Tungsten and Other Refractory Metals for VLSI Applications II

MATERIALS RESEARCH SOCIETY SYMPOSIA PROCEEDINGS

ISSN 0272 - 9172

Volume 1—Laser and Electron-Beam Solid Interactions and Materials Processing, J. F. Gibbons, L. D. Hess, T. W. Sigmon, 1981

Volume 2—Defects in Semiconductors, J. Narayan, T. Y. Tan, 1981

Volume 3—Nuclear and Electron Resonance Spectroscopies Applied to Materials Science, E. N. Kaufmann, G. K. Shenoy, 1981

Volume 4—Laser and Electron-Beam Interactions with Solids, B. R. Appleton, G. K. Celler, 1982

Volume 5—Grain Boundaries in Semiconductors, H. J. Leamy, G. E. Pike, C. H. Seager, 1982

Volume 6—Scientific Basis for Nuclear Waste Management IV, S. V. Topp, 1982

Volume 7—Metastable Materials Formation by Ion Implantation, S. T. Picraux, W. J. Choyke, 1982

Volume 8—Rapidly Solidified Amorphous and Crystalline Alloys, B. H. Kear, B. C. Giessen, M. Cohen, 1982

Volume 9—Materials Processing in the Reduced Gravity Environment of Space, G. E. Rindone, 1982

Volume 10—Thin Films and Interfaces, P. S. Ho, K.-N. Tu, 1982

Volume 11—Scientific Basis for Nuclear Waste Management V, W. Lutze, 1982

Volume 12—In Situ Composites IV, F. D. Lemkey, H. E. Cline, M. McLean, 1982

Volume 13—Laser-Solid Interactions and Transient Thermal Processing of Materials, J. Narayan, W. L. Brown, R. A. Lemons, 1983

Volume 14—Defects in Semiconductors II, S. Mahajan, J. W. Corbett, 1983

Volume 15—Scientific Basis for Nuclear Waste Management VI, D. G. Brookins, 1983

Volume 16—Nuclear Radiation Detector Materials, E. E. Haller, H. W. Kraner, W. A. Higinbotham, 1983

Volume 17—Laser Diagnostics and Photochemical Processing for Semiconductor Devices, R. M. Osgood, S. R. J. Brueck, H. R. Schlossberg, 1983

Volume 18—Interfaces and Contacts, R. Ludeke, K. Rose, 1983

Volume 19—Alloy Phase Diagrams, L. H. Bennett, T. B. Massalski, B. C. Giessen, 1983

Volume 20—Intercalated Graphite, M. S. Dresselhaus, G. Dresselhaus, J. E. Fischer, M. J. Moran, 1983

Volume 21—Phase Transformations in Solids, T. Tsakalakos, 1984

Volume 22—High Pressure in Science and Technology, C. Homan, R. K. MacCrone, E. Whalley, 1984

Volume 23—Energy Beam-Solid Interactions and Transient Thermal Processing, J. C. C. Fan, N. M. Johnson, 1984

Volume 24—Defect Properties and Processing of High-Technology Nonmetallic Materials, J. H. Crawford, Jr., Y. Chen, W. A. Sibley, 1984

MATERIALS RESEARCH SOCIETY SYMPOSIA PROCEEDINGS

Volume 25—Thin Films and Interfaces II, J. E. E. Baglin, D. R. Campbell, W. K. Chu, 1984

Volume 26—Scientific Basis for Nuclear Waste Management VII, G. L. McVay, 1984

Volume 27—Ion Implantation and Ion Beam Processing of Materials, G. K. Hubler, O. W. Holland, C. R. Clayton, C. W. White, 1984

Volume 28—Rapidly Solidified Metastable Materials, B. H. Kear, B. C. Giessen, 1984

Volume 29—Laser-Controlled Chemical Processing of Surfaces, A. W. Johnson, D. J. Ehrlich, H. R. Schlossberg, 1984.

Volume 30—Plasma Processsing and Synthesis of Materials, J. Szekely, D. Apelian, 1984

Volume 31—Electron Microscopy of Materials, W. Krakow, D. A. Smith, L. W. Hobbs, 1984

Volume 32—Better Ceramics Through Chemistry, C. J. Brinker, D. E. Clark, D. R. Ulrich, 1984

Volume 33—Comparison of Thin Film Transistor and SOI Technologies, H. W. Lam, M. J. Thompson, 1984

Volume 34—Physical Metallurgy of Cast Iron, H. Fredriksson, M. Hillerts, 1985

Volume 35—Energy Beam-Solid Interactions and Transient Thermal Processing/1984, D. K. Biegelsen, G. A. Rozgonyi, C. V. Shank, 1985

Volume 36—Impurity Diffusion and Gettering in Silicon, R. B. Fair, C. W. Pearce, J. Washburn, 1985

Volume 37—Layered Structures, Epitaxy, and Interfaces, J. M. Gibson, L. R. Dawson, 1985

Volume 38—Plasma Synthesis and Etching of Electronic Materials, R. P. H. Chang, B. Abeles, 1985

Volume 39—High-Temperature Ordered Intermetallic Alloys, C. C. Koch, C. T. Liu, N. S. Stoloff, 1985

Volume 40—Electronic Packaging Materials Science, E. A. Giess, K.-N. Tu, D. R. Uhlmann, 1985

Volume 41—Advanced Photon and Particle Techniques for the Characterization of Defects in Solids, J. B. Roberto, R. W. Carpenter, M. C. Wittels, 1985

Volume 42—Very High Strength Cement-Based Materials, J. F. Young, 1985.

Volume 43—Fly Ash and Coal Conversion By-Products: Characterization, Utilization, and Disposal I, G. J. McCarthy, R. J. Lauf, 1985

Volume 44—Scientific Basis for Nuclear Waste Management VIII, C. M. Jantzen, J. A. Stone, R. C. Ewing, 1985

Volume 45—Ion Beam Processes in Advanced Electronic Materials and Device Technology, B. R. Appleton, F. H. Eisen, T. W. Sigmon, 1985

Volume 46—Microscopic Identification of Electronic Defects in Semiconductors, N. M. Johnson, S. G. Bishop, G. D. Watkins, 1985

MATERIALS RESEARCH SOCIETY SYMPOSIA PROCEEDINGS

Volume 47—Thin Films: The Relationship of Structure to Properties, C. R. Aita, K. S. SreeHarsha, 1985

Volume 48—Applied Materials Characterization, W. Katz, P. Williams, 1985

Volume 49—Materials Issues in Applications of Amorphous Silicon Technology, D. Adler, A. Madan, M. J. Thompson, 1985

Volume 50—Scientific Basis for Nuclear Waste Management IX, L. O. Werme, 1986

Volume 51—Beam-Solid Interactions and Phase Transformations, H. Kurz, G. L. Olson, J. M. Poate, 1986

Volume 52—Rapid Thermal Processing, T. O. Sedgwick, T. E. Seidel, B.-Y. Tsaur, 1986

Volume 53—Semiconductor-on-Insulator and Thin Film Transistor Technology, A. Chiang. M. W. Geis, L. Pfeiffer, 1986

Volume 54—Thin Films—Interfaces and Phenomena, R. J. Nemanich, P. S. Ho, S. S. Lau, 1986

Volume 55—Biomedical Materials, J. M. Williams, M. F. Nichols, W. Zingg, 1986

Volume 56—Layered Structures and Epitaxy, J. M. Gibson, G. C. Osbourn, R. M. Tromp, 1986

Volume 57—Phase Transitions in Condensed Systems—Experiments and Theory, G. S. Cargill III, F. Spaepen, K.-N. Tu, 1987

Volume 58—Rapidly Solidified Alloys and Their Mechanical and Magnetic Properties, B. C. Giessen, D. E. Polk, A. I. Taub, 1986

Volume 59—Oxygen, Carbon, Hydrogen, and Nitrogen in Crystalline Silicon, J. C. Mikkelsen, Jr., S. J. Pearton, J. W. Corbett, S. J. Pennycook, 1986

Volume 60—Defect Properties and Processing of High-Technology Nonmetallic Materials, Y. Chen, W. D. Kingery, R. J. Stokes, 1986

Volume 61—Defects in Glasses, F. L. Galeener, D. L. Griscom, M. J. Weber, 1986

Volume 62—Materials Problem Solving with the Transmission Electron Microscope, L. W. Hobbs, K. H. Westmacott, D. B. Williams, 1986

Volume 63—Computer-Based Microscopic Description of the Structure and Properties of Materials, J. Broughton, W. Krakow, S. T. Pantelides, 1986

Volume 64—Cement-Based Composites: Strain Rate Effects on Fracture, S. Mindess, S. P. Shah, 1986

Volume 65—Fly Ash and Coal Conversion By-Products: Characterization, Utilization and Disposal II, G. J. McCarthy, F. P. Glasser, D. M. Roy, 1986

Volume 66—Frontiers in Materials Education, L. W. Hobbs, G. L. Liedl, 1986

Volume 67—Heteroepitaxy on Silicon, J. C. C. Fan, J. M. Poate, 1986

Volume 68—Plasma Processing, J. W. Coburn, R. A. Gottscho, D. W. Hess, 1986

Volume 69—Materials Characterization, N. W. Cheung, M.-A. Nicolet, 1986

Volume 70—Materials Issues in Amorphous-Semiconductor Technology, D. Adler, Y. Hamakawa, A. Madan, 1986

Volume 71—Materials Issues in Silicon Integrated Circuit Processing, M. Wittmer, J. Stimmell, M. Strathman, 1986

Volume 72—Electronic Packaging Materials Science II, K. A. Jackson, R. C. Pohanka, D. R. Uhlmann, D. R. Ulrich, 1986

Volume 73—Better Ceramics Through Chemistry II, C. J. Brinker, D. E. Clark, D. R. Ulrich, 1986

Volume 74—Beam-Solid Interactions and Transient Processes, S. T. Picraux, M. O. Thompson, J. S. Williams, 1987

Volume 75—Photon, Beam and Plasma Stimulated Chemical Processes at Surfaces, V. M. Donnelly, I. P. Herman, M. Hirose, 1987

Volume 76—Science and Technology of Microfabrication, R. E. Howard, E. L. Hu, S. Pang, S. Namba, 1987

Volume 77—Interfaces, Superlattices, and Thin Flims, J. D. Dow, I. K. Schuller, 1987

Volume 78—Advances in Structural Ceramics, P. F. Becher, M. V. Swain, S. Somiya, 1987

Volume 79—Scattering, Deformation and Fracture in Polymers, B. Crist, T. P. Russell, E. L. Thomas, G. D. Wignall, 1987

Volume 80—Science and Technology of Rapidly Quenched Alloys, M. Tenhover, L. E. Tanner, W. L. Johnson, 1987

Volume 81—High-Temperature Ordered Intermetallic Alloys, II, C. C. Koch, C. T. Liu, N. S. Stoloff, O. Izumi, 1987

Volume 82—Characterization of Defects in Materials, R. W. Siegel, R. Sinclair, J. R. Weertman, 1987

Volume 83—Physical and Chemical Properties of Thin Metal Overlayers and Alloy Surfaces, D. M. Zehner, D. W. Goodman, 1987

Volume 84—Scientific Basis for Nuclear Waste Management X, J. K. Bates, W. B. Seefeldt, 1987

Volume 85—Microstructural Development During the Hydration of Cement, L. Struble, P. Brown, 1987

Volume 86—Fly Ash and Coal Conversion By-Products Characterization, Utilization and Disposal III, G. J. McCarthy, F. P. Glasser, D. M. Roy, S. Diamond, 1987

Volume 87—Materials Processing in the Reduced Gravity Environment of Space, R. H. Doremus, P. C. Nordine, 1987

Volume 88—Optical Fiber Materials and Properties, S. R. Nagel, G. Sigel, J. W. Fleming, D. A. Thompson, 1987

Volume 89—Diluted Magnetic (Semimagnetic) Semiconductors, J. K. Furdyna, R. L. Aggarwal, S. von Molnar, 1987

Volume 90—Materials for Infrared Detectors and Sources, R. F. C. Farrow, J. F. Schetzina, J. T. Cheung, 1987

Tungsten and Other Refractory Metals for VLSI Applications II

Proceedings of the 1986 Workshop held November 12-14, 1986, Palo Alto, California, U.S.A. Sponsored by Continuing Education in Engineering, University Extension, University of California, Berkeley.

EDITOR:

Eliot K. Broadbent

Philips Research Laboratories Sunnyvale, Signetics Corporation, Sunnyvale, California, U.S.A.

ASSOCIATE EDITORS:

Robert S. Blewer
Sandia National Laboratories, Albuquerque, New Mexico, U.S.A.

Carol M. McConica
Colorado State University, Fort Collins, Colorado, U.S.A.

John E. Sanchez, Jr.
Xerox Palo Alto Research Center, Palo Alto, California, U.S.A.

Krishna C. Saraswat
Stanford University, Stanford, California, U.S.A.

Michael D. Strathman
Charles Evans and Associates, Redwood City, California, U.S.A.

Victor A. Wells
Unisys Corporation, Egan, Minnesota, U.S.A.

Ronald H. Wilson
General Electric Corporation, Schenectady, New York, U.S.A.

MRS

MATERIALS RESEARCH SOCIETY

Pittsburgh, Pennsylvania

Published by:

Materials Research Society
9800 McKnight Road, Suite 327
Pittsburgh, Pennsylvania 15237
telephone (412) 367-3003

ISSN 0886-7860

ISBN 0-931837-66-9

Printed in the United States of America

Manufactured by Publishers Choice Book Mfg. Co.
Mars, Pennsylvania 16046

Contents

PREFACE xv

Acknowledgements xvii

KEYNOTE PRESENTATION

*TUNGSTEN - A SPECTATOR'S VIEW 3
D.B. Fraser

PRACTICAL AND DEVICE CONSIDERATIONS FOR REFRACTORY METAL USAGE IN VLSI

*DEVICE CONSIDERATIONS FOR USING TUNGSTEN AND OTHER REFRACTORY METALS IN CMOS VLSI 9
J.Y. Chen

*THIN - FILM DIFFUSION BARRIERS FOR METAL-SEMICONDUCTOR CONTACTS 19
M.-A. Nicolet

*THE PHYSICS AND CHEMISTRY OF THIN "NATIVE" OXIDE FILMS ON SILICON 27
B.E. Deal and D.-B. Kao

SELECTIVE TUNGSTEN: SELECTIVITY LOSS AND NUCLEATION ON INSULATORS

OBSERVATIONS ON SELECTIVITY LOSS DURING TUNGSTEN CVD 43
J.R. Creighton

A MODEL FOR TUNGSTEN NUCLEATION ON OXIDE 51
C.M. McConica

EVALUATION ON SELECTIVE DEPOSITION OF CVD W FILMS BY MEASUREMENT OF SURFACE TEMPERATURE 59
T. Ohba, Y. Ohyama, S. Inoue, and M. Maeda

*Invited Paper

CVD TUNGSTEN: FUNDAMENTAL REACTION AND GROWTH STUDIES

TEMPERATURE PROGRAMMED DESORPTIONS OF TUNGSTEN CVD MATERIALS — 69
M.A. Heiny and C.M. McConica

THE REACTION KINETICS OF THE H_2 REDUCTION OF WF_6 IN THE CHEMICAL VAPOR DEPOSITION OF TUNGSTEN FILMS — 77
P. van der Putte

THE MICROSTRUCTURE OF CVD W FILMS PRODUCED BY THE Si REDUCTION OF WF_6 — 85
M.L. Green, Y.S. Ali, T. Boone, B.A. Davidson, L.C. Feldman, and S. Nakahara

INTERACTION OF CVD TUNGSTEN WITH UNDERLYING METAL LAYERS — 93
S.L. Ng, S.J. Rosner, S.S. Laderman, T.I. Kamins, D.R. Bradbury, and J. Amano

GROWTH OF SELECTIVE TUNGSTEN ON SELF-ALIGNED $CoSi_2$ BY LOW PRESSURE CHEMICAL VAPOR DEPOSITION — 101
P. van der Putte, D.K. Sadana, E.K. Broadbent, and A.E. Morgan

RECENT CVD PROCESS AND EQUIPMENT DEVELOPMENTS

SELECTIVE CVD OF $TaSi_2$ — 111
T.P.H.F. Wendling, C. Wieczorek, and K. Hieber

FLOW CONFIGURATION EFFECTS ON THE DEPOSITION PROPERTIES OF SELECTIVE TUNGSTEN — 123
Y. Kusumoto, K. Takakuwa, T. Ikuta, and I. Nakayama

SELECTIVE AND BLANKET TUNGSTEN BY APCVD — 131
N. Gralenski and L. Bartholomew

SELECTIVE TUNGSTEN DEPOSITION IN A BATCH COLD WALL CVD SYSTEM — 137
R. Chow, S. Kang, W.R. Harshbarger, and M. Susoeff

SELECTIVE TUNGSTEN DEPOSITION IN A VARIAN/TORREX 5101 COLD WALL CVD REACTOR — 147
R.F. Foster, D.L. Brors, and S. Tseng

BLANKET CVD TUNGSTEN: METAL GATE APPLICATIONS AND ADHESION STUDIES

*HIGHLY RELIABLE TUNGSTEN GATE TECHNOLOGY — 159
N. Kobayashi, S. Iwata, N. Yamamoto, and N. Hara

*Invited Paper

A REFRACTORY METAL GATE APPROACH FOR MICRONIC CMOS TECHNOLOGY 169
V. Lubowiecki, J.-L. Ledys, C. Plossu, and B. Balland

ADHESION OF BLANKET TUNGSTEN TO OXIDES 177
K.C. Ray Chiu and N.E. Zetterquist

THIN LAYERS OF TiN AND Al AS GLUE LAYERS FOR BLANKET TUNGSTEN DEPOSITION 187
V.V.S. Rana, J.A. Taylor, L.H. Holschwandner, and N.S. Tsai

CHEMICAL VAPOR DEPOSITION OF MOLYBDENUM FROM MoF_6

LOW PRESSURE CHEMICAL VAPOR DEPOSITION OF MOLYBDENUM: KINETICS AND REACTION MECHANISMS 199
T. Sahin, E.J. Flanigan, and J.T. Sears

CVD MOLYBDENUM FROM MoF_6 207
D.W. Woodruff and R.A. Sanchez-Martinez

MICROSTRUCTURAL CHARACTERIZATION OF MOLYBDENUM FILMS DEPOSITED BY LPCVD 215
N. Lifshitz, J.M. Brown, and D.S. Williams

CHARACTERISTICS OF CVD TUNGSTEN FILMS IN DEVICE STRUCTURES

THE INFLUENCE OF SELECTIVE TUNGSTEN DEPOSITION ON SHALLOW n^+(As or P)/p AND p^+(BF_2)/n JUNCTION LEAKAGE 227
G.E. Georgiou, J.M. Brown, M.L. Green, R. Liu, and D.S. Williams

DETRIMENTAL EFFECTS OF RESIDUAL SILICON OXIDES ON LPCVD TUNGSTEN DEPOSITIONS IN SHALLOW JUNCTION DEVICES 235
R.S. Blewer and M.E. Tracy

ELECTROMIGRATION-INDUCED SHORT CIRCUIT FAILURE IN ALUMINUM/TUNGSTEN (CVD) CONDUCTORS 247
E.K. Broadbent and J.M. Towner

SELECTIVE CVD TUNGSTEN FOR BIPOLAR MULTILEVEL INTERCONNECTION 257
N. Tsuzuki, M. Ichikawa, K. Kurita, K. Watanabe, and K. Inayoshi

THERMAL STABILITY OF Al/W/p-n Si CONTACTS DURING POST-METALLIZATION ANNEALING 265
L. Gutai, M. Delfino, and J.M. DeBlasi

PROPERTIES OF REFRACTORY METALS DEPOSITED BY SPUTTERING OR EVAPORATION

THE CHARACTERIZATION OF TITANIUM-NITRIDE THIN FILM AS DIFFUSION BARRIER FOR VLSI METALLIZATION — 275
K.-H. Park, S. Mihara, Y. Sato, H. Tsuchikawa, and M. Yoshida

THE DEPOSITION, CONTACT AND ELECTROMIGRATION PROPERTIES OF REACTIVELY SPUTTERED TITANIUM NITRIDE FILMS — 283
R.C. Ellwanger and J.M. Towner

FORMATION KINETICS AND PROPERTIES OF TITANIUM NITRIDE FORMED BY RAPID THERMAL ANNEALING — 291
R.K. Shukla and S.A. Yoshikawa

THERMAL STABILITY AND NITROGEN REDISTRIBUTION IN THE Ti/W-N BILAYER — 301
F.C.T. So, C.-D. Lien, E. Kolawa, X.-A. Zhao, and M.-A. Nicolet

PROPERTIES OF REACTIVELY SPUTTERED WN_x FILMS — 311
E. Kolawa, F.C.T. So, X.-A. Zhao, and M.-A. Nicolet

TiW AS A BARRIER METAL FOR SMALL CONTACTS — 319
S. Saito, K. Matsuda, K. Nishizawa, and K. Sakiyama

THE STABILITY OF THE ELECTRICAL AND STRUCTURAL PROPERTIES OF ELECTRON BEAM DEPOSITED AND SPUTTER DEPOSITED Ta AND Mo SCHOTTKY CONTACTS TO GaAs — 325
A.K. Kulkarni, M.G. Thompson, and G.H. Whipple

PRACTICAL PREPARATION OF SPUTTERED REFRACTORY METAL DIFFUSION BARRIERS — 333
R.S. Nowicki

DEVELOPMENT OF REFRACTORY METALS AND SILICIDES TARGETS, AND THEIR CHARACTERISTICS — 339
M. Suzuki

REACTIVE ION ETCHING OF REFRACTORY METAL FILMS

REACTIVE ION ETCHING OF REFRACTORY METALS FOR GATE AND INTERCONNECT APPLICATIONS — 349
R.J. Saia and B. Gorowitz

PLASMA ETCHING OF CVD TUNGSTEN FILMS FOR VLSI APPLICATIONS — 357
C.-H. Chen, L.C. Watson, and D.W. Schlosser

REACTIVE ION ETCHING OF SUBMICROMETER SIZE FEATURES IN TUNGSTEN THIN FILMS — 363
T.J. Whetten, N.Y. Whetten, and E.D. Wolf

REACTIVE ION ETCHING OF AN AL-Si/TiW MULTILAYER 371
N. Takenaka, D. Kimura, K. Nishizawa, and K. Sakiyama

RECENT NEW PAPERS

THE HF PARTIAL PRESSURE AS A SELECTIVITY DETERMINANT DURING THE SELECTIVE DEPOSITION OF TUNGSTEN 377
L.F.Tz. Kwakman, W.J.C. Vermeulen, E.H.A. Granneman, and M.L. Hitchman

THE CONTACT PROPERTIES TO $TiSi_2$ AND THE ADHESION WITHIN SUB-MICRON CONTACT HOLES OF ETCHED BACK CVD W/ADHESION LAYER FILMS 385
R.C. Ellwanger, J.E.J. Schmitz, R.A.M. Wolters, and A.J.M. van Dijk

MECHANISMS FOR WHISKER GROWTH IN THIN FILMS 395
J.E. Sanchez, Jr.

TUNGSTEN DEPOSITION IN POROUS SILICON FOR THE FORMATION OF BURIED LAYER CONDUCTORS 401
R.S. Blewer, S.S. Tsao, and M. Tracy

PANEL DISCUSSION SUMMARY

RELIABILITY OF DEVICES WITH TUNGSTEN ELEMENTS 411
S. Mehta

DEVICE RELIABILITY WHEN USING TUNGSTEN INTERCONNECTS 419
J.P. Roland

AUTHOR INDEX 423

SUBJECT INDEX 425

Preface

This Materials Research Society volume contains forty-seven papers presented at the Third Workshop on Tungsten and Other Refractory Metals For VLSI Application held November 12-14, 1986, in Palo Alto, California. The Workshop series recognizes the ever increasing importance of refractory metal materials in present and future IC device applications, and provides a unique annual forum for presentation and discussion of the most up-to-date results.

This second volume continues an agreement between conference organizers, University Extension, University of California, Berkeley, and the Materials Research Society to document the proceedings of the workshop in a series of special books. The first volume, published in 1986 by the MRS, contained the joint proceedings from both the 1984 and 1985 workshops which were held in Albuquerque, New Mexico. As with its predecessor, this volume contains the majority of what has been published during the course of the past year on CVD tungsten.

From an informal survey conducted at this years workshop, it was clear that the near-term priority application for CVD tungsten films is that of volume filling for contacts and vias in multilevel IC metallization. Both selective and blanket deposition techniques appear available to match the specific requirements of each merchant and captive chip manufacturer. Concurrently, focused interest and research activities continue with the application of CVD tungsten directly onto single and polycrystalline Si as a current shunt and diffusion barrier, above silicides as a self-aligned diffusion barrier, and as a high-reliability pure interconnect material. Ultimately, the work function value and low resistivity of tungsten, coupled with the high-purity material aspects of chemical vapor deposition, make it an attractive candidate for use as a gate material in sub-micron feature size CMOS devices.

It has proven most appropriate in this symposium to complement the purview in CVD of metals (papers on tungsten, molybdenum and tantalum silicide were presented this year) with that of refractory metals prepared by sputtering and evaporation techniques. Several papers covering reactive ion etching of refractory metals are also contained in this volume. This volume should serve well as a convenient reference source to the papers contained herein for many years to come.

Future workshops on tungsten and other refractory metals will continue to be held in the fall of each year. Information on the date and location of upcoming workshops can be obtained from Ms. Linda Reid, University Extension, University of California, Berkeley, (415) 642-4151, who is in charge of workshop administration.

Eliot K. Broadbent
1986 Symposium Chariman

January 1987

Acknowledgments

Executive Committee

Robert S. Blewer, Sandia National Laboratories
Eliot K. Broadbent (Chairman), Philips Research Laboratories Sunnyvale
Werner Metz, Kodak Research Laboratories
Takahiko Moriya, Toshiba Corporation
Krishna C. Saraswat, Stanford University
*Michael D. Strathman, Charles Evans and Associates
*Thomas E. Wade, University of South Florida
Victor A. Wells, Unisys Corporation

Executive Secretaries: Linda Reid and Karen Anderson, University of California, Berkeley

Local Arrangements

John E. Sanchez, Jr., Xerox Palo Alto Research Center
Gloria Warner, Xerox Palo Alto Research Center

1986 Session Chairman

Robert S. Blewer, Sandia National Laboratories
Carol M. McConica, Colorado State University
Marc-A. Nicolet, California Institute of Technology
John E. Sanchez, Jr., Xerox Palo Alto Research Center
Krishna C. Saraswat, Stanford University
Arjun N. Saxena, Gould AMI Semiconductors
Thomas E. Wade, University of South Florida
Ronald H. Wilson, General Electric Corporate Research & Development
Victor A. Wells, Unisys Corporation

*Ex officio

Keynote Presentation

TUNGSTEN-A SPECTATOR'S VIEW

DAVID B. FRASER
Intel Corporation, 250 North Mines Road, Livermore, CA 94550
Components Research & Development

INTRODUCTION

Tungsten, deposited by CVD processes, has been very attractive for application in silicon integrated circuits [1] where the dimensions are in the micron or submicron range. The bulk resistivity of tungsten is 5.3u Ω cm and the coefficient of thermal expansion is 4.5 ppm°C^{-1} (similar to 3.1 ppm°C^{-1} (of silicon). Tungsten forms a stable silicide and can be made to form low resistance, of the order of 10-7 cm^{-2}, contacts to silicon. In addition, the refractory nature of tungsten or molybdenum [2] make them attractive because of electromigration resistance and the ability to withstand higher process temperatures. However, tungsten does oxidize readily and the adherence to dielectric surfaces is poor.

The alternative to CVD is physical vapor deposition such as electron beam evaporation and sputter deposition. The use of electron beam evaporation leads to radiation damage due to the high energy electrons creating X-rays and the vapor from the source travels in straight lines resulting in poor step coverage. In sputtering, the radiation damage can be reduced by use of magnetron sputter sources (low voltage in comparison with the e-beam) but step coverage is still poor unless a bias deposition process is used. The adherence of the physically deposited films is poor and this is compounded by the high compressive stress that can result during sputter deposition. [3]

CURRENT VIEW

Two approaches are being taken in CVD - a blanket process [4] and a selective process [5]. Selective tungsten deposition has been attractive because of the advantages it offers. For example, it allows the planarizing of contact and via structures and it can act as a contact/ barrier between a primary metallization and the substrate. The feasibility of the process has been well demonstrated. In actual application though, there are reports of "tunnels" and "encroachment"[6] [7] due to the displacement reaction occurring between the silicon substrate and the WF6; junction leakage has occurred ($J >$ 1nA cm^{-2} or $>$ .01 nA cm^{-1}) and there is loss of selectivity in cases where thick deposits are required.

From the point of view of one who would like to produce chips that have 500,000 and more contacts, it is essential that the process be reliable. In reviewing some of the selective deposition literature, the experimental prior processing has not been very well documented. To clarify our understanding, we need more information than is often offered. For example, is the surrounding dielectric a deposited film or is it a thermal oxide; what is the implanted species and the dose; what heat cycle did the implanted silicon see prior to the CVD step; if there was a masking layer during the implant, how thick was it, and how was it removed; what was the configuration of the contact relative to the field oxide or the junction edge; and, what was the thermal history following the CVD step? If N+ contacts are worse than P+, what can we conclude? Is there more damage associated with the N+? Do shallow junctions with minimal heat treatment aggravate the problems?

Coming from a metallization background where sputtering equipment is now capable of yielding base pressure in the 10^{-8} torr range, it seems that the bulk of CVD stations are underpumped. I base that opinion on the fact that the deposition rates are not that different in sputter equipment and the effects of residual gas can not be all that different either. These concerns are further compounded if one considers a batch versus a single wafer load locked reactor. The batch reactor may require more pumping and purging because the internal surfaces are exposed to the atmosphere each cycle and the reactive deposits may "getter" water vapor. The load locked system should minimize these problems and lead to shorter process times. Intuitively, it appears simpler to optimize the process for a single wafer system since the volume is less and it would be possible to have a larger dynamic range for flow rate and residence time values. However, the batch system can compensate by increasing the load size. I look forward to the presentation, at this conference, of the paper by Kusomoto et al describing their load locked selective reactor.

What if you do not have a selective process? The alternative appears to be a blanket deposition process followed by an etch to yield filled and planar contacts and vias. This approach will also permit the implementation of a selective process when it is production worthy. For contacts in the submicron, this blanket-etchback plus a different metal "1" layer looks attractive. Because of the restrictions on real estate, nested contacts are a luxury we will not have and dissimilar etching characteristics will be necessary in order to preserve contact integrity. I like the scheme proposed by Brown [8] in which a tungsten-filled contact is used in conjunction with a sputtered molybdenum layer. The use of a molybdenum first layer also would lead to the use of a selective tungsten process to fill the vias [9]. Where the contacts can be nested a blanket tungsten process could fill the contact and provide metal "1".

Where blanket tungsten can be used, there are other issues that must be considered such as adherence and mechanical stress. The stress for CVD tungsten films on 100nm of sputtered titanium is plotted as a function of the tungsten thickness. For the thinnest layer the stress is high, 1500 MPa and tensile, and film stress drops to about 1200 MPa for the thickest tungsten. The stress as a function of temperature for the thinnest film is shown in the next figure. Since the films were deposited at 300°C the latter plot tells us that the tensile stress in the film is primarily intrinsic. Such high intrinsic stress levels indicate the need for adhesive layers beneath the tungsten. It may be that the preferential etching observed in contact perimeters during the tungsten etchback is due to the high stress levels.

CONCLUSION

Clearly, there is more work to be done in both selective and blanket tungsten deposition. In spite of the difficulties encountered in practical application, CVD tungsten remains attractive. At this time selective deposition may "buy time" through implementation of blanket-etchback processes. It is clear that the problems in this technology are formidable and that some of the solutions will come from collaborative efforts between equipment vendors and equipment users.

ACKNOWLEDGEMENTS

The author would like to thank R. Shukla, S. Sivaram, and M. Okano for providing data on the stress in the CVD tungsten films.

REFERENCES

1. D.M. Brown et al, Solid State Electronics, Vol. 11, p. 1105 (1968).
2. D.M. Brown, J. Electrochem Soc. 114, p. 730 (1967).
3. R.S. Wagner, A.K. Sinha, T.T. Sheng, H.J. Levinstein and F.B. Alexander, J. Vac. Sci. Tech., Vol. 11, p. 582 (1974).
4. J. Amick and J. Shaw in Proceedings of the Conference on Chemical Vapor Deposition of Refractory Metal Alloys, and Compounds. A.C. Schaffhauser, Editor, Gatlinburg, Tennessee, September 12-14, 1967, American Nuclear Society.
5. J.J. Cuomo, "Selective Chemical Vapor Deposition of Tungsten", Proceedings of the Third International Conference on Chemical Vapor Deposition, April 24-27, 1972, American Nuclear Society, pp. 270-291.
6. M.L. Green and R.A. Levy "Tungsten and Other Refractory Metals for VLSI Applications" R.S. Blewer Editor, 1984 Workshop, p. 423, 1986.
7. E.K. Broadbent, A.E. Morgan, J.M. Deblasi, P. Van der Putte, A. Reader, B. Coulman, B.J. Burrow and D.K. Sodana, "Tungsten and Other Refractory Metals for VLSI Applications", R.S. Blewer Editor, 1985 Workshop, p. 63, 1986.
8. D.M. Brown private communication.
9. R.H. Wilson, R.W. Stoll and M.A. Calacone, Editor, 1985 "Tungsten and Other Refractory Metals for VLSI Applications", R.S. Blewer Editor, 1985 Workshop, p. 35, 1986.

CVD W Film Stress versus Thickness (CVD W over 100 nm Ti)

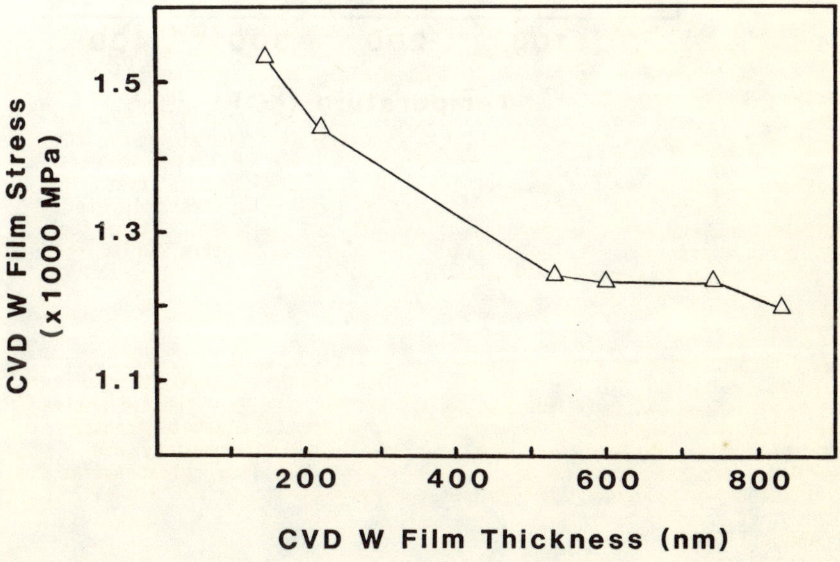

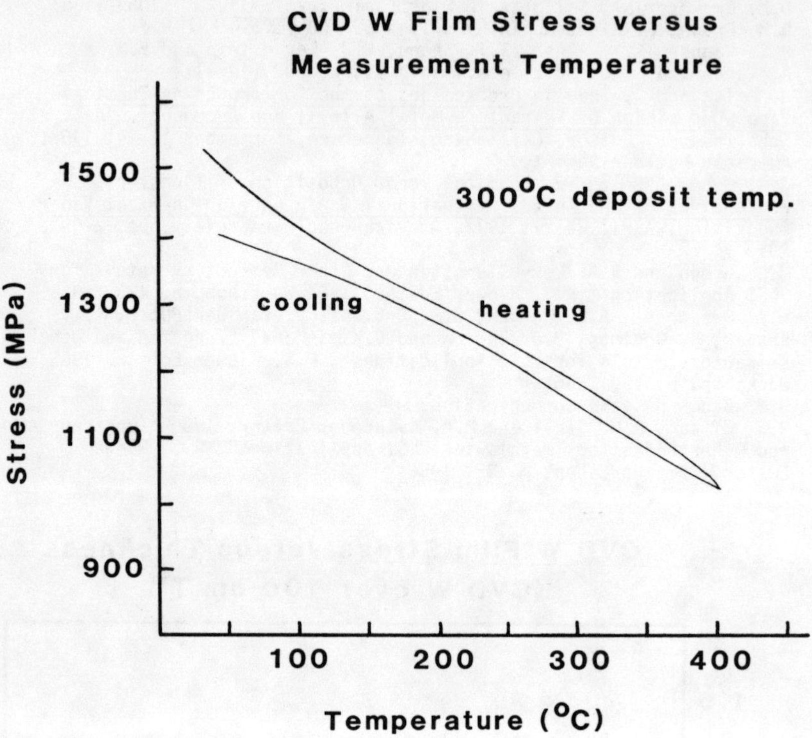
CVD W Film Stress versus
Measurement Temperature
300°C deposit temp.
cooling
heating
Stress (MPa)
1500
1300
1100
900
100
200
300
400
Temperature (°C)

Practical and Device Considerations for Refractory Metal Usage in VLSI

DEVICE CONSIDERATIONS FOR USING TUNGSTEN AND OTHER REFRACTORY METALS IN CMOS VLSI

JOHN Y. CHEN
Xerox Palo Alto Research Center, 3333 Coyote Hill Road, Palo Alto, CA 94304

ABSTRACT

Tungsten and other refractory metals can be applied to a CMOS IC in gate electrodes, metallized or silicided source/drain regions and interconnects. Besides the well-known advantage in reducing intra- and inter-transistor parasitic resistances, hence increasing circuit speed, other advantages in CMOS include improved turn-off characteristics in pMOS FETs, enhanced latch-up immunity and reduced electromigration failure. Meanwhile a new reliability concern has been raised for nMOS FETs with silicided source/drains.

INTRODUCTION

CMOS has become the dominant VLSI technology due to its inherent low power characteristics. The speed of a CMOS VLSI circuit is also rapidly catching up with its NMOS counterpart because as the transistor is scaled down, a p-channel FET is no longer much slower than an n-channel FET with the same size [1]. In fact, the speed of a VLSI circuit is becoming more and more sensitive to the intra- and inter-device parasitic resistances and less dependent on the intrinsic transistor performance since a large number of small-size transistors are used in a circuit [2]. A small transistor offers very low intrinsic resistance, hence any additional parasitic resistance becomes a significant portion for the entire device performance. The large number of transistors requires enormous interconnects among them. This is particularly true for CMOS circuits due to the difficulty of implementing buried contacts and the need of a p-channel load for each n-channel driver in a static CMOS design. The inter-device interconnects increase RC delay which plays a dominant role in determining circuit speed. Delay time associated with transmitting signals from one end of the chip to the other end is set by the dimension and the material property of the long-distance interconnects instead of transistor performance [3].

Tungsten or other refractory metals can be applied to CMOS ICs in several places such as gate electrodes, metallized or silicided source/drains, and interconnects. In general, these applications reduce intra- or inter-device parasitic resistances and increase device/circuit speed. They also have other effects on device performance and the effects are different in each application. These effects must be completely understood before one can take full advantage of using tungsten and other refractory metals in CMOS VLSI.

Gate Electrode Application

In a standard MOS technology, MOS gates are made of n^+ polysilicon (poly-Si) or silicide on n^+ poly-Si, called polycide. The n^+ poly-Si is adequate for n-channel transistors because the work function difference (ϕ_{ms}) between the n^+ poly-Si and a p-type substrate tends to promote inversion, thereby resulting in a threshold reduction which is in favor of

MOSFET scaling for VLSI. This material however is not desirable for p-channel transistors because the associated ϕ_{ms} causes band bending in an opposite way for inversion to occur. In order to turn the transistor on, a large gate voltage must be applied to compensate the effect of ϕ_{ms}. Such a large threshold voltage is undesirable for MOS scaling and circuit speed.

In typical CMOS IC fabrication, boron atoms are implanted into the pMOS channel area near the Si surface. The idea is to use ionized boron impurity atoms as a negative interface charge (Q_{FC}) which can decrease the magnitude of the p-channel threshold. Due to the implantation process and the subsequent high-temperature IC process steps, the boron atoms at the end of the IC processing always result in a p-type layer with finite thickness. Though depending on the gate voltage, this p-type layer changes the channel potential. More specifically, the potential minimum is shifted from the Si surface into bulk Si shown as the dashed line in Fig. 1. Thinning p layer can reduce this effect (solid line in Fig. 1) but is limited by the number of boron atoms needed for the threshold change. Since mobile carriers, holes in this case, follow the potential minimum, the majority of the free carriers flow below the Si surface rather than at the surface as exhibited in a conventional surface-channel MOSFET. This type of conduction is referred to as buried-channel conduction, and there are specific problems associated with buried-channel MOSFETs.

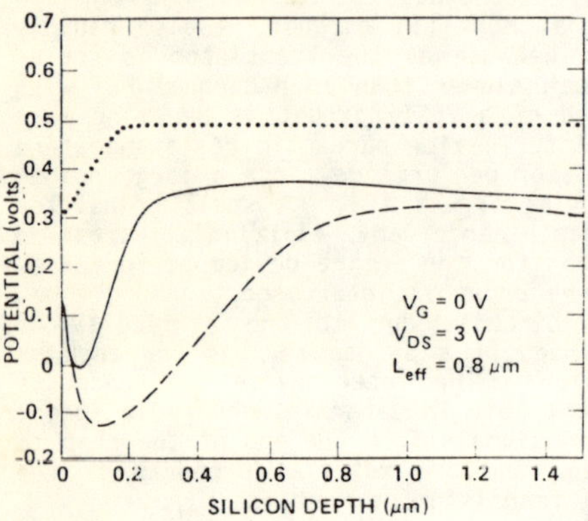

Fig.1 Potential distribution for surface- and buried-channel MOSFETs; dotted line: surface-channel nMOS, dashed line:buried-channel pMOS, solid line:buried-channel pMOS with a thinner buried p layer (ref. 2 & 5).

The main problem in a buried-channel device is the turn-off characteristic [4,5]. Carriers in this type of device are located farther away from the gate, hence are harder to control by changing the gate voltage. As a result, turning current off by decreasing gate voltage is not as sharp as the one normally obtained in a surface-channel device Fig. 2 shows that a slow turn-off rate and a leakage in the range of pA to nA at V_{GS}=0 V are commonly observed in a buried-channel pMOS FET. An obvious possibility is to replace n^+ poly-Si by p^+ poly-Si for CMOS gate electrodes, but then the problem will appear in nMOS FETs, though possibly to a lesser degree. Another option is to use n^+ poly-Si and p^+ poly-Si for nMOS and pMOS FETs respectively. This approach however would greatly complicate the CMOS process. Refractory metals (tungsten and molybdenum) [6,7] and their silicides [8] have been directly applied on CMOS gate oxide. A pure tungsten or molybdenum gate offers a work function value near the middle of the silicon bandgap, resulting in nearly identical work function differences (ϕ_{ms} values) for both an n- and a p-type substrate [9]. Consequently, symmetric and adequate threshold voltages can be obtained for scaled n- and p-channel transistors while buried-channel operation is eliminated or greatly reduced. Fig. 3 shows the subthreshold characteristics of half-micron Mo-gate nMOS FETs with various channel doping concentrations (N_C). With reasonable doping concentrations($N_C \geq 8 \times 10^{15} cm^{-3}$), good turn-off rate with adequate threshold voltage (0.6-0.8 V) can be obtained [7]. Experimental CMOS transistors have also been made using $MoSi_2$ gates. A significant reduction in buried-channel effect can be seen from Fig. 4a. Notice the thinner buried p-layer associated with the $MoSi_2$ gate when compared with the n^+ poly-Si gate. Consequently, short-channel effects such as threshold lowering in the saturation region is improved (Fig. 4b). It was also shown [8] that with careful device design, $MoSi_2$ gate can provide current driving ability higher than that offered by n^+ or p^+ gate for both nMOS and pMOS FETs.

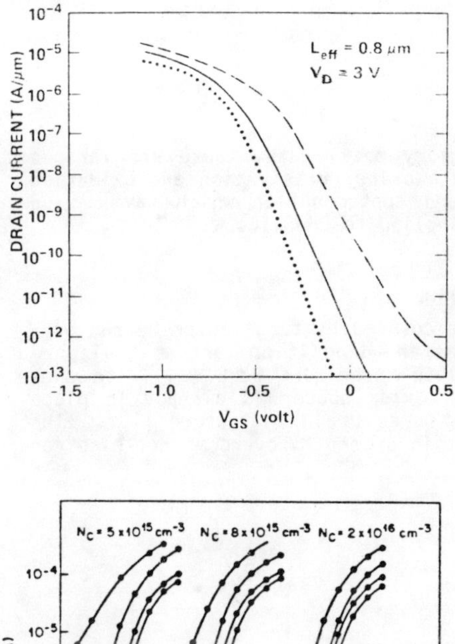

Fig.2 Subthreshold leakage for surface- and buried-channel MOSFETs; dotted line: surface-channel nMOS, dashed line:buried-channel pMOS, solid line:buried-channel pMOS with a thinner buried p layer (ref. 2 & 5).

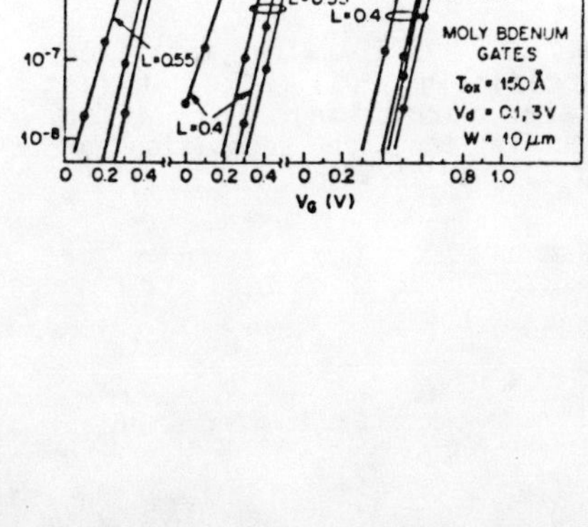

Fig.3 Simulated subthreshold characteristics for half-micron FETs with Mo gate (ref. 7).

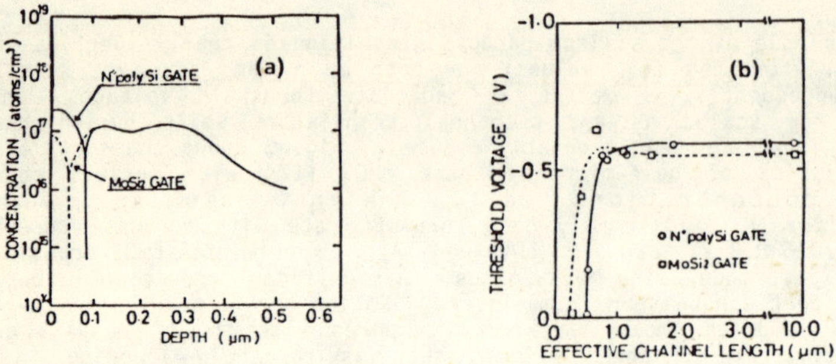

Fig.4 Buried-channel pMOS FETs with $MoSi_2$ (dashed lines) and n^+ poly-Si (solid lines) gates; (a) net impurity profiles in the channel region, (b) threshold lowering (ref. 8).

In spite of the advantages of a refractory metal gate, there are various process issues including adhesion on gate oxide, passivation and oxidation of the pure metal or silicide gate, and contamination which have to be resolved before this technology can be applied to CMOS ICs.

Tungstenized Or Silicided Source And Drain

CMOS source/drain (S/D) areas can be covered by tungsten or refractory-metal silicides using selective tungsten deposition or self-aligned silicidation technique respectively. Both techniques work well with one-micron MOS technology in which sidewall oxide spacer is already in place (Fig. 5a) for implementing LDD (Lightly-Doped Drain) structure [10]. The LDD structure is needed for 1 μm MOSFETs in order to overcome hot-electron

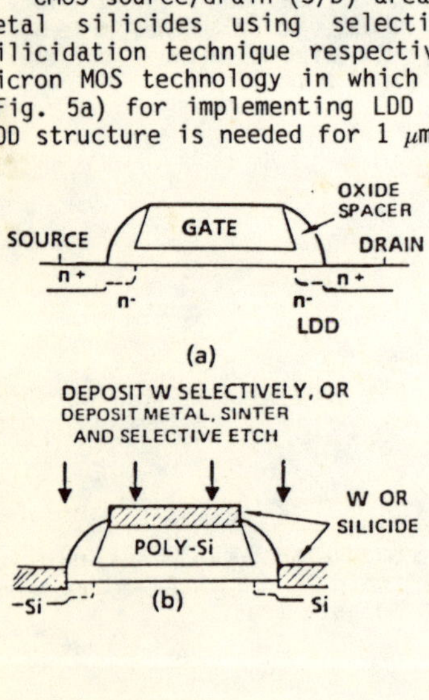

Fig.5 A MOSFET with metalized source/drain: (a) after oxide spacer formation, (b) after selective tungsten deposition or self-aligned silicidation.

induced device degradation. The sidewall oxide spacer separates the gate electrode from S/D regions and allows selective W deposition or selective silicidation. As a result, W or silicide only forms on poly- and monocrystalline Si surface, not on the oxide spacer (Fig. 5b). The associated advantage of reducing parasitic series resistance by an order of magnitude is well-known. Metal to S/D contact resistance is also reduced because the contact resistivity for metal to tungsten (or silicide) contact is low and the contact area between tungsten (or silicide) to Si is the entire S/D surface which is large. For example, if tungsten (W) or titanium/tungsten (TiW), with or without aluminum (Al) above it, is used for first-level metal and the entire S/D is covered by W, then the contact resistance at the contact hole is negligible. Meanwhile, W to Si contact resistance is low due to the large contact area between them. The reduction of both S/D series resistance and metal-Si contact resistance by using tungstenized or silicided S/D technique greatly decreases the intra-device parasitic resistance, allowing CMOS down-scaling below half-micron. As shown in Fig. 6 [11], with S/D silicidation, CMOS transistor gains scale well with L^{-1} until L reaches 0.3-0.4 μm. Notice the dramatic difference for pMOS gains with and without silicide.

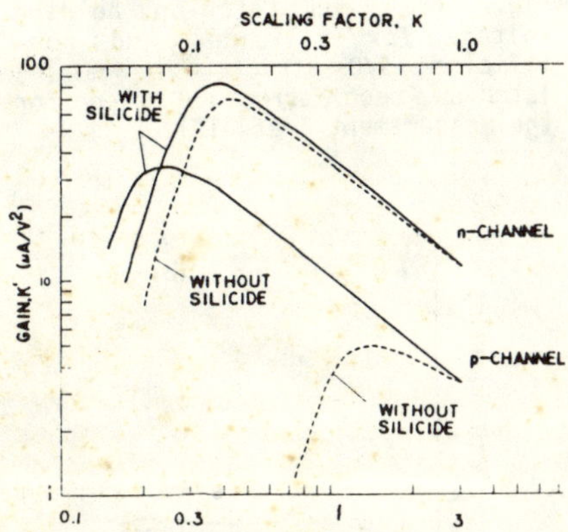

Fig.6 Scaling CMOS transistor gains with and without S/D silicidation (ref. 11).

Another advantage which has not yet been greatly appreciated is the improvement of latch-up immunity with S/D areas silicided in a CMOS circuit [12]. Fig. 7a and 7b show the measured holding voltages versus n^+-to-p^+ spacing(s) and the test site used for measurement. For the silicided case, Si surface of the entire n^+ and p^+ regions (shown as the bold lines in Fig. 7b) are converted to $TiSi_2$. It is clear that the holding voltage increases significantly by using $TiSi_2$. The higher the holding voltage is, the better the latch-up immunity is with a total latch-up free condition existing if the holding voltage is higher than the supply voltage [13]. This improvement is attributed to the poor emitter efficiency caused by reduced emitter width and enlarged effective emitter area when $TiSi_2$ is applied. The emitter width is reduced due to Si consumption during the silicidation process. The effective emitter area is enlarged because the surface recombination velocity is increased when the entire n+ or p+ surface is covered by $TiSi_2$ instead of SiO_2. Poor emitter efficiency spoils bipolar gains for both n/p/n and p/n/p transistors, thereby making latch-up difficult to occur [14].

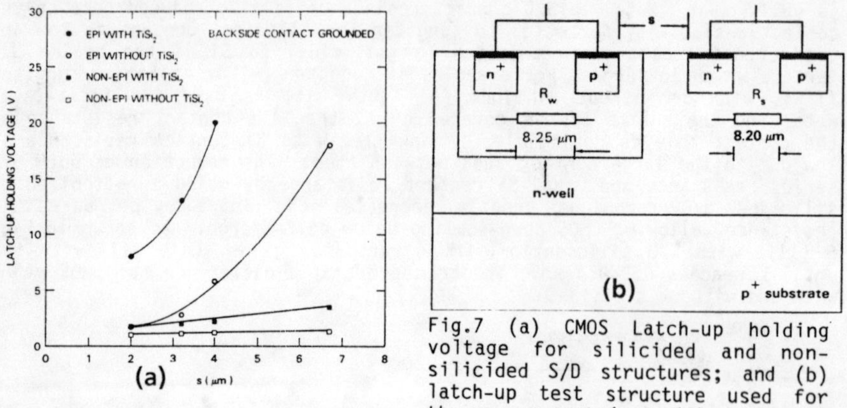

Fig.7 (a) CMOS Latch-up holding voltage for silicided and non-silicided S/D structures; and (b) latch-up test structure used for the measurement (ref. 12).

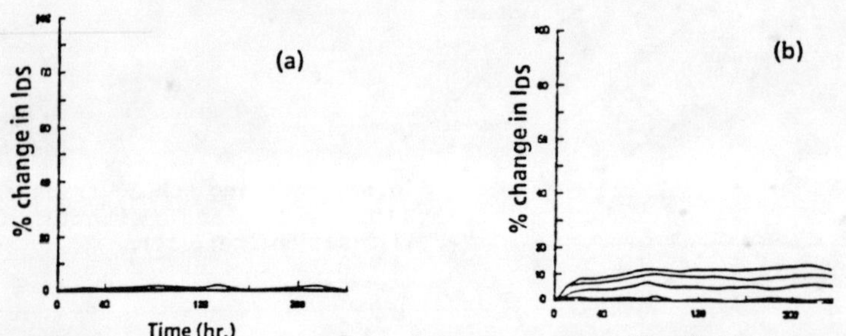

Fig.8 Hot-electron induced device degradation (a) without silicided S/D and (b) with silicided S/D (ref. 11).

S/D silicidation has, however, caused additional concern in long-term MOSFET reliability. As mentioned before, hot-electron effects in a short-channel MOSFET are mitigated by the use of LDD structure. Consequently, long-term device instability commonly expressed as threshold shift and current degradation is minimum. This is indeed true for the experimental LDD devices shown in Fig. 8a [11]. However, current reduction after long-term stressing is observed when these devices are made with silicided S/D (Fig. 8b) [11]. The exact reason for this silicidation-induced degradation has not yet been made clear. One possible explanation is a current crowding effect. In the silicided devices, the majority of the drain current flows within the highly-conductive and very thin silicided layer at the S/D surface. Since the electric field (E) is highest at the Si surface near the drain, impact ionization and hence hot electron generation is enhanced due to the increase **J•E** product where **J** is current density. Another possibility is that additional oxide traps are generated in the gate or spacer oxide during the silicidation process in which sputtering damage and/or silicide film stress may be introduced. While the current crowding is associated with device physics and is basically unavoidable if S/D is silicided, the process-induced damage probably can be reduced or eliminated. Both effects however have to be investigated in more detail before solutions are sought.

Tungsten Or Titanium/Tungsten Alloy For Interconnects

Lastly, the use of CVD tungsten [15] or sputtered titanium/tungsten (TiW with 10 wt.% Ti) [16] alloy for contacts and interconnects offers contact resistivities to n^+ and p^+ Si in the range of 10^{-7} Ω-cm, lower than Al-Si contacts especially for contacting n^+ Si. Moreover, the W or TiW to Si contact is stable even after higher temperature anneal. Fig. 9 shows the Auger depth profiles for as-deposited and annealed Al-Si contacts and annealed TiW-Si contact [17]. As can be seen from the data in Figs. 9a and 9b, a 450 °C anneal has resulted in a thorough intermixing of Al and Si. In contrast, Fig. 9c shows that for the TiW-Si contact, a 550 °C anneal has

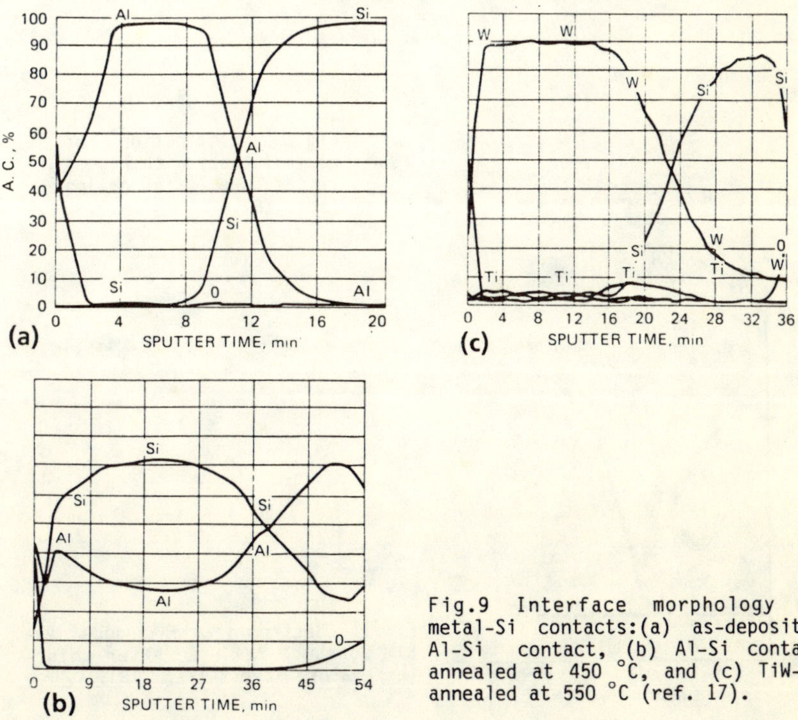

Fig.9 Interface morphology at metal-Si contacts:(a) as-deposited Al-Si contact, (b) Al-Si contact annealed at 450 °C, and (c) TiW-Si annealed at 550 °C (ref. 17).

produced no detectable inhomogenous diffusion. Higher temperature anneal is desirable to remove neutral traps in MOS gate oxide. As shown in Fig. 10, neutral traps are first induced by electron-beam exposure and subsequently filled by injected electrons, causing a threshold shift [18]. The number of neutral traps however can be reduced by annealing the e-beam exposed devices at a temperature between 400-550 °C . The removal of neutral traps can improve device reliability because otherwise they can capture hot electrons which exist in micron and submicron MOSFETs. Higher temperature annealing, allowed only by using refractory metal contact, is a more effective means to remove neutral traps.

Another reliability consideration is electromigration-induced contact failure. The electromigration related problem can be very severe and may place a limit in CMOS scaling [19]. It has been shown [20] that leakage current from an n^+-to-p-substrate diode with Al-Si contact increases rapidly after the device has been stressed at 225 °C for 10 hours (Fig. 11). When this problem occurs in NMOS technology it is commonly solved by an n-type implant after the contact holes are opened. The additional implant (known as plug implant) is difficult to implement in a CMOS process because both n^+ and p^+ exist under the contact holes. However, with TiW underneath Al for interconnect on shallow n^+ junctions, no leakage is detected after several hundreds of hours stress at 180 °C [21].

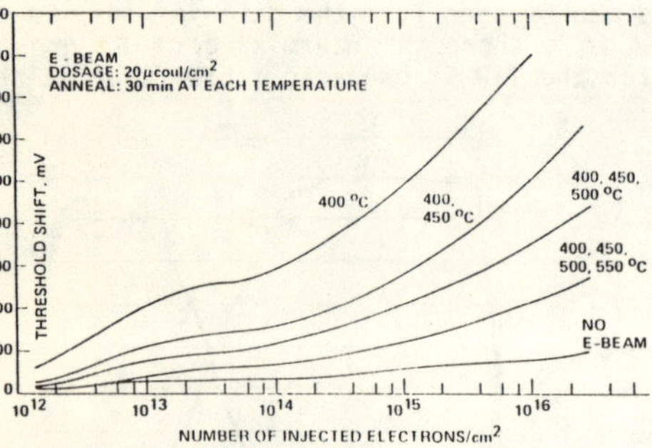

Fig.10 Elimination of neutral traps with higher temperature annealing (ref. 18).

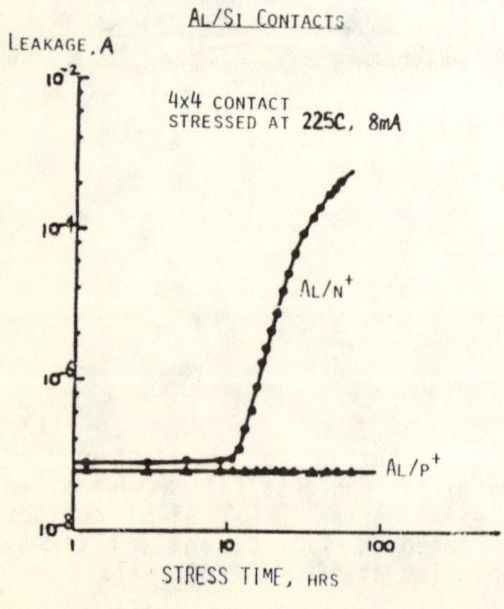

Fig.11 Electromigration-induced Al-Si contact failure after high-temperature stress (ref. 20).

SUMMARY

The use of tungsten and other refractory metals not only improves circuit speed through the reduction of parasitic resistances, it also offers other advantages for CMOS devices in VLSI. Replacing doped polysilicon by tungsten or molybdenum or its silicide for MOS gate electrodes can provide adequate threshold voltages and sharp turn-off characteristics for both nMOS and pMOS FETs in a CMOS IC. When silicided source and drain are applied to a CMOS circuit, latch-up immunity is also enhanced due to the deterioration of parasitic bipolar transistors. On the other hand, silicided nMOS FETs are found less reliable due to the increase hot electron effect. Tungsten or titanium/tungsten interconnects allow higher temperature annealing which is more effective in removing oxide neutral traps. Consequently, less hot electron trapping and better device stability can be expected. Lastly, with the refractory metal contacting shallow CMOS junction, electromigration failure is greatly reduced.

REFERENCES

[1] R. Chwang and K. Yu, VLSI Design, 4 42, (1981).

[2] J. Y. Chen, IEEE Circuits and Devices Mag. 2 (2), 16 (1986).

[3] J. D. Meindl, IEEE Trans. Electron Dev. 31 (11), 1555, (1984).

[4] T. N. Nguyen and J. D. Plummer, presented at 1981, Device Research Conf., abstract published on IEEE Trans. on Electron Dev. 29, 1663, (1982).

[5] G. J. Hu, C. Y. Ting, Y. Taur and R. H. Dennard, in IEDM Digest 1982, p. 710.

[6] N. Kobayashi, S. Iwata, N. Yamamoto, and T. Terada, in the Dig. of Symp. on VLSI Technology, 1983, p. 94.

[7] S. J. Hillenius and W. T. Lynch, in the Proc. of IEEE Inter'l Conf. on Computer Design: VLSI in Computers, Port Chester, NY, 1985, p. 147.

[8] T. Noguchi, Y. Asahi, M. Nakahara, K. Maeguchi and K. Kanzaki, in the Dig. of Symp. on VLSI Technology, 1986, p. 19.

[9] J. Y. Chen and L. B. Roth, Solid State Technology, 145, (1984).

[10] P. J. Tsang, S. Ogura, W. W. Walker, J. F. Shepard, and D. L. Critchlow, IEEE Trans. Electron Dev. 29, 590 (1982).

[11] R. Haken, R. Chapman, T. Holloway, C-F Wan, D. Bell, B. Gale, and T. Tang, in the Proc. of IEEE Inter'l Conf. on Computer Design: VLSI in Computers, Port Chester, NY, 1985, p. 151.

[12] F. S. J. Lai, L.-K. Wang, Y. Taur, J. Y.-C. Sun, K. E. Petrillo, S. K. Chicotka, E. J. Petrillo, M. R. Polcari, T. J. Bucelot, and D. S. Zicherman, IEEE Trans. on Electron Dev. 33, 1308 (1986).

[13] G. J. Hu and R. Bruce, IEEE Electron Dev. Lett. 5 (6), 211 (1984).

[14] D. B. Estreich and R. W. Dutton, IEEE Trans. on Computer-Aided Design of Integrated Circuits and Systems, CAD-1, 157 (1982).

[15] S. Swirhun, K. C. Saraswat, and R. M. Swanson, IEEE Electron Dev. Lett. 5 (6), 209, (1984).

[16] J. Y. Chen, and D. B. Rensch, IEEE Trans. Electron Dev., 30 (11), 1542 (1983).

[17] J. Y. Chen, G. Eckhardt and L. D. Hess, in Semiconductor Silicon 1981, edited by H. Huff, R. Kreigler and Y. Takeishi, The Electrochem. Soc. Inc., 81-5, 694 (1981).

[18] J. M. Aitken, IEEE J. of Solid State Circuits, 14, 294 (1979).

[19] M. H. Woods and B. L. Euzent, in the Dig. of IEDM, 50 (1984).

[20] J. Chern, W. G. Oldham, and N. Cheung, in the Extended Abstract of 1984 ECS Fall Meeting, New Orleans, 84-2, 855 (1984).

[21] H. VanderPlas, (private communication).

THIN-FILM DIFFUSION BARRIERS FOR METAL-SEMICONDUCTOR CONTACTS

M-A. NICOLET
California Institute of Technology, Pasadena, CA 91125

ABSTRACT

The notion of a diffusion barrier is discussed as it applies to thin films, and to metal-semiconductor contacts in particular. The electrical requirements on the barrier material are easily met; those of the thermodynamic stability of the barrier and of the low atomic diffusivity in the barrier are not. The critical influence of the deposition process on the latter property is pointed out. Several different ways to solve the problem are discussed by examples.

INTRODUCTION

To impart stability to a metal-semiconductor contact by inserting a kinetic barrier in the form of a thin film is difficult, because the barrier layer has to be limited to a thickness of the order of 100 nm or less. The time required to penetrate through a given thickness by a diffusion process scales with the square of the thickness. A 1 mm thick diffusion barrier that will fail after 10 years does so already after 3 sec when reduced to 100 nm. Very low atomic diffusivities are thus essential for thin-film diffusion barriers. In addition, a thin-film diffusion barrier must also meet conditions that are imposed on its physical and chemical characteristics by virtue of its function in the contact and on a chip, as well as by practical requirements of manufacturing. Compromises are therefore unavoidable. To find acceptable ones requires an appreciation of the issues and trade-offs involved.

This paper considers briefly some of these questions. For comprehensive treatments and references, the reader is referred to recent reviews [1-3].

A diffusion barrier does not eliminate the thermodynamic instability of a metal-semiconductor combination. A diffusion barrier is a "kinetic barrier", that is, it retards the rate of progression of a degradation process and thereby provides stability which is necessarily only temporary. Whether a diffusion barrier should be included in the design of a device is an economic issue. This paper concerns itself only with the technical aspects of thin-film barriers.

CONCEPTS

An electrical contact to a semiconductor must fulfill two major conditions: (i) It must satisfy a prescribed electrical characteristic, and (ii) this characteristic must be stable in time. In the implementation of Fig. 1, both these conditions must be met by the same layer X. Although the scheme is simple, it is easier by far to find acceptable solutions to both requirements by solving each condition separately.

This engineering approach naturally leads to the generic contact scheme shown in Fig. 2. The <u>contacting layer</u> determines the electrical characteristics of the contact (ohmic or rectifying). The material of this layer is selected to <u>realize a prescribed electrical interface with the semiconductor</u> that satisfies the requirements imposed on the contact by device considerations. In addition, this contacting layer/semiconductor

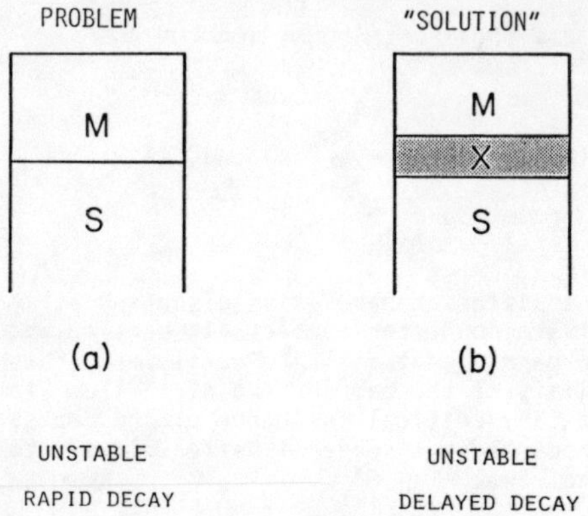

Fig.1 A thin-film diffusion barrier has a typical thickness of 100 nm or less. It does not remove an existing instability but only retards its rate of progression. When the layer X is in immediate contact with the semiconductor, X must satisfy electronic requirements at the barrier/semiconductor interface, as well as requirements imposed on its structure to fulfill its role as a diffusion barrier.

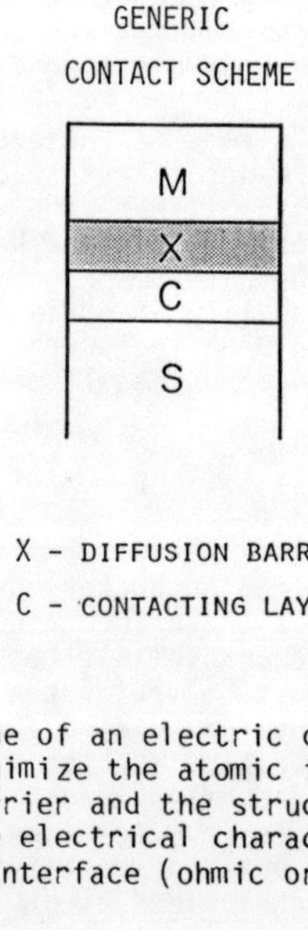

Fig.2 In this generic scheme of an electric contact, the diffusion barrier layer X serves to minimize the atomic interaction between the metal layer M above the barrier and the structure below it. The contacting layer C optimized the electrical characteristics of the contacting layer/semiconductor interface (ohmic or rectifying).

configuration has to be able also to sustain subsequent processing steps without significant degradation. In silicon VLSI circuits, this contacting layer is typically made of a transition metal silicide. For GaAs and other III-V compounds, satisfactory contacting layers are presently lacking; to find some is the subject of ongoing searches. The thin-film diffusion barrier minimizes the interaction between the metallic overlayer and the contact configuration below. Ideally, the diffusion barrier is electronically transparent and atomically opaque. Conceivably, a good thin-film diffusion barrier can serve well on devices involving different semiconductors and metal overlayers. Contacting layers, in contrast, are semiconductor-specific.

"Electrically transparent" means that the voltage drop across the barrier layer should be negligibly small. Assuming a current flow of 10E6 A/cm^2 perpendicularly to the plane of a 100 nm film, and an acceptable voltage drop of 25 mV (=kT/q), resistivities on the order of 1 mOhm-cm or less are sufficient to qualify a material electrically. This condition is met by many substances, and includes heavily doped semiconductors, amorphous metallic alloys, and conducting oxides. To meet the electrical requirement of a diffusion barrier is thus easy; even poor electrical conductors are admissible for this purpose.

To appreciate when and how atoms diffuse through solid films is essential to the design of effective thin-film diffusion barriers. There are many possible diffusion mechanisms in solids. Their relationship to atomic diffusion in thin films is depicted schematically in Fig. 3. At high temperatures, diffusion mechanisms through the bulk of the material dominate. At low temperatures, diffusion along extended defects of the material (such as grain boundaries or dislocations) is faster than through the bulk. The activation energy for these processes typically scales with the melting point of the element [4]. The temperature of transition from bulk to defect mechanisms depends on the density of these extended defects, and lies typically near half the absolute melting point for polycrystalline materials. More important than the actual location of this transition point is the recognition that if defects determine the diffusivity, the diffusivity becomes a function of the manner in which the film is deposited, because it is the deposition process that determines the defect structure of a film. In working with and discussing thin-film diffusion barriers, it is therefore essential to pay as much attention to the methods of film deposition as to the choice of the material.

IMPLEMENTATIONS

Stuffed Barrier

How important the deposition method is in determining the performance of a thin-film barrier was demonstrated by an experiment with a Ti-W film deposited between a Si substrate and a Au overlayer by rf sputtering with argon [5]. Two depositions were made in which all sample processing steps were identical, except for the reduced base pressure achieved in one case over the other by a longer pump-down period before back-filling with argon (see Fig.4). A subsequent thermal annealing of the two samples under an SEM microscope revealed that the barrier deposited with the lower base pressure (samples shown on the left side of the frames in Fig. 4) failed to prevent Au-Si intermixing at temperatures at which the other sample showed no signs of degradation. An AES analysis indicated that the good barrier contained about 5 at.% oxygen impurity and the poor one only 2 at.%. The better performance of the good barrier may thus have had its origin in the elevated content of oxygen (although other impurities - not monitored in the experiment - may have been at cause as well).

If one assumes that a polycrystalline material consists of cubic grains

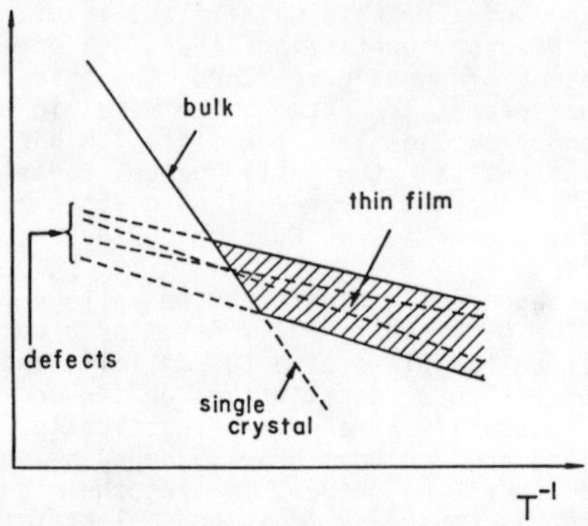

Fig.3 Typical dependence of atomic diffusivity in a solid as a function of reciprocal absolute temperature. At elevated temperatures, bulk processes dominate. At low temperatures, diffusion is primarily along extended defects. Their number and nature depend on the deposition process of the film. Diffusivity in thin films at low temperatures thus is a process-dependent quantity.

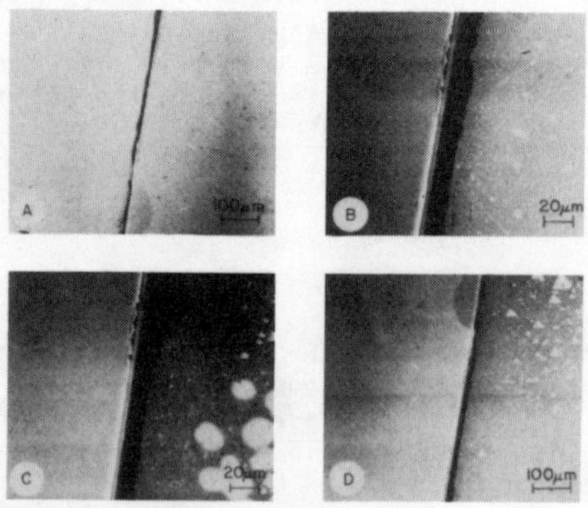

Fig.4 The notion of a stuffed diffusion barrier explains why seemingly small variations in a thin-film deposition process can lead to major differences in behavior (from [5]).

50 nm on edge, it takes only 1.6 at.% of impurity atoms to completely decorate all grain boundaries. A full decoration may not be necessary to impede the transport of atoms along fast diffusion paths, and there may not be as many extended defects to plug as this simple geometric model assumes. The main point of the argument is that a total of less than 1 at.% of impurities may have a big effect on the atomic diffusivities of thin films.

A number of experiments performed on films deposited in UHV systems with and without intentionally added impurities suggests that thin films of elemental metals are poor diffusion barriers when they are pure, because fast diffusion paths will promote rapid interdiffusion. With impurities, however, the same elemental film may serve well. Films whose good diffusion barrier performance depends on the presence of minor foreign constituents have been referred to as "stuffed barriers" to indicate that their effectiveness rests on impurities. Stuffed barriers are valid solutions to the thin-film diffusion barrier problem. The danger lies in not recognizing that a barrier is of this type.

Tungsten seems to be an exception to the rule that pure elemental thin films are poor diffusion barriers. How little impurity is needed to alter the diffusivities along fast diffusion paths is not normally known. Tungsten may be particularly sensitive in this respect. To clarify this issue, in general, requires careful experiments in UHV systems with ultra-pure metallic charges and an accurate tracing of impurities. Conclusive experimental evidence that a solute segregates along defects and alters diffusivities there has so far been obtained only in a few instances [4]. Until such experiments are carried out for W, the reason for its good diffusion barrier properties remains uncertain.

The notion of the stuffed barrier is a useful concept that explains experimental observations simply, but one should be aware that it has yet to be validated in most instances by direct experimental proofs relating cause and effect.

Refractory Compound Barriers

The only alternative to elemental films is that of compounds and alloys. These materials are usually polycrystalline as well, so that fast diffusion paths exist there also. What is new here is that compounds can have very high melting points. Since the activation energy for diffusion along extended defects tends to scale with the melting point, polycrystalline materials with a high melting point are advantageous for diffusion barrier applications.

Interstitial alloys form a group of refractory compounds that have shown much promise as thin-film diffusion barriers. They are binary alloys of the early transistion metals (Ti, V, and Cr goups) with the metalloids B, C, and P [6,7], they are metallic in character and have extremely elevated melting points. They must therefore be synthesized as part of the deposition process. Deposition by chemical vapors and reactive sputtering are the most common methods.

TiN has been studied most for thin-film diffusion barrier applications. Its commercial use in electrical contacts goes back to the beam-lead technology of the 60's, but interest in its systematic investigation and application is of recent date. Successful applications have been reported in III-V solid-state injection lasers, GaAs microwave devices, and Si-based semiconductor devices and VLSI circuits (see Fig. 5). TiN also forms upon thermal annealing of Ti films in a nitrogen ambiant if oxygen is absent. This fact is exploited in combination with rapid thermal annealing to simultaneously form a self-aligned $TiSi_2$ contacting layer to a shallow junction in a contact opening, and a TiN layer above it as a diffusion barrier [9]. Over the field oxide, a conducting layer (of a general composition Tin_xO_y) forms that is subsequently removed, or that can serve as

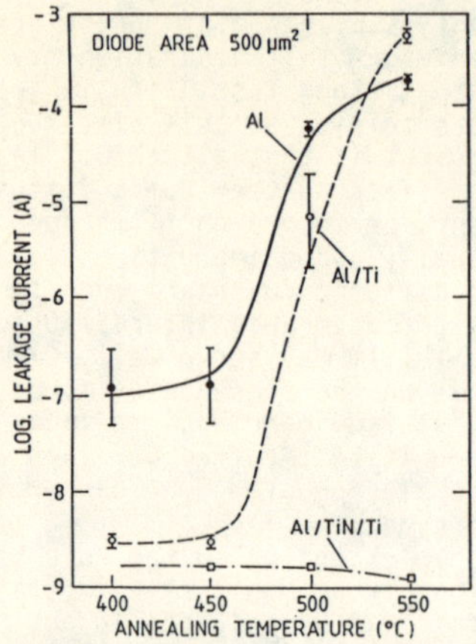

Fig.5 A successful application of the reactively sputtered refractory compound TiN as a thin-film diffusion barrier (from [8]). The figure shows the leakage current of a reverse biased Si n^+p junction with Al/Si, Al/Ti/Si and Al/TiN/Ti/Si contact systems as a function of the temperature of an annealing treatment in flowing argon for 15 min.

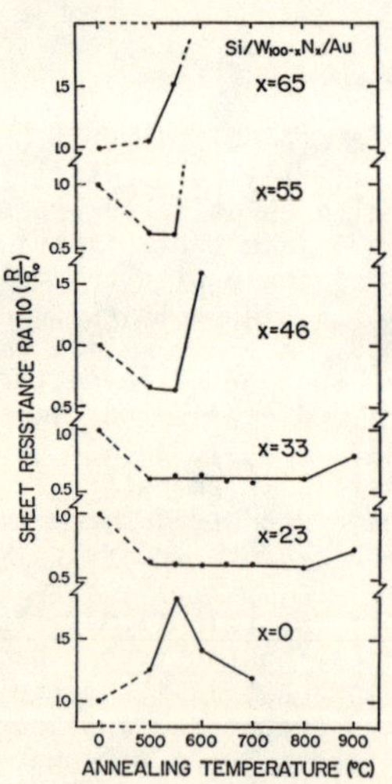

Fig.6 A successful application of the amorphous alloy $W_{100-x}N_x$ as a thin-film diffusion barrier between a Au overlayer and a Si substrate (from [11]).

a local interconnection. On Si with an Al overlayer, a TiN barrier of about 100 nm can prevent degradation upto about 500 °C for 30 min annealing.

The TiN/Al system is thermally unstable; the dominant failure mode is a locally triggered reaction. One difficulty of TiN films that are reactively sputtered in a argon/nitrogen gas mixture is the large compressive stress (on the order of GPa). Kanamori has recently reviewed the properties of TiN thin films for diffusion barrier applications [10]. Solute atoms and/or non-stoichiometric composition can of course lead to segregation along grain boundaries as well; the distinction between compound barriers and stuffed barriers may then be difficult to make.

Amorphous Alloy Barriers

Amorphous metallic alloys are attractive as diffusion barriers because they have no extended structural defects of the kind that produce fast diffusion paths in cyrstalline materials. (Recent research indicates, however, that not all amorphous materials have necessarily low atomic diffusivities.) Amorphous materials are also metastable. This is not a serious obstacle; some amorphous metallic alloys are known to crystallize above 600 to 800°C. One difficulty is that amourphous alloys tend to react, so that it is important to minimize their reactivity.

Pure metals are not amorphous at room temperature. Amorphous metallic alloys are therefore necessarily multi-elemental. One class of amorphous metallic alloys combines two transistion metals. Because near-noble transistion metals react readily at fairly low temperatures with Si, binary alloys that do not contain these metals are preferable for contacts to Si devices [11]. Another class of amorphous metallic binaries combines transistion metals with metalloids (B,C,P,Si). Current research suggests that some of these alloys have promising properties for diffusion barrier applications.

An example is N-W which, in its crystalline form, is an interstitial compound. In the composition range of 10 to 30 at.% N, reactive sputtering produces the amorphous phase [12]. Such a film, about 100 nm thick, has prevented metallurgical interactions between a Au overlayer and a Si substrate above 700 °C annealing for 30 min (see Fig.6). The Au-Si eutectic temperature is 360 °C. A highly pinhole-free film is required to endure this treatment.

CONCLUSION

As this sketchy review indicates, the subject of thin-film diffusion barriers for metal-semiconductor contacts combines structural, chemical and electronic aspects of materials. When the two basic requirements of a contact, meaning its electrical characteristics and its thermal stability are met by two distinct layers, as shown in Fig. 2, the constraints on the electrical resistivity of the diffusion barrier material are weak. This opens the way to many possible choices of materials, including doped semiconductors, metallic alloys, and conducting oxides in crystalline or amorphous forms. Only few have been investigated so far.

A thorough experimental investigation of any particular system is a demanding undertaking that normally has to include details of the film deposition process. These particularities of the subject make it likely that continued good progress can be expected from investigations of new materials and/or deposition techniques inspired by ideas and concepts outlined in this paper. The clear understanding of these concepts on a fundamental level will advance at a moderate pace only. Progress there requires carefully planned experiments, executed with the advanced tools of thin-film synthesis and analysis that modern materials research now has, and dedicated financial backing.

ACKNOWLEDGEMENTS

This work was supported by the Army Research Office under contract number DAAG29-85-K-0192.

REFERENCES

1. M-A. Nicolet, Thin Solid Films 136, 415 (1978).
2. M. Bartur and M-A. Nicolet, J. Electrochem. Soc. 131, 1118 (1984).
3. M. Wittmer, J. Vac. Sci. Technol. A2, 273 (1984).
5. J.E. Baker, R.J. Blattner, S. Nadel, C.A. Evans Jr., R.S. Nowicki, Thin Solid Films, 69, 53 (1980).
6. H.J. Goldschmidt, Interstitial Alloys, (Plenum Press, New York, 1967).
7. L.E. Toth, Transition Metal Carbides and Nitrides, (Academic Press, New York, 1971).
8. I. Suni, M. Blomberg, J. Saarilahti, J. Vac. Sci. Tech. A3, 2233 (1985).
9. M. Delfino, E.K. Broadbent, A.E. Morgan, B.J. Burrow and M.H. Norcott, IEEE Electron Dev. Lett. EDL-6, no 11, 591 (1985).
10. S. Kanamori, Thin Solid Films 136, 195 (1986).
11. R.E. Thomas, J.H. Perepezko, J.D. Wiley, Mat. Res. Soc. Symp. Proc., Vol. 55 (1985).
12. H.P. Kattalus, E. Kolawa, K. Affolter, M-A. Nicolet, J. Vac. Sci. Technol. A3, 2246 (1985).

THE PHYSICS AND CHEMISTRY OF THIN "NATIVE" OXIDE FILMS ON SILICON

BRUCE E. DEAL AND DAH-BIN KAO
Fairchild Research Center, 4001 Miranda Ave., Palo Alto, CA 94304

ABSTRACT

Thin "native" oxide films on silicon, resulting from various surface treatments or even exposure to room ambient, play a critical role in the fabrication of high performance, reliable VLSI semiconductor devices. The nature of these 0.5 to 2.0 nm films is discussed with respect to chemical, physical, and electrical properties. Particular emphasis is placed on the effects of individual chemical species, used in common cleaning processes, along with those of various impurities contained therein. Also, effects of these films on the characteristics of specific device structures, such as MOS gate dielectrics, metal-semiconductor contacts, and poly-Si emitters, are briefly reviewed.

INTRODUCTION

The preparation of bare, silicon surfaces has been an important aspect of silicon device technology since the diffused transistor was first developed in the 1950's [1], and even more so with the currently used Planar structure - first invented in 1960 [2]. In the latter, the active junctions are passivated by the masking thermal oxide after the dopants diffuse laterally under this oxide. In addition, openings have to be etched through the passivating oxide to establish contacts to the active device region. It was observed early on that thin layers of oxide material form very rapidly on the silicon surface after removing the thermal oxide, and that these so-called "native" oxides critically affect the electrical properties of the metal-silicon contacts. It was also noted that the nature and composition of these thin oxides vary considerably, depending on the methods used for etching and cleaning the silicon, and also depending on the ambient conditions such as humidity, illumination and temperature.

The problems of achieving reproducible silicon surfaces with minimum oxide thicknesses for obtaining reliable, high performance devices have continued up to the present era of VLSI. In fact, various components in today's device structures, such as MOS gate oxides, poly-Si emitters in bipolar circuits, and all types of interconnect contacts are extremely dependent on the nature of the native oxide films produced due to the various chemical treatments used in each application.

In this paper we will briefly review the current understanding of thin, native or surface oxide films on silicon and how they can affect device fabrication. Included will be a description of chemical and other cleaning processes for silicon, a discussion of thin surface films - their types, properties, effects of impurities and also deionized water - and finally, examples of surface effects on specific device structures.

CLEANING PROCESSES FOR SILICON

Various chemical treatments have generally been used for a number of deposition for forming contacts to the silicon. These chemicals include solvents such as TCE or acetone to remove organic impurities, strong oxidizing agents such as sulfuric peroxide, ammonium peroxide, aqua regia ($HCl:HNO_3$), or other acidic mixtures to remove organic or metal residues, and hydrofluoric acid-water mixtures to remove silicon oxide species [3-5]. In some cases mechanical scrubbing procedures are also employed. The final drying step of the silicon wafer is also important with respect to composition and thickness of the surface film. Various drying techniques have included isopropanol vapor (isodry), nitrogen blow-dry, spin-dry in air, and inert gas furnace dry. Typical pre-oxidation cleaning techniques are listed in Table I. In general, native or surface oxide films resulting from these cleaning treatments are 1.0 to 1.5 nm thick without a final HF dip, and approximately 0.5 nm with the HF dip. It should be noted that many current procedures for pre-oxidation cleans, where very thin (10-40 nm) gate oxides are involved, utilize "sacrificial" oxides. This means that a thin 20 nm or so oxide is first grown and then removed in an HF solution immediately prior to the final oxidation process. Significant improvements in gate oxide reliability have been observed using this procedure. It should also be noted that proper DI water rinses between the various chemical treatments are critical, as is the control of impurities in the DI water. This subject is discussed below.

As device technology has advanced, and feature sizes have approached one micrometer or less, other more sophisticated cleaning processes have been developed. Some of these are tabulated in Table II [6-14]. Several of these were developed to satisfy the more demanding requirements of such techniques as Si MBE (molecular beam epitaxy), and can take advantage of the high vacuum atmosphere employed in the latter. One cleaning process which can be used under more conventional applications is the use of ultraviolet illumination in the presence of oxygen [10,12-13]. The latter is converted to ozone (O_3) and acts to oxidize surface impurities. The process also results in a very dense, thin oxide to help passivate the silicon surface until further processing is carried out. An example of a U.V. reactor is shown in Fig. 1, along with a similar microwave discharge system.

As indicated in Table II, an alternate cleaning procedure is to anneal the silicon wafer at elevated temperature in an inert gas or in vacuum. An example of surface oxide removal in a high vacuum anneal is shown in Fig. 2, where oxide thickness removal is plotted against heating time in vacuum at 710°C. This technique is generally suitable only if subsequent processing can be carried out immediately in situ - otherwise a native oxide will be reformed. Of course for certain applications, this might still be satisfactory if a thin, clean native oxide can be tolerated.

Table I. Common chemical pre-oxidation cleans for silicon.

RCA CLEAN	MPR CLEAN
$NH_4OH : H_2O_2 : H_2O$	$H_2SO_4 : H_2O_2$
$HCl : H_2O_2 : H_2O$	$HCl : HNO_3$ (A.R.)
W, W/O $HF : H_2O$	W, W/O $HF : H_2O$
APPROP. DI RINSE	APPROP. DI RINSE

FSC NO. 2A CLEAN	FINAL DRYING TECHNIQUES
HOT TCA	ISOPROPANOL VAPOR
$CH_3CH(OH)CH_3$	BLOW DRY (N_2)
HOT $HNO_3 : H_2SO_4$	SPIN DRY (AIR)
W, W/O $HF : H_2O$	FURNACE DRY (N_2, AR)
APPROP. DI RINSES	

SACRIFICIAL PRE-OXIDATION OPTIONAL IN ALL CASES

Table II. Other silicon cleaning techniques.

- HIGH TEMPERATURE ANNEAL (Vacuum, Inert Gas)
- FLASH HEATING (RTA)
- SPUTTER CLEAN, PLASMA
- UV + O_2 (O_3)
- IN-SITU GAS CLEAN (HCl)
- LASER CLEAN
- Si-BEAM ETCH
- GALLIUM BEAM TREATMENT
- CLEAVAGE

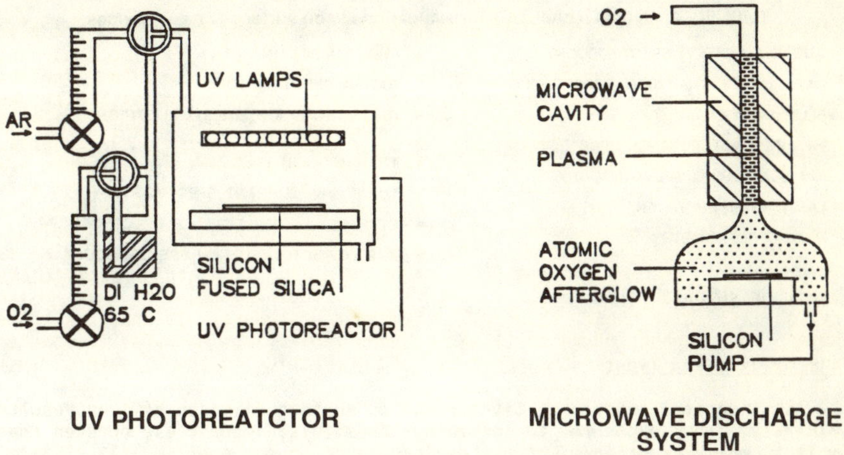

Fig. 1. Apparatus used to generate atomic oxygen for cleaning silicon wafers [13].

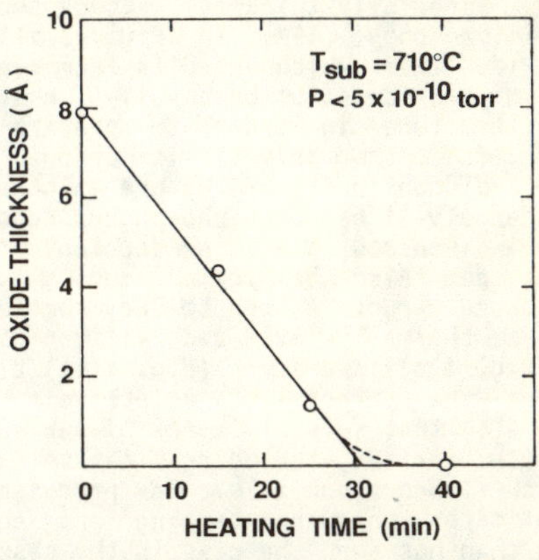

Fig. 2. Native oxide removal from (111) silicon surface as a function of time in ultra-high vacuum [14].

SURFACE ANALYSIS TECHNIQUES

Many analytical techniques have been employed to examine the silicon surface before and after various cleaning procedures. Some, but undoubtedly not all, are listed in Table III. Such analyses have been used to identify species and residues remaining on the surface after cleaning, to determine composition of surface films, and to investigate the location and incorporation of these species in subsequently deposited oxide or metal films. Depending on the analytical technique used, different types of data can be obtained, and it is unlikely that any one method will provide sufficient information to properly characterize a given surface treatment. Some of the more interesting analyses reported include ellipsometry [15-17], x-ray photoelectron spectroscopy [18-20], Auger spectroscopy [21-23], transmission electron spectroscopy [24-25], and ultraviolet reflectance spectroscopy [26].

Table III. Analytical techniques for evaluationsilicon surfaces / native oxides.

- AUGER ELECTRON SPECTROSCOPY
- CAPACITANCE TRANSIENT SPECTROSCOPY
- ELLIPSOMETRY
- INFRARED REFLECTANCE SPECTROSCOPY
- ION SCATTERING SPECTROSCOPY
- LASER ATOM PROBE ANALYSIS
- MASS SPECTROMETRY
- PHOTOINJECTION
- RAMAN SPECTROSCOPY
- REFLECTION HIGH ENERGY ELECTRON DIFFRACTION
- TRANSMISSION ELECTRON SPECTROSCOPY
- TUNNELING CURRENT SPECTROSCOPY
- ULTRAVIOLET REFLECTANCE SPECTROSCOPY
- X-RAY PHOTOELECTRON SPECTROSCOPY

NATURE OF SURFACE FILMS

Film Origins and Types

As indicated above, native oxide or surface films on silicon result primarily from exposure to cleaning chemicals, impurities, or even the ambient which contains water, oxygen, and other species [19,27-31]. Common film origins are listed in Table IV. Also, Table IV indicates that other variables such as temperature and pressure will influence the properties of the films, especially thickness. Cross-sections obtained by transmission electron microscopy (TEM) in Fig. 3 illustrate a typical native or chemical oxide film which results from cleaning the silicon surface in an ammonium peroxide solution. Poly-Si has been deposited at 620°C on the film structure in order to prepare the sample for evaluation. In Fig. 3(a), the poly-Si is undoped which leaves a continuous film, while a discontinuity or rupture is observed in the oxide in Fig. 3(b), where the poly-Si has been phosphorus doped and annealed at 1000°C. Such effects are discussed in a later section.

Native oxide films can also be formed during gaseous or "dry" processing. In fact these structures can be very complex as is indicated in Fig. 4. Here, a thermally oxidized silicon surface is shown schematically before (Fig. 4(a)) and after (Fig. 4(b)) plasma etching in a CF_4/H_2 gas mixture for the purpose of selectively removing the SiO_2 layer. Note in Fig. 4(b) that several layers of various types remain in the silicon and on its surface, including a native oxide-type layer as well as a fairly thick fluorocarbon film. Dry processes, common in VLSI technology, will complicate the understanding and control of silicon surface films even more than has been the case in the past.

Film Properties

Growth kinetics of native oxide films at room temperature are not well understood, since film compositions vary considerably. However, typical kinetic data are shown in Fig. 5 where oxide thickness is plotted as a function of HF-dipped-silicon air exposure time at room temperature. Both x-ray photoelectron spectroscopy (XPS) and neutron activated analysis (NAA) were used to determine film thickness. Note that after 1000 min (16.6 hr), film thickness is 0.6 nm or 6 A.

Composition and stoichiometry of thin native oxides have been determined by a number of analytical methods [15-26]. In most cases, it has been reported that the oxide part of the film is represented by SiO_x. In Fig. 6, pulsed laser atom probe has been used to characterize a native oxide film on silicon (formed after an HF rinse). Most of the 1 to 1.5 nm film has a composition of SiO, with a very thin layer next to the silicon having a composition of $Si_{0.7}O_{0.3}$. These results compare favorably with others obtained by XPS and Auger spectroscopy [36,37]. As might be expected, other impurities such as sulfur and carbon are also

Table IV. Origins of silicon surface films.

- ROOM AMBIENT (H_2O, O_2)
- CLEANING CHEMICALS (HNO_3, NH_4OH, H_2O_2, etc.)
- D I WATER
- IMPURITIES (Carbon, Sulfur, etc.)
- DRY PROCESSING (Plasma, Sputter, etc.)

* *

OTHER VARIABLES - TEMPERATURE
- PRESSURE
- ILLUMINATION
- RADIATION

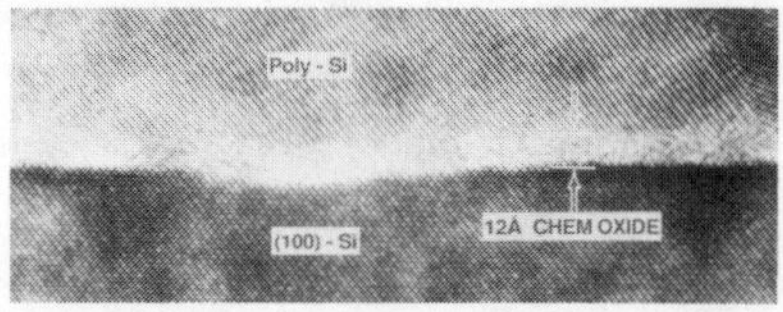

(a) NO PHOSPHORUS DIFFUSION IN POLY-Si

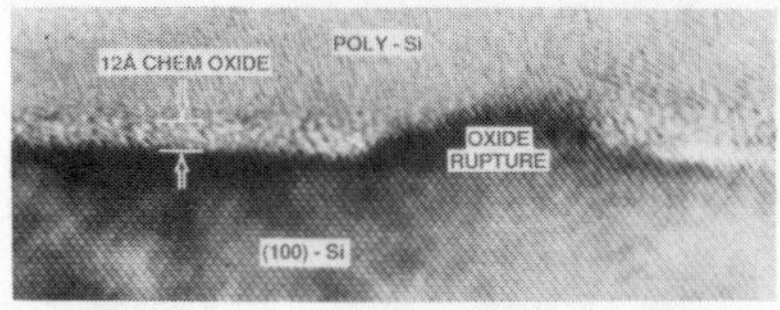

(b) PHOSPHORUS DIFFUSION AND ANNEAL

Fig. 3. Chemical oxide ($NH_4OH : H_2O_2$) on silicon [32].

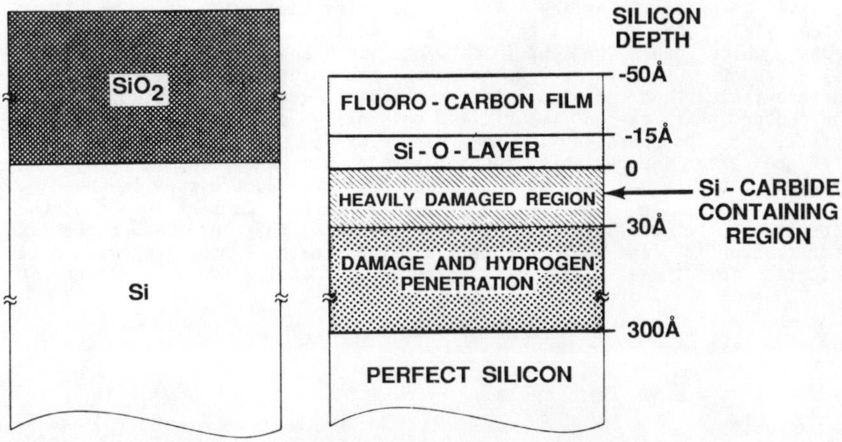

(a) BEFORE ETCHING (b) REACTIVE ION ETCHED

Fig. 4. Surface film due to reactive ion etching of SiO_2 in CF_4 / 40% H_2 [33].

found in these oxides, especially if no HF rinse is employed after the chemical cleaning procedure. Effects of process variables on surface properties, structure, and roughness have been reported [38].

One other important property of native silicon-oxide films is that of wetting or non-wetting (hydrophilic vs hydrophobic) of the oxide surface by water. Investigators have shown that for bare silicon and for thin oxide layers up to 2.5 nm or so, the surface is hydrophobic. Above this thickness, the oxide is hydrophilic. This phenomenon has been used to measure oxide thickness since the wetting contact angle is a function of thickness [39]. This property has also been proposed to be due to oxide composition and charge effects [40], and to the chemical composition [41].

Effects of Impurities

It has already been noted above that not only chemical species such as fluorides, sulfur, etc., from the cleaning solutions can be incorporated

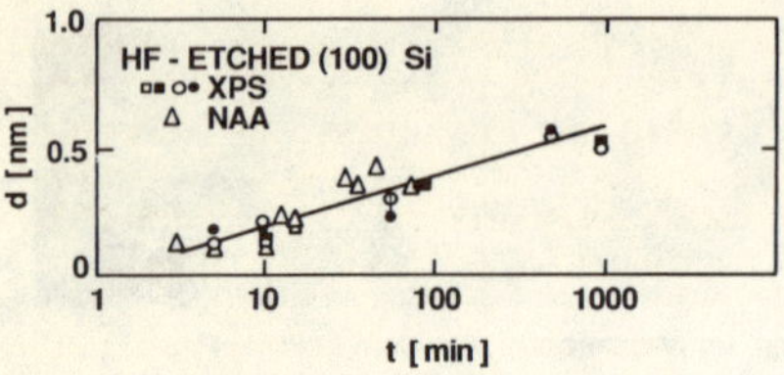

Fig. 5. Silicon oxide thickness (XPS and NAA) vs air exposure time at 25°C [34].

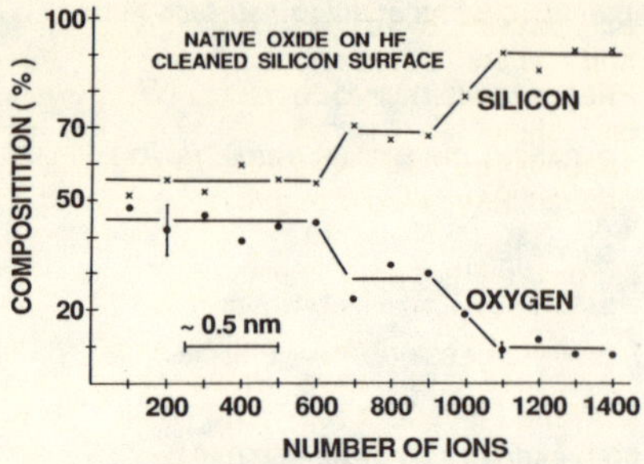

Fig. 6. Pulsed laser atom probe analysis of native oxide film (~1.5 nm) on silicon (HF pretreatment [35].

in the native oxide films formed on silicon, but impurities from these solutions may also be contained in the thermal oxide or even segregated at the silicon-oxide interface. These impurities then quite often result in device yield and reliability problems [42-47]. One likely candidate is carbon, which can be present in many chemicals and in D.I. water as well. Small amounts of carbon, like nitrogen, can change the $Si-SiO_2$ interface characteristics and can also retard subsequent oxidation kinetics. It has been noted that carbon impurities originally present on the silicon surface, or incorporated in the native oxide, will stay at the $Si-SiO_2$ interface after appreciable thermal oxide has been formed [48]. An example of the carbon pile-up determined by Auger spectroscopy is shown in Fig. 7, indicating considerable impurity carbon content in one of the reagents. The possible subsequent effect of this interface carbon is demonstrated in Fig. 8. Here native oxide growth at room temperature is indicated for three different surfaces. Silicon fractured in either air

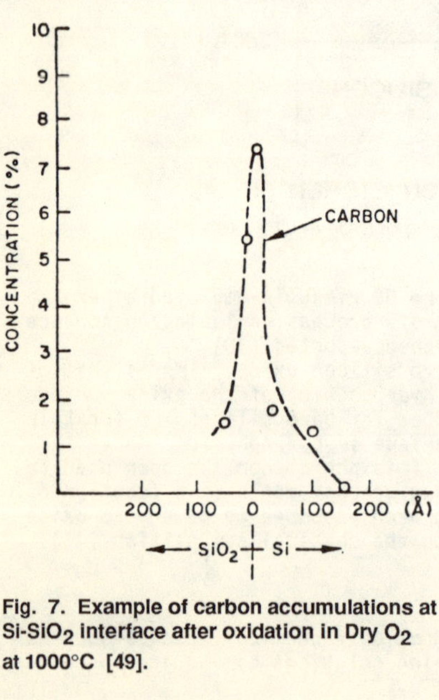

Fig. 7. Example of carbon accumulations at $Si-SiO_2$ interface after oxidation in Dry O_2 at 1000°C [49].

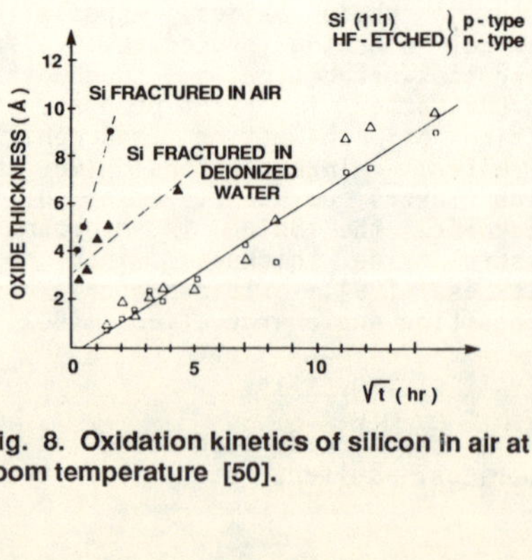

Fig. 8. Oxidation kinetics of silicon in air at room temperature [50].

or water oxidizes much faster than for silicon etched in an HF solution and then exposed to air. Note that after 16 hours ($t^{1/2} = 4$), the HF-treated surface oxide thickness was only 0.2 nm, much less than the 0.6 nm shown in Fig. 5 for the same time. The authors [50] assume that carbon impurities in the HF cause the retardation. They may be right, and this carbon may also retard subsequent thermal oxidation growth rate.

Deionized Water Effects

Up till now, only processing chemicals (or gasses) and impurities contained in them have been discussed with respect to effects on silicon surface properties and the formation of native oxide layers. Mention should be made, however, of possible effects due to impurities in deionized water. Several types of impurities are generally monitored in semiconductor fabrication areas and these acceptable limits are tabulated in Table V. Excess amounts of any of these impurity types can cause detrimental effects on device properties, primarily through initial incorporation in the native oxide. One of these, bacteria, is less understood at present and consequently leads to the most problems regarding analysis and control [52-54]. An example of a recent experiment designed to evaluate the effects of excess bacteria in DI water is shown in Fig. 9. The accumulation of excess numbers of bacteria in the DI water was sufficient to cause considerable yield loss in power transistors through degradation of reverse breakdown. Unfortunately, not too much such data are available in the literature, although similar results which correlate high concentrations of carbon in DI water to yield loss have also been reported [56-57].

SURFACE OXIDE EFFECTS ON DEVICE PROCESSING

Other than the scientific interest in native or surface oxide films on silicon, many important device applications depend on a better understanding of the subject. These include thin thermal oxides for advanced MOS and CMOS circuits, metal-silicon contacts in all types of device structures, poly-Si emitters for VLSI bipolar devices, and various types of "next-generation" structures which involve such technology as molecular beam epitaxy. Each of these examples are briefly discussed below.

Table V. Pure water specifications for VLSI applications [51].

			1985	
TEST	1975	1980	ACCEPTABLE	ATTAINABLE
RESIDUE (ppm)	0.5	0.3	< 0.3	0.1
TOTAL OXIDIZABLE CARBON (ppb)	200	100	< 50	< 20
PARTICULATES (per liter)	4000	1000	<1000	<500
BACTERIA (per 100 ml)	20	< 20	< 6	0
DISSOLVED SiO_2 (ppb)	5	< 3	< 5	<3
RESISTIVITY (meg - ohm)	16.0	17.0	17.9	18.3
IONS - CATIONS (ppb)	-	-	0.1 - 2.0	0.02 - 0.2
IONS - ANIONS (ppb)	-	-	0.1 - 0.3	0.05 - 0.1

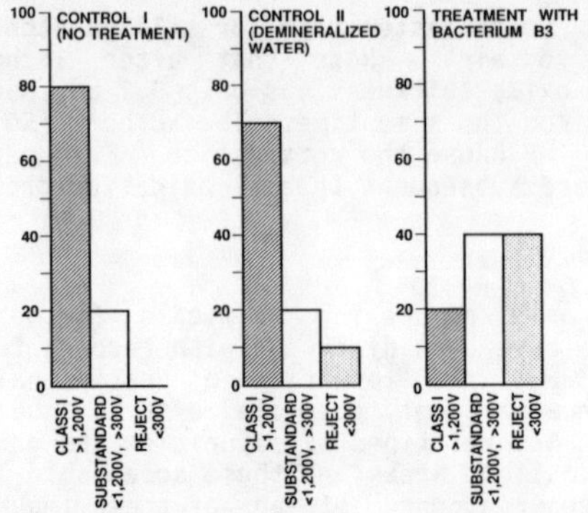

Fig. 9. Dependence of device quality (yield) on D.I. water purity [55].

Thermal Oxidation

As gate dielectric thicknesses in VLSI MOS devices approach the 10-20 nm range, much more attention is given to control of surface films and impurities which can have considerable effect on gate oxide properties. In fact, even oxide growth kinetics have been shown to depend directly on the nature of the pre-oxidation cleaning procedure [48,58]. An example is shown in Fig. 10, where oxide thickness is plotted against oxidation time for three different cleaning processes. In the case of the "RCA" clean, it has been proposed that either nitrogen from the ammonia, or carbon impurities, react with the silicon to form a retarding barrier which slows

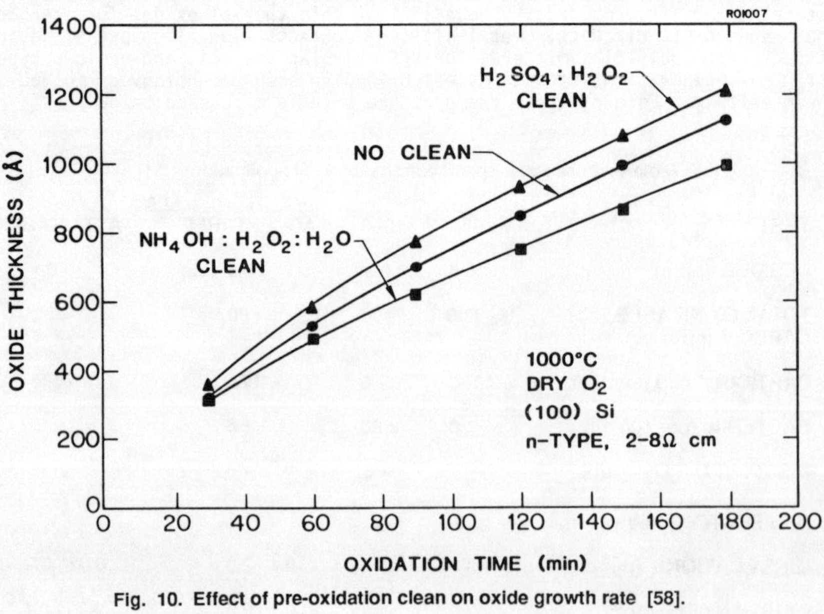

Fig. 10. Effect of pre-oxidation clean on oxide growth rate [58].

down the oxidation. In addition, other thin gate oxide properties, such as dielectric integrity or breakdown, are extremely dependent on impurities and particulates in the processing environment [59].

Metal-Silicon Contacts

Thin surface or chemical oxide films in contact openings have caused problems in contacting metal leads or aluminum interconnects to silicon for more than twenty five years. As new interconnect materials have been utilized and as geometries have decreased, the problem has intensified. One appropriate example involves the selective deposition of tungsten onto silicon for contact purposes. The problem is more complex than for previous systems, since both chemical and physical phenomena are involved in this particular CVD process. One reasonable model for the reaction, proposed by Stacy et. al [60], is shown in Fig. 11. If a thin, uniform surface oxide is initially present on the silicon (or if no oxide is present) the WF_6 reactant reacts with the silicon to form a thin (up to 10 nm) tungsten film, which then stops growing since no more silicon is available. Hydrogen is then added to the reactants which allows the vapor deposition of tungsten to continue. If, on the other hand, the surface oxide is defective or thick and non-uniform, the tungsten continues to react with the silicon and forms non-uniform, irregular deposits of tungsten. Thus, the control of the original native oxide is very critical. However, it has also been reported [62] that under certain conditions, thicker tungsten can be deposited during the initial silicon - reduction phase, which has been proposed to be due to the catalysis action of tungsten nodules. It is obvious, therefore, that more information is needed to understand the complete mechanism.

Similar effects on Schottky barrier heights in metal-semiconductor contact structures due to surface oxides and different cleaning procedures have been reported [63].

Poly-Silicon Emitters

An important development in bipolar device technology has been the use of poly-Si emitter structures. It has been determined that the use of polycrystalline silicon for both a diffusion source and a contacting layer in the emitter region results in considerable improvement in device characteristics [64-66], especially current gain. Typical results are shown in Fig. 12. The figure also includes a cross-section of the poly-Si emitter region with the interfacial oxide layer. Various models for the improved performance have been proposed, which includes the possibility that the thin oxide layer at the poly-Si/single crystal silicon interface acts as a tunnel barrier to hole transport. Indeed, it has been found that the nature of this interface layer does affect the device characteristics considerably. As a result, control of the interfacial oxide uniformity and integrity is critical for this particular device structure, even though the exact model for the effects observed has not been established.

Silicon Molecular Beam Epitaxy

One of the advanced techniques which quite likely will be used to fabricate certain VLSI and ULSI advanced silicon structures is molecular beam epitaxy (MBE). Requirements for depositing these atomic thickness layers will be such that no native oxide films can be tolerated. Therefore, MBE technology will undoubtedly lead the way in controlling and eliminating such films. The fact that MBE generally involves very high vacuum will make it easier for oxide removal. In fact, a typical

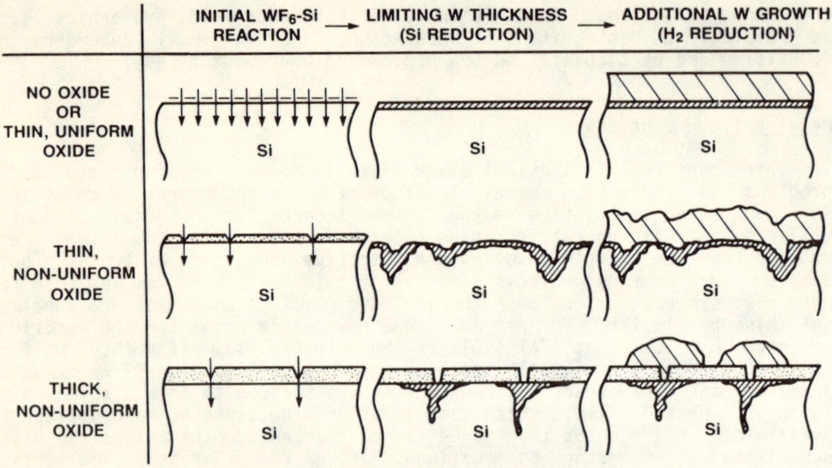

Fig. 11. Effect of surface oxide on tungsten growth on silicon [61].

procedure already followed for preparing silicon surfaces for MBE includes: (1) wet chemical treatment to minimize carbon and other impurities, (2) thin oxide formation to protect the clean surface, and (3) desorption of oxide in UHV and high temperature to provide a clean, bare silicon surface. Steps (1) and (2) might be accomplished using ultraviolet light and oxygen as mentioned earlier [6-14].

The use of silicon-beam radiation to remove the remaining oxide was mentioned earlier. Figure 13 depicts the effect of various processes, including silicon-beam radiation on etch pit densities in MBE silicon. It can be observed that such a cleaning process is quite effective.

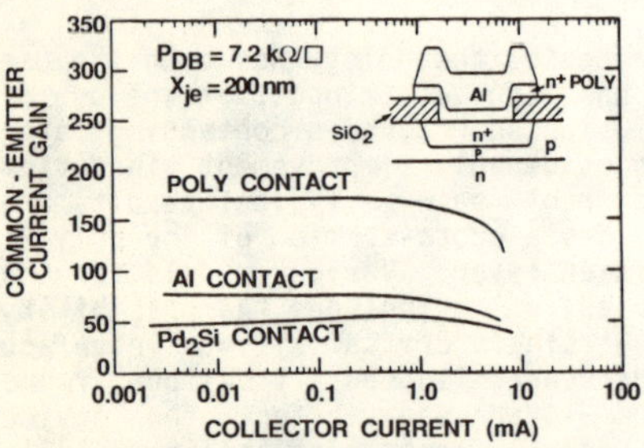

Fig. 12. Use of poly-Si emitters to improve bipolar transistor characteristics [67].

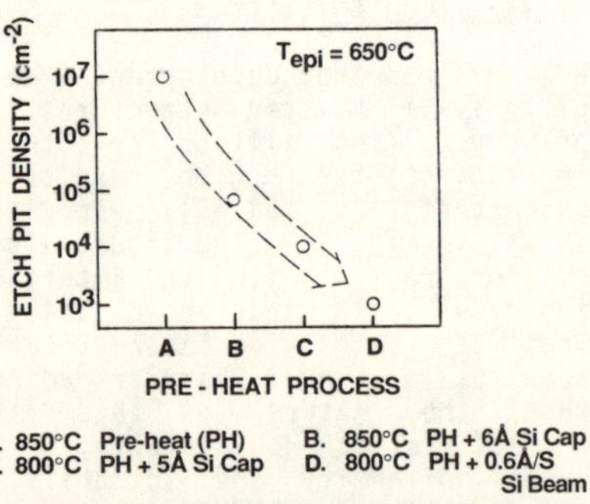

Fig. 13. Effects of four cleaning processes on etch pit densities of one micrometer silicon MBE films [68].

CONCLUSIONS AND FUTURE DIRECTIONS

We have reviewed the physics and chemistry of thin native or surface oxide films on silicon - especially with respect to their relationship to device applications. It has been noted that these films are primarily silicon oxide (probably SiO), but also include chemical species from the cleaning solutions employed as well as impurity atoms, such as carbon, that may have been contained in the chemicals. Analytical techniques are slowly being developed to analyze the nature and composition of these layers, which range from 0.5 to 1.5 nm in thickness. Control of native oxide films can be a critical factor in producing advanced VLSI or ULSI devices.

Some general conclusions and comments regarding future directions can be made. First, the current understanding of native oxide film formation on silicon needs to be improved. This is because VLSI-ULSI device requirements are such that control of these films will be critical if such devices are to be successfully fabricated. Likewise, analysis and impurity control of process chemicals, gases, and DI water must be improved. Finally, advanced technologies such as silicon MBE and tungsten-silicon contact structures are providing the incentive for obtaining a better understanding of thin surface oxide layers, as has been the case with other similar problems and related device requirements for the past twenty five years.

REFERENCES

1. M. Tannenbaum and D.E. Thomas, Bell Syst. Tech. J., 35, 1 (1956); J.F. Aschner, C.A. Bittmann, W.F.J. Hare, and J.J. Kleimack, J. Electrochem. Soc., 106, 415 (1959).
2. J.A. Hoerni, presented at The IRE Electron Devices Meeting, Washington, D.C., October 1960; U.S. Patents 3,025,589 (1962) and 3,064,167 (1962).
3. W. Kern and D.A. Puotinen, RCA Review, 31, 187 (1970).
4. R.C. Henderson, J. Electrochem. Soc., 119, 772 (1972).
5. J.A. Amick, Solid State Tech., 19(11), 47 (1976).
6. B.A. Joyce, R.R. Bradley, and G.R. Booker, Phil. Mag., 15, 1167 (1967).
7. M.A. Hopper, R.A. Clarke, and L. Young, Surface Sci., 56, 472 (1976).
8. S. Wright and H. Kroemer, Appl. Phys. Lett., 36, 210 (1980).
9. J.L. Vossen, J.H. Thomas, III, J.-S. Maa, and J.J. O'Neill, J. Vac. Sci. Technol., A2, 212 (1984).
10. M. Tabe, Appl. Phys. Lett., 45, 1073 (1984).
11. Y.H. Xie, K.L. Wang, and Y.C. Kao, J. Vac. Sci. Technol., A-3, 1035 (1985).
12. J.R. Vig, J. Vac. Sci. Technol., A-3, 1027 (1985).
13. J. Ruzyllo, A.M. Hoff, and G. Duranko, in Semiconductor Silicon 1986, edited by H.R. Huff, T. Abe, and B. Kolbesen (The Electrochemical Society, Pennington, NJ, 1986), p. 398.
14. A. Ishizaka and Y. Shiraki, J. Electrochem. Soc., 133, 666 (1986).
15. R.J. Archer, J. Opt. Soc. Am., 52, 970 (1962).
16. Y.J. van der Meulen and N.C. Hien, J. Opt. Soc. Am., 64, 804 (1974).
17. D.E. Aspnes, in Microelectronics Processing: Inorganic Materials Characterization, edited by L.A. Casper (American Chemical Society, Washington, DC, 1986) p. 192.
18. R.A. Clarke, R.L. Tapping, M.A. Hopper, and L. Young, J. Electrochem. Soc., 122, 1347 (1975).
19. S.I. Raider, R. Flitsch, and M.J. Palmer, J. Electrochem. Soc., 122, 413 (1975).
20. F.J. Grunthaner, B.F. Lewis, N. Zamini, J. Maserjian, and A. Madhukar, IEEE Trans. Nuclear Sci., NS-27, 1640 (1980).
21. M.G. Yang, K.M. Koliwad, and G.E. McGuire, J. Electrochem. Soc., 122, 675 (1975).
22. B.A. Joyce and J.H. Neave, Surface Sci., 27, 499 (1971).
23. C.R. Helms, N.M. Johnson, S.A. Schwarz, and W.E. Spicer, J. Appl. Phys., 50, 7007 (1979).
24. O.L. Krivanek and J.H. Mazur, Appl. Phys. Lett., 37, 392 (1980).
25. J.C. Bravman and R. Sinclair, Thin Solid Films, 104, 153 (1983).
26. H.R. Philipp and E.A. Taft, J. Appl. Phys., 53, 5224 (1982).

27. R.J. Archer, J. Electrochem. Soc., 104, 619 (1957).
28. A.C. Adams, T.E. Smith, and C.C. Chang, J. Electrochem. Soc., 127, 1787 (1980).
29. R.E. Oakley and G.A. Godber, Thin Solid Films, 9, 287 (1972).
30. J.A. Aboaf, J. Electrochem. Soc., 118, 1370 (1971).
31. J. Ruzyllo, J. Electrochem. Soc., 133, 1677 (1986).
32. Courtesy of J.C. Bravman.
33. G.S. Oehrlein, G.J. Coyle, J.G. Clabes, and Y.H. Lee, Surface and Interface Anal., 9, 275 (1986).
34. G. Mende, J. Finster, D. Flamm, and D. Schulze, Surface Sci., 128, 169 (1983).
35. C.R.M. Grovenor, A. Cerezo, and G.D.W. Smith, Mats. Rsch. Soc. Proc., 37, 199 (1985).
36. E.H. Poindexter and P.J. Caplan, Prog. Surface Sci., 14, 201 (1983).
37. F.J. Grunthaner, P.J. Grunthaner, M.M. Hecht, and D. Lawson, in Insulating Films on Semiconductors, edited by J.J. Simonne and J. Buxo (Elsevier, Amsterdam, 1986) p. 1.
38. P.O. Hahn, S. Yokohamu, and M. Henzler, Surface Sci., 142, 545 (1984).
39. R. Williams and A.M. Goodman, Appl. Phys. Lett., 25, 531 (1974).
40. A.M. Stoneham and P.W. Tasker, Semi. Sci. Tech., 1, 93 (1986).
41. M. Grundner and H. Jacob, Appl. Phys. A, 39, 73 (1986).
42. G.B. Larrabee, K.G. Heinen, and S.A. Harrell, J. Electrochem. Soc., 114, 867 (1967).
43. S. Kawado, T. Tanigaki, and T. Maruyama, presented at The 1986 Electrochemical Society Spring Meeting, Abs. 200, Boston, MA, May 1986.
44. L.A. Fergason, Microcontamination, 4(4), 33 (1986).
45. E.M. Juleff, Microelectronics, 7(4), 17 (1976)
46. M.A. Taubenblatt and C.R. Helms, J. Appl. Phys., 59, 1992 (1986).
47. I.K. Bansal, Solid State Tech., 29(7), 75 (1986).
48. F.J. Grunthaner and J. Maserjian, IEEE Trans. Nucl. Sci., NS-24, 2108 (1977).
49. G.A. Haas and H.F. Gray, J. Appl. Phys., 46, 3885 (1975).
50. A. Licciardello, P. Puglisi, and S. Pignataro, Appl. Phys. Lett., 48, 41 (1986).
51. Courtesy M.K. Balazs.
52. W. Harned, J. Environ. Sci., 29(3), 32 (1986).
53. M.W. Mittleman, Microcontamination, 3(10), 51 (1985).
54. J.W. Costerton and G.G. Gessey, in Surface Contamination, edited by K.L. Mittal (Plenum Press, NY, 1979) p. 211.
55. A. Gough, R.W. Attwell, D.F.D. Hardy, and R. Caldwell, Solid State Tech., 29(2), 139 (1986).
56. P.A. McConnelee, S.J. Poirier, and R. Hanselka, Semi. Int., 9(8), 82 (1986).
57. C. Mistry and J. Gililland, Microcontamination, 3(5), 71 (1985).
58. F.N. Schwettmann, K.L. Chiang, and W.A. Brown, presented at The 1978 Electrochemical Society Spring Meeting, Abs. 276, Seattle, WA, May 1978.
59. J.R. Monkowski and R.J. Zahour, IEEE Proc. Int. Rel. Phys. Symp., 244 (1982).
60. W.T. Stacy, E.K. Broadbent, and M.H. Norcott, J. Electrochem. Soc., 132, 444 (1985).
61. E.K. Broadbent and W.T. Stacy, Solid State Tech., 28(12), 51 (1985).
62. M.L. Green, Y.S. Ali, T. Boone, B.A. Davidson, L.C. Feldman, and S. Nakahara, presented at The 1986 Fall Meeting of The Electrochemical Society, Abs. 332, San Diego, CA, October 1986.
63. M.A. Taubenblatt, Tech. Report No. G-740-1, Stanford University, June 1985.
64. T.H. Ning and R.D. Isaac, IEEE Trans. Electron Devices, ED-27, 2051 (1980).

65. H.C. deGraaff and J.D. deGroot, IEEE Trans. Electron Devices, ED-26, 1771 (1979).
66. G.L. Patton, J.C. Bravman, and J.D. Plummer, IEEE Trans. Electron Devices, ED-33, 1754 (1986).
67. J.D. Plummer, private communication.
68. K. Kugimiya, Y. Hirofuji, and N. Matsuo, Jap. J. Appl. Phys., 24, 564 (1985.)

Selective Tungsten: Selectivity Loss and Nucleation on Insulators

OBSERVATIONS ON SELECTIVITY LOSS DURING TUNGSTEN CVD

J. R. CREIGHTON
Sandia National Laboratories, Division 1126, P. O. Box 5800, Albuquerque, NM 87185

ABSTRACT

Possible mechanisms for the loss of selectivity during tungsten CVD (i.e., growth on SiO_2) were investigated using a dual vacuum system composed of a growth chamber and a UHV analysis chamber equipped with Auger and mass spectroscopy. An oxide-covered silicon sample could be transferred between the two chambers while under vacuum to maintain surface composition integrity. With this experimental system, we have found evidence for a selectivity loss mechanism which involves desorption of tungsten subfluorides from tungsten surfaces. The disproportionation of the subfluorides then produces a reactive state of tungsten on the SiO_2 surfaces that leads directly to selectivity loss.

INTRODUCTION

Low pressure CVD of tungsten by reduction of WF_6 has a number of applications for VLSI fabrication [1-6]. These applications include formation of diffusion barriers and planarization by filling up contact or via holes. The process may be selective in that tungsten deposits only where there is exposed silicon, silicides, or metal, but not on adjacent silicon dioxide or other insulator surfaces. The selective nature of the deposition has proven to be extremely useful, as it significantly reduces the number of steps in the metallization process. Unfortunately, it is not always possible to maintain selectivity when desired and tungsten may nucleate on the insulator surfaces. The purpose of this study is to determine the mechanism(s) of selectivity loss with hopes that this information will make it possible to devise ways to enhance selectivity.

Several researchers have noted that selectivity loss usually first occurs near areas of tungsten deposition [5, 7, 8]. This observation strongly suggests that a reaction product, by-product, or intermediate is initiating selectivity loss. For many applications, the first reaction occurring is the silicon reduction of tungsten hexafluoride; i.e., $3/2\ Si(s) + WF_6(g) \rightarrow 3/2\ SiF_4(g) + W(s)$. Several authors have postulated that the silicon tetrafluoride or other silicon-containing species produced by this initial reaction may contribute to selectivity loss [5, 8]. We presently have no data to prove or disprove such a mechanism, but note that it could not explain all observations because selectivity loss is known to occur in systems where there is no exposed silicon. The main tungsten CVD reaction is the hydrogen reduction reaction, i.e., $3H_2(g) + WF_6(g) \rightarrow 6HF(g) + W(s)$, in which a large amount of hydrogen fluoride is produced. This hydrogen fluoride has also been suspected of initiating selectivity loss [7]. However, our experiments investigating the role of HF failed to establish a definitive link between the HF partial pressure and the degree of selectivity loss. We have instead found convincing evidence for a mechanism that first involves desorption of tungsten subfluorides (WF_x, $x<6$) from tungsten surfaces. The adsorption and disproportionation of the tungsten subfluorides on SiO_2 surfaces then produces a reactive state of tungsten that can immediately nucleate tungsten CVD.

EXPERIMENTAL

These experiments were performed in a dual chamber vacuum system composed of a reaction chamber and a UHV analysis chamber (see Fig. 1). The analysis chamber was equipped with a cylindrical mirror analyzer for Auger analysis and quadrupole mass spectrometer for TPD (temperature programmed desorption) measurements. This chamber was pumped with a turbomolecular pump to a base pressure of less than 2×10^{-10} torr. The reaction chamber was also pumped by a turbomolecular pump and the base pressure was approximately 5×10^{-10} torr. However, due to the large number of samples studied, the working base pressure was higher; normally $\sim 2 \times 10^{-7}$ torr. Because pressures as high as 300 millitorr were used during CVD experiments, the turbopump for the reaction chamber was throttled to give a nitrogen pumping speed of 1.3 ℓ/sec.

The samples were 2 cm by 1 cm rectangular slices of 0.4-Ω-cm Si(100) which were covered with 400 Å of thermally-grown SiO_2. These samples were mounted on a long-throw manipulator by means of tantalum clips which allowed direct resistive heating to above 800°C. For temperature measurements, a chromel-alumel thermocouple was attached to the top edge of the samples with a small drop of ceramic cement (Aremco #516). By opening a straight-through valve, a sample could be moved from the reaction chamber into the UHV chamber for surface analysis.

A polycrystalline tungsten foil was also mounted in the reaction chamber about 3 cm from the silicon sample and served as an independent source of reaction products and intermediates. The foil was heated resistively and the temperature was measured with a chromel-alumel thermocouple spot-welded to the middle of the foil.

VACUUM SYSTEM SCHEMATIC

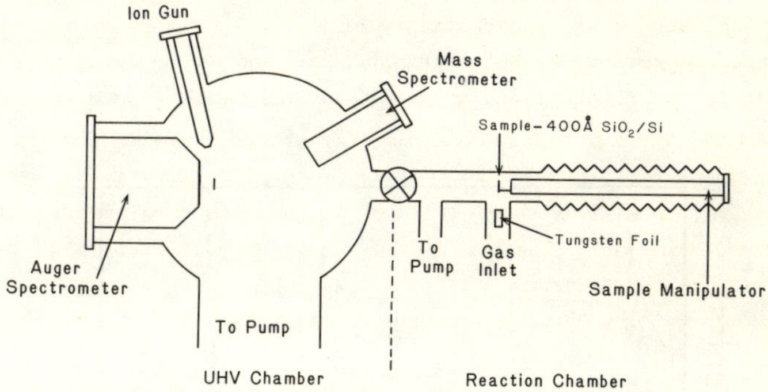

Fig. 1 Schematic of dual chamber vacuum system

RESULTS

Our initial experiments focused on the role of HF as a possible promoter of selectivity. These experiments involved flowing additional amounts of HF through the reactor under normal CVD conditions to see if there was any noticeable increase in nucleation on SiO_2. No effort was made to quantify low densities of tungsten nuclei, but high density of nuclei is measurable with Auger spectroscopy. All experiments using additional HF failed to give measurable tungsten nuclei densities using this method. The following experiment exemplifies this point. The SiO_2/Si sample was heated to 400°C in the presence of 5-millitorr WF_6, 40-millitorr H_2, and 100 millitorr HF. After 60 minutes, tungsten coverage was still below Auger detection limits.

We then searched for possible reaction intermediates or by-products generated by tungsten surfaces in contact with the reactant gases. For these experiments, the additional tungsten foil in the reaction chamber proved invaluable. Before performing any experiments on the SiO_2/Si sample, an Auger spectrum of the surface was recorded (see curve (a), Fig. 2). The silicon and oxygen peaks are indicative of SiO_2, and the absence of carbon indicates the lack of significant hydrocarbon contamination. A very small fluorine signal is probably residual from the HF/NH_4F etch performed just before sample installation in the vacuum system. With the SiO_2/Si sample held at room temperature in the reaction chamber, WF_6 was introduced to a pressure of 5 millitorr. The tungsten foil was heated to 600°C for 90 minutes, then cooled and the chamber evacuated. Within five minutes, the sample was transferred into the UHV chamber and an Auger spectrum taken (curve (b), Fig. 2). Note the presence of tungsten and fluorine as well as the complete attenuation of the silicon signal. Quantification of the Auger signals indicates a WF_x stoichiometry of X = 3.7 ± 0.5. This quantification was done by measuring the WO_3 Auger spectrum and using the relative sensitivities of fluorine and oxygen to obtain the F/W ratio [9].

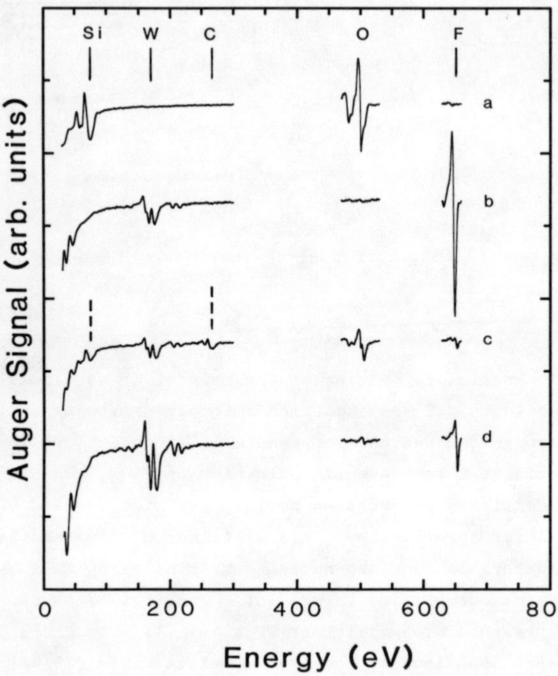

Fig. 2 Auger spectra of sample: (a) initially clean SiO_2, (b) after deposition of WF_x, (c) after heating to 500°C (d) after tungsten CVD

If the WF_x-covered SiO_2 surface is then heated to 500°C, significant chemical changes in the overlayer occur and some tungsten subfluoride is observed desorbing around 160 and 310°C (see Fig. 3). Curve (c) of Fig. 2 shows that the remaining tungsten is much less fluorinated and now contains some carbon and oxygen. We believe that this tungsten is formed by thermally induced disproportionation; i.e., $2WF_4(s) \rightarrow WF_3(s) + WF_5(g)$. This remaining tungsten is in a lower oxidation state and more reactive towards background gases, thus explaining the carbon and oxygen Auger signals.

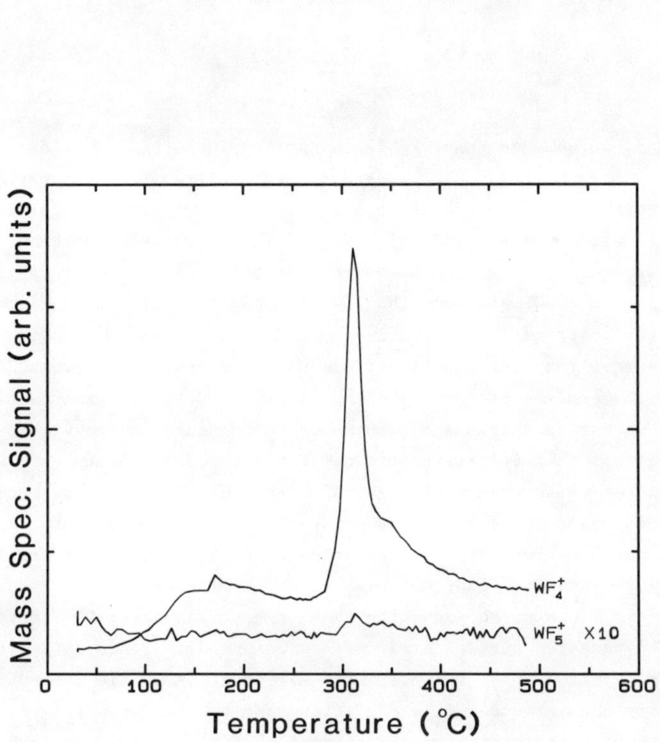

Fig. 3 TPD spectrum of WF_x-covered SiO_2 surface. Heating rate was 5°C/sec.

If the sample is then moved back into the reaction chamber and heated to 400°C in the presence of 5-millitorr WF_6 and 100-millitorr H_2, very rapid deposition of tungsten occurs. Changes in the surface optical properties are seen within two and a half minutes of deposition, and by six minutes the entire surface is covered with a continuous metallic tungsten film. After ten minutes of deposition, the Auger spectrum (curve (d), Fig. 2) shows a large signal indicative of metallic tungsten and the silicon signal is again completely attenuated.

TPD experiments were performed to examine the reactions occurring on the WF_x-covered SiO_2 surface. Starting with the surface described above (curve (b), Fig. 2), the sample was heated at 5°C/sec while the WF_4^+ and WF_5^+ mass spectrometer signals were recorded. The resulting TPD spectrum is displayed in Fig. 3. The WF_4^+ ion shows a small peak at 160°C and a larger peak at 310°C. Other similar experiments show that the WF_2^+ and WF_3^+ ions track the WF_4^+ response indicating that they are all fragments of the same species. The WF_5^+ signal shows virtually no response indicating that WF_6 is not the primary desorbing species since this ion is most intense in the WF_6 mass spectrum [10]. The only possible remaining species would be WF_4 and WF_5. We have not been able to find the mass spectral fragmentation pattern for WF_4, but since most metal fluorides have very weak or non-existant parent ions, we believe that WF_5 is the primary desorbing species responsible for the TPD spectrum seen in Fig. 3 [11, 12]. Since the initial surface stoichiometry is approximately $WF_{3.7}$, the WF_5 must be forming by disproportionation, i.e., $2WF_4(s) \rightarrow WF_3(s) + WF_5(g)$. We also know that the desorbed WF_5 cannot be due to a simple desorption mechanism since the heat of vaporization of WF_5 is much too low to account for the high desorption temperatures observed [13].

Since most tungsten CVD is performed at temperatures between 250 and 400°C and the experiments described above produced the tungsten subfluorides at 600°C, we have measured the temperature dependence of the WF_x formation rate to determine if it is significant at the lower temperatures. Since time dependent measurements of Auger signals would be very time consuming, we used the steady-state gas-phase pressure drop of WF_6 to measure the formation rate. A fundamental assumption is that formation of the low volatility tungsten subfluorides removes WF_6 from the system. Ignoring stoichiometry factors, the reaction rate, $R = S\Delta P/A$, where S = pumping speed of WF_6, A = tungsten surface area, ΔP = pressure difference before reaction and during reaction. For example, if 5 millitorr of WF_6 is introduced into the reaction chamber under flow conditions and the tungsten foil is then heated to 940°C, the system pressure drops to 4 millitorr. An Arrhenius plot of the results (see Fig. 4) gives a nice, straight line with an activation energy $E_a = 23.9 \pm 0.9$ kcal/mole. Extrapolation of the subfluoride formation rate to 300°C and 400°C gives values of 1×10^{11} molecules/cm^2-sec and 3×10^{12} molecules/cm^2-sec, respectively. Although these values may not seem large at first glance, selectivity loss is considered severe with tungsten nuclei densities of less than 10^{10}/cm^2. With this perspective, it is seen that the tungsten subfluoride formation rate is very important at normal CVD temperatures.

CONCLUSIONS

We have shown that tungsten subfluorides desorb from tungsten surfaces exposed to WF_6 at significant rates at normal CVD temperatures. These subfluorides condense on room temperature SiO_2 to yield a WF_x stoichiometry of $X = 3.7 \pm 0.5$. Upon heating, the $WF_{3.7}$ disproportionates to gas phase WF_5 and an even lower oxidation state tungsten. This remaining tungsten is very reactive and immediately nucleates further tungsten growth under normal CVD conditions. We believe that this tungsten subfluoride desorption-disproportionation mechanism is the primary cause of selectivity loss during tungsten CVD by hydrogen reduction of WF_6. Now that we have some degree of understanding of the process, it may be possible to devise ways to enhance selectivity.

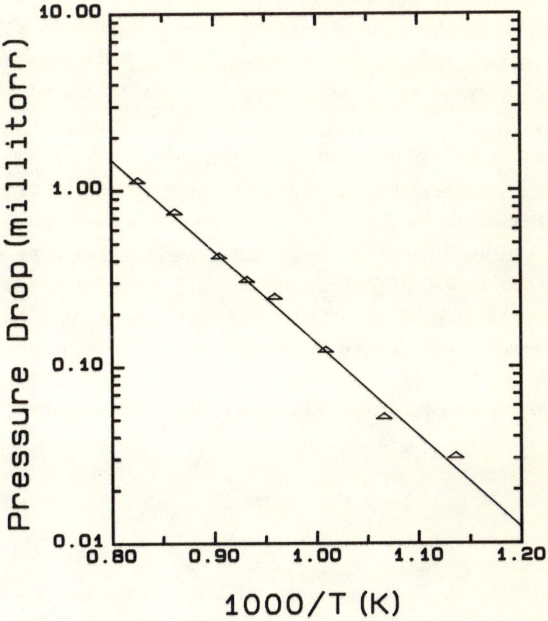

Fig. 4 Arrhenius plot of WF_6 pressure drop (proportional to WF_x formation rate) for an initial WF_6 pressure of 5 millitorr. $E_a = 23.9 \pm 0.9$ kcal/mole.

ACKNOWLEDGMENTS

The author would like to thank John Medernach for supplying the oxide-covered silicon samples and Robert Blewer for valuable discussions.

This work performed at Sandia National Laboratories supported by the U.S. Department of Energy under contract number DE-AC04-76DP00789 for the Office of Basic Energy Sciences.

REFERENCES

[1]J. M. Shaw and J. A. Amic, RCA Rev. 31, 306 (1970).

[2]E. K. Broadbent and C. L. Ramiller, J. Electrochem. Soc. 131, 1427 (1984).

[3]N. E. Miller and I. Beinglass, Solid State Technology 23, 79 (1980).

[4]N. E. Miller and I. Beinglass, ibid. 25, 85 (1982).

[5]E. K. Broadbent and W. T. Stacy, ibid. 28, 51 (1985).

[6]R. S. Blewer, ibid. 29, 117 (1985).

[7]Y. Paleau and Ph. Lami, J. Electrochem. Soc. 132, 2779 (1985).

[8]T. Moriya and H. Itoh, in proceedings of the 1985 Workshop on Tungsten and Other Refractory Metals for VLSI Applications (MRS, Pittsburgh, 1986), p. 21.

[9]L. E. Davis, N. C. Macdonald, P. W. Palmberg, G. E. Riach, and R. E. Welser, Handbook of Auger Electron Spectroscopy (Physical Electronics Industries, Minnesota, 1986).

[10]S. R. Heller and G. W. Milne, EPA-NIH Mass Spectral Data Base, Vol. 3 (NBS, Washington, 1978), p. 2223.

[11]A. V. Gusarov and L. N. Gorokhov, Adv. Mass Spectrom. 7a, 666 (1978).

[12]W. E. Falconer, G. R. Jones, W. A. Sunder, M. J. Vasile, A. A. Muenter, T. B. Dyke, and W. Klemperer, J. Fluorine Chem. 4, 213 (1974).

[13]J. Schroder and F. J. Grewe, Chem. Ber. 103, 1536 (1970).

A MODEL FOR TUNGSTEN NUCLEATION ON OXIDE

CAROL M. McCONICA
Department of Chemical Engineering, Colorado State University,
Fort Collins, Colorado 80523

ABSTRACT

The rate of initiation of tungsten nuclei on thermal oxide has been determined as a function of time, temperature and distance from areas of tungsten growth. The creation rate of infinitely small nuclei increases exponentially with time and decays exponentially with distance from large regions of tungsten growth. As time passes, the nucleation front progresses away from regions of tungsten deposition. The results suggest an autocatalytic nucleation phenomenon coupled with either gas phase or surface diffusion of the reactive intermediate. An activation energy for nucleation of 25 kcal mol^{-1} was determined using three independent methods.

It is hypothesized that HF attacks the SiO_2 to form SiF_x in the presence of tungsten catalyst. Solution of the reactor continuity equation shows that the steady state HF pressure is most sensitive to reactive area (hot metallic surfaces) in the reactor.

INTRODUCTION

The initiation of tungsten nucleation in the absence of silicon on thermal oxide from WF_6 and H_2 has been observed to be primarily dependent upon temperature, and Cooper showed it to be autocatalytic in nature [1,2]. The purpose of this years work has been to accurately determine the activation energy for tungsten nucleation on oxide, to propose a model describing the chemistry of nucleation and to solve the continuity equations to find HF concentrations as dependent upon reactor design variables.

EQUIPMENT AND PROCEDURE

The equipment is described in detail elsewhere, and can be summarized as a high vacuum stainless steel cold wall reactor with a tungsten coated stainless steel substrate heater [2]. The 2 inch n type wafers used in the study were oxidized on all sides to 6000 Å prior to nucleation studies. A wafer was loaded into the cold reactor and then baked at less than 10^{-5} Torr and 400°C for 1 hour prior to deposition. The detailed procedure is reported by Cooper [3]. In all depositions the partial pressure of hydrogen was kept at 0.70 Torr and WF_6 at 0.05 Torr, and the deposition times were less than 10 minutes. No silicon surfaces were ever exposed in the reactor.

The tungsten islands were analyzed using two independent techniques. A surfscan particle counter made by Tencor was used to obtain the total number of islands and their size distribution as a function of location on the wafer. Each wafer was also analyzed for total tungsten deposited using Neutron Activation Analysis NAA (accurate to the nanogram range).

RESULTS

Three independent methods were used for analyzing the Tencor data. In the first method plots were made of the number of islands versus each island size for various edge exclusions (figure 1). The intercept gives the number of infinitely small sites at any radius (figure 2.) Because this is an integral curve, the local density of infinitely small islands can only be found by $\frac{1}{2\pi r}\frac{dN}{dr}$, where $\frac{dN}{dr}$ is the slope of fig. 2 at any r. The results are plotted in figure 3.

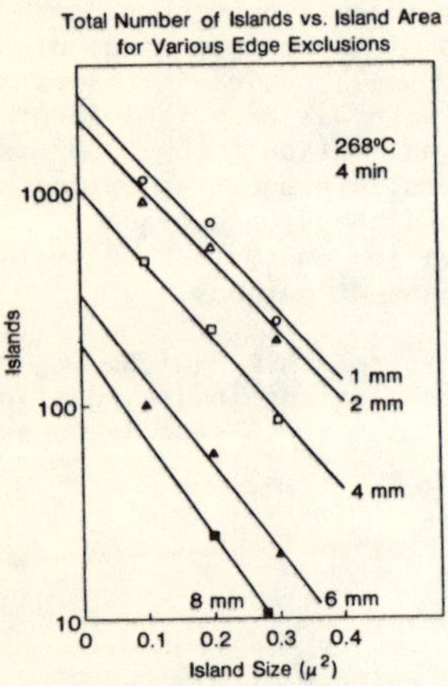

Figure 1

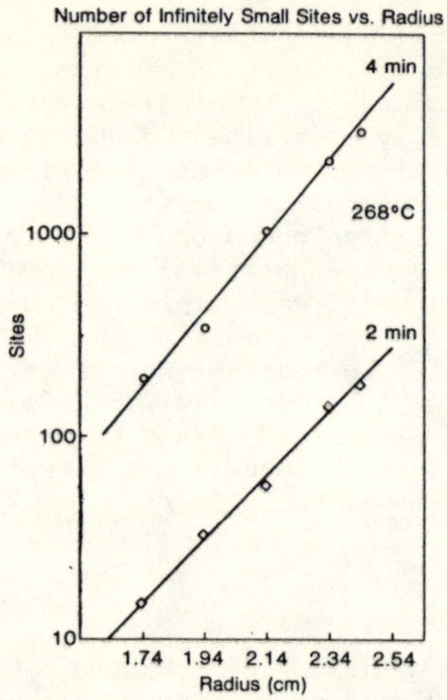

Figure 2

Because the data for infinitely small sites were valid only at low temperatures, the same method was applied to find the local density of $0.1\mu^2$ sites for all temperatures. This evaluation of the Tencor data resulted in a local rate of change in density of $0.1\mu^2$ sites at each location on the wafer (figure 4).

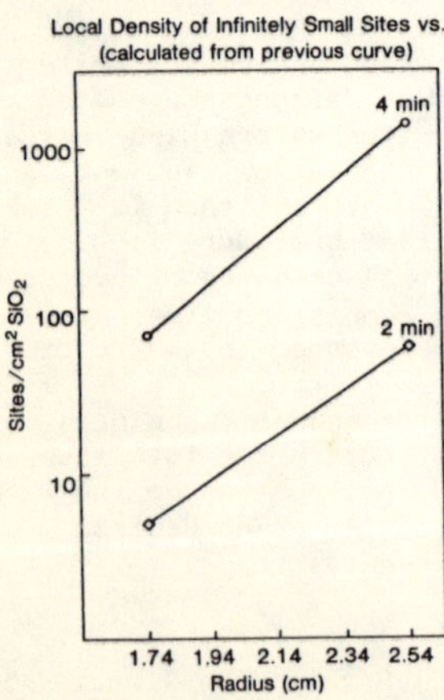

Figure 3

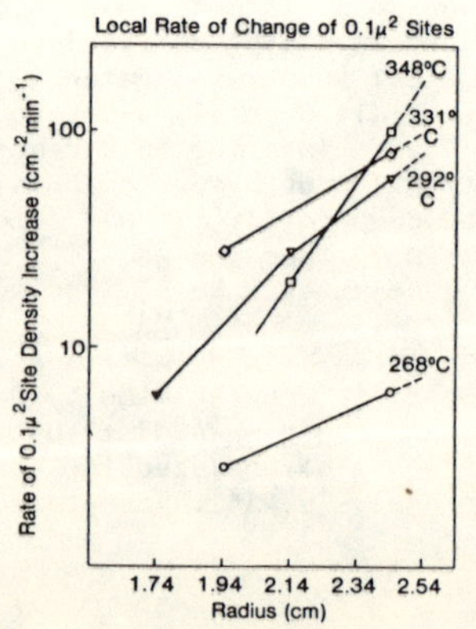

Figure 4

The third method involved integrating the area under the island number *vs* size curve generated by Tencor to find the total area of tungsten on each wafer. Again, the area is a function of radius, time and temperature. The resulting data can be summarized by figure 5 as the local rate of tungsten area growth versus radius.

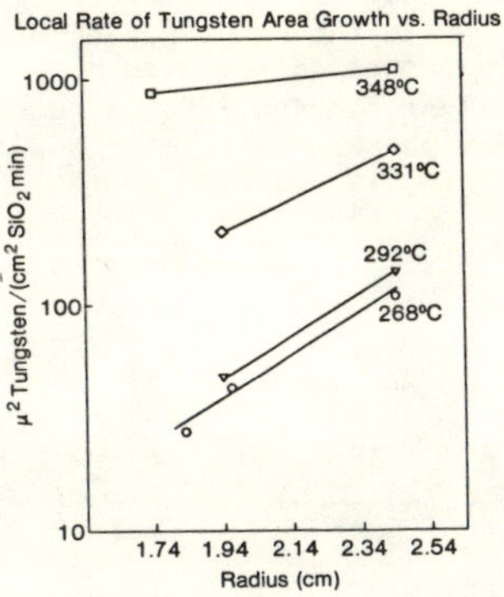

Figure 5

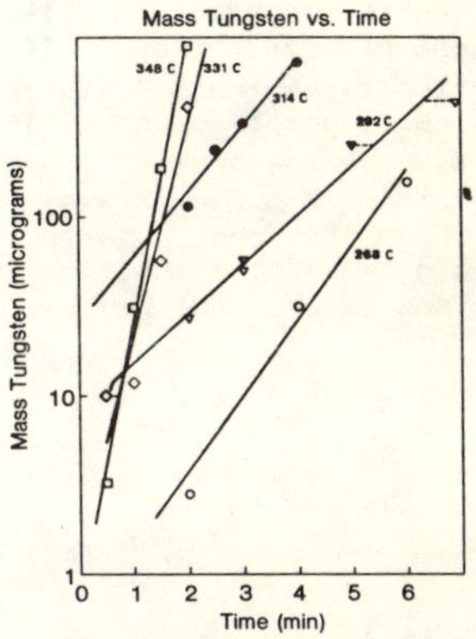

Figure 6

Figure 6 shows the results of NAA evaluation of the total mass of tungsten on each wafer. The mass of tungsten on the wafer is exponentially dependent upon time and the rate of mass addition is linearly dependent upon the mass currently there (figure 7).

Three independent methods were used to generate Arrhenius plots of nucleation rate on thermal oxide (figure 8). The three methods resulted in activation energies for nucleation of 25, 24 and 27 kJ mol^{-1}.

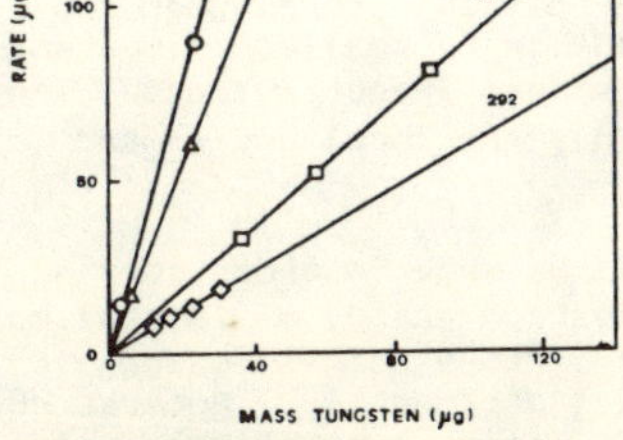

Figure 7

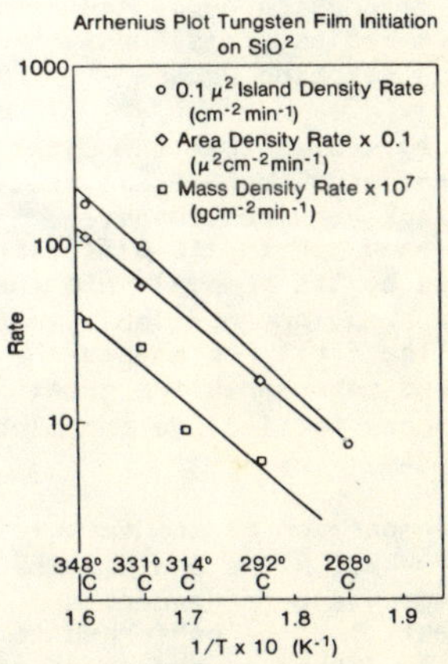

Figure 8

DISCUSSION OF NUCLEATION EXPERIMENTS

Because the local nuclei density was found to consistently decay exponentially with distance from the tungsten heater block (wafer edge) and increase exponentially with time, it is believed that the nucleation reaction is autocatalytic. Because the activation energy for nucleation is in the middle of the regime for chemical reaction (25k cal mol^{-1}), it is concluded that nucleation on oxide results from a chemical attack of the oxide rather than random defects. The results of Heiny imply that HF in the presence of a catalyst attacks the SiO_2 to form SiF_x [4]. While the exact series of chemical reactions is unknown, it is hypothesized that dissociated HF attacks SiO_2 in the proximity of the metal catalyst forming SiF_x, a reactive intermediate. This SiF_x travels through the gas phase, reacts with WF_6, leaving WF_y which forms a surface nuclei. Once WF_y is adsorbed on the oxide, it grows into a tungsten island. A model consistent with the observations is proposed below:

$$\frac{d\,[S_o]}{dt} = k_1\,[SiF_x]^n$$

$$\frac{\partial\,[SiF_x]}{\partial t} = k_2\,[HF]^m\,[A_W] + \mathcal{D}_{SiF_x}\left\{\frac{1}{r}\frac{\partial}{\partial r}\left(r\,\frac{\partial SiF_x}{\partial r}\right) - \partial^2\,\frac{SiF_x}{\partial Z^2}\right\}$$

$$\frac{\partial\,[A_W]}{\partial t} = A_o\,\frac{\partial\,[S_o]}{\partial t} + 4\pi R\int^{a} N(a)\,da$$

S_o = Surface density of infinitely small sites [sites m^{-2}]
k_1&k_2 = Reaction rate constants
$[SiF_x]$ = Gas phase or surface concentration of SiF_x
[HF] = Gas phase concentration of HF [mol ℓ^{-1}]
$[A_W]$ = Area of tungsten/area of SiO_2 [$m^2_W\ m^{-2}_{SiO_2}$]
$\mathcal{D}_{SiF_x}$ = Diffusivity of SiF_x
r = Radial coordinate on wafer
Z = Dimension perpendicular to the wafer
A_o = Area of the smallest nucleation site
R = Tungsten deposition rate on tungsten
a = Radius of islands
n,m = Reaction Orders

This model states that the sites are initiated by a locally high SiF_x concentration. The SiF_x is created by a catalyzed attack of SiO_2 by HF. This attack requires tungsten. The local SiF_x concentration on the wafer is increased due to its diffusion from regions of higher concentration and decreased by its removal into the Z direction. The local rate of growth of tungsten area, A_W, available to catalyze the HF attack depends upon two terms. The first represents the rate of initiation of smallest sites and the second represents the growth of all islands into larger islands. The A_W term does not include the block area, but only the local nuclei and island area.

A comparison of the data to the model will be made in order to find values for k_1, k_2 n and m. The observed activation energy of 25 kcal mol^{-1} is a lumped value dependent upon both k_1 and k_2. More sophisticated experiments must be performed to observe these independently. Because HF is inherently linked to the creation of nuclei, it became necessary to solve the continuity equation for HF and observe its local concentration dependence upon reactor design variables.

STEADY STATE HF CONCENTRATION

At steady state, assuming no HF flow into the reactor, and initially assuming a gradientless (single wafer) reactor:

$$\text{HF creation rate by reaction} = \text{exit flow rate of HF}$$
$$6\,(r_W)\,A_r = F \qquad (1)$$

r_W = tungsten deposition rate = $0.495 \exp(-8800/T)\, H_2^{1/2} \; \dfrac{\text{moles}}{\text{cm}^2\ \text{min}}$

A_r = reactor area available for deposition [cm^2]

F = HF exiting flow rate [moles min^{-1}]

The exiting flow rate can be related to the inlet flow rate by:

$$F_{out} = F_{in}\,(1+\epsilon\, x)$$

ϵ = volume expansion coefficient: $\left[\dfrac{\text{moles}_{x=1} - \text{moles}_{x=0}}{\text{moles}_{x=0}}\right]$

X = WF_6 conversion in the reactor.

Insertion into (1) gives:

$$6\,(r_W)\,A_r = \left[V_F\,(1+\epsilon\, x)\frac{P}{RT}\right] y_{HF} \qquad (2)$$

V_F = Inlet total volumetric flow rate [cm^3 min^{-1}]
P = Atmospheric Pressure
T = Standard Temperature [298°K]
R = Universal gas constant [cm^3 atm mol^{-1} $°k^{-1}$]
y_{HF} = Mole fraction of exiting stream which is HF

The mole fraction of the exiting stream which is HF can be given in terms of WF_6 conversion by:

$$y = \frac{6x}{E + 4 + 2x} \qquad (3)$$

Where E = hydrogen in excess of the stoichiometric value of 3.

The rate equation must consider the actual hydrogen concentration within the reactor, as diluted by the HF:

$$r_W = 0.495 \exp(-8800/T)\,\left(P_E + 3P_{WF_{6_0}}\,(1-x)\right)^{1/2} \qquad (4)$$

P_E = Pressure of H_2 in excess of the stoichometric value
$P_{WF_{6_0}}$ = Inlet pressure of WF_6

Inserting (3) and (4) into (2) gives the relationship between conversion and the reactor variables:

$$\frac{298.6\ k\ A_r\ R}{V_F} = (1 - \epsilon x)\left\{\frac{6x}{E+4+2x}\right\}\left(P_E + 3P_{WF_{6_o}}(1-x)\right)^{-\frac{1}{2}} \quad (5)$$

where $k = 0.495 \exp(-8800/T)$

Using (5), the conversion in the reactor can be determined as dependent upon reactive area, flow rate, temperature, and excess hydrogen. It is then possible to find the pressure of HF in the reactor by:

$$P_{HF} = Y_{HF}P_T.$$

P_T Total reactor pressure

If it is assumed that the volume expansion is small ($\epsilon \rightarrow 0$) due to large excess hydrogen ($E \gg 1$) then the above equation can be solved analytically with the results plotted in figures 9 and 10.

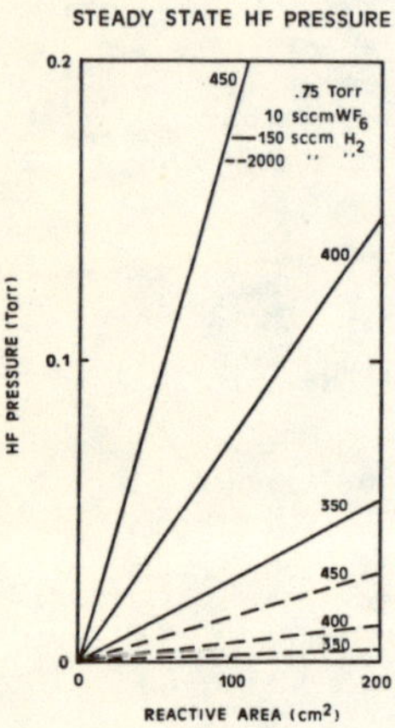

Figure 9

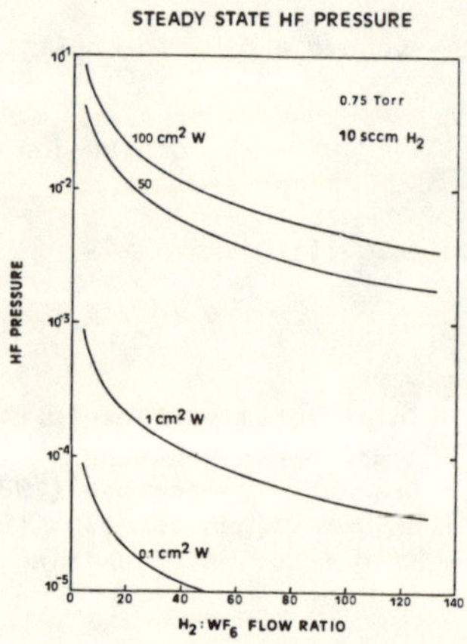

Figure 10

DISCUSSION ON HF CALCULATION

The primary variables dominating HF partial pressure are the reactive area, gas flow rate and reactive surface temperature. Lowering the reactive area has a much greater effect upon HF concentration than increasing flow rate. Reactors with hot walls and tungsten coated liners will have HF partial pressures greater than 0.2 Torr and would be undergoing a significant rate of the reverse tungsten deposition reaction (observation of a lower tungsten deposition rate). Cold wall, single wafer reactors should minimize nucleation by minimizing HF and catalyst available to dissociate the HF.

CONCLUSION

The initiation of tungsten island growth on thermal oxide could be the result of a catalyzed chemical attack of the oxide by some reactive intermediate. This reactive intermediate is probably monatomic fluorine and the species responsible for initiating growth is hypothesized to be SiF_x. Nucleation proceeds radially inward due to the diffusion of SiF_x from regions of high concentration to low concentration.

The observation that selectivity is enhanced by cold wall single wafer reactors is supported by this model (5). Authors with hot wall reactors, but high flow rates and low tungsten deposition areas also report good selectivity, in agreement with our model (6). Elimination of tungsten nucleation on thermal oxide requires minimization of the reactive area available to catalyze the formation of some reactive intermediate, in this case hypothesized to be SiF_x.

ACKNOWLEDGEMENTS

This research was supported by Sandia National Labs. The author acknowledges Kent Cooper, Kit Bauman and Kirk Anderson for performing the experiments and Tencor for donating time on their surfscan.

REFERENCES

1. C.M. McConica and K. Krishnamani, J. Electrochem. Soc., 2542 (1986).

2. C.M. McConica and K.J. Cooper, in Tungsten and other Refractory Metals for VLSI Applications, edited by R.S. Blewer (Materials Research Society, Pittsburgh, PA, 1986) p. 197.

3. K.J. Cooper, M.S. Thesis, Colorado State University, 1987.

4. M. Heiny and C.M. McConica, in Tungsten and Other Refractory Metals for VLSI Applications, in press (1987).

5. R.H. Wilson, R.W. Stoll and M.A. Calacone, in Tungsten and Other Refractory Metals for VLSI Applications, edited by R.S. Blewer (Materials Research Society, Pittsburgh, PA, 1986) p. 35.

6. R.S. Blewer, M.E. Tracy and V.A. Wells, Ibid p. 407.

EVALUATION ON SELECTIVE DEPOSITION OF CVD W FILMS BY MEASUREMENT OF SURFACE TEMPERATURE

T. Ohba, Y. Ohyama, S. Inoue, and M. Maeda
Fujitsu Limited, 1015, Kamikodanaka Nakahara-ku, Kawasaki, 211, JAPAN

ABSTRACT

The CVD W deposition process is observed while changing the substrate temperature, with IR thermometer measurement. The temperature indicated changes with emissivity depending on the surface morphology and materials. As the response time of this measurement is short, it is possible to perform in situ observation of changing surface morphology. This is a valid method to evaluate the deposition process. On a silicon substrate, the indicated temperature increases rapidly as soon as the gas flow starts, i.e. as soon as tungsten is deposited on the substrate. The indicated temperature increases in two steps. One is for silicon reduction and the other is for hydrogen reduction. Furthermore, it is found that the start of deposition on insulators is delayed. This phenomenon is related to the selective deposition. The thickness of deposited tungsten films can on insulatars be estimated by measuring the delay time and the thickness of tungsten on silicon.

The reaction process for silicon and hydrogen reduction is evaluated by this method. We found that within 10 seconds the silicon reduction was finished and the hydrogen reduction started. We also found that optimum silicon reduction enhances the selectivity.

INTRODUCTION

Chemical vapor deposition (CVD) is an important process in the fabrication of VLSI devices. Selective tungsten films deposited by low pressure CVD (LPCVD) are essential to fill vias and diffusion barriers. Selective deposition has been discussed widely regarding via filling [1-3], interface phenomenon [4-11], thermodynamics (e.g., chemical kinetics) [12-14], and electrical characteristics [8, 15, 16].

Selective deposition is capable of depositing on silicon or metal surfaces preferentially without depositing on insulator surfaces, because the deposition processes are different between the silicon and insulator surfaces due to their different chemical or physical properties. Evaluation of selectivity loss is estimated by observing the deposited wafer surface with SEM photomicrographs or measurement of nucleation. However it is impossible to define the process during the deposition on silicon or insulators. Particularly, as the reactions are important parameters to the selectivity, it is necessary to define the processes of the silicon and hydrogen reduction of WF_6 and the deposition on insulators (i.e., selectivity loss). The infrared (IR) radiation thermometer performs integrating measurement characterized by the emissivity, since the emissivity depends on materials and surface morphology [9]. Temperature measurement using the IR thermometer is capable of determining the process in real time, e.g. when deposition or nucleation starts on the silicon or insulators, and of determining changes in surface morphology during the deposition. This paper discusses relationships between the reduction reaction process and nucleation or deposition on insulators and selectivity.

EXPERIMENTAL

Tungsten films were deposited by using a parallel plate load-lock LPCVD system reactor. A schematic diagram of the deposition system is shown in Figure 1. The substrate on the susceptor was heated by a resistance heater. The gas flow and pumping portion of the system are similar to that of conventional LPCVD systems.

The gases used were tungsten hexafluoride (WF_6), hydrogen (H_2) and inert gases (He, Ar, and N_2). Only the WF_6 and H_2 were used for the deposition by hydrogen reduction. For the study of substrate Si reduction (i.e., WF_6-Si system), WF_6 and inert gases were used. There are two types of deposition sequences, the one step and the two step deposition processes. The one step deposition process used only $WF_6 + H_2$. The two step deposition process used WF_6 + inert gas and $WF_6 + H_2$.

The conditions during deposition ranged from 60 to 200 Pa total pressure; 3 to 900 seconds deposition time and temperature from 250 to 450°C. The temperature of the substrate was measured by an IR radiation thermometer calibrated initially with a thermocouple attached to a test substrate surface and corrected for emissivity. Figure 2 shows a model of the indicated surface temperature during deposition, therefore the indicated temperature changed when the emissivity changed. The indicated temperature was increased by deposited mirror surface metal and decreased by rough surface morphology according to the increase in grain size.

The substrates used in this study were 4 inch diameter, p-type (100) oriented single crystal silicon wafers (Figure 3.); patterned oxide or nitride coated. The substrates were precleaned by immersion in dilute HF, rinsed and dried and immediately placed in the load lock chamber. In the experiment, the substrate was heated under an inert or H_2 ambient prior to deposition in a reactor. The temperature near the boundary between silicon substrate and insulators, measured by IR radiation thermometer, was monitored. The thickness of tungsten was determined by step height measurement or SEM photomicrograph.

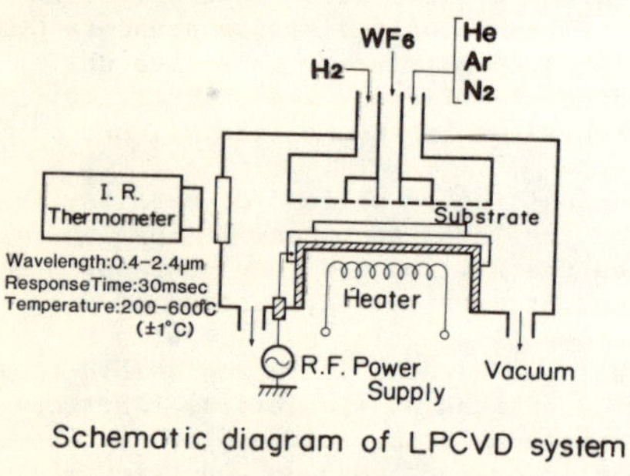

Fig. 1. Schematic diagram of LPCVD system.

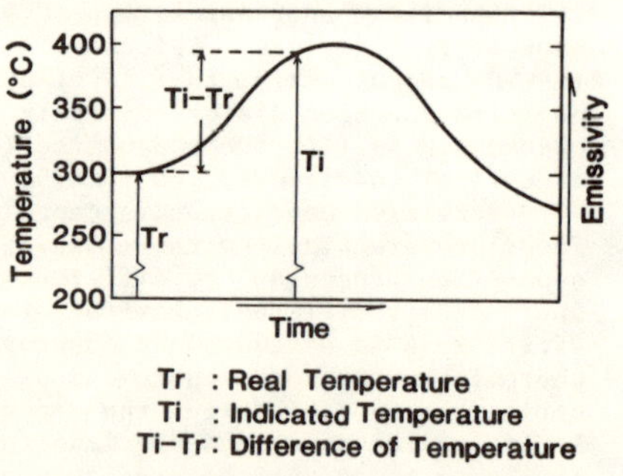

Fig. 2. A model of temperature and emissivity as a function of deposition time for CVD.

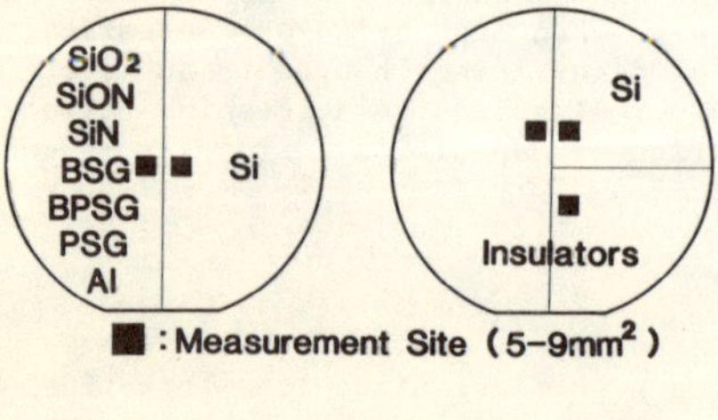

Fig. 3. Schematic diagram of substrate.

RESULTS AND DISCUSSION

I. TEMPERATURE AND SURFACE MORPHOLOGY

Figure 4 shows the temperature difference (Ti-Tr), percentage of reflectance and SEM photomicrographs of the silicon substrate as a function of deposition time in the one step deposition process and percentage of reflectance compared with a 0.2 um thick sputtered Al film at 800 um wavelength. Initially, the reflectance of silicon substrate is about 42%. Reflectance increases rapidly for about 30 seconds and then decreases slowly. As compared with the surface morphology seen in the SEM photomicrographs, the surface is smooth at 30 seconds and subsequently changes into a rough surface. The increasing and decreasing reflectance suggests the changes in surface morphology, the increase indicates that the tungsten has formed a thin layer, and the decrease signifies that the surface is becoming rough due to increasing grain size. The indicated temperature almost parallels the reflectance, because the emissivity is related to the reflectance. Thus, the surface morphology or the starting time of deposition can be evaluated by the indicated temperature in this measurement method (response time 30 seconds minimum).

II. EVALUATION OF SELECTIVITY

Figure 5 shows the temperature difference as a function of deposition time on several substrates. On silicon, the indicated temperature increases rapidly as soon as gases are supplied. However, on SiO_2 and PSG, the indicated temperature increase is delayed, the temperature curves have broad peaks and the heights are different. The thickness of tungsten film on several substrates was obtained as a function of deposition time (Figure 6). The tungsten film deposition on silicon substrates starts as soon as the gases flow starts, but the deposition on insulators is delayed. The start of deposition is coincident with the first temperature change, as expected, On insulators, it means that until the indicated surface temperature changes, tungsten is not deposited. The time that deposition starts on the insulator can be known by the increase in indicated temperature, and the time can be used to evaluate the thickness on silicon (i.e., selective thickness on silicon vs. insulators). Thus, the delay time on insulators is the index to the selectivity on silicon.

Figure 7 shows the correlation between the delay time for the initial temperature change and for the start of deposition, from Figures 5 and 6, for several substrates. It shows that the two delay times are coincident. The delay times are different for several kinds of insulators; delay increasing in the order,

SiON <SiO_2 <SiN <BSG <BPSG <PSG order (5 wt% <8 wt%).

Particularly, delays for doped oxides are markedly greater than for oxide or oxynitride. The selectivity increases for the doped oxides depending on the kind of dopant (i.e., boron, boron-phosphorus, and phosphorus) and higher dopant content (i.e., from 5 to 8 wt% phosphorus-doped oxide) [11]. The marked effect of doped oxides on the selectivity suggests that having boron and phosphorus on the surface can inhibit nucleation sites and deposition. It is necessary to inquire further into whether reaction between solid-gas phase and absorption involves chemisorption or physical absorption (i.e., by-product molecular; SiF_4, HF, reactant gases; WF_6, H_2, surface structure; Si, oxide, nitride etc.) [8, 12, 13].

In our method, Figure 8 shows the selective thickness of tungsten using silicon vs. silicon dioxide as functions of deposition temperature for different gas flow ratios (i.e., H_2/WF_6 at 20 and 500). When the resulting selective thickness is compared with our method and by observing SEM photomicrographs, we see that these thickness coincide, as expected. The selectivity is increased by lower deposition temperature and high gas flow ratio (e.g., selective thickness for lower and higher gas flow ratio at 300°C is 360 nm and 540 nm, respectively) [11]. Other insulators show the same tendency. It is concluded that our method is useful for evaluation of selectivity.

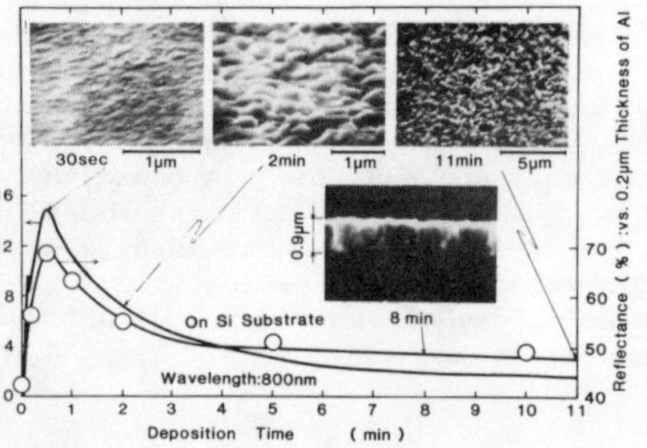

Fig. 4. Temperature difference, percentage of reflectance, and surface morphology as a function of deposition time.

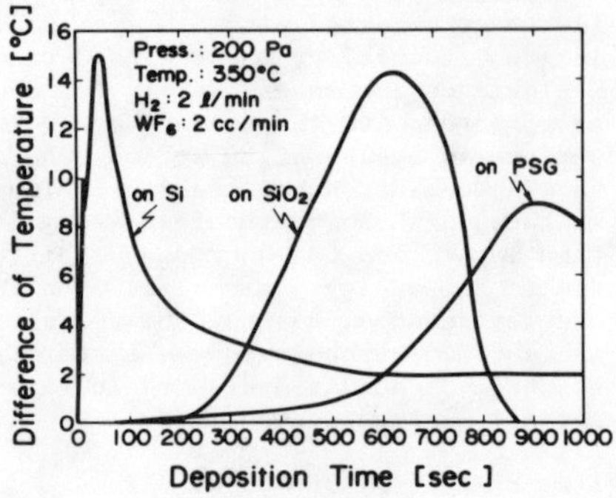

Fig. 5. Temperature difference as a function of deposition time for Si, SiO_2, and PSG.

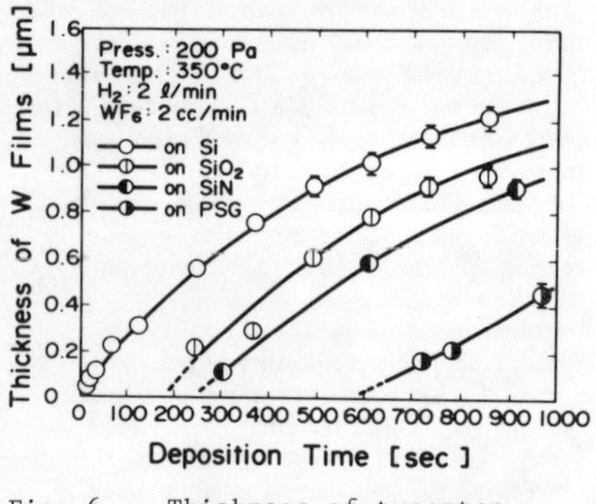

Fig. 6. Thickness of tungsten films as a function of deposition time for several kinds of substrates.

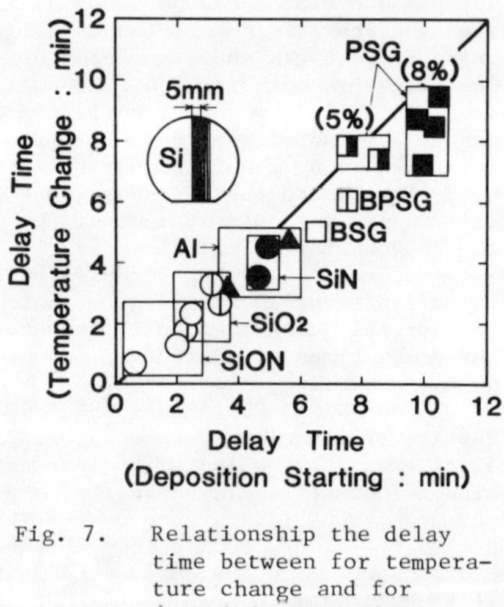

Fig. 7. Relationship the delay time between for temperature change and for deposition starting.

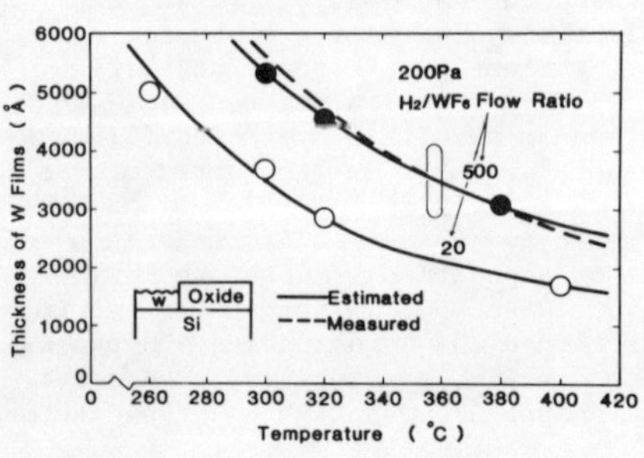

Fig. 8. Selective thickness of tungsten films as functions of deposition temperature for different gas flow ratios.

III. Si AND H_2 REDUCTION PROCESS

Figure 9 shows temperature difference and sheet resistivity as functions of deposition time in the one step deposition process (WF_6 + H_2 system). The temperature is similar to that in Figure 4 and the deposition time is on a coordinate logarithm base to explain the short time (-100 seconds) for change in the temperature. There are two peaks of temperature change within 30 seconds. Those peaks are at about 10 seconds (sharp and small) and 30 seconds (large). Further, sheet resistivity decreases rapidly at the peak at 10 seconds. Figure 10 shows the temperature difference at two deposition temperatures, when Ar was used instead of H_2 ambient (WF_6 + Ar system), for comparison with Figure 9. The temperature changed after only a few seconds, the same as with the hydrogen ambient, and after that, temperature difference is saturated. Similar results were obtained in high or low temperature deposition. The tungsten layer was formed by silicon reduction, because under the inert ambient, WF_6 must be reduced chemically only by the silicon substrate. For WF_6 + Ar ambient, no further reduction occurs after the tungsten has covered the original silicon surface. Hence the limiting thickness theory for a conventional LPCVD of tungsten is reasonable. Consequently, the reaction time for silicon reduction can be evaluated as about 10 seconds. Comparing the reactions with reducing (H_2) and inert (Ar) ambient shown in Figures 9 and 10 respectively, it is found that the reaction between silicon and hydrogen reduction even reacting simultaneously under hydrogen ambient can be discriminated by our own technique, which can be indicated in real time. Apparently, when silicon reduction is completed, hydrogen reduction starts.

For silicon reduction, the initial peak and saturation height of the temperatures shown in Figure 10 are different for different deposition temperatures. For two different deposition temperatures in a patterned oxide, surface morphology and thickness are shown in Figure 11. It shows the differences in the resulting layer thickness and surface roughness. The thickness of the low temperature deposition is less than that of the high temperature deposition. Also, the surface of the low temperature deposition is smoother than the surface of the high temperature deposition. In Figure 10, at high deposition temperature (400°C), the final surface can be speculated to be smoother than the surface at first peak temperature. The tungsten layer deposits nonuniformly at the first peak temperature because the tungsten deposits locally on the original silicon surface, as WF_6 and silicon react extremely at places when the temperature is higher. Then, as WF_6 reacts with the remaining silicon, the surface becomes smooth. But, the surface at lower temperature during deposition is smoother than at higher deposition temperature. Figure 12 shows the sheet resistivity and thickness of tungsten as functions of deposition temperature for several inert gases (i.e., He, Ar and N_2; 300 seconds deposit time). Note that the thickness appears to be almost independent of temperature from 300 to 400°C [5-7]. The thicknesses are dependent on inert gases; however, details of these are not evaluated, yet. Especially, using Ar gas, the thickness of tungsten is uniform for the temperatures examined; the limiting deposit thickness is 10 nm.

IV. SELECTIVITY AND Si REDUCTION

The following discusses the selective deposition properties for the different deposition processes. Figure 13 shows the topography of deposited wafer surfaces using two types of deposition processes. A wafer is patterned as shown (Si, SiO_2, and PSG). To compare the selectivity loss on insulators, tungsten is deposited for 15 min which is a long time in excess of loss of selectivity for WF_6 + H_2 system. Therefore, deposited tungsten is observed even on the insulators. Light lines in Figure 13 show tungsten thickness contrours. Comparing the deposited regions on the insulators for one and two

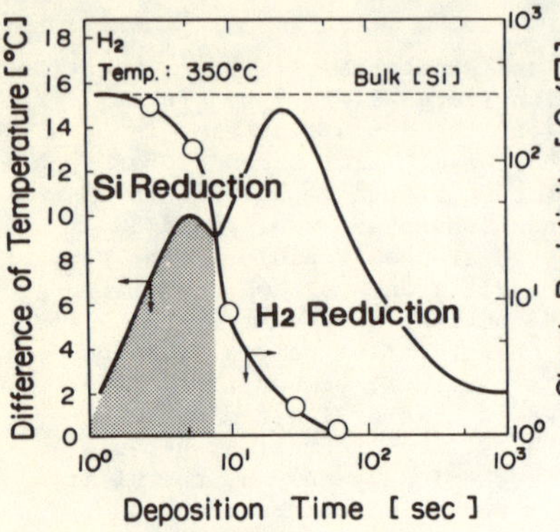

Fig. 9. Temperature difference and sheet resistivity as a function of deposition time. (WF_6 + H_2 system)

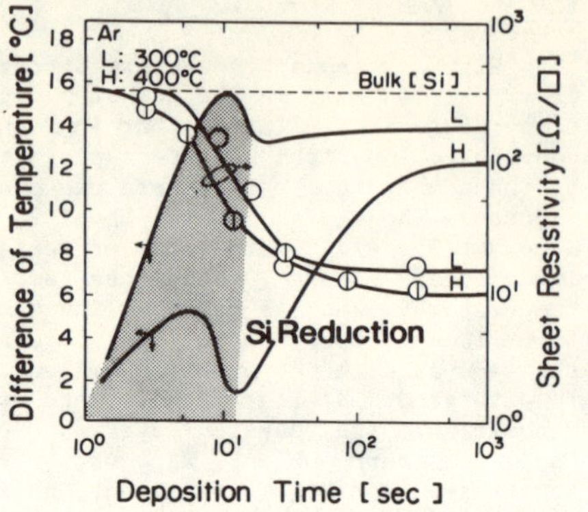

Fig. 10. Temperature difference and sheet resistivity as a function of deposition time. (WF_6 + Ar system)

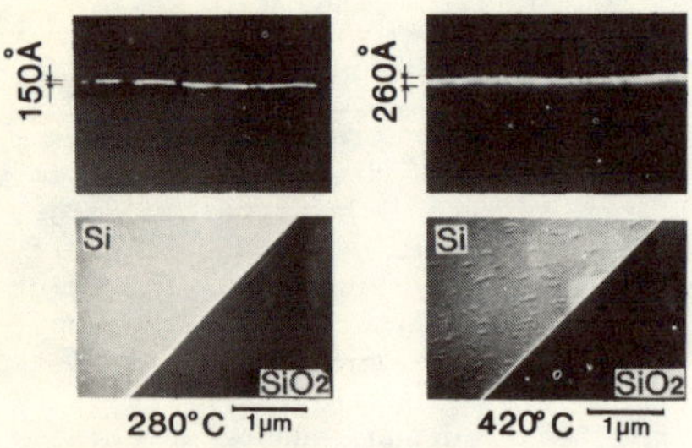

Fig. 11. SEM photomicrograph; deposition temperature at 280 and 420°C. (WF_6+Ar system)

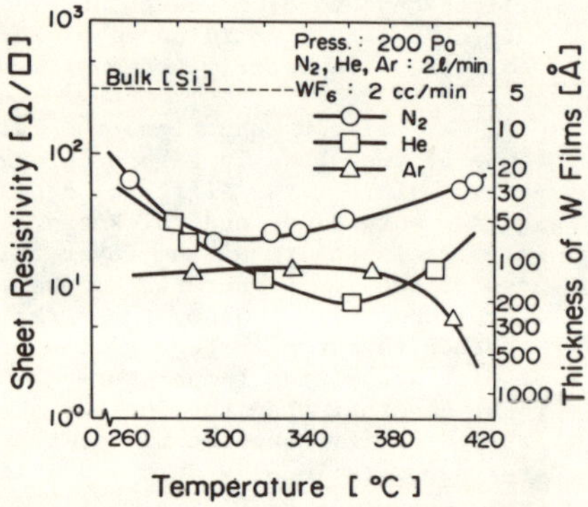

Fig. 12. Sheet resistivity and thickness as functions of deposition temperature for several inert gases. (WF_6+inert gas)

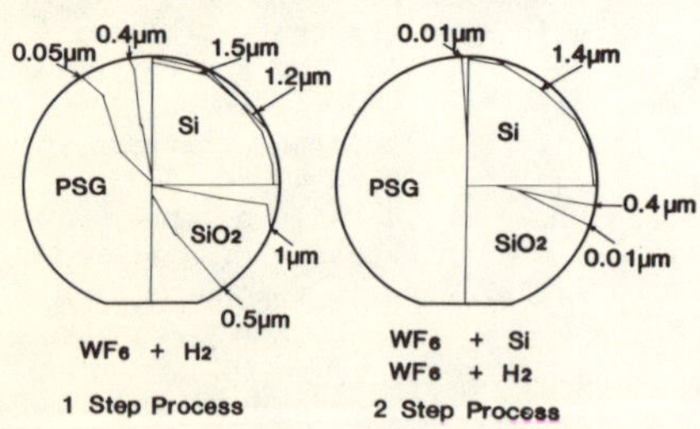

Fig. 13. Topography of deposited wafer surfaces; light lines are tungsten thickness contours.

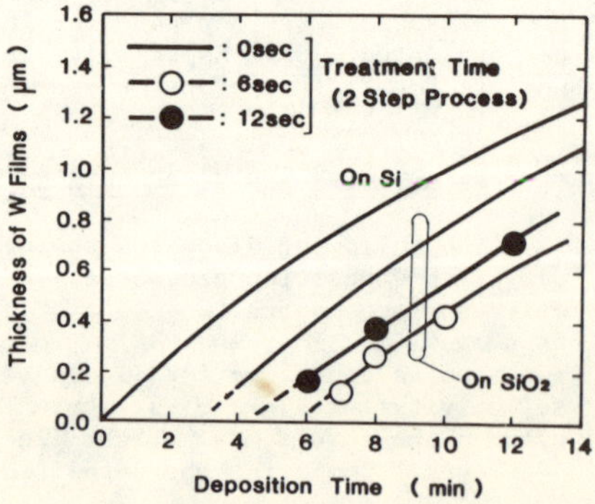

Fig. 14. Thickness of tungsten as a function of deposition time using the two step process.

step deposition processes, it is found that the selectivity loss is reduced in the two step deposition process (e.g., thickness of tungsten on SiO_2 at the boundary of Si is 1 μm in the one step process and 0.4 μm in the two step process). Thus the selectivity is enhanced by the two step process in which the first step is controlled silicon reduction. Furthermore, when we compare the deposited regions, we see that the selectivity is defferent between SiO_2 and PSG.

Figure 14 shows the thickness of tungsten as a function of deposition time using the two step deposition process. The treatment time shown in Figure 14 is the time for the silicon reduction process (i.e., WF_6 + Ar system). When treatment time is zero, it becomes a one step deposition process. Comparing the selective thickness of tungsten with the treatment time for silicon reduction, the selectivity is enhanced by silicon reduction when the time is increased (e.g., selective thickness is enhanced two times which is 0.38 μm for no treatment and 0.7 μm for 6 seconds silicon reduction). In addition, as silicon reduction yields the highest selectivity, it seems that 6 seconds must be an optimum condition in this experiment (i.e., selective thickness for 6 seconds treatment (0.7 μm) is thicker than for 12 seconds treatment (0.52 μm). We found that an over-supply of WF_6 for silicon reduction (i.e., the same as over-treatment time) produces worse selectivity and the optimum time is the efficient time for the first stage of reduction reaction. Further, this treatment time is related to the reaction time for silicon reduction evaluated by indicated temperature. Using this measurement method, we can be evaluate the optimum treatment time for silicon reduction. The reason for selectivity loss is the forming of nucleation sites on insulators by the by-product (i.e., $WF_6 + 3/2Si = W + 3/2\ SiF_4$). Therefore, using optimum treatment time for silicon reduction in the two step deposition process reduces selectivity loss.

CONCLUSIONS

Substrate surface temperature measurement using the IR thermometer has been estimated for selective deposition of CVD tungsten films. The selectivity can be evaluated by this measurement method and the deposition processes are observed in real time. Measurement of selectivities are evaluated as functions of parameters in tungsten deposition over several kinds of substrates, deposition temperature, gas flow ratio, and reaction time of silicon reduction. The selectivity is enhanced by the low deposition temperature, higher gas flow ratio, and using doped oxides. Especially, the doped oxides markedly reduce selectivity loss. The reaction between silicon and hydrogen reduction even reacting simultaneously under hydrogen ambient can be discriminated. The reaction times for silicon reduction are about 10 seconds and the time is an optimum condition for selectivity using the two step deposition process. The 10 seconds is enough time for the first stage deposition in which the original silicon surface is covered uniformly and properly. Therefore, no reaction occurs under the layer of tungsten. The limited thickness layer of tungsten is an interface for interconnection of VLSI devices, and influences the electrical characteristics (e.g., encroachment or wormholes). It is necessary to study this process and results further.

ACKNOWLEDGEMENTS

The authors wish to thank K. Yanagida and H. Ino for encouragement.

REFERENCES

1. R.W. Stoll and R.H. Wilson, Extend Abstracts of the 170th Meeting of the E.C.S., p. 531, October (1986).
2. R.H. Wilson, R.W. Stoll, and M.A. Calacone, Proc. 1985 Workshop on LPCVD Tungsten in VLSI Applications, 2nd ed., 35 (1985).
3. Y. Kusumoto, K. Takanashi, T. Ikuta, A. Suzuki, H. Nawa, and I. Nakayama, Extended Abstracts of the 47th Autumn Meeting, The Japan Soc. of Applied Physics, 509 (1986).
4. N.E. Miller and I. Beinglass, Solid State Technol., 25, December, 85 (1982); ibid., December, 79 (1980).
5. E.K. Broadbent, Proc. 1984 Workshop on LPCVD Tungsten in VLSI Applications, 2nd ed., 365 (1985).
6. E.K. Broadbent and W.T. Stacy, Solid State Technol., 28, December, 51 (1985).
7. E.K. Broadbent and C.L. Ramiller, J. Electrochem. Soc., 131(6), 1427 (1984).
8. T. Moriya and H. Itoh, Proc. 1985 Workshop on LPCVD Tungsten in VLSI Applications, 2nd ed., 21 (1985).
9. T. Ohba, Y. Ohyama, S. Inoue, and M. Maeda, Extended Abstracts of the 47th Autumn Meeting, The Japan Soc. of Applied Physics, 509 (1986).
10. T. Moriya, K. Yamada, Y. Tsunashima, S. Nakata, and M. Kashiwagi, Extended Abstracts of the 15th Conference on Solid State Devices and Materials, Tokyo, 225 (1983).
11. D.R. Bradbury and T.I. Kamins, J. Electrochem. Soc., 133(6), 1214 (1986).
12. Y. Pauleau and Ph. Lami, ibid., 132(11), 2779 (1985).
13. Y. Pauleau, Thin Solid Films, 122, 243 (1984).
14. Jan-Otto Carlsson and M. Boman, J. Vac. Sci. Technol. A, 3(6), 2298 (1985).
15. I. Beinglass, Proc. 1985 Workshop on LPCVD Tungsten in VLSI Applications, 2nd ed., 13 (1985).
16. R.A. Levy, M.L. Green, P.K. Gallagher, and Y.S. Ali, J. Electrochem. Soc., 133, 1905 (1986).

CVD Tungsten: Fundamental Reaction and Growth Studies

TEMPERATURE PROGRAMMED DESORPTIONS OF TUNGSTEN CVD MATERIALS

MICHAEL A. HEINY* AND CAROL M. McCONICA**
*Colorado State University, Department of Agricultural and Chemical Engineering, Fort Collins, CO 80523
**Colorado State University, Department of Agricultural and Chemical Engineering, Fort Collins, CO 80523

ABSTRACT

Temperature programmed desorption (TPD) studies were conducted on high surface area tungsten and silicon dioxide materials following exposure to a typical low pressure chemical vapor deposition (LPCVD) of tungsten in the absence of silicon. The depositions were run at 235°C, and the desorptions were conducted between 245°C and 470°C. It was found that significant amounts of HF desorbed from the tungsten and that SiF_4 and HF desorbed from the silicon dioxide. Peak desorption temperature ranges for HF are 375°C to 425°C from SiO_2 and above 450°C from W. Peak SiF_4 desorption from SiO_2 occurred between 325°C and 375°C. Related studies on the interaction of gas phase HF with oxidized wafers indicated that a reactive surface was needed for the attack of the oxide by the HF to proceed.

Studies on the effects of reactive surfaces and HF concentrations on tungsten nucleation initiation on thermal oxide were also performed. It was found that nucleation can be prevented by the elimination of all reactive surfaces. HF, when controlled amounts were added to the reactor, did not initiate nucleation on oxide surfaces, but appeared to inhibit the H_2 reduction of WF_6. Silicon produces SiF_4 when exposed to WF_6. These conditions also initiated nucleation on the wafer and on the inert walls of the non-reactive reactor.

INTRODUCTION

The initiation of tungsten nucleation on insulators such as SiO_2 is not a well understood process. It has been shown to be autocatalytic[1], but the exact mechanism and the species responsible are unknown. To gain a better understanding of the kinetics and mechanism of the loss of selectivity, TPD studies and other experiments were performed.

In an effort to determine which species are responsible for nucleation initiation, high surface area materials were used to simulate tungsten and silicon dioxide surfaces and were subjected to TPD following standard LPCVD runs. Commercially available hydrophilic SiO_2 powder (340 m^2/gm) was used to simulate the surface of an oxidized wafer, and tungsten coated Ni powder (estimated 100 to 300 cm^2/gm) was used to simulate tungsten surfaces. These materials were chosen for their superior surface area compared to actual wafer surfaces. Similar tests were performed on wafers to justify the use of these materials, but absolute correlation cannot be guaranteed.

The cold walled, single wafer reactor was also modified to test several theories on the cause of nucleation initiation on thermal oxide. A non-reactive liner was developed for the reactor which eliminated all reactive surfaces from the reaction chamber, and high purity hydrogen fluoride was plumbed to the system. The effect of nucleating surfaces and HF concentrations upon nucleation could then be observed.

EQUIPMENT

All experiments were conducted in a single wafer, pseudo-cold wall, high vacuum reactor where the gases were analyzed in-situ with a mass spectrometer (Fig. 1). The reactor was heated with a tungsten coated copper substrate heater and the temperature was controlled between 30°C and 550°C within 2°C by a programmable temperature controller. Actual wafer and powder temperatures were experimentally correlated to the controller reading. He, H_2, and WF_6 entered the system through mass flow controllers at rates of up to 200 sccm, while the flow rate of HF was controlled manually. Reactor pressure was controlled to within 0.5%, separately and independently from the gas flow rates, at pressures up to 10 torr with an exhaust valve control unit. Pumping is provided by a diffusion pump backed up with a mechanical pump. Maximum vacuum attainable is below 10^{-5} torr. There is a liquid nitrogen trap between the reaction chamber and the diffusion pump, and an afterburner and zeolite filter between the reactor and the mechanical pump.

The Residual Gas Analyzer (RGA) is a UTi 100C mass analyzer coupled with a UTi Q-30C mass spectrometer/pumping station. It is capable of measuring m/e ratios up to 300 m/e over the entire range of operating conditions.

The stainless steel reactor chamber is 21.7 cm in diameter, with 1 cm of clearance between the wafer and the reactor wall, and has a volume of about 75 cm^3. To eliminate the possible nucleating surfaces of the stainless steel walls and the tungsten coated heating block, an inert insert was designed and machined to cover all reactive surfaces in the reaction chamber. This insert reduced the reactor volume by approximately 50% and limited the maximum operating temperature.

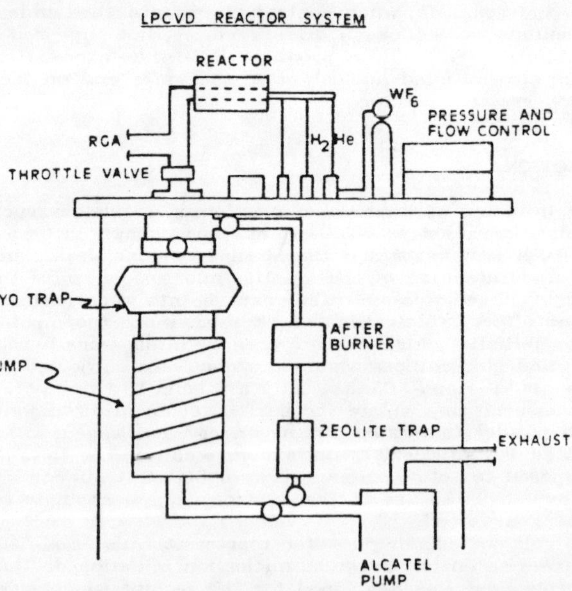

Figure 1. LPCVD reactor system.

EXPERIMENTAL PROCEDURES

Temperature Programmed Desorptions

The material to be studied was placed in a specially made tungsten coated copper sample box and covered with a fine tungsten coated wire screen. The sample box was then placed on the substrate heater and the reactor was sealed. The reactor was purged with He for five minutes and then evacuated with the diffusion pump while the substrate heater was controlled at 550°C. This bake was continued until the pressure in the reactor was less than 10^{-5} torr, usually from 2 to 4 hours. During the bake, the composition in the reactor was monitored with the RGA for evidence of leaks, and the run was terminated if any were found.

Following the bake out, the reactor was exposed to typical deposition conditions of 235°C and 5.00 torr while flowing H_2 and WF_6 at 195 sccm and 13 sccm respectively for 60 minutes. After the deposition the system was again evacuated with the diffusion pump to 10^{-5} torr at 10°C above the deposition temperature. The reactor exhaust was then completely sealed and the reactor was heated in approximately 50°C temperature increments and allowed to stabilize at each new temperature. Recordings of pressure and composition were taken at the end of each increment.

Selectivity Experiments

Cleaned wafers, either patterned oxide, field oxide, or bare silicon, were placed on the substrate heater and the reactor was sealed. When the reactive wall system was used, the system was baked and evacuated to 10^{-5} torr at 550°C. If the non-reactive reactor was used, the system was baked at a lower temperature for 4 hours or until a pressure of less than 10^{-5} torr was reached. Depositions were then made under various conditions and the wafers were examined visually.

RESULTS

Temperature Programmed Desorptions

The TPD experiments showed that SiF_4 and HF are chemisorbed by SiO_2 and that tungsten chemisorbs HF. Figure 2 shows the relative amounts of HF and SiF_4 that desorbed from the different materials and from a control run on the sample box alone. The relative amounts desorbed were obtained from correlations experimentally derived for pressure as a function of m/e peak height. A certain amount of error is inevitable. While the total amount desorbed from the SiO_2 is greater than that desorbed from the tungsten powder, the amount desorbed per unit surface area is actually much greater for the tungsten, which had about 2500 to 10,000 times less surface area than the SiO_2 used.

Figure 3 is a plot of the temperature derivative of the amount desorbed from SiO_2 powder in relative units versus temperature. The individual points each represent a 50°C temperature interval and should not be considered as absolute values. From these plots and others like them, the peak desorption temperature for HF from SiO_2 was found to be between 375°C and 425°C and the peak for SiF_4 desorption between 325°C and 375°C. Figure 4 is a similar plot for desorption from W powder. Although the graph clearly indicates a peak for HF desorption in the range of 450°C to 500°C, it should noted that the desorption is still rising and the actual desorption temperature and the amount desorbed may be greater than shown.

In related studies, 4 inch oxidized wafers were exposed to static vapor phase HF at 10 torr for 30 minutes in both the reactive and the

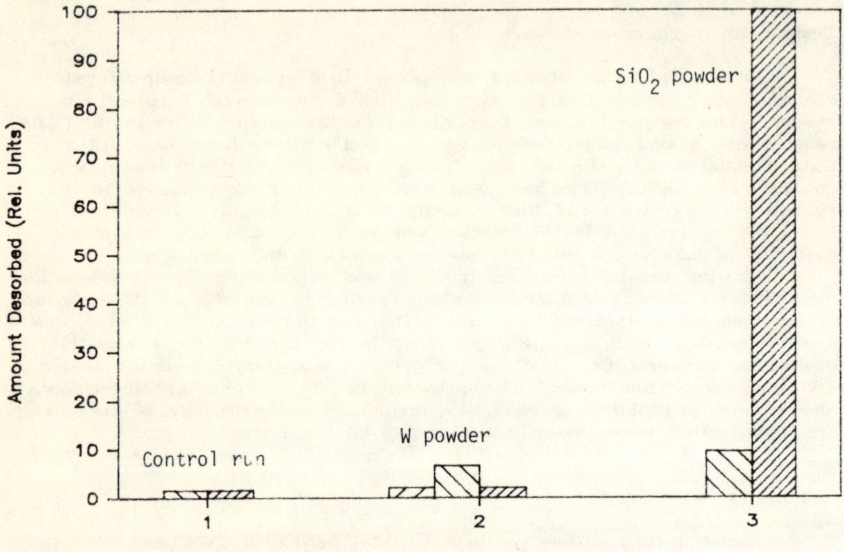

Figure 2. Total amounts desorbed.

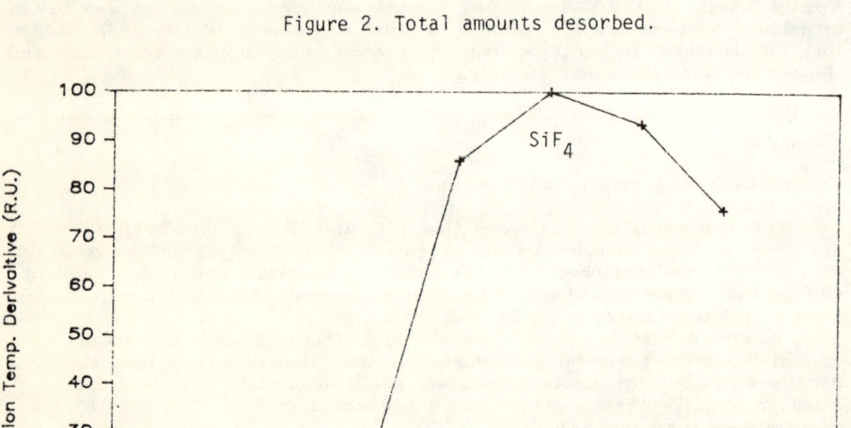

Figure 3. HF and SiF_4 desorption from SiO_2 powder.

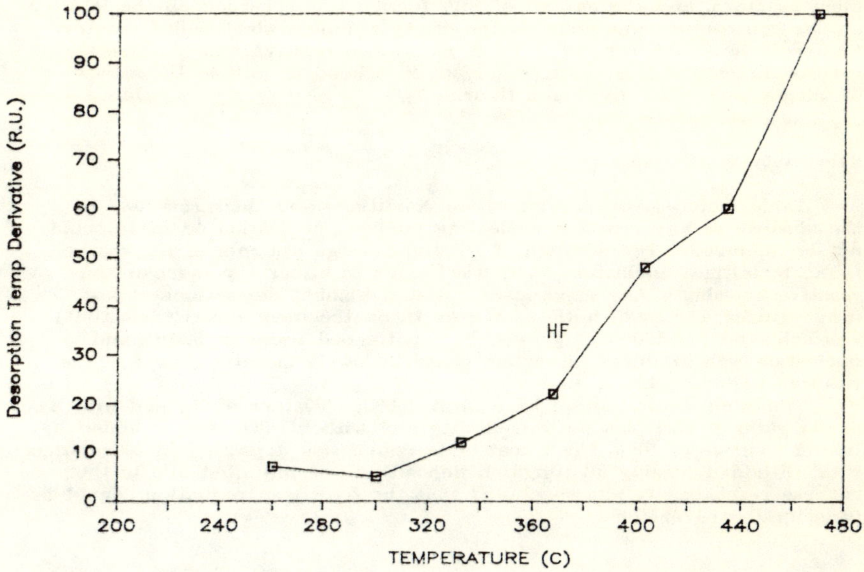

Figure 4. HF desorption from W powder.

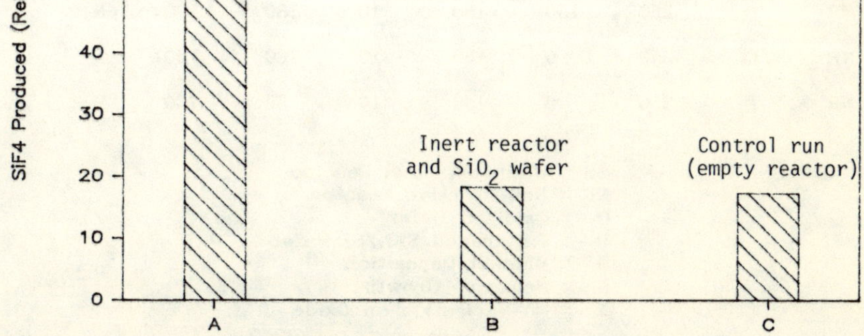

Figure 5. SiF_4 produced by oxidized wafers exposed to gas phase HF at 10 torr and 250°C.

inert reactors, and the amount of SiF_4 formed was recorded on the RGA. A similar control run was done on the empty stainless steel walled reactor. Figure 5 shows the results of these runs on a relative scale. These results indicate that a reactive surface is needed to initiate the attack of silicon dioxide by hydrogen fluoride.

Selectivity Experiments

Table I shows various deposition conditions and their results. In the absence of any reactive nucleating surface, nucleation on oxide could not be achieved. The addition of relatively large amounts of HF was not found to initiate nucleation on thermal oxide in either the inert or the reactive chamber. Any exposed Si initiated blanket depositions at low temperatures and even initiated W growth on the inert insert. Selectivity was achieved for 5000A of growth on a patterned wafer at 500°C, and nucleation was inhibited in certain cases in low temperature runs in the reactive walled reactor.

Tungsten deposition on bare Si at 350°C, .300 torr of H_2, and with a H_2:WF_6 ratio of 15:1 was performed with and without .300 torr of added HF. In the absence of HF a thick coat of tungsten was deposited in 20 minutes. With HF added, nearly all tungsten deposition was inhibited due to the reverse reaction. It was concluded that the hydrogen reduction cannot be considered irreversible.

Table I. Deposition Results

Reactor Type	Wafer Type	Temp. (°C)	HF (sccm)	H_2 (sccm)	WF_6 (sccm)	Time (min)	Press (torr)	Results
SS	O	250	150	150	10	120	1.500	N
SS	O	250	0	150	10	120	.750	N,B
SS	P	250	0	150	10	60	.750	B
SS	P	520	0	100	40	2	.500	S
NR	O	250	150	150	10	180	1.500	N
NR	O	250	0	150	10	180	.750	N
NR	P	250	0	150	10	60	.750	B

SS = Stainless Steel Reactor
NR = Non-Reactive Reactor
O = Oxidized wafer
P = Patterened SiO_2/Si Wafer
B = Blanket Deposition
S = Selective Growth
N = No Nucleation on Oxide

DISCUSSION AND CONCLUSIONS

The TPD studies and the experiments with wafers indicate that nucleation on oxide is catalyzed by reactive surfaces (metal walls and tungsten area). The loss of selectivity is probably due to a catalyzed attack of the SiO_2 by some reactive intermediate (possibly HF). If HF dissociates in the presence of a catalyst to form monatomic fluorine, it could readily attack the SiO_2 to form SiF_4 and water. The SiF_4 or subfluorides could then act as nucleators as has been demonstrated with experiments on Si surfaces. The addition of large amounts of HF in the reactive walled reactor resulted in no nucleation because the hydrogen reduction was totally suppressed under these conditions.

The most certain method to prevent nucleation initiation on thermal oxide is to eliminate all reactive surfaces in the reactor. Elimination of all quartz surfaces is advised because HF attacks it forming SiF_4 and water. The water can then catalyze the further attack of the quartz.[2]

Depositing at temperatures above the peak desorption temperature of SiF_4 (375°C) may help to eliminate nucleation. Further research is in progress to quantify the relationship between chemisorbed SiF_4, subfluorides, and nucleation rate.

ACKNOWLEDGEMENTS

We are very pleased to thank Sandia National Laboratories for funding this project and for the extended loan of the UTi mass spectrometer system. We also wish to thank the GENUS corporation for the donation of 99.999% pure WF_6 and Dr. J. Boyd and Delco Electronics for the donation of high quality oxidized wafers. Without these generous contributions this project would not have been possible.

REFERENCES

1. Kent J. Cooper, MS thesis, Colorado State University, 1986.

2. D.F. Weston and R.J. Mattox, J. Vac. Sci. Technol. 17(1), 466 (1980).

THE REACTION KINETICS OF THE H_2 REDUCTION OF WF_6 IN THE CHEMICAL VAPOR DEPOSITION OF TUNGSTEN FILMS

P. VAN DER PUTTE
Philips Research Laboratories Sunnyvale, Signetics Corporation, Sunnyvale, CA. 94088

ABSTRACT

The possible rate limiting steps in the deposition reaction of tungsten onto silicon are reviewed. The experimental results for the deposition of tungsten at various WF_6 partial pressures, and the influence of the addition of PH_3 to the gas phase, show the rate limiting step to be the desorption of HF from the tungsten surface, rather than the hydrogen dissociation reaction.

INTRODUCTION

In recent years the application of tungsten in very large scale integrated (VLSI) circuit devices has gained more and more importance. The reason for this is that tungsten as part of the metallization has many advantages [1].

The reaction of tungsten hexafluoride and hydrogen in the presence of silicon oxide and silicon has been studied by different authors [2-4]. The general observation is that the activation energy for the reaction between WF_6 and hydrogen on a tungsten surface is 0.71 eV/atom; the reaction rate is of order 0.5 with respect to the hydrogen partial pressure, and the reaction rate does not depend on the partial pressure of the WF_6. It is pointed out that the rate limiting step in the reduction sequence is either the dissociation reaction of adsorbed hydrogen molecules, or the desorption reaction of HF, with most authors in favor of the former.

It is the purpose of this work to evaluate the reaction mechanisms involved in the low temperature and low pressure regime of the chemical vapor deposition of tungsten. The effect of the addition of PH_3 to the gas is studied, and shown to offer the opportunity to gain more information regarding the rate limiting step in the deposition reaction.

EXPERIMENTAL PROCEDURE

The experiments were conducted in a commercially available Anicon reactor. In this reactor the wafers are positioned parallel to the gas flow direction and the effective reactor length equals the diameter of the wafer, minimizing depletion effects.

The gases used in the experiments were H_2, purified with a palladium diffuser element, WF_6 (99.95%), and PH_3 (0.1%, 100 and 5 ppm PH_3 in H_2).

The substrates used were 100 mm diam. 17-33 ohm-cm p-type <100> oriented silicon wafers. Prior to deposition the wafers were etched for two minutes in a (3:400) $HF:H_2O$ solution to remove the native oxide, followed by a deionized water rinse, and spin dry. The thickness of the deposited layer was measured using the beta-backscatter technique.

EXPERIMENTAL RESULTS

Deposition using WF_6 and H_2

The deposition rate of W was studied as a function of the temperature between 260 and 350°C. During the deposition the ratio of H_2 and WF_6 as well as the total flow was kept constant. The activation energy for the H_2 reduction reaction of WF_6 on a W surface was found to be 0.71 eV/atom in agreement with earlier work [2-4].

The dependence of the deposition rate on the H_2 concentration was studied at H_2 partial pressures varying between 200 and 2000 mtorr. It was found that the deposition rate depended on the square root of the H_2 partial pressure. This again is in accordance with the value reported previously [4]. Apart from one reference [5] the dependence of the deposition rate of W on the partial pressure of WF_6 is reported to be zero [2-4]. This is confirmed only partially in this work. The dependence of the deposition rate on the partial pressure of WF_6 was determined at different partial pressures of H_2 for WF_6 partial pressures ranging between 5 and 90 mtorr. The results are shown in Figure 1. It can be seen that indeed above a certain value of the WF_6 partial pressure the deposition rate becomes independent of the WF_6 partial pressure.

Deposition of Tungsten with PH_3 added

When concentrations in the order of 1 ppm PH_3 are added to the gas stream, the deposition rate decreases with increasing PH_3 concentration (Figure 2). The PH_3 concentration not only has a strong effect on the tungsten growth rate, it also shows a very pronounced depletion effect. Therefore the deposition rate of tungsten is plotted for three regions on the silicon wafer: a region close to the gas injectors, a region halfway the wafer, and a region closer to the gas exit. All three regions show a linear decrease of the deposition rate with increasing PH_3 input concentration.

For the same regions on the wafer, the activation energy of the deposition reaction was determined for a PH_3 concentration of 1.2 ppm, at a pressure of 550 mtorr and a WF_6/H_2 flow ratio of 60/2640. In Figure 3 the deposition rate of tungsten from the reaction between WF_6 and H_2 is plotted as a function of temperature in an Arrhenius plot. It can be seen that the activation energy is strongly influenced by the addition of

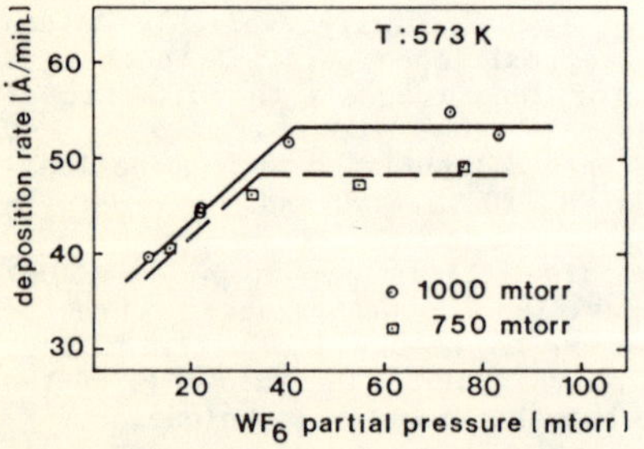

Fig. 1 Dependence of the deposition rate of W on the partial pressure of WF_6.

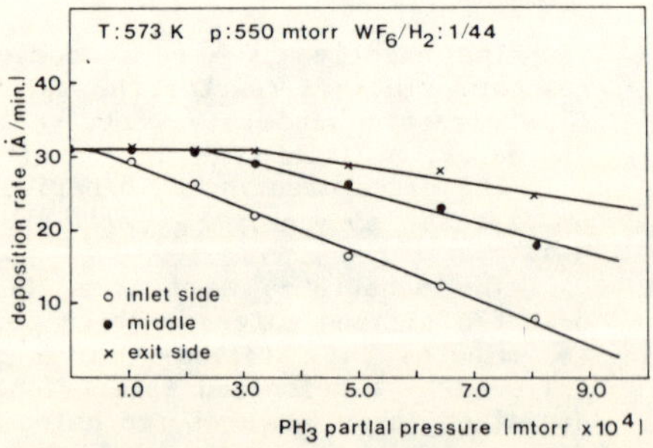

Fig. 2 Deposition rate of W as a function of the PH_3 concentration at 550 mtorr, 300°C.

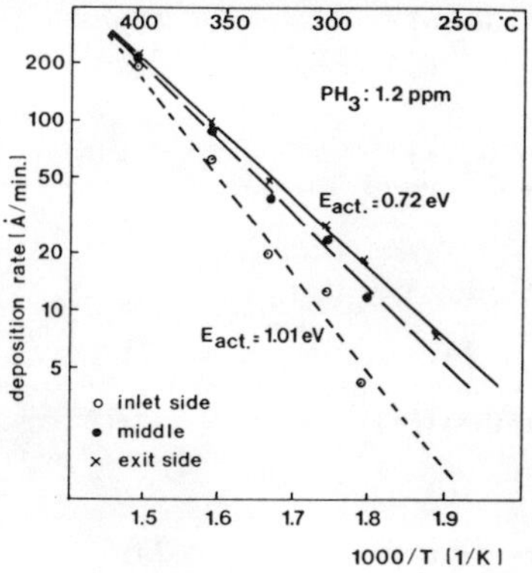

Fig. 3 Arrhenius plot of the deposition rate versus temperature for a PH_3 concentration of 1.2 ppm.

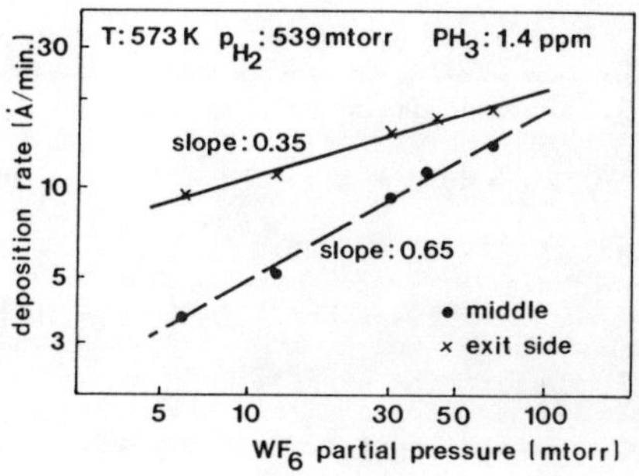

Fig. 4 Deposition rate as a function of the WF_6 partial pressure for a PH_3 concentration of 1.40 ppm.

the PH_3. Depending on the actual PH_3 concentration, the activation energy varies between 1.01 eV/atom for the region closest to the gas inlet, and 0.72 eV/atom for the area closest to the gas exit, a value close to the one found when no PH_3 is added to the gas.

PH_3 was added to the gas to influence the reaction so, that more information could be obtained about the rate limiting step in the reaction path. The information needed is the dependence of the deposition rate on the WF_6 concentration when PH_3 is added to the gas. In Figure 4 this dependence is plotted for a PH_3 concentration of 1.4 ppm for WF_6 partial pressures between 8 and 70 mtorr. The deposition rate appears to depend strongly on the WF_6 partial pressure.

DISCUSSION

It is known that the hydrogen reduction reaction of WF_6 can be expressed according to the following overall reaction [e.g. 6]:

$$WF_6 + 3\ H_2 \longrightarrow W + 6\ HF \qquad (1)$$

However, to gain more information on the reaction mechanism, and in particular the rate limiting step in the reaction, this overall reaction has to be divided into a series of steps, each of which has to be considered a possible candidate for the rate limiting step.
For this purpose the overall reaction will be divided into the following steps:

1. Adsorption of WF_6 on the surface:

$$* + WF_6(g) == WF_6* \qquad K_1 = (WF_6*)/(WF_6)(*) \qquad (2)$$

2. Decomposition of WF_6*:

$$WF_6* + 5\ * == W + 6F* \qquad K_2 = (F*)^6/(WF_6*)(*)^5 \qquad (3)$$

3. Adsorption of H_2 on the surface:

$$H_2(g) + * == H_2* \qquad K_3 = (H_2*)/(*)(H_2(g)) \qquad (4)$$

4. Double adsorption of H_2 on the surface:

$$H_2* + * == H_2** \qquad K_4 = (H_2**)/(H_2*)(*) \qquad (5)$$

5. Decomposition of H_2**

$$1/2\ H_2** + * == H* + * \qquad K_5 = (H*)(*)/(H_2**) \quad (*) \qquad (6)$$

6. Formation of HF*:

$$H* + F* == HF* + * \qquad K_6 = (HF*)(*)/(H*)(F*) \qquad (7)$$

7. Desorption of HF*:

$$HF* == HF(g) + * \qquad K_7 = (HF(g))(*)/(HF*) \qquad (8)$$

In the equations above the following notation is used:
* means a free surface site, F* means an adsorbed F atom, (g) means a species in the gas phase. K_1 through K_7 are the equilibrium constants for the reaction shown in the same lines.

In this approach other reactions will be treated fast compared to the rate limiting one, and those preceding the rate limiting one are thus in equilibrium. This implies that the adsorption of the different gas species is a competitive process. Using this model, an expression can be derived for the deposition rate for each of the above given reactions being rate limiting.

The results of the derivations are given in Table I, where for each rate limiting reaction the corresponding overall reaction rate is given.

All the values in the table are known, except for the concentration of free surface sites (*). The number of free surface sites can be found as follows. The surface is at least partly covered by species that are adsorbed on the surface, like WF_6*, F* and H*. Thus, the number of free surface sites per unit area is reduced, and can be expressed as:

$$(*) = C - F* - WF_6* - H_2* \ldots\ldots \qquad (9)$$

with C being the constant number of surface sites, or, using eq. 2 to 8:

$$(*) = C\ /\ (\ 1 + K_1K_2.p_{WF_6} + k_1.p_{WF_6}^{1/6}.(*) + k_3.p_{H_2}.(*) + \ldots) \qquad (10)$$

In the case of the reaction between H_2 and WF_6 it can be argued that the principle surface covering species is adsorbed F atoms, so equation 10 can be reduced to

$$(*) = C\ /\ (1 + K_1K_2.p_{WF_6}^{1/6}) \qquad (11)$$

To find out which of the different steps in the reaction mechanism can be rate limiting, the two extreme cases are considered. First the surface is assumed free of any adsorbed species, and second when the whole surface is covered with F atoms. In the first case equation 12 will be reduced to

$$(*) = C \qquad (12)$$

TABLE I: Expressions for reacton rates assuming different rate limiting steps in the H_2 reduction of WF_6.

For rate limiting step:	the reaction rate is:
1. adsorption of WF_6	$R=k_1 \cdot p_{WF_6} \cdot (*)$
2. decomposition of WF_6	$R=k_2K_1 \cdot p_{WF_6} (*)^6$
3. adsorption of H_2	$R=k_3 \cdot p_{H_2} \cdot (*)$
4. double adsorption of H_2	$R=k_4K_3 \cdot p_{H_2} \cdot (*)^2$
5. decomposition of H_2**	$R=k_5(K_4K_3)^{1/2} \cdot p_{H_2}^{1/2} \cdot (*)^2$
6. formation of HF	$R=k_6K_5(K_4K_3)^{1/2}(K_2K_1)^{1/6} p_{H_2}^{1/2} p_{WF_6}^{1/6} \cdot (*)^2$
7. desorption of HF	$R=k_7K_6K_5(K_4K_3)^{1/2}(K_2K_1)^{1/6} p_{H_2}^{1/2} p_{WF_6}^{1/6} \cdot (*)$

and in the second case it will be reduced to:

$$(*) = C \;/\; K_1K_2 \cdot p_{WF_6}^{1/6} \tag{13}$$

Both expressions for the amount of free surface sites can be introduced in the expressions for the reaction rates when considering each of the steps to be rate limiting, as given in Table I. Then, the resulting rate equation can be compared with the experimental results, in the region where the deposition rate only depends on the square root of the H_2 concentration. The expression for the deposition rate in that region is of the form:

$$R = k_{tot} \cdot p_{H_2}^{1/2} \tag{14}$$

where k_{tot} stands for any combination of constants.

It appears that only two steps are possibly rate limiting:

1) the decomposition of double adsorbed H_2 on the surface resulting in:

$$R= C^2 \; k_5(K_4K_3)^{1/2} \; p_{H_2}^{1/2} \tag{15}$$

in the case that the surface is free of adsorbants, and

2) the desorption of HF from the surface, resulting in:

$$R = C\, k_7 K_6 K_5 (K_4 K_3)^{1/2}\, p_{H_2}^{1/2} \qquad (16)$$

in the case that the surface is covered with F.

To discriminate further between the two possible rate limiting steps left, the dependence of the deposition rate on the WF_6 concentration in the presence of PH_3 can be taken into account.

In the presence of PH_3 the surface coverage changes drastically. Because of the effect of the addition of PH_3 on the deposition rate, we assume that the surface for the largest part is covered with PH_3. This means that besides the reactions given in equations 2 to 8, an additional reaction has to be introduced:

8. Adsorption of PH_3 on the surface:

$$PH_3(g) + * == PH_3^* \qquad K_8 = (PH_3^*)/(PH_3)(*) \qquad (17)$$

The adsorption of the phosphine on the surface will clearly have an effect on the expression of the concentration of free surface sites. In the presence of PH_3 equations will change into:

$$(*) = C \,/\, (1 + K_1 K_2 \cdot p_{WF_6}^{1/6} + K_8 \cdot p_{PH_3}) \qquad (18)$$

and can be reduced, since the surface coverage by the PH_3 is so large, to

$$(*) = C \,/\, K_8 \cdot p_{PH_3} \qquad (19)$$

However, it must be taken into account, that the amount of PH_3 on the surface is not solely a function of the PH_3 concentration in the gas, but that WF_6 plays an active role in the removal of PH_3 from the surface. In that case the following equation must be added:

9. Removal of PH_3 by WF_6:

$$PH_3^* + WF_6^* == PWH_xF_y + z\,HF \qquad (20)$$

(While it does not have to be this particular reaction that is rate limiting in the removal of PH_3, this reaction represents correctly a group of removal reactions of possible complexes of PH_3, which ultimately result in the similar kinetic expressions for the removal rate of PH_3).

Assuming a predominant active removal of PH_3 by WF_6 results in a different expression for the PH_3 surface concentration. The PH_3 flux to the surface still is regulated by the adsorption reaction:

$$(dPH_3/dt) = k_8 p_{PH_3} \cdot (*) \qquad (21)$$

but the PH_3 removal is determined predominatly by the reaction described in eq. 20:

$$(dPH_3/dt) = k_9\, PH_3^* \cdot WF_6^* \qquad (22)$$

Assuming steady state conditions these two expressions must be equal. The expression for the adsorbed PH_3 becomes (using eq. 2):

$$PH_3^* = k_8 \cdot p_{PH_3} \,/\, k_9 \cdot p_{WF_6} \qquad (23)$$

and the expression for the amount of free surface sites, for the case that the surface is mainly covered with PH_3 becomes (cf. eq. 9)

$$(*) = C - k_8 . p_{PH_3} / k_9 . p_{WF_6} \quad (24)$$

The expressions for the reaction rates for the steps that were possibly rate limiting now become (all constrants combined):

a) for the H_2 dissociation rate limiting, and no WF_6 involved in the removal of PH_3:

$$R = K_a . p_{H_2}^{1/2} / p_{PH_3}^{2} \quad (25)$$

b) for the H_2 dissociation rate limiting, and WF_6 involved in the removal of PH_3:

$$R = K_b \, p_{H_2}^{1/2} \, (1 - K_c \, p_{PH_3} / p_{WF_6}) \quad (26)$$

c) for the HF desorption being rate limiting, and no WF_6 involved in the removal of PH_3:

$$R = K_d . \, p_{H_2}^{1/2} \, p_{WF_6}^{1/6} / p_{PH_3} \quad (27)$$

d) and for HF desporption rate limiting and WF_6 involved in the removal of PH_3:

$$R = K_e . \, p_{H_2}^{1/2} . p_{WF_6}^{1/6} (1 - K_f \, p_{PH_3} / p_{WF_6}) \quad (28)$$

According to the experiments the correct description of the deposition reaction has to:

a. be linear in the dependence on PH_3 when the concentrations of WF_6 and H_2 constant,
b. allow for a varying dependence of the deposition rate on WF_6, with an order that can be larger than 1/6 for constant H_2 and PH_3 concentrations
c. allow for a change in activation energy depending on the PH_3 concentration

The only reaction mechanism that fits this description is the one where it is assumed that HF desorption is rate limiting in the deposition of tungsten from H_2 and WF_6, and that WF_6 plays an active role in the removal of WF_6 from the surface.

The experimentally found dependences of the deposition rate on the WF_6 partial pressure and the H_2 partial pressure for low WF_6 concentrations support this conclusion.

When the WF_6 concentration is low, the surface coverage with F will be low, and the equation for the amount of free surface sites is described by eq. 12.

The expressions for the deposition rates for H_2 decomposition and HF desorption will become of the form (all constant gathered into one):

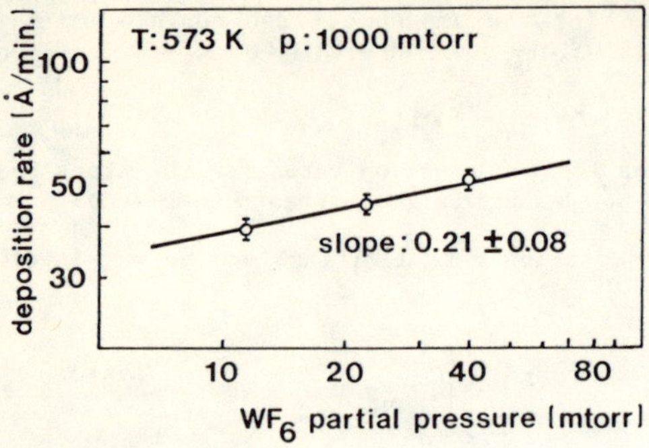

Fig. 5 Log. -log. plot of the deposition rate vs. the partial pressure of WF_6 for the lower WF_6 concentrations at 300°C and 750 mtorr.

$$R = K'.p_{H_2}^{1/2} \qquad \text{and} \qquad R = K''.p_{H_2}^{1/2}.p_{WF_6}^{1/6}$$

respectively. Only when the desorption of HF is rate limiting can the dependence of the deposition rate on the WF_6 partial pressure be explained. In Figure 5 the dependence of the deposition rate on the WF_6 partial pressure for low WF_6 concentrations at a total pressure of 750 mtorr and a temperature of 300°C is plotted on a log-log scale. A slope of 0.20 is found, which is reasonably close to the expected value of 0.17, corresponding to the power 1/6 in equation 33.

CONCLUSIONS

It has been shown that the deposition rate of W onto W surfaces depends for low WF_6 concentrations on the partial pressure of WF_6 in the gas phase. For higher WF_6 concentrations the deposition rate is independent of the WF_6 partial pressure, and only dependent on the H_2 partial pressure. When PH_3 is added to the gas stream, the deposition rate becomes dependent on the WF_6 partial pressure also for the higher WF_6 partial pressures. The experimental results can be explained by assuming that the rate limiting step in the deposition reaction is the desorption of HF from the surface.

REFERENCES

1. See, for example, the review and references provided in E.K. Broadbent and W.T. Stacy, Solid State Technol. 28, (12) p. 51 (1985).
2. W.A. Bryant, J. Electochem., Soc., 125, 1534 (1978).
3. H. Chueng, in "The Third International Conference on Chemical Vapor Deposition", ed. F.A. Glaski, p. 136, The American Nuclear Society, Hindsdale IL, (1972).
4. E.K. Broadbent and C.L. Ramiller, J. Electrochem. Soc., 131, 1427 (1984).
5. M.L. Green, R.A. Levy, J. Electrochem. Soc. 132, 1243 (1985).
6. W.A. Bryant and G.M. Meier, J. Electrochem Soc. 120, 559 (1973).

The Microstructures of CVD W Films Produced by the Si Reduction of WF_6

M. L. Green, Y. S. Ali, T. Boone,
B. A. Davidson, L. C. Feldman and S. Nakahara

AT&T Bell Laboratories
Murray Hill, New Jersey 07974

ABSTRACT

In this paper, we present data on the formation and structure of CVD W films deposited by the Si reduction of WF_6. Although the great majority of CVD W films deposited for IC applications are ostensibly deposited by H_2 reduction, the Si reduction of WF_6 always occurs first, even in the presence of copious amounts of H_2. Therefore, it is the Si reduction reaction that determines the nature of W/Si interface, and, in turn, such important properties as contact resistance and leakage current.

We have found that Si reduced W films deposited between 210°C and 700°C are porous and discontinuous, and are probably not effective barriers to the further diffusion of WF_6. The discontinuous structure of the films provides a simple mechanism for the growth of thick W films by Si reduction; such thick films have been observed with increasing frequency as of late and have been enigmatic, because the reaction between WF_6 and Si was thought to be self-limiting.

An unusual temperature dependence of W film growth is reported. Films deposited below 310°C self-limit at about 0.3×10^{17} atoms/cm^2 coverage (~60Å of full dense W). In the temperature range 320°C to 450°C, film coverages up to 6.1×10^{17} atoms/cm^2 (~950Å, at 340°C), are observed. In the higher temperature range, 500°C to 700°C, film coverages of about 1.5×10^{17} atoms/cm^2 (200Å) are observed. The temperature dependence and the self-limiting behavior of film growth (regardless of the ultimate film thickness at which the film self-limits), are believed to be controlled by some surface-catalyzed reaction. Thus, the temperature dependence of the sticking coefficient of a reactive intermediate probably controls the reaction. This kind of model has the potential to explain why minor changes in the Si surface due to cleaning variations, for example, can result in radically different W film growth behavior.

Introduction

The interest in chemical vapor deposition (CVD) of W films for a variety of integrated circuit (IC) metallization applications has increased steadily in the past five years. While most papers are concerned with the application of selective CVD W as a diffusion barrier between Al and Si at source/drain areas (1-4), other CVD W applications, i.e., as an interconnect, are also gaining popularity (5,6). Much of the latest work in the CVD W field is published in the proceedings of the 1984 and 1985 Albuquerque (Sandia) meetings on CVD refractory metals (7).

The chemical vapor deposition of W on Si, using WF_6 as the source, can involve two reactions. The first reaction, the Si reduction of WF_6, also known as the displacement reaction, involves consumption of Si:

$$2WF_6 + 3Si \rightarrow 2W + 3SiF_4 \uparrow \quad . \qquad (1)$$

For every angstrom of W produced, approximately two angstroms of Si are consumed (3). Even when hydrogen is present (the H_2 reduction of WF_6 will be described next), this reaction will always occur first. The Si reduction of WF_6 can be prevented through the use of excess SiF_4 in the reactor (8), although it is not possible to eliminate Si reduction entirely without poisoning the H_2 reduction reaction. Judicious use of SiF_4, however, can minimize the reduction and ensuing erosion of Si due to Si reduction.

Whether SiF_4 is used, the displacement reaction usually stops once W films of about 100-200Å have formed, ostensibly due to the lack of contact between WF_6 and Si. Thicker W films can be grown on the Si-reduced W film by H_2 reduction of WF_6 according to a second reaction:

$$WF_6 + 3H_2 \rightarrow W + 6HF \uparrow \qquad (2)$$

It is believed that the rate limiting step in this reaction is the dissociation of H_2 into atomic hydrogen (9), which is probably catalyzed by the freshly deposited displaced W film. W films of unlimited thickness can be deposited by H_2 reduction.

Although the great majority of CVD W films deposited for IC applications are deposited by H_2 reduction, it is important to emphasize that even in the the presence of copious amounts of H_2 (typical flows for selective CVD W film deposition are 2000 sccm H_2, 10 sccm WF_6), the displacement reaction, Eq. 1, will always take place first. This fact is dictated by the higher free energy of reaction (1) as compared to reaction (2) (−147 vs. −15 Kcal/mole of W, at 327°C), and can readily be verified by selectively etching the W from source/drain windows and observing the erosion of the Si. Therefore, it is the Si reduction reaction, and not the H_2 reduction reaction, that determines the nature of the W/Si interface. In turn, this interface will determine such important device characteristics as contact resistance and leakage current, and such physical characteristics as interfacial cleanliness and silicide formation kinetics. Thus, the purpose of this paper is to exclusively study the nature of the displacement reaction, and the structure and properties of the resultant W films.

Background

A previous paper (10) reported on studies of Si-reduced W films as a function of growth temperature and crystal perfection. Figure 1 illustrates the observed temperature dependence of W film growth. The unusual temperature dependence suggests that kinetic limitations, as opposed to nucleation limitations, influence growth, because even at very low temperatures, e.g. 210-290°C, W films can nucleate, although they become self-limiting at thicknesses of less than 100Å. (Assuming full density of W, 10^{17} atoms/cm^2 coverage equals 159.3Å of W.)

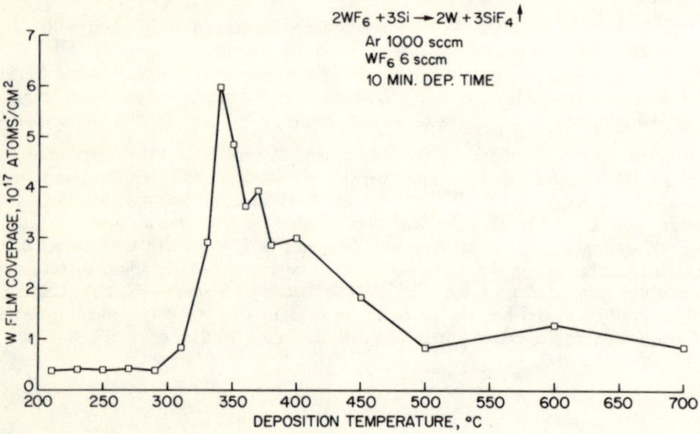

Fig. 1 W film coverage as a function of deposition temperature for lightly doped Si (100) wafers. Assuming full density of W, 10^{17} atoms/cm^2 coverage equals 159.3 Å of W.

Other researchers have observed similar temperature dependences of W growth, as is shown in Fig. 2. The data of Fig. 1 is shown for comparison, replotted in terms of thickness, assuming full W density. Tracy (11) sees a peak in W growth at about the same temperature as the present data, 350°C. Tracy's cleaning and related processing techniques are the same as in this paper, although the deposition time is only 1 min. Itoh's data (12), generated using an unreported cleaning technique, but a protected (N_2) wafer introduction technique, reveals a strong growth peak, but at 450°C instead of the 350°C observed here and by Tracy (11). On the other hand, Broadbent's data (13), generated on samples that were introduced into a cold furnace and then heated under vacuum, show almost no temperature dependence. Thin films are always observed. They note, however, (14), that introduction into a hot (T > 300°C) furnace under air ambient results in thicker W films.

With the exception of Itoh's data (12), whose film thickness measurement technique is unknown, the other data on Fig. 2 were deduced from either RBS or β-backscattering measurements. Therefore, the thicknesses plotted for these data are derived from the measured areal coverage of W and the assumption of full W density. We will shortly show that the assumption of full density, and therefore the calculation of film thickness, is often erroneous.

Results

Thick and thin self-limiting W films by Si reduction of WF_6

Conventional wisdom in the CVD W field has stated many times that the Si displacement reaction, Eq. 1, results in thin (~100Å), self-limiting films of W. The data of Figs. 1 and 2 have now completely upset this simple picture. In addition, Tsao et al. (15) have seen Si-reduced W films as thick as ~2000Å and Mianowski et al. (16) have seen similar films up to 6000Å thick when deposited in an N_2 ambient at atmospheric pressure. Also, Mo films as thick as several microns have been deposited by Si reduction of MoF_6, a reaction analogous to Eq. 1 (17).

The Si reduction reaction does in fact result in self-limiting films, but not all films are self-limiting at the same thickness. Figure 3, for example, illustrates the effect of time on the growth of various

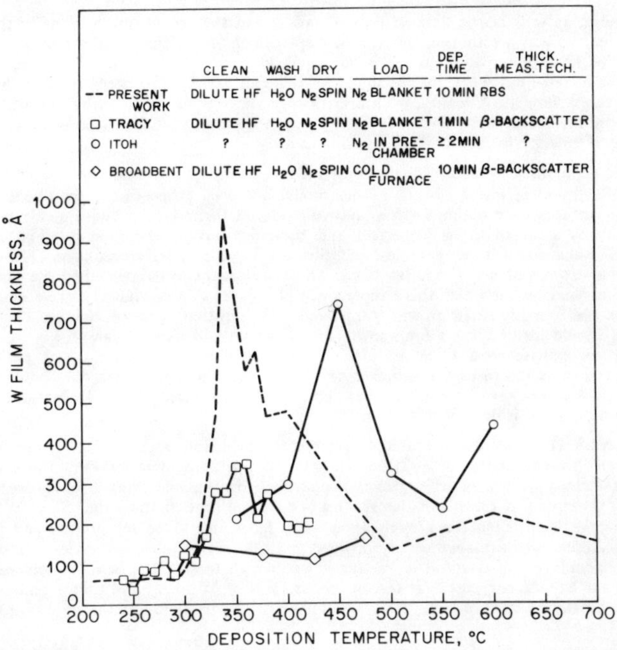

Fig. 2 W film thickness as a function of deposition temperature. Data from a variety of sources is plotted along with the data of Fig. 1.

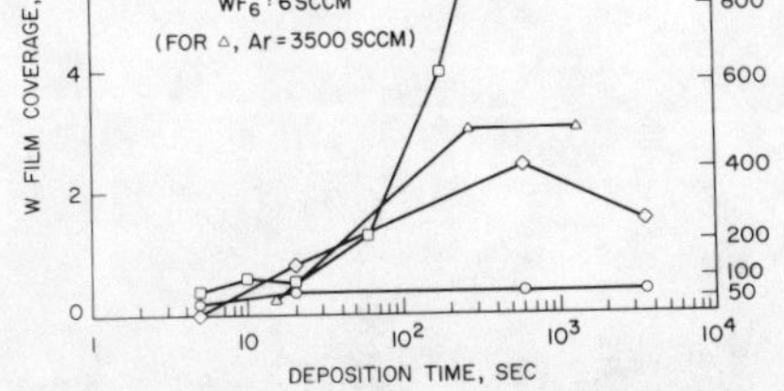

Fig. 3 W film coverage and thickness as a function of deposition time for deposition temperatures of 290°C, 340°C and 550°C.

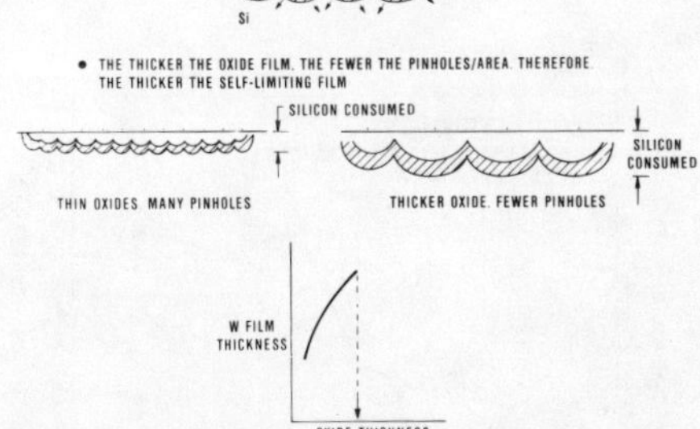

Fig. 4 The pinhole model of W film growth by Si reduction of WF_6.

Si-reduced W films. Each data point represents an individual deposition run, and the run-to-run variations can be seen. However, the self-limiting trend is clear, in that somewhere between 3 and 10 min, probably closer to 3 min, the films attain their ultimate thickness. Films deposited at 340°C self-limit at t ~ 1300Å, whereas those deposited at 290°C self-limit at t < 100Å.

These new findings, as well as the data in Figs. 1 and 2, and the recent observations in the literature (15-17), must change the way we think about the Si displacement reaction. First, we must ask how it is possible for thick W films to grow by what is supposed to be a diffusion controlled reaction that is traditionally thought to self-limit when the W thickness is about 100Å. Furthermore, now that it is well established that thick W films are a reality, we must ask why all W films produced by Si reduction are not thick, especially those formed at T > 340°C. Finally, we should ask why the thick W films do not grow thicker with increased time. Why are they self-limiting at all?

One model to explain thick W film growth by Si reduction of WF_6 is what we will call the pinhole theory of growth, illustrated in Fig. 4. Much of this model has been proposed by Broadbent et al. (14). It is known that Si surfaces have a thin (~5Å) native oxide on them (18), which quickly regrows after cleaning. This oxide is assumed to be imperfect and therefore have pinholes in it. The Si reduction reaction is assumed to nucleate at these pinholes, and advance into the underlying silicon. The self-limiting thickness is the thickness at which the reaction stops, which is thought to happen when the growing fronts coalesce. The W film thickness at which this happens will be related to the original pinhole density. Many pinholes should give rise to many small growth fronts, and a W film that coalesces quickly, i.e., a thin film. Few pinholes, then, should lead to just a few growth fronts, which will have to consume a lot of Si before coalescing, thereby giving rise to a thick W film. Furthermore, if one assumes that pinhole density decreases as oxide thickness increases (i.e., thicker oxides are more perfect), then one should see W film thickness vs. oxide thickness behavior similar to that depicted at the bottom of Fig. 4. Naturally, when the oxide gets thick enough, no W is deposited by Si reduction.

In fact, recent work (19) has demonstrated this predicted behavior, with a peak W film thickness of about 1000Å for an oxide thickness of ~20Å. In spite of the agreement between the predicted and experimentally determined forms of the growth behavior, the pinhole theory of growth is rather unsatisfactory. For one thing, if growth of the W film is diffusion limited (by either Si or WF_6 diffusion through the growing W film), then the growth along each front should be self-limited and the growing fronts should not coalesce. Growth should only proceed laterally under the native oxide. In other words, perhaps just a few monolayers of continuous W should be enough to stop the Si reduction reaction. The fact that this does not happen suggests that discontinuous films of W form, and that a pinhole argument may only partially explain the data of Ref. (19). We will show shortly that Si-reduced W films are in fact very discontinuous.

Furthermore, RBS and spectroscopic ellipsometry (20,21) were used to characterize the surface of cleaned Si wafers that underwent simulated W deposition runs, i.e., were removed from the furnace just prior to W deposition. Figure 5, which illustrates these results, reveals that very little change in native oxide thickness is noted in the temperature range 300°-350°C, where, according to Fig. 1, very significant changes in W deposition behavior were noted. That a major change in W deposition behavior can be precipitated with only negligible change in oxide thickness suggests that the pinhole theory of growth does not have much validity. Note from Fig. 5 that a simulation temperature greater than 350°C is required before significant process-related oxide growth is seen. Even then, it is observed that this oxide growth is not associated with enhanced W film growth, as can be seen in Fig. 1.

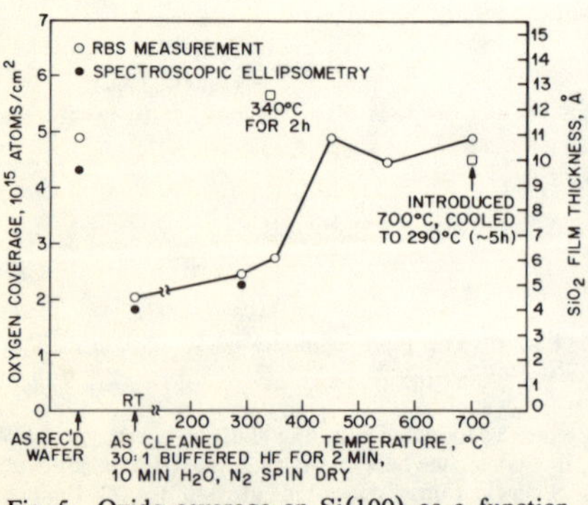

Fig. 5 Oxide coverage on Si(100) as a function of deposition (simulated) temperature.

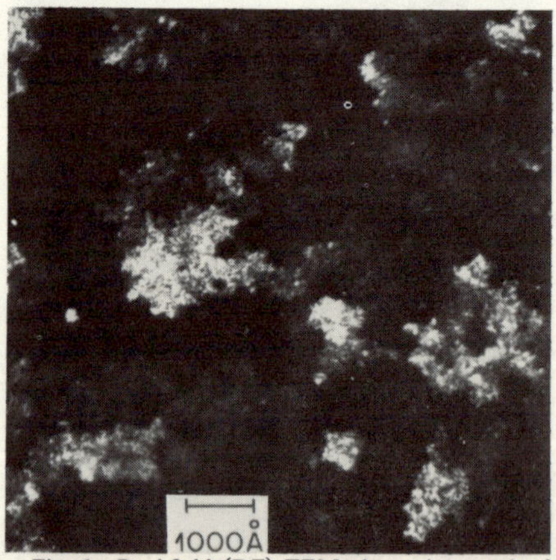

Fig. 6 Darkfield (DF) TEM photomicrograph of a Si-reduced W film, in plan view. W film was deposited at 340°C, Ar = 1000 sccm, WF_6 = 5 sccm, t = 5 sec.

A final aspect of the pinhole theory that is troublesome is its assumption that the native SiO_2 layer remains intact, and that attack of the Si takes place through and underneath it. In fact, the native oxide is not stable in the presence of WF_6:

$$3SiO_2 + 2WF_6 \rightarrow 2WO_3 + 3SiF_4 \tag{3}$$

with $\Delta G = -48$ Kcal/mole of WO_3 at 327°C. Due to the destruction of the SiO_2 film, a model as simple as the pinhole model probably only has limited validity.

The preceding discussions, and the evidence of Fig. 1, suggest that there is no kinetic limitation to the formation of CVD W by Si reduction. That a very thick film forms at 340°C but not at 310°C indicates that growth is not a diffusion limited process, since it is unlikely that the temperature dependence of the diffusion mechanism is that strong. Even assuming it were that strong, why would growth suddenly decrease after only a slight further increase in temperature to, e.g., 375°C? It is more likely that the temperature dependence of CVD W film growth by Si reduction is controlled by factors such as the temperature dependence of nucleation rate, sticking coefficient, atomic reduction mechanism, and other mechanisms. To study these effects, sophisticated surface chemistry studies, performed in high vacuum with very good diagnostic techniques such as Auger spectroscopy, RHEED, XPS and mass spectrometry need to be carried out.

Film Structure

Table 1 summarizes an experimental fact that was repeatedly observed. W film thicknesses, calculated from RBS measurements using W full density, were consistently smaller than actual film thicknesses as observed by TEM cross-section. The ratio of the two thickness measurements is equal to the density of the W film, as expressed as a fraction of its full density. Thus, CVD W films, deposited at a variety of temperatures, are all less than full density. It is interesting to note that the W film density peaks at about the same temperature as the growth maximum, (~340°C), and that it follows the behavior of Fig. 1 in general. The percentage of space not occupied by W (i.e. trapped gases, pores, inclusions) in the CVD W films varies from 40 to 58.8%. Careful scrutiny of the RBS spectra reveals less than 2% Si in the W films, the detectability limit by RBS.

The discontinuous, porous nature of the Si-reduced W film is illustrated in Fig. 6, which is a plan view TEM photomicrograph of a W film deposited on Si. It is seen that each grain is composed of tiny particles separated by a fine pore network. The grains are therefore "spongy," and reminiscent of a partially sintered mass of particles. From Fig. 6, the particles that form the grains can be seen to be about 25-200Å in size.

The origin of the porous structure of the W films seems quite obvious. Since a gas (SiF_4) is evolved during the formation of the W deposits, disruption of the film is to be expected. For example, if a

TABLE 1

Volume Percent W in Si Reduced W Films

(100) Si Wafers Exposed to WF_6 for 10 minutes with 6 sccm WF_6 and 1000 sccm Ar.

Deposition Temperature, °C	t_{RBS}, Å	t_{TEM}, Å (Avg)	$\frac{t_{RBS}}{t_{TEM}}$ = % Full Density of Film
290	59	129	45.9
350	741	1236	60.0
550	392	737	53.2
700	200	485	41.2

nucleating SiF_4 "bubble" is occupying a site on the growing W film, somewhere near the W/Si interface, another W particle cannot nucleate there. The pores, therefore, may be the remnants of bubbles of SiF_4.

These new observations of microporosity in Si-reduced W films are consistent with two other observations of these films. First, since the porous W film may not be a barrier to further WF_6 in-diffusion, the porosity provides a mechanism for the growth of thick Si-reduced W films. In other words, such films are not self-limiting at small W coverages because they are not barriers to further reaction. Second, we have observed very high film resistivities ($50-200\times10^{-6}$ Ω-cm, 10 to 40 times the bulk W value) for Si-reduced W films. A significant portion of this increase is explained by the lack of interconnectivity of the porous W films.

Knowledge of CVD W film morphology as a function of time might offer some insight into the formation of the porosity. Figure 7 illustrates, in TEM cross-section, the morphology of films deposited for 5 sec (a), and for 10 min (b), at a temperature of 350°C. The TEM samples were tilted in the microscope to various degrees, to reveal more of the surface detail of the W films.

Even after only 5 sec deposition time, a significant W film has formed. This film exists as blotchy patches covering the Si surface, about 204Å average size. The patches are composed of smaller W particles, and both the inter and intrapatch porosity are evident. The film, after 10 min of growth, looks remarkably similar to the initial stages of growth. The film is porous, the blotchy patches have enlarged to 457Å average size, and the particles comprising the patches are now discernible and are 37Å average size.

Discussion

The observation that Si-reduced W films are porous and discontinuous provides an obvious model for thick W film growth. The W film, previously thought to be a barrier to further contact between Si and WF_6, may not be a barrier at all. Upon first considering the data of Fig. 1, the question one asks is how, under certain conditions, can such thick W films grow? However, given the discontinuous, porous nature of these films, one should ask why all Si-reduced W films are not thick, like the Mo films produced by Si reduction of MoF_6. These films (17) have been found to be porous to about the same extent as the present W films, and are capable of forming to a thickness of several microns. We believe that both W and Mo films form by the same mechanism, that of fluoride percolation through the porous metal film. Why the Mo films tend to grow an order of magnitude greater than the W films is not understood.

We believe that the answer to this question will not be resolved until the detailed mechanism of the WF_6-Si reaction is understood. The temperature dependence of W film growth, as observed in Fig. 1, is probably due to the temperature dependence of sticking coefficient, desorption coefficient, etc., of one of the reaction intermediates. The appeal of such a catalytic model is that it has the potential to explain the effects of minor surface modifications on W growth behavior, Fig. 2, and also to explain the data of Fig. 1 in terms of "poisoning" phenomena. Self-limiting behavior, independent of the W film thickness at which

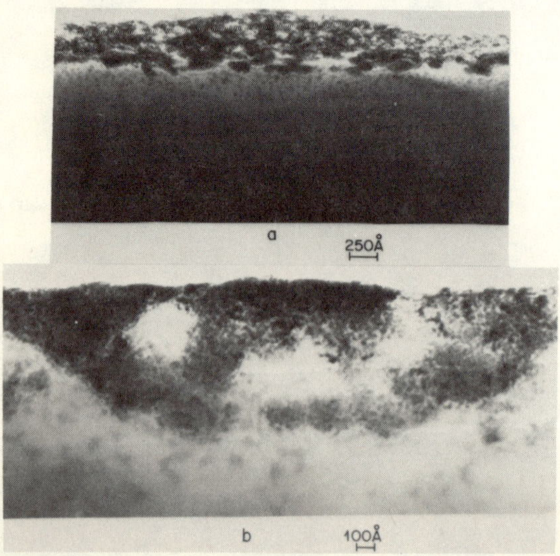

Fig. 7 Brightfield (BF) TEM cross-section photomicrographs of a Si-reduced W film, deposited at 350°C, on a (100) Si wafer. Ar = 1000 sccm, WF_6 = 5 sccm, a) t_{dep} = 5 sec and b) t_{dep} = 10 min.

it occurs, is consistent with such a mechanism.

Studies are now underway (24) to understand the reaction of WF_6 with Si clusters of various sizes. Such studies may eventually lead to the determination of the rate limiting step of the reaction between WF_6 and bulk Si. High vacuum, in-situ reaction studies on carefully cleaned and characterized Si surfaces are also needed to solve this problem. High-vacuum diagnostics such as RHEED, Auger spectroscopy, mass spectrometry and X-ray photoelectron spectroscopy are needed to understand the initial stages of the WF_6—Si reaction.

Finally, we would like to comment on the lack of "wormhole," or "tunnel" defects (3,4,25) in the Si-reduced W films. Due to their similarity with defects generated during the catalytic hydrogenation (26) or oxidation (27) of graphite, we have ascribed (4) the production of these defects to the catalytic gasification of Si by HF,

$$Si + 4HF \rightarrow SiF_4 + 2H_2 \tag{4}$$

for which $\Delta G = -90$ Kcal/mole of Si at 300°C. Since HF is necessary for this reaction to occur, and since HF is a by-product of the H_2 reduction of WF_6, one should not expect to see wormhole defects in Si-reduced W films. Stacy et al. (25) claim to see them in such films, but the feature they observe may be nothing more than a W particle formed by Si reduction and embedded in the Si matrix.

Summary

We have studied the formation and structure of CVD W films deposited by the Si reduction of WF_6, in the temperature range 210°C-700°C. Rutherford backscattering spectroscopy and transmission electron microscopy were the primary diagnostic techniques used. Although the great majority of CVD W films deposited for IC applications are deposited by H_2 reduction, the Si reduction of WF_6 always occurs first, even in the presence of copious amounts of H_2. Therefore, it is the Si reduction reaction that determines the nature of the W/Si interface, and, in turn, such important properties as contact resistance and leakage current.

We had previously reported that in this temperature range, the growth of the films exhibited an unusual temperature dependence. Between 210°C and 310°C, the films self-limit at about 0.3×10^{17} atoms/cm^2 (~60Å, assuming full dense W) coverage of W. In the temperature range 320°C-450°C, coverages up to 6.1×10^{17} atoms/cm^2 (~950Å) (at 340°C), are observed. In the higher temperature range, 500°C-700°C, film coverages of about 1.5×10^{17} atoms/cm^2 (200Å) are observed. All films, thick or thin, are found to be self-limiting. That is, once a certain deposition time (about 3 min) is past, the films achieve their ultimate thickness; further increases in deposition time cause negligible changes in film thickness.

TEM and RBS results show that the films are discontinuous and porous. The discontinuous nature of the films provides a simple mechanism for the growth of thick W films by Si reduction, i.e., the in-diffusion of WF_6 to the Si. Such thick films have been observed with increasing frequency as of late and have been enigmatic, because it has long been assumed that a thin W film (~100Å) could prevent Si and WF_6 from reacting, thus self-limiting the film. The lack of interconnectivity of these films also explains their high resistivity.

It is suggested that since thick W films by Si reduction of WF_6 are a reality, and that the discontinuous nature of the W films is a likely mechanism for the growth of such films, that the temperature dependence of W film growth is controlled by some surface catalyzed reaction. Thus, the temperature dependence of the sticking coefficient of some reactive intermediate probably controls the reaction. Such a model has the potential to explain the self-limiting behavior of the reaction, e.g. by catalytic poisoning at some stage of film growth, and the dramatic effects of minor changes of the Si surface (i.e., due to cleaning variations) on W film growth behavior. We believe that high-vacuum, in-situ reaction studies have to be carried out using a variety of diagnostic techniques before an understanding of the reaction mechanism will be reached.

Acknowledgments

The authors are indebted to D. Aspnes, R. A. Levy, L. Manchanda and N. Lifshitz for interesting discussions.

REFERENCES

1. N. E. Miller and I. Beinglass, Solid State Technology, **25**, 85 (1982).
2. P. A. Gargini, Ind. Res. Dev. **25** (3), 141 (1983).
3. M. L. Green and R. A. Levy, J. Electrochem. Soc., **132**, 1243 (1985).
4. M. L. Green and R. A. Levy, Tech. Proc. Semicon East, Boston, Sept. 17-19, 1985, SEMI, Mountain View, CA, p. 57.

5. C. M. Melliar-Smith, A. C. Adams, R. H. Kaiser and R. H. Kushner, J. Electrochem. Soc. **121** (2), 298 (1974).

6. C. Fuhs, E. J. McInerney, L. Watson and N. Zetterquist, in Ref. 7, p. 257.

7. Proceedings of the Workshop on W and Other Refractory Metals for VLSI Applications, Vol. I/II, R. S. Blewer, ed., MRS, Pittsburgh, Pa. 1986.

8. R. A. Levy, M. L. Green, P. K. Gallagher and Y. S. Ali, J. Electrochem. Soc., **133** (9), 1905 (1986).

9. W. A. Bryant, J. Electrochem. Soc. **125**, 1534 (1978).

10. M. L. Green, Y. S. Ali, B. A. Davidson, L. C. Feldman and S. Nakahara, in Materials Research Society Symposia Proceedings, Vol. 54: Thin Films – Interfaces and Phenomena, R. J. Nemanich, P. S. Ho and S. S. Lau, eds., MRS, Pittsburgh, Pa., 1986, pp. 723-728.

11. M. E. Tracy, in Ref. 7, p. 211.

12. H. Itoh, R. Nakata and T. Moriya, Proc. IEDM '85, 25.6, p. 606, 1985.

13. E. K. Broadbent and W. T. Stacy, Solid State Technology **28**, 51 (1985).

14. E. K. Broadbent and C. L. Ramiller, J. Electrochem. Soc. **131**, 1427 (1984).

15. K. Y. Tsao and H. H. Busta, J. Electrochem. Soc. **131**, 2702 (1984).

16. R. J. Mianowski, K. Y. Tsao and H. A. Waggener, Ref. 7, p. 145.

17. N. Lifshitz, to be published.

18. A. Licciardello, O. Puglisi and S. Pignataro, Appl. Phys. Lett. **48**, 41, 1986.

19. H. H. Busta and C. H. Tang, J. Electrochem. Soc. **133** (6), 1195 (1986).

20. D. E. Aspnes and A. A. Studna, Phys. Rev. B, **27**, 985 (1983).

21. D. Aspnes and M. L. Green, to be published.

22. P. Petroff, T. T. Sheng, A. K. Sinha, G. A. Rozgonyi and F. B. Alexander, J. Appl. Phys. **44**, 2545 (1973).

23. D. Hess, in Ref. 7, p. 105.

24. W. D. Reents, Jr., M. L. Mandich and V. E. Bondybey, submitted to Chem. Phys. Lett.

25. W. T. Stacy, E. K. Broadbent and M. H. Norcott, J. Electrochem. Soc. **132**, 444 (1985).

26. R. T. K. Baker, R. D. Sherwood and J. A. Dumesic, J. Catal. **66**, 56 (1980).

27. R. T. K. Baker, J. A. France, L. Rouse and R. J. Waite, J. Catal. **41**, 22 (1976).

INTERACTION OF CVD TUNGSTEN WITH UNDERLYING METAL LAYERS

S.L. NG, S.J. ROSNER, S.S. LADERMAN, T.I. KAMINS,
D.R. BRADBURY, AND JUN AMANO
Hewlett-Packard Laboratories, P O Box 10350, Palo Alto CA 94303-0867

ABSTRACT

The deposition of Chemical Vapor Deposition(CVD) Tungsten on aluminum- and titanium-containing surfaces and the interaction of the tungsten and the metal underlayers have been investigated. CVD tungsten readily deposits on freshly etched aluminum surfaces, but does not nucleate when the aluminum is covered with an oxide layer formed by treating in an oxygen plasma. When the surface is covered with a native oxide, considerable fluorine is found at the interface. A sputtered tungsten layer thicker than 10 nm allows reproducible deposition of CVD tungsten and minimizes fluorine incorporation. Fluorine was also detected when CVD tungsten was deposited on titanium or $TiSi_2$, but not when it was deposited on a Ti-W film containing 5 weight percent titanium.

INTRODUCTION

Recently, tungsten has received much attention because of its potential applications in VLSI devices. It has low resistivity, high electromigration resistance and acts as a diffusion barrier. Using chemical vapor deposition (CVD) to deposit tungsten interconnections provides good step coverage. In addition, it has been demonstrated that selectively deposited CVD tungsten can be used as a diffusion barrier [1], as a contact material [2,3], as a polysilicon shunt to reduce interconnection resistance [4], and as a multi-level metal layer interconnection [3,5]. It should be especially important for filling vias between layers of metal to provide a planar surface on which the subsequent metal layer can be readily deposited [6].

To take advantage of these characteristics, it is desirable to be able to deposit CVD tungsten on a wide variety of materials. Since CVD tungsten deposition involves fluorine chemistry, the chemically-vapor-deposited film may interact differently with underlying metals than does sputter-deposited tungsten. Indeed, it has been observed that the CVD tungsten process can result in fluorine accumulation at the underlying interface, that metal fluorides such as TiF_3 can be formed on $TiSi_2$, and that the morphology of the CVD tungsten interfaces can be strongly influenced by surface preparation [7-10]. Previous workers have also concluded that Al is a poor choice for an underlying metal surface because of a high contact resistance postulated to be related to AlF_3 formation [6,8].

The objective of this work was to increase our understanding of, and control over, such effects by studying the interfacial reactions between CVD tungsten and a number of underlying metals. Because of the potential importance of CVD tungsten for filling vias between aluminum metalization layers to form a planar surface for subsequent metal deposition, the deposition of CVD tungsten on aluminum was again investigated here. Titanium-containing layers were also studied because of their importance in VLSI technology. The following underlying metals are discussed in this paper: $TiSi_2$, Ti, Ti-W, Al, Al-capped layered Ti-Al metallization, and Al capped with various thicknesses of sputtered W. CVD tungsten deposition conditions suitable for selective tungsten formation were used in this study.

EXPERIMENTAL TECHNIQUE

All the underlying metals were deposited by sputtering. $TiSi_2$ films were formed by sputtering 50 nm of Ti on Si and reacting in a nitrogen ambient at 675°C for 20 min. The film was immersed in a 2:1 H_2SO_4:H_2O_2 solution for 5 min to remove unreacted Ti and then annealed at 800°C for 30 min. Al-capped layered metal consisted of 50 nm Al-2.0%Si/30 nm Ti/250 nm Al-Si/30 nm Ti/250 nm Al-Si on oxidized Si wafers. The TiW films consisted of 5 weight percent Ti.

Most of the CVD tungsten films were deposited at 400°C in a hot-wall LPCVD reactor constructed in this laboratory. The deposition pressure was 210 mtorr while the WF_6 and H_2 flow rates were 10 and 1300 sccm, respectively. Some of the CVD W films deposited onto $TiSi_2$ and Ti-W were formed in a cold-wall LPCVD reactor at 450°C or 600°C with WF_6 and H_2 flow rates of 4 and 40 sccm, respectively and a total pressure of 140 mtorr.

All of the films were characterized with Rutherford backscattering (RBS), which was performed using 2.0 MeV He^+ ions. The backscattered particle energies were measured using a silicon surface barrier detector. The total flux for each measurement was 30 μC. The RBS sensitivity to fluorine is about 2×10^{16} cm^{-2} in a very thin film or, equivalently, 15 atomic percent in a film 20 nm or more thick.

Some of the films were further characterized with x-ray diffraction analysis. The x-ray diffraction data were collected using a Huber thin-film camera in a Seemann-Bohlin geometry. Copper $K_{\alpha1}$ radiation was employed, and the angular range studied covered d-spacings from 0.44 to 0.081 nm. Generally, x-ray diffraction data were collected for the as-deposited films and then after successive etchings in warm H_2O_2, which removed some CVD tungsten and thereby improved the sensitivity to underlying layers. In addition, selected samples were examined by secondary ion mass spectroscopy (SIMS) using a 10 KV primary oxygen beam.

EXPERIMENTAL RESULTS

Because of the possible effect of fluorine-containing layers on contact resistance, the most important interaction investigated during this study was the possible formation of fluorine-containing layers between CVD tungsten and the underlying metal layers.

Al-containing surface layers

The behavior of CVD tungsten on aluminum depended critically on the surface condition of the aluminum layer. When the aluminum was subjected to an oxygen plasma treatment, such as that used to remove photoresist, and not subsequently cleaned before CVD tungsten deposition, no tungsten was deposited and no fluorine was found on the surface of the remaining film. To show that tungsten does not nucleate on aluminum oxide, an attempt was made to deposit tungsten films on bulk Al_2O_3 (sapphire) wafers. No deposit was formed.

When no plasma treatment was used and a brief, hot, concentrated HCl etch was employed to clean the surface before tungsten deposition, uniform tungsten layers were reproducibly formed, and fluorine could be observed by RBS at levels close to the detection limit, implying a flourine concentration of about $2x10^{16}$ cm^{-2}. A similar low level of fluorine was found for Al-capped multilayers, when the surface was cleaned with HCl. These levels are close to the detection limit of the x-ray technique used here, and no fluoride-containing compound was found for these samples. However, SIMS analysis did detect a small concentration of fluorine at the interface.

In many IC processes, however, using an HCl etch before tungsten deposition may not be practical, and a method of obtaining reproducible CVD tungsten deposition without any fluorine accumulation is important. It has been proposed that fluorine accumulation at Al/CVD tungsten interfaces results from the Al reduction of WF_6, leading to the formation of the non-volatile metal fluoride AlF_3 [6,8]. No such fluorine accumulation is expected on W surfaces. Therefore, the use of a thin "nucleating" layer of sputtered W on Al was investigated to determine if it would act as a barrier to suppress AlF_3 formation and allow deposition of a uniform tungsten layer.

The samples used in this portion of the study consisted of sputtered W on Al. They were neither exposed to an oxygen plasma treatment nor etched with HCl before CVD tungsten deposition. When the sputtered W layer was 10 nm thick or thicker, thick CVD tungsten films were reproducibly formed, and no fluorine was detected by RBS, as shown in Fig. 1. When the sputtered W was thinner than 10 nm, thick CVD tungsten films still formed, but fluorine was readily seen by RBS, as shown in Fig. 2; the concentration was higher for thinner sputtered W layers. In the case of a 3.5 nm sputtered W film, RBS analysis indicates a fluorine concentration at the CVD tungsten/Al interface of about $2x10^{17}$ cm^{-2}. If the fluorine were all contained in a solid AlF_3 phase, this corresponds to more than 100 nm of AlF_3. Careful x-ray analysis revealed only Al, W, and a very small quantity of polycrystalline silicon (from the Al-Si metallization). The levels of fluorine found by RBS imply that, if the fluorine were contained in AlF_3, there would be sufficient material to observe by x-ray diffraction. We conclude, therefore, that the fluorine-containing compound is either extraordinarily fine-grained, nearly amorphous AlF_3 or some fluorine-containing material other than AlF_3.

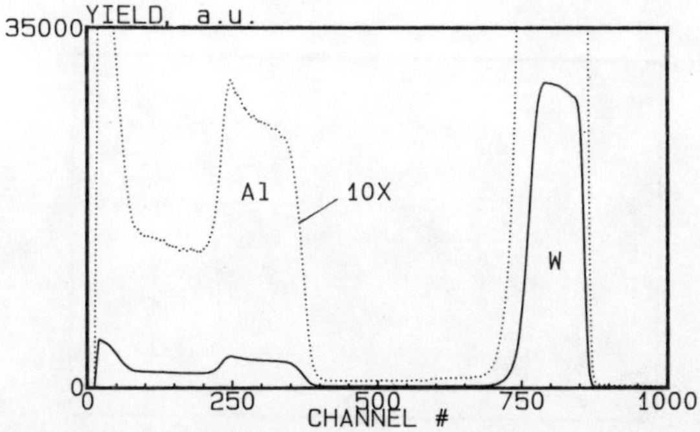

FIG. 1. RBS data of CVD W on 10 nm sputtered W on Al

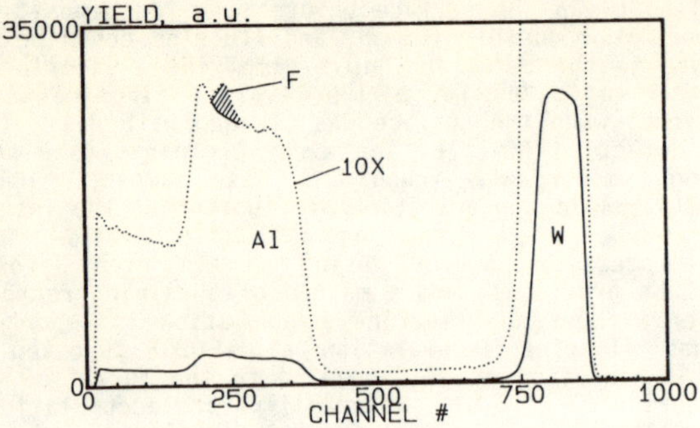

FIG. 2. RBS data of CVD W on 3.5 nm sputtered W on Al

Finally, when CVD W, with or without an intervening sputtered W film, was deposited on Al, x-ray diffraction analysis revealed that the compound WAl_{12} was formed.

Ti-containing surface layers

When CVD tungsten was deposited directly on Ti at 400°C in a hot-wall reactor, fluorine was detected by RBS at the interface (Fig. 3), and the presence of TiF_3 was confirmed by x-ray diffraction (Table 1). However, when CVD tungsten was deposited on $TiSi_2$ at 600°C in a cold-wall reactor, lower, but still detectable, quantities of fluorine appeared in the RBS spectra; no TiF_3 was detected by x-ray analysis in this case.

When CVD tungsten was deposited on layers of sputtered 5wt%Ti-95wt%W, no subsurface fluorine was detected by RBS analysis. Similar behavior was observed even if a $TiSi_2$ layer was beneath the Ti-W film. The Ti content of these Ti-W films does not cause any significant fluorine incorporation, and the deposition resembles deposition onto W, rather than deposition onto Ti.

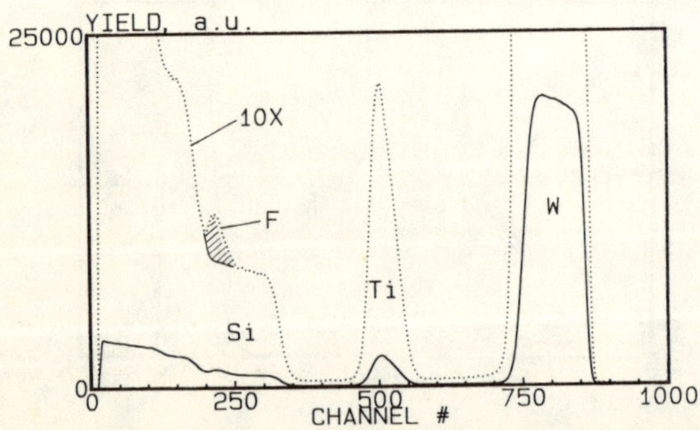

FIG. 3. RBS data of CVD W on Ti

Table 1. Phases identified by x-ray diffraction after deposition of CVD tungsten on titanium. The table lists the d-spacings corresponding to the Bragg reflections observed and the assignments based on comparison to the JCPDS standards. [12]

observed d spacing (Angstroms)	Identification (phase and reflecting planes)		
3.880	TiF3 (110)		
2.545		Ti (010)	
2.238	TiF3 (222)	Ti (002)	W (110)
1.936	TiF3 (220)		
1.742		Ti (012)	
1.744	TiF3 (321)		
1.582		Ti (110)	W (200)
1.352	TiF3 (220)		
1.329		Ti (103)	
1.292			W (211)
1.119			W (220)
1.001			W (310)
0.9137			W (222)
0.8459			W (321)

DISCUSSION

The most important set of results are those concerning Al surfaces. Previous reports have postulated that the formation of insulating AlF_3 is a direct consequence of Al reduction of WF_6 [6,8]. This implies that unacceptably high contact resistances are inevitable unless a non-Al nucleating layer is used.

It is important to note, therefore, that we find substantially more fluorine when a very thin sputtered W film is deposited as a nucleation layer than when the Al surface is HCl etched just before CVD tungsten deposition. The thinnest sputtered W films are probably not continuous, so that both W and Al are exposed on the surface. Since these samples had not been etched in HCl, the surface could be expected to be, at least partially, covered with native aluminum oxide, providing a non-homogeneous surface for CVD tungsten nucleation. A high fluorine concentration was found with the thinnest sputtered tungsten layers. As the thickness of the sputtered W film increases and it becomes continuous, covering the entire aluminum surface, the fluorine accumulation decreases. No fluorine could be detected for samples with thick, sputtered tungsten layers. This behavior implies that fluorine accumulation is related to the exposed Al surfaces. However, since considerably less fluorine is found when the Al surface is HCl etched to remove the surface oxide, it appears that, under the deposition conditions used here, fluorine accumulation at the CVD tungsten/Al interface is correlated with aluminum oxide, not with bare Al metal.

The correlation of fluorine accumulation with aluminum oxide can be understood in the following way: Since we have shown that the presence of thick aluminum oxides inhibits CVD tungsten deposition, the Al surfaces on which CVD tungsten does deposit must differ. Furthermore, CVD tungsten did deposit on a sapphire surface partially covered with a thin, sputtered W layer, but no fluorine accumulated. Both results suggest that the native aluminum oxide may differ from thick Al_2O_3.

A typical native oxide may be discontinuous, chemically different from Al_2O_3 or too thin to inhibit diffusion to unoxidized metal underneath. If the native oxide is discontinuous, the tungsten can deposit quickly over the unoxidized metal surface, freeing reactive fluorine species. These species either diffuse through the oxide covering other regions and are trapped below the oxide or react directly with the oxide, perhaps because it is non-stoichiometric. If the native oxide is continuous, the WF_6 may possibly diffuse through the oxide to react with the Al underneath. Alternatively, since thin native oxides of Al are Al rich [11], the non-stoichiometric oxide may promote deposition and fluorine accumulation.

Regardless of the detailed mechanisms involved, we have shown that large fluorine accumulations are not intrinsic to clean Al surfaces. However, it is important to note that, in this study, H_2 was present in the gas phase. It is possible that bare Al, and possibly the native oxide as well, catalyzes the hydrogen reduction of WF_6. The relative rates of hydrogen reduction and Al reduction of WF_6 may be crucial in achieving low fluoride accumulation at the Al/CVD tungsten interface. Therefore, the gases present during the deposition may be important.

Fluorine was also detected when CVD tungsten was deposited on titanium or $TiSi_2$, but not when it was deposited on Ti-W films containing a low concentration of titanium. Etching in dilute HF before insertion into the reactor did not reduce the fluorine accumulation at titanium-rich surfaces. However, as noted in the results section, less fluorine accumulated when CVD tungsten was deposited onto $TiSi_2$ at 600^oC instead of 400^oC. While temperature was not the only variable in the comparison, we believe it was the most important difference. This is consistent with thermodynamic data which imply that the tendency to form TiF_3 from WF_6 and Ti decreases as the temperature increases, while the competing reaction, which forms SiF_4, tends to increase as temperature increases.

SUMMARY

The deposition of CVD tungsten on aluminum- and titanium-containing surfaces and the interaction of the tungsten and with metal underlayers have been investigated. CVD tungsten readily deposits on freshly etched aluminum surfaces, but does not nucleate when the aluminum is covered with an oxide layer formed by treating in an oxygen plasma. When the surface is partially covered with a native oxide, considerable fluorine is found at the interface. Fluorine is also detected when CVD tungsten was deposited on titanium or $TiSi_2$, but not when it is deposited on a Ti-W film containing a low concentration of titanium.

The lower levels of fluorine observed on clean aluminum surfaces in this study suggest that low contact resistances can possibly be obtained on aluminum, but that the contact resistance will be very sensitive to the condition of the aluminum surface on which the tungsten is being deposited by CVD. A sputtered tungsten layer only about 10 nm thick between the aluminum and the CVD tungsten prevents fluorine accumulation and should lead to low contact resistance.

ACKNOWLEDGMENT

We are grateful for the technical assistance of Bob Smith for x-ray analysis, Hope Keller for RBS measurements, John Turner for SIMS analysis, Jo Soto for tungsten deposition, and to Terry Nordstrom, Tom Anthony, C. H. Lin, and Paul Merchant for supplying experimental samples.

REFERENCES

1. M.L. Green and R.A. Levy in "Semicon/East 1985 Technical Program Proc." pp. 57, Boston, Mass. (Sept. 17-19, 1985).

2. R.A. Levy, M.L. Green, P.K. Gallagher, and Y.S. Ali, J. Electrochem. Soc., 133, 1905 (1986).

3. N.E. Miller and I. Beinglass, Solid State Tech., 25, 85 (Dec. 1982).

4. W.A. Metz and N.J. Szluk, G.W. Miller, and H.O. Hayworth, IEEE Electron Device Letters, EDL-6, 372 (1985).

5. W.A. Metz and E.A. Beam, in "Proc. Second Int'l. IEEE VLSI Multilevel Interconnect Conference," pp. 357, Santa Clara, CA (June 9-10, 1985).

6. T. Moriya, S. Shima, Y. Hazuki, M. Chiba, and M. Kashiwagi, IEDM Tech. Dig., CH1973-7/83, 550 (1983).

7. E.K. Broadbent and C.L. Ramiller, J. Electrochem. Soc., 131, 1427 (1984).

8. W.T. Stacy, E.K. Broadbent, and M.H. Norcott, J. Electrochem. Soc., 132, 144 (1985).

9. E.K. Broadbent and W.T. Stacy, Solid State Tech., 28, 51 (Dec. 1985).

10. E.K. Broadbent, A.E. Morgan, J.M. DeBlasi, P. van der Putte, B. Coulman, B.J. Burrow, and D.K. Sadana, J. Electrochem. Soc., 133, 1715 (1986).

11. J. Grimblot and J.M. Eldridge, J. Electrochem. Soc., 129, 2369 (1982).

12. "ASTM Powder Diffraction File", Joint Committe on Powder Diffraction Standards, Swarthmore, Pennyslvania.

GROWTH OF SELECTIVE TUNGSTEN ON SELF-ALIGNED $CoSi_2$ BY LOW PRESSURE CHEMICAL VAPOR DEPOSITION

P. VAN DER PUTTE, D.K. SADANA, E.K. BROADBENT and A.E. MORGAN
Philips Research Laboratories Sunnyvale, Signetics Corporation, Sunnyvale CA 94088

ABSTRACT

The selective deposition of tungsten films onto $CoSi_2$ and onto Co by low pressure chemical vapor deposition and their material properties have been investigated with Auger electron spectroscopy, transmission electron microscopy and Rutherford backscattering. When using WF_6 and H_2, uniformly thick tungsten films can be deposited onto $CoSi_2$ without substrate alteration. In patterned structures however, void formation was found at the perimeters of $CoSi_2$ contacts to silicon, indicating encroachment of WF_6 down the edge of the silicide-Si interface. In WF_6 and Ar the film thickness was limited to 10 nm and some Si was locally consumed from the upper part of the $CoSi_2$ film. Transmission electron diffraction showed evidence of Co_2Si formation in these areas. High deposition temperatures result in enhanced $WF_6/CoSi_2$ interaction.

INTRODUCTION

Chemical vapor deposition of tungsten has been studied extensively during the past few years since it provides a very promising technique for use in large scale integrated circuit metallization [1]. Direct application of tungsten onto single crystal silicon suffers from the formation of tunnel defects, a side effect of the reaction between WF_6 and silicon [2,3].

In an earlier publication, we examined an alternative metallization scheme in which tungsten is used as a barrier and/or contact fill material above regions where a silicide has been formed first [4,5]. For W on $TiSi_2$, it was found that Si was consumed from the upper 40-50 nm of silicide, and that polycrystalline TiF_3 was formed as a by-product. For W on PtNiSi, void formation was observed at the perimeter of the self-aligned PtNiSi contacts, a result of WF_6 encroachment down the edge of the silicide-Si interface.

In this paper, we report on the deposition of W onto pure Co and $CoSi_2$ structures. $CoSi_2$ has a layer resistivity comparable to $TiSi_2$ [6] and can be formed in a self aligned manner [7].

EXPERIMENTAL

Tungsten films were deposited onto self-aligned $CoSi_2$ structures and onto unpatterned $CoSi_2$ and Co layers. The substrates used were 17-33 Ω-cm, 100 mm diameter, p-type <100> silicon wafers, sometimes covered with 150 nm thick thermally grown SiO_2. A number of oxide-coated wafers were patterned into repetitive bar structures and contacts. A conical magnetron sputter source was used to deposit 28.5 nm of cobalt. All samples received a dilute HF immersion preclean and an in-situ sputter etch before deposition.

The $CoSi_2$ was formed by heating samples in an halogen lamp rapid thermal annealer at 850°C for 10 seconds in a nitrogen ambient. The unreacted cobalt was then removed in 3:1 HCl:30%H_2O_2 at room temperature for 2 minutes. The resulting $CoSi_2$ layer had a thickness of about 110 nm.

Tungsten was deposited by low pressure chemical vapor deposition (CVD) in a vertical flow, hot wall deposition system. WF_6 (99.98%), and purified hydrogen or argon (99.998%) were used as reaction gases at a temperature of 300°C and 450°C and a pressure of 550 mtorr. The WF_6/H_2 or WF_6/Ar flow ratio was 1/44. Tungsten depositions were performed both with and without precleaning of the $CoSi_2$ surface by dilute HF. At a temperature of 600°C, the experiments were performed in a lamp heated, cold wall deposition system. The WF_6/H_2 flow ratio was 1/10, and the total pressure 124 mtorr. Before the deposition of tungsten the surface was cleaned by dilute HF.

RESULTS AND DISCUSSION

Deposition of W onto $CoSi_2$ by hydrogen reduction of WF_6

In contrast to W on $TiSi_2$, the deposition of W onto $CoSi_2$ can be performed without difficulty in a gas mixture of H_2 and WF_6. In this surface catalysed reaction, WF_6 is reduced by hydrogen to form W and HF [1]. Auger depth profiling of a 30 nm thick tungsten layer on top of $CoSi_2$ does not reveal any fluorine at the $CoSi_2/W$ interface or within the silicide and W layers themselves, indicating that no non-volatile fluorides were formed. Some interfacial oxygen was noted which originated from the silicide surface since the native oxide was not removed prior to deposition. No consumption of either Co or Si from the $CoSi_2$ layer was apparent.

A tungsten layer of about 60 nm thickness was deposited onto self-aligned $CoSi_2$ contact structures using WF_6 and H_2 gases. Figure 1 shows a transmission electron micrograph of the metallized contact in a perimeter region. Void formation within the Si is found at the perimeter of the contact, as WF_6 has apparently encroached down the edge of the original silicide-Si interface and reacted with the silicon underneath the SiO_2 layer. As the reaction between WF_6 and Si results in a 2:1 volumetric ratio between the amount of silicon consumed and the amount of

Fig. 1 TEM micrograph showing void formation near perimeter of self aligned $CoSi_2$ structure following selective W deposition.

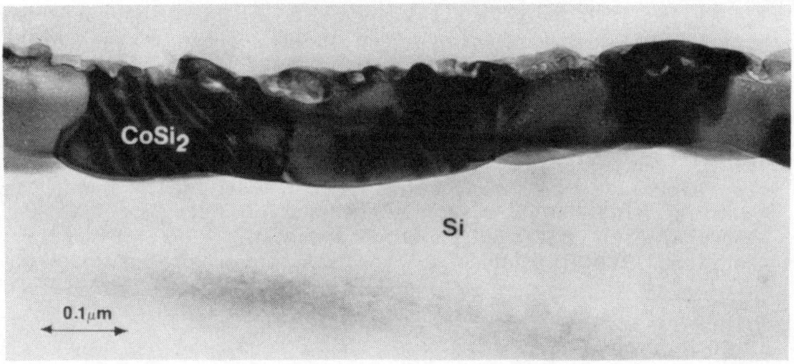

Fig. 2 TEM micrograph of a 0.11 μm thick $CoSi_2$ layer formed by RTA in nitrogen (28.5 nm Co, 850°C, 10 sec).

W formed, void formation results [5]. Application of tungsten onto $CoSi_2$ contact structures would therefore be restricted to those situations where the contact perimeter would not be exposed to WF_6 vapor (e.g. nested contact openings).

Substrate reduction of WF_6 by $CoSi_2$

Without the presence of hydrogen, direct reduction of WF_6 by certain surfaces can take place. For example, the substrate reduction of WF_6 by Si is well known from the literature [1]. Substrate reduction of WF_6 by $CoSi_2$ was studied by exposing $CoSi_2$ samples to a mixture of argon and WF_6. Again, no surface cleaning was used prior to tungsten deposition. Transmission electron microscopy (TEM), combined with Auger and Rutherford backscattering (RBS) analyses, indicated a very non-uniform layer of tungsten formed on the surface of the $CoSi_2$ with an average thickness of about 7 nm. The non-uniform nature of the deposit can be traced to the original structure of the $CoSi_2$ layer (figure 2); viz. a layer consisting

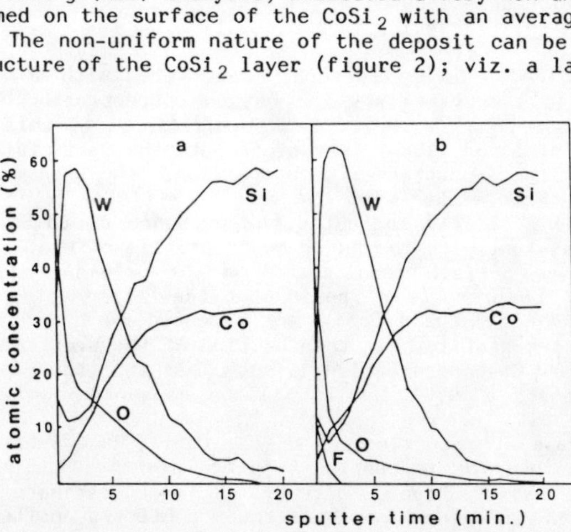

Fig. 3 Auger depth profile of W/$CoSi_2$ structure, after deposition (300°C, WF_6 + Ar, 0.55 torr): a) without cleaning, b) with cleaning prior to W deposition.

of large grains (150 nm laterally) arranged in a columnar structure with a voided upper region containing a few atomic percent oxygen. The oxide on the $CoSi_2$ clearly impedes the formation of a uniform tungsten layer and is responsible for the large amount of oxygen seen in the upper portion of the Auger depth profile from the W/$CoSi_2$ structure (Fig. 3a).

After removal of the uppermost portion of the $CoSi_2$ in dilute HF, exposure of the silicide to WF_6 and Ar produced a more uniform and slightly thicker (10 nm) tungsten layer. Auger depth profiling (Fig. 3b) showed a decrease in oxygen concentration near the W-$CoSi_2$ interface. According to RBS, the reduction amounted to 35%. Figure 4 shows a cross-sectional TEM micrograph of the structure. The remainder of the

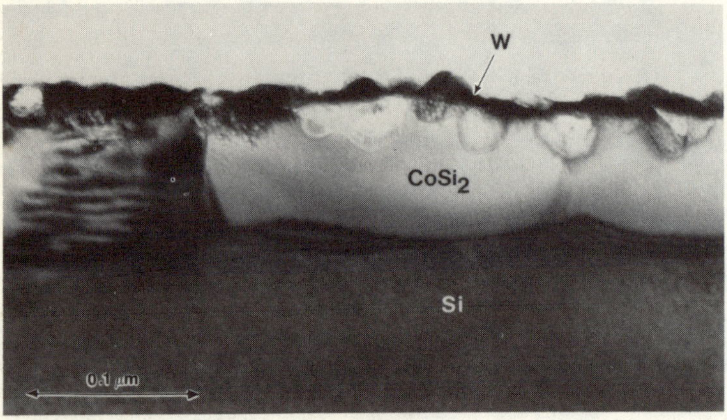

Fig. 4 TEM micrograph of a thin W layer deposited on $CoSi_2$ by substrate reduction only (300°C, WF_6 + Ar, 0.55 torr, 5 min.) No cleaning of the silicide prior to W deposition.

original silicide surface structure, now covered with a W layer, can be recognized. This explains why the oxygen concentration found by Auger and RBS analyses remains high. A striking aspect of this micrograph is the formation of large, local intrusions into the $CoSi_2$ layer, apparently as a result of the substrate-WF_6 interaction. The transmission electron diffraction pattern, taken from the reacted silicide layer after removal of the tungsten, clearly indicated the presence of Co_2Si. The results are not inconsistent with the Auger depth profile of Fig. 3b. Because of the very uneven surface resulting from the original silicide surface roughness and the presence of residual oxide-filled voids (Fig. 2), the Auger signals of W, O and $CoSi_2$ are expected to be intermixed in the upper part of the profile. In this section of the profile, the ratio between the Si and Co concentration is much less than the expected ratio of 2. This suggests a reduction in silicon concentration at the W/$CoSi_2$ interface.

For the case of reduction of WF_6 by $CoSi_2$, two reduction reactions may be responsible for initiating tungsten growth. The reaction of WF_6 with silicon was mentioned already. To investigate the possible reduction of WF_6 by cobalt, a 35 nm thick sputter-deposited cobalt layer was exposed to WF_6 and argon gases for 5 minutes. To avoid oxidation of the cobalt layer, the substrates were loaded into the reactor at room temperature and then ramped up to temperature in an inert ambient. RBS analyses showed an amount of W present on the surface equivalent to a 1.3Å thick uniform layer, i.e. much less than a monolayer. In addition, a high concentration of fluorine was found. This indicates that cobalt itself does not reduce WF_6. The amount of W found on the surface can be explained by assuming that absorbed WF_6 is oxidized when the reactor was opened and the hot surface exposed to air, thus allowing the liberated fluorine to react with cobalt. It can therefore be assumed that the reduction of WF_6 by $CoSi_2$ results from WF_6-Si interactions only. Thus, to form the tungsten layer on $CoSi_2$, silicon is removed from the $CoSi_2$ layer at areas not covered by oxide. Intrusion-like structures are produced within the silicide which are found to be silicon deficient and at least partly consist of Co_2Si.

Influence of temperature on the WF_6-$CoSi_2$ interaction

The formation of intrusion at 300°C indicates that an interaction between the WF_6 and the $CoSi_2$ layer is occurring which, if taking place to a large extend may be damaging to the W/$CoSi_2$/Si contact structure. When the $CoSi_2$ surface is exposed to WF_6 and argon at a temperature of 450°C indeed such a reaction results: the $CoSi_2$ layer brakes up locally and regions are formed of a much lower density where even void formation can take place (Figure 5). Also a reaction takes place at the

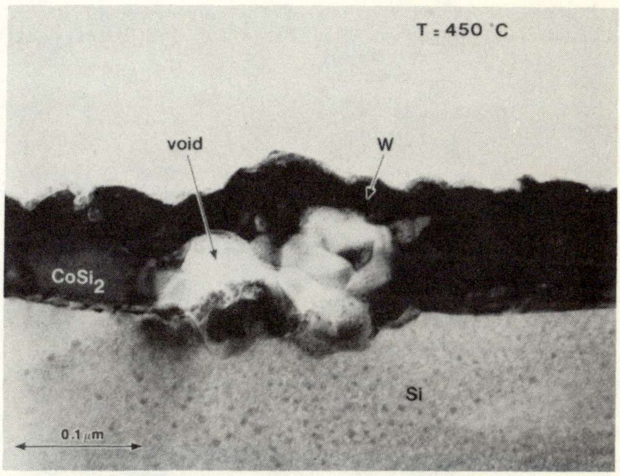

Fig. 5 TEM micrograph showing WF_6-Si_2 interaction after exposing $CoSi_2$ to an Ar/WF_6 mixture at 450°C for 5 minutes.

$CoSi_2$/Si interface resulting in intrusions into the Si substrate. Similar structures are found abundantly when tungsten was deposited onto $CoSi_2$ structures at 600°C. Using STEM/EDAX, in those structures the presence of W could be detected, both in the $CoSi_2$ layer and at the $CoSi_2$/Si interface. The occurence of these structures can possibly be explained when it is assumed that WF_6 diffuses along the grain boundaries through the CoSi layer and reacts with the silicon substrate underneath the $CoSi_2$ layer. This explanation is supported by the observation that already at 300°C, where no such structures are found, RBS shows evidence of the presence of W at the $CoSi_2$/Si interface (Figure 6).

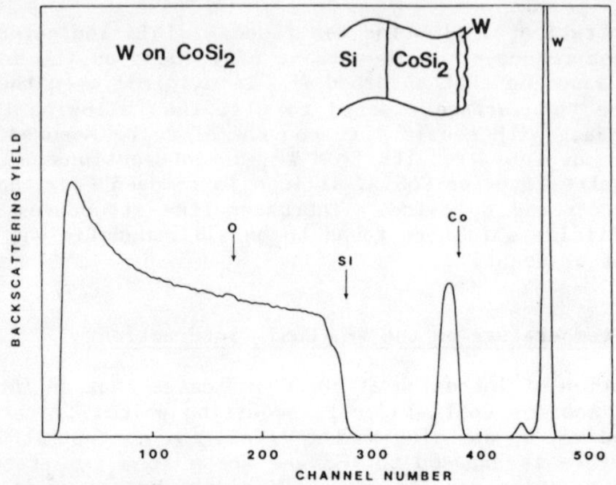

Fig. 6 RBS spectrum of W on $CoSi_2$ after exposing $CoSi_2$ to an Ar/WF_6 mixture for 5 minutes.

Figure 7 shows a TEM micrograph of the perimeter of a contact after deposition at 600°C. Increased void formation and local destruction of $CoSi_2$ layer are visible.

CONCLUSIONS

Tungsten can be deposited successfully onto $CoSi_2$ by both the hydrogen and Si reduction of WF_6. The latter is accomplished via reduction of WF_6 by silicon from the $CoSi_2$ layer, leaving behind Si-deficient regions (intrusions) within the silicide. Substrate reduction

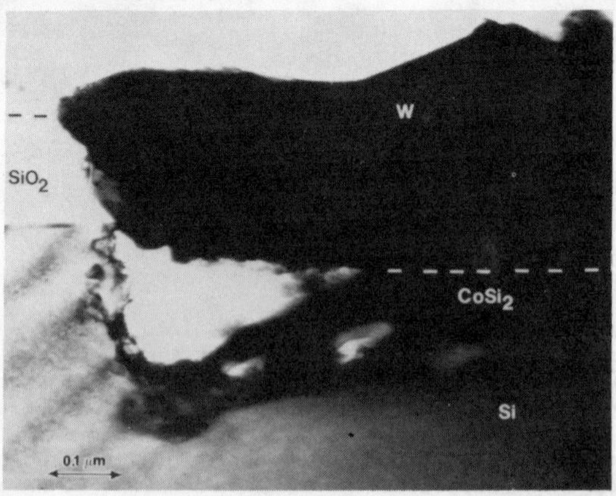

Fig. 7 TEM micrograph showing the interaction of WF_6 and $CoSi_2$ at the perimeter of a self aligned CoSi structure after selective deposition at 600°C.

of WF_6 by $CoSi_2$ is hampered when the silicide surface is covered with an oxide layer. Void formation occurs at the perimeters of the contacts when selective tungsten is applied to self-aligned $CoSi_2$ structures.

High temperatures should be avoided during deposition of W onto $CoSi_2$. At 300°C some reaction already is noticable, temperatures of 450°C and higher result in destructive interaction between WF_6 and $CoSi_2$.

ACKNOWLEDGEMENT

The authors would like to thank H. Shishido for the Auger measurements, D. Stadtler for TEM sample preparation, K. Ritz for the STEM/EDAX measurements and Dr. A.E.T. Kuiper of Philips Research Labs Eindhoven (the Netherlands) for the RBS measurements.

REFERENCES

1. See, for example, the review and references provided in: E.K. Broadbent and W.T. Stacy, Solid State Technol. 28, (12), 51 (1985) and the papers contained in "Tungsten and Other Refractory Metals for VLSI Applications". R.S. Blewer, Ed. Materials Research Society, Pittsburgh, Penn. (1986).
2. W.T. Stacy, E.K. Broadbent and M.H. Norcott. J. Electrochem. Soc. 132, 444 (1985).
3. D.M. DeBlasi, D.K. Sadana and M.H. Norcott, Mat. Res. Soc. Symp. Proc. 71, p. 303, 1986.
4. Saraswat, R.M. Swanson, IEEE Trans., Electron Device Letters, EDL 5, 209 (1984).
5. E.K. Broadbent, A.E. Morgan, J.M. De Blasi, P. van der Putte, B. Coulman, B.J. Burrow, D.K. Sadana and A. Reader, J. Electrochem. Soc. 133, 1715 (1986).
6. See, for example, S.P. Murarka, Silicides for VLSI Applications, Academic Press, New York (1983).
7. A.E. Morgan, E.K. Broadbent, M. Delfino, B. Coulman and D.K. Sadana, accepted for publication in J. Electrochem. Soc.

Recent CVD Process and Equipment Developments

SELECTIVE CVD OF $TaSi_2$

T.P.H.F. WENDLING, C. WIECZOREK, K. HIEBER
CORPORATE RESEARCH & TECHNOLOGY, SIEMENS AG

ABSTRACT

A selective CVD $TaSi_2$ deposition process has been observed in a Siemens built Low Pressure CVD-System [1]. The deposition was performed according to the reaction chemistry:

$$4\ SiH_2Cl_2 + 5\ H_2 + 2\ TaCl_5 \text{ ---> } 2\ TaSi_2 + 18\ HCl \qquad (1)$$

For some deposition conditions this $TaSi_2$ does not consume underlying silicon and the TEM cross sections indicate a smooth interface, no encroachment under the SiO_2 and no wormholes as it was observed for e.g. selective W.

These facts promise to be of great potential in the metallization of shallow junctions in submicron IC structures and will lead to new and denser metallization schemes.

I. Introduction

The development of denser circuit integration requires new metallization schemes which use materials that show low bulk and contact resistances and compatibility with subsequent processing. Our interest focuses on a low resistance contact material, which stands up to high temperature processes, shows good RIE selectivity vs SiO_2 and does not encroach much into the underlying silicon and under the silicon oxide. Titanium silicide [2], cobalt silicide [3], molybdenum silicide [4], tungsten [5] and molybdenum [6] are currently under investigation for this purpose. We have observed the selective LPCVD of $TaSi_2$ and planned to investigate its qualifications as a low resistance contact metallization material for very large, high density and high speed integrated circuits.

II. Process

The self-aligned LPCVD $TaSi_2$ process is pictured in figure 1. Before the metallization we have a silicon/silicon oxide structure as examplified by the MOS structure in figure 1.

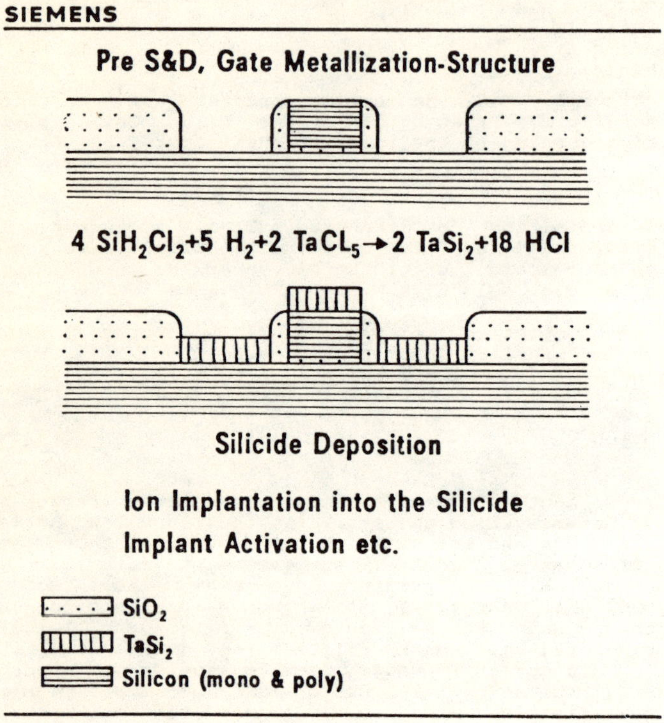

Figure 1: The self-aligned LPCVD $TaSi_2$ process.

According to the reaction equation and as previously described by Hieber, et al [1]

$$4\ SiH_2Cl_2 + 5\ H_2 + 2\ TaCl_5 \dashrightarrow 2\ TaSi_2 + 18\ HCl, \qquad (1)$$

we now selectively deposit $TaSi_2$ on the exposed silicon. The deposition can be several hundred nanometers thick for good sheet resistance R_s of the deposited $TaSi_2$ film. Following a masking step the source & drain regions are now implanted with the respective ions into the p-MOS and n-MOS regions. In the subsequent CMOS processing during a high temperature anneal, some dopant ions are driven into the silicon and activated to form conformal and shallow junctions dependent on the temperature of the anneal.

III. Equipment

The equipment used for this selective LPVCD of $TaSi_2$ has been described before by M. Stolz, et al [1]. The principal vacuum reactor set up is represented in figure 2.

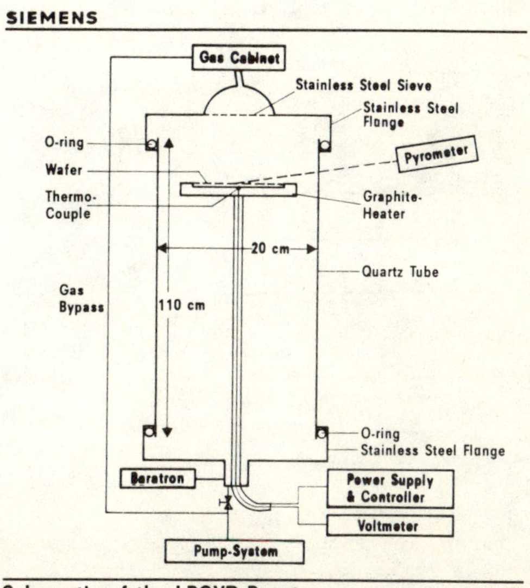

Figure 2: **Schematic of the LPCVD-Reactor**

The graphite heater used in this system is shown in figure 3. The heating element which winds back and forth is locally of variable thickness. This arrangement serves to optimize the temperature uniformity across a wafer in the neighborhood of 600°C.

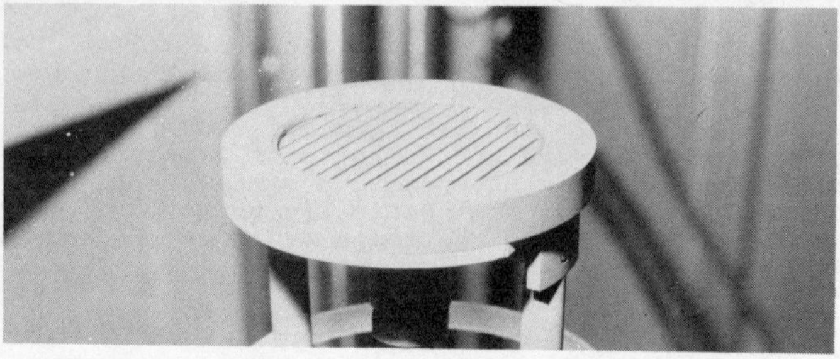

Figure 3: The graphite wafer heater

The sublimator, which provides the system with $TaCl_5$ is shown in figure 4.

SIEMENS

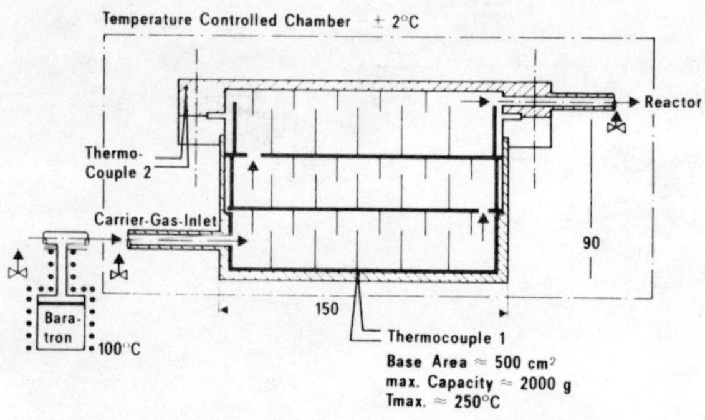

Figure4 :$TaCl_5$-Sublimator

ZT MTZ 233

The $TaCl_5$ is spread out on a large surface and H_2 is purged over it. The $TaCl_5$ sublimator can be temperature controlled to ±2°C. All gas pipes between the sublimator and reactor are temperature controlled to prevent the $TaCl_5$ from condensing in the system.

IV. Experiments

100mm diameter wafer with a 5x5mm² SiO_2 chessboard structure or a 10mm wide SiO_2 band across its center and in some cases with SiO_2 device pattern were used. Before the $TaSi_2$ deposition the patterned wafers received a 100:1 HF clean. Within 10 min of this HF clean, the wafers were inserted into the LPCVD system. We first observed selective LPCVD of $TaSi_2$ at a heater temperature of about 640 C, a SiH_2Cl_2 flow of 20sccm, a H_2 flow through the sublimator of 20sccm, an additional direct H_2 flow of 1,200sccm and a reactor pressure of about 650mtorr. The L_9 orthogonal array experiment [8] shown in table I was set up to explore the optimum deposition conditions in the neighborhood of the above deposition. To conduct the L_9 orthogonal array experiments, the wafers with the 10mm wide SiO_2 band across their center were used.

#	T_{Heater} /°C	$SiH_2Cl_2/H_2(TaCl_5)$	T_{TaCl_5} /°C	$H_2 + SiH_2Cl_2 + H_2(TaCl_5)$ = total flow through reactor in sccm
1	630	1 / .75	75	1200+20 +15 = 1235
2	630	1 /1.00	80	1200+20 +20 = 1240
3	630	1 /1.25	85	1200+20 +25 = 1245
4	650	1 / .75	80	1200+25.7+19.3 = 1245
5	650	1 /1.00	85	1200+17.5+17.5 = 1235
6	650	1 /1.25	75	1200+17.8+22.2 = 1240
7	670	1 / .75	85	1200+22.9+17.1 = 1240
8	670	1 /1.00	75	1200+22.5+22.5 = 1245
9	670	1 /1.25	80	1200+15.6+19.4 = 1235

Reactor pressure ~ 630 mtorr
Deposition time ~ 15 min
The SiH_2Cl_2 flow is delayed in time by 30 sec vs. the $TaCl_5$ flow.

Table I: The orthogonal array after Taguchi, et al [8].

Independent of this array experiment we explored the effects of the heater temperature, the total H_2 flow as well as the H_2 flow through the sublimator, the SiH_2Cl_2 flow, the delay between the SiH_2Cl_2 flow and the $TaCl_5$ flow, the reactor pressure, the sublimator temperature and two step processes.

Two kinds of multiple step processes were investigated:

(a) For 5 to 7 min the deposition took place at certain flows and pressures. Then the flows and pressures were changed for the remainder of the experiment.

(b) In the second case of multiple step processes, the experiment was interrupted after the first 5 to 7 minutes and the heater temperature was changed to a higher level. The reactor pressure and gas flows were changed as well.

The averaged thickness of the $TaSi_2$ films was evaluated by secondary X-ray emission spectroscopy. The selectivity and the structure of the $TaSi_2$ films were evaluated by optical microscopy, scanning electron microscopy (SEM) and in some cases by transmission electron microscopy (TEM). When possible we also measured the resistivity of the $TaSi_2$ films with a 4 point probe.

V. Results

The deposition time of 15 min for each experiment of the orthogonal array L_9 was in no case sufficient to form a closed conducting $TaSi_2$ film on the mono-silicon. As example a SEM of matrix-experiment #8 is shown in figure 5.

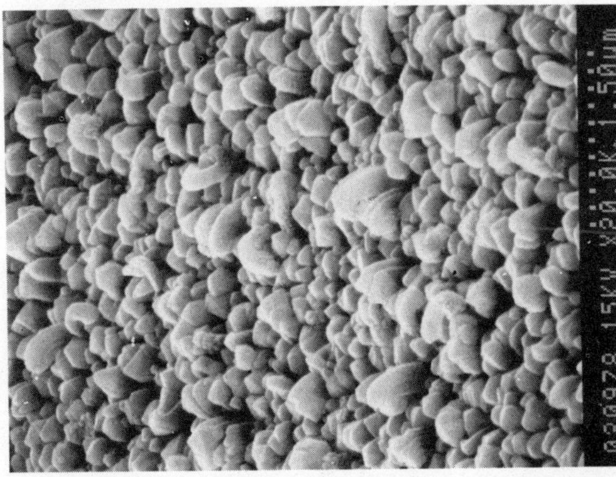

Figure 5: $TaSi_2$ deposition as obtained from matrix-experiment #8.

The selectivity was generally good and only a few relatively small $TaSi_2$ crystals were seen on the SiO_2 area. The beginning of whiskers growth, similar to that described by Williams, et al [9] for CVD $TiSi_2$ and W, was observed for some of the samples treated at heater temperatures at and above 630°C.

Whiskers growth was more predominant for higher SiH_2Cl_2 flows eq. 20 sccm in the Siemens system. This observation about whiskers was true for all experiments even those conducted outside the described L_9 matrix.

The averaged $TaSi_2$ thickness of each L_9 experiment was measured with secondary X-ray emission spectroscopy. The statistical evaluation of these measurements suggested optimum growth conditions within the limits of this L_9 at:

Heater temperature:	670°C
Sublimator temperature:	85°C
SiH_2Cl_2 flow:	17.5 sccm
H_2 flow through $TaCl_5$ sublimator:	17.5 sccm
H_2 flow direct through reactor:	1,200 sccm

A 15 min confirmation run with these settings gave us an averaged growth rate of ~ 9.5 nm/min of CVD $TaSi_2$. The increase of the $TaSi_2$ deposition rate for higher heater temperatures was confirmed by depositions conducted outside the L_9 array.

The selectivity was determined by a convolution of the crystal size and crystal density on the silicon and silicon oxide. The statistical evaluation here points towards lower heater temperatures. This trend is confirmed for the experiments conducted outside the L_9 array, where the selectivity improves for the lower heater temperatures and becomes perfect for temperatures below ~610°C. For these low temperatures, the deposits do not show the crystal planes as for the higher temperature depositions. The crystal facetting was less pronounced at lower temperatures. For the higher heater temperatures, not only the selectivity suffers, but the interactions with the Silicon interface increase. Depending on the gas composition these interactions become severe for temperatures of about 720°C and above. At temperatures as high as 800°C the $TaSi_2$ encroaches deep into the silicon as shown on the TEM in figure 6.

The selectivity did not depend on the $TaSi_2$ film thickness up to an observed thickness of 2500 nm.

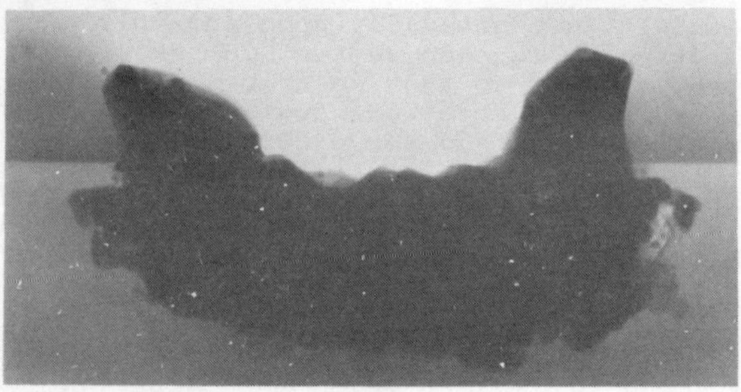

0.1 µm

Figure 6: $TaSi_2$ deposited at 800°C

For the two kinds of multiple step processes investigated, we obtained in case:

(a) continuous films with rough surfaces and polycrystalline microstructures, good selectivity and no encroachment into the mono-silicon, as shown in figures 7 and 8.

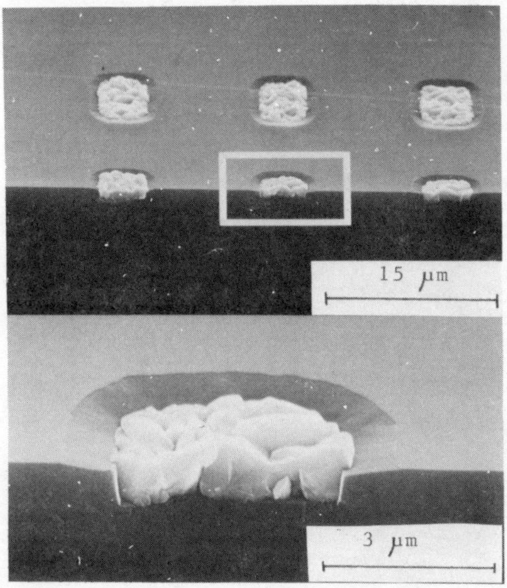

Figure 7: 2 step $TaSi_2$ (SEM)

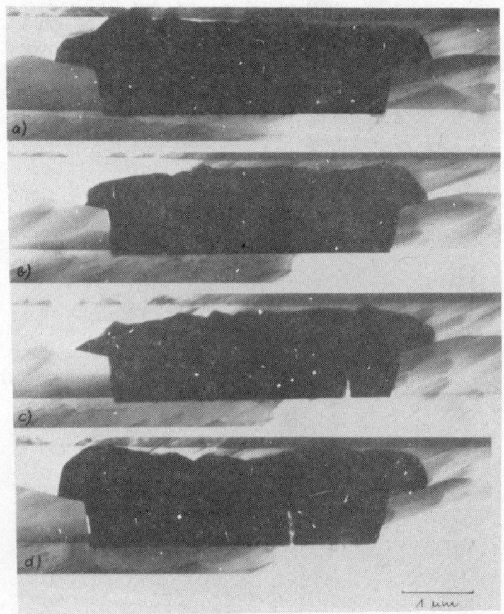

Figure 8: 2 step $TaSi_2$ (TEM)

The encroachment of about 50nm derives from the previous RIE and was also seen on blank wafers. After a 2 min in situ etch with 40 sccm HF and 1,200 sccm H_2 at 640°C the deposition conditions in this case were:

Heater temperature:	640°C
Sublimator temperature:	80°C
SiH_2Cl_2 flow:	20 sccm
H_2 flow through $TaCl_5$ sublimator:	20 sccm
H_2 flow direct through reactor:	400 sccm
Reactor pressure:	450 mtorr

These conditions were kept for ~ 5 min, then the H_2 flow direct through the reactor was increased to 1,200 sccm and the reactor pressure was raised to 650 mtorr for the following 25 min.

As seen in figures 7 and 8, the films from this experiment have some voids between the grains. Some voids can also be seen between the $TaSi_2$ and the underlying mono-silicon. The thinnest part of the film is about 500nm and the thickest part is about 1,900nm. The sheet resistance measured with a 4 point probe was 2.5 ohm/sq., which would result in a specific resisitivity of 125 μohm-cm for a 500nm film.

In the other kind b) of the 2 step deposition processes, the heater temperature was increased from 670°C to 750°C after the initial 4 min, in addition to an increase in H_2 flow and reactor pressure.

The resulting films showed a closed $TaSi_2$ film of about 210nm thickness encroached by as much into the mono-silicon. Cuts through the big almost spherical crystals embedded into this film showed encroachments into the mono-silicon of up to 280 nm.(Fig.9)

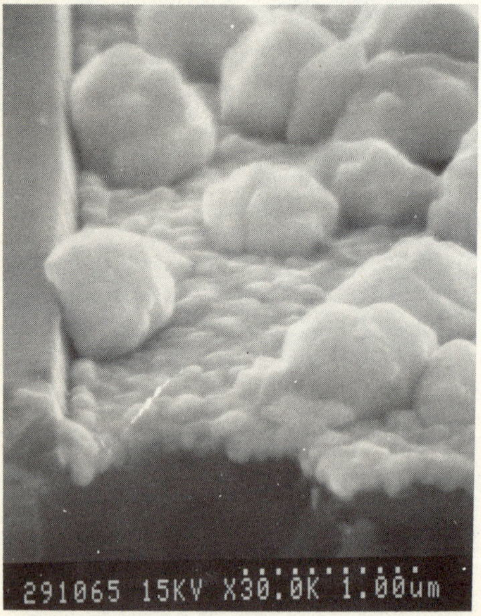

Figure 9: 2 step $TaSi_2$ deposition

X-ray analysis of several samples showed that the crystalline films consisted of stochiometric $TaSi_2$ and that Cl incorporation was below 0.3%. Tape tests showed good adhesion of all films.

VI. Discussion and Outlook

The principle method for the investigation of the LPCVD of selective $TaSi_2$ followed the orthogonal array method described by Taguchi, et al [8]. Since one changes several parameters simultaneously, it is difficult and often impossible to extract the exact behavior of a certain parameter. Therefore, we can only give rough estimates about the behavior of some parameters. For a more careful investigation the matrix has to be increased. The 15 min deposition time in the described L_9 matrix experiment was in no case sufficient to produce a continuous film. We have, however confirmed that a longer deposition time gives a continuous and thicker film. These films showed some voids and not always continuous contact with the underlying mono-silicon. Furthermore, the surfaces of these films were very rough. We believe that the relatively high resistivity of our $TaSi_2$ films of 125 µohm-cm can be attributed to the voids between the crystals and the roughness of the films. Murarka [10] states specific resistivities of 35 µohm-cm to 65 µohm-cm for $TaSi_2$. Since the components of our films are $TaSi_2$ crystals, which themselves have optimum resistivity, we have to reduce the voids and improve the contact between the crystals as well as the surface roughness to obtain better sheet resistances. These improvements could also prevent aluminum from diffusing through the $TaSi_2$ films and thereby causing junction shorts. To improve the $TaSi_2$ film properties and the run to run reproducibility a better control of the substrate temperature (e.g. by pyrometer) is needed. Better control for the start up of gases e.g. through computer control would substantially improve the control over the crucial first few minutes of the deposition. A load lock system would create reproducible clean conditions for the LPCVD of $TaSi_2$.

ACKNOWLEDGEMENTS

The authors wish to thank Dr. Meyer and his department for the analytical support and Mr. Endres and Mr. Stolz for their help and many fruitful discussions:

References

[1] M. Stolz, K. Hieber and C. Wieczorek, Thin Solid Films, 100 (1983) p. 209-218.

[2] M.E. Alperin, T.C. Hollaway, R.A. Haken, C.D. Gosmeyer, R. V. Karnaugh and W. D. Parmantie, IEEE Journal of Solid-State Circuits, Vol. SC-20, No.1 (Feb. 1985) p. 61.

[3] L. Van den hove, R. Wolters, K. Maex, R. De Keersmaecker and and G. Declerck, "A self-aligned cobalt silicide technology using rapid thermal processing" presented at the "Workshop on Refractory Metal silicides for VLSI IV" May 1986, San Juan Bautista, CA.

[4] A. Tabuchi, S. Inoue, M. Maeda and M. Takagi, Jpn., Semicond. Technol. News, (Feb. 1983) p. 43.

[5] E. K. Broadbent and W. T. Stacy, Solid State Technology, (Dec. 1985) p. 51.

[6] N. Lifshitz, D. S. Williams and C. D. Capio, "Selective Molybdenum Deposition by LPCVD", this workshop.

[7] K. Hieber, F. Neppl, Thin Solid Films, 140 (1986)p. 131-135.

[8] Genichi Taguchi & Yu-In Wu: Introduction to off line quality control; Central Japan Quality Control Association.

[9] D. S. Williams, E. Coleman ad J. M. Brown, "Filament Growth during LPCVD of Refractory Metals and Silicides" in Tungsten and other Refractory Metals for VLSI Applications, R. S. Blewer, eds., MRS Pittsburgh, PA 1985, p. 125.

[10] S. P. Murarka, Silicides for VLSI Applications, Academic Press, Inc., 1983.

FLOW CONFIGURATION EFFECTS ON THE DEPOSITION PROPERTIES OF SELECTIVE TUNGSTEN

Y.KUSUMOTO, K.TAKAKUWA, T.IKUTA AND I.NAKAYAMA
ULVAC Corp.,2500 Hagisono,Chigasaki 253,JAPAN

ABSTRACT

The deposition properties of selective tungsten have been investigated as a function of deposition temperature,pressure and a flow configuration, using a novel CVD equipment ERA-1000. The selective deposition is found to be rate-limited process with an activation energy of 0.38eV, whose value is significantly lower than 0.7eV observed in the previous works. It is also found that the deposition properties are affected by flow configuration. Turbulence favors the silicon consumption in filling contacts, or leads to the selectivity breakdown in filling vias after glow discharge treatment. These observations and the discrepancy in activation energies will be discussed.

1. INTRODUCTION

Much attention has been paid to the CVD technique of selective tungsten because of its extensive application to VLSI manufacturing, especially to contact filling or via filling. However there is no commercial CVD equipment available for mass production of VLSI devices. A novel CVD equipment designated as "ERA-1000" has been developed in ULVAC. The ERA-1000 has experienced more than 700 deposition runs for contact filling, and more than 100 runs for via filling, by which the following performance has been established.

1) High deposition rate ranging from 500 to 3000 A/min allows high throughput 5-15 pcs/hr,when filling contacts or vias of 1 micron depth with total selectivity(see Fig.8).
2) Lateral encroachment and silicon consumption are suppressed to be less than 800A(see Fig.9 where a non-doped Si contact is filled with selective tungsten after 1 minute immersion into dilute HF solution).

2. EQUIPMENT

The specifications of ERA-1000 are listed below.

i) Capable of handling a wafer up to 8 inches in diameter.
ii) Cold wall reactor.
iii) Single wafer system.
iv) High vacuum system eliminating residual gas contamination.
v) Fully automatic CPU control equipped with touch sensitive CRT.

Figure 0 shows a schematic sketch of the reactor consisting of an infrared lamp heater,a rotatable susceptor,a gas nozzle,a quartz window, and a quartz mesh plate suspended from the top lid. Hydrogen or argon gas can be led through the mesh plate into a reaction zone in order to stabilize the two dimensional jet of reactive gases emanating from the nozzle. By adjusting the respective flow-rates, the flow configuration can be altered from turbulent to laminar and vice versa.

3. CONTACT FILLING

3.1 Deposition Rate

An example of Arrhenius plot for contact filling is illustrated in Fig.1, where the depositions were carried out under the constraints that the lateral encroachment and the silicon consumption were suppressed to be less than 1000A,and that the total selectivity was preserved. The activation energy was found to be 0.38 eV/atom, which agree well with the value observed in a lamp heating system[1], but differ from 0.7eV observed in otherwise heating system[2,3]. This discrepancy may be mainly due to the difficulties in measuring the temperature of thin films directly irradiated by a lamp. We will not go further into this problem here,which will be discussed later.

The reaction orders for WF_6 and H_2 are found to be zero and one half respectively,which are in accord with the previous results[2,3]

Mass transport mechanism does not have a significant influence on the deposition rate, but on the silicon consumption.

3.2 Selectivity

We have investigated the dependence of selectivity on pressure and temperature. The SiO_2 coated wafers of 6 inches in diameter were deposited under various conditions, where the ratio H_2/WF_6 was fixed at 100. The increment of particles of size ranging from 0.3 to 3 microns on post-deposition wafer was determined using laser beam scattering counter HLD-300 (HITACHI DECO). From each incremental value, we have subtracted the reference value which was evaluated from the wafers deposited in the absence of WF_6,in order to exclude the number of particles other than tungsten islands. Moreover to reduce a statistical error,two deposition runs have been repeated for each case.

The selectivity mapping thus obtained is shown in Fig.2. The tungsten islands were distributed homogeneously all over the wafer surface. Over the pressure range of 0.2 to 0.8 Torr,the nucleation density increases monotonously as temperature increases. At higher temperature(450°C) the nucleation density shows monotonous increase with respect to the pressure, while at lower temperature(350°C or 400°C) it seems to decrease as the pressure increases. This implies that the nucleation density cannot be expressed in a simple form of separation of variables P, T. Though the mapping data are not completely free from statistical error because of small sampling number, we have found again the same tendency in reproducing the mapping for a turbulent flow,which was very similar to Fig.2. At present we have no satisfactory model compatible with these experimental data.

3.3 Encroachment

It is well known that the sizes of the lateral encroachment and the silicon consumption depend on a number of variables; deposition temperature,the partial pressure of WF_6,[4] and the dopant species and concentration. The lateral encroachment lengths after 10 minutes deposition are plotted versus the temperature in Fig.3. By adopting a multi-step recipe,we can suppress the encroachment to be less than 800A irrespective of the deposition temperature up to 500°C. At the initial stage of deposition, the barrier layer which prevents the silicon substrate from further reaction with WF_6 ,has been formed at relatively low temperature . Then the high-rate deposition stages succeeded. There may exist an optimal temperature of the initial stage, below which the barrier layer initially formed is too thin to protect the underlying junction, and beyond which the initial damage is too serious. The optimal temperature is strongly dependent on the dopant species and concentration.

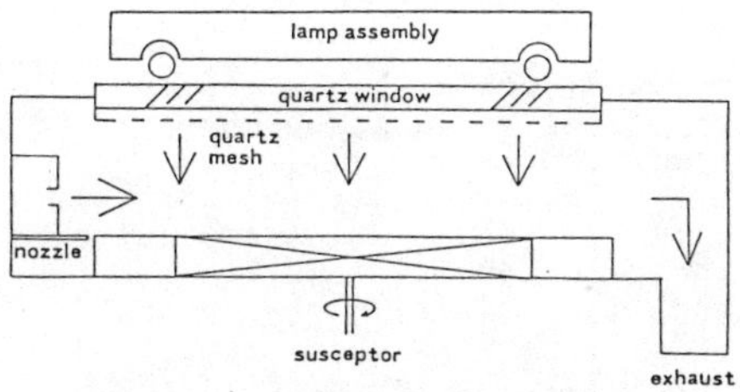

Fig.0
Schematic sketch of reactor.

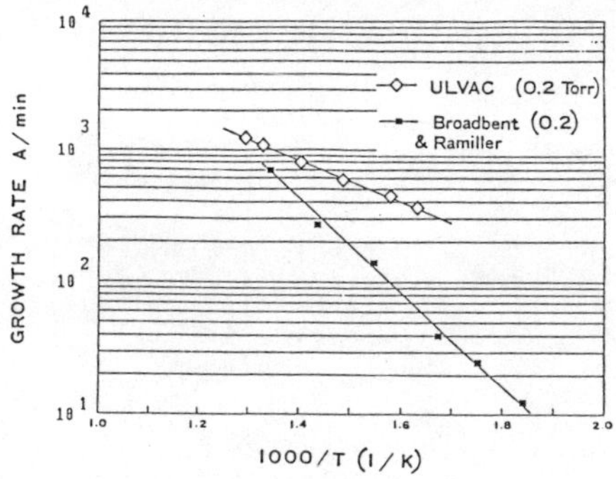

Fig.1
Arrhenius plot of selective deposition of tungsten. (Contact filling)

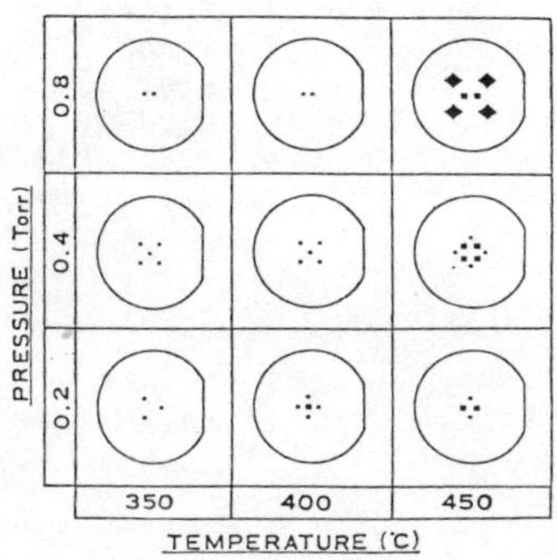

Fig.2
Number density of tungsten nucleation on SiO_2 surface after 10 min deposition. The 6-inches wafers coated with thermal SiO2 were used with no pre-deposition treatment. (Laminar jet)

◆ 10 ptcls/cm²
▪ 1 "
· 0.1 "

$H_2/WF_6 = 100$

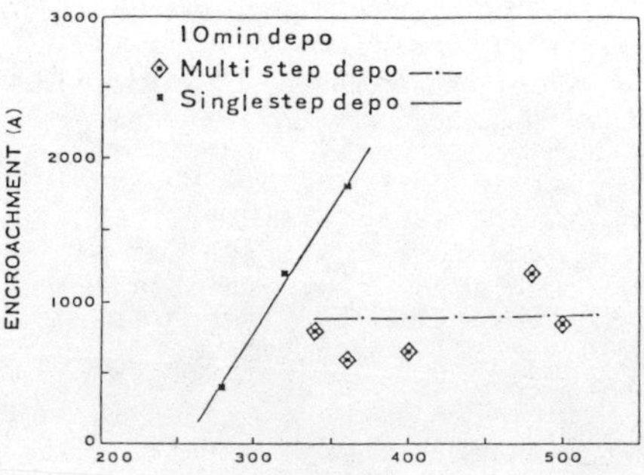

Fig.3
Temperature dependence of encroachment length after 10min. deposition onto non-doped Si contacts at 0.2 Torr.
In multi-step deposition, the abscissa indicates the temperature at final step. (Turbulent jet)

4. FLOW CONFIGURATION

4.1 Laminar and Turbulent Jet

It is of particular interest to reveal how the mass transport mechanism affects the deposition of selective tungsten,which is considered as a surface-reaction-limited process.

The two dimensional jet of reactive gases ejected from the slit nozzle can be laminarized by downward flow which uniformly emanates from the upper mesh plate. Then the reactants are confined to the vicinity of deposition zone. Without the stabilizing downward flow, it freely diverges into a reactor resulting in a quasi-turbulent flow, where the reactants diffuse rapidly with turbulent diffusivity.

The alternation of flow configuration from one to the other was confirmed both by a visualization experiment and by a numerical simulation (see Figs 4).

4.2 Effect on Si Consumption

Figures 5a and 5b respectively show the silicon consumption resulting from laminar and turbulent flow, where argon gas was employed as a carrier gas to maintain sufficiently large Reynolds number. The carrier gas was ejected from the upper mesh in laminar case ,or from the slit nozzle in turbulent case, while the other parameters were kept constant. The contacts of the deposited sample have original overetched depth about 2300A. In Fig.5a the silicon consumption is less than 500A,while in Fig.5b the isotropic erosion over 2000A was observed.

We can conceive the following process to elucidate how the turbulent flow favors silicon consumption. Turbulent eddies could transport an excessive amount of WF_6 to silicon surface or rapidly pump out the by-products such as SiF_x in an intermittent manner. And the tungsten film formed at silicon reduction step would be rough and nonuniform, which does not work as good barrier at subsequent hydrogen reduction step. This speculation is based on the assumption that the silicon reduction step is strongly affected by the flow. It is necessary to investigate how the erosion evolves at each reduction step.

4.3 Effect on Selectivity

The flow configuration has no significant influence on the selectivity in filling contacts after usual HF treatment (e.g. 1min. immersion into dilute HF). However glow discharge treatment leads to the selectivity breakdown only in the turbulent flow (see Figs.6). The sample of aluminum via hole underwent H_2 plasma treatment for 3 minutes before the deposition. The WF_6-H_2 gas mixtures was used with no additional inert gas. Figure 6 suggests that the PSG surface activated by glow discharge easily causes the nucleation of tungsten islands, and that the transport properties of turbulence strikingly differ from those of laminar flow. Without glow discharge treatment,it was found that the turbulence induces only a side-wall growth, which is not shown here.

Figure captions:

Fig.4 Laminar and turbulent jets.
Fig.5 Effect of flow configuration on Si consumption.
Fig.6 Effect of flow configuration on selectivity (with H2 glow discharge treatment).

LAMINAR	TURBULENT

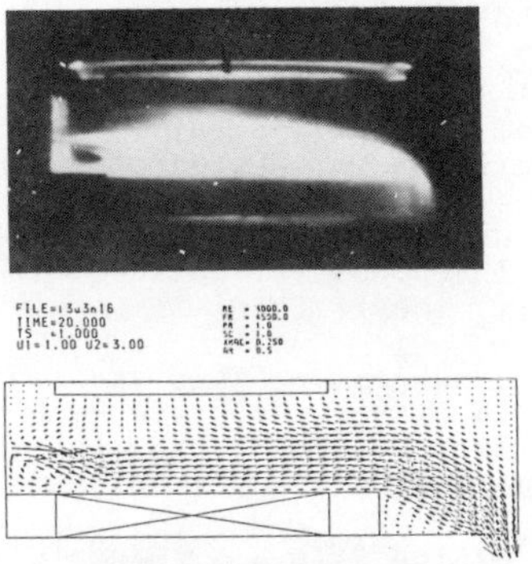

FIG. 4a

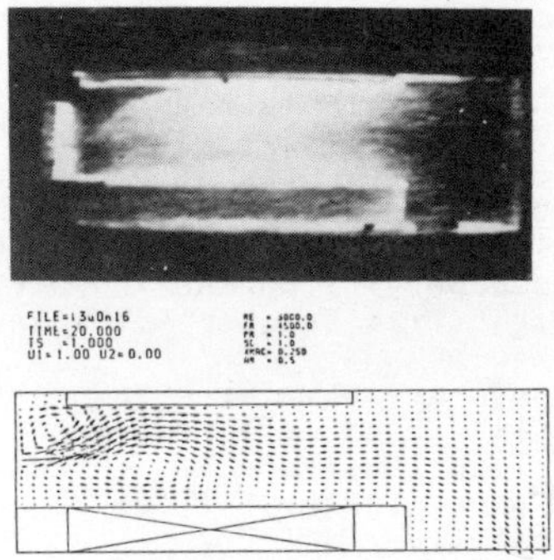

FIG. 4b

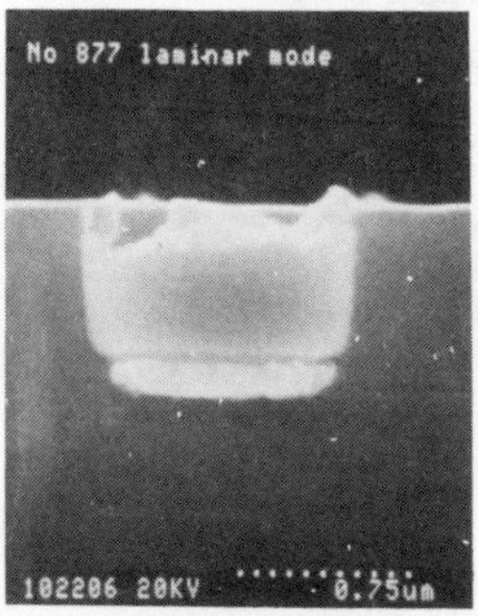

FIG. 5a

400 °C
0.4 Torr
10 min

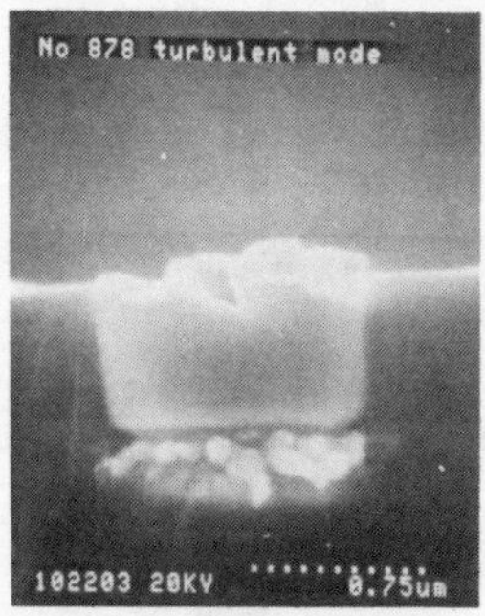

FIG. 5b

CONTACT FILLING

overetch ~ 2300A

FIG. 6a

460 °C
0.2 Torr
10 min

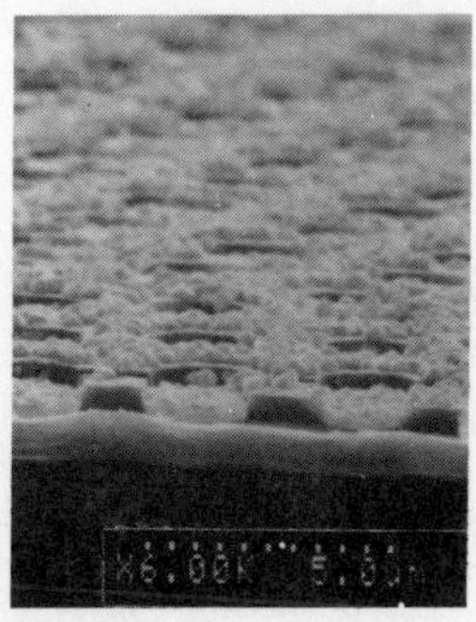

FIG. 6b

AL-VIA FILLING

after H_2 glow discharge treatment

5. DISCUSSION

When a silicon wafer is directly irradiated by a lamp,the deposition parameter of "temperature" only specifies the temperature of the wafer on which no film is deposited. The tungsten film has relatively low reflectance (less than 60%) to the infrared whose wavelength is shorter than 3 microns[5]. On the other hand,the silicon substrate is almost transparent to that of wavelength longer than 1.1 microns. If the temperature of a lamp filament is kept at 2200K or so, the peak of black body radiation is located at 1.3 microns wavelength accordong to Wien's displacement law. Consequently the balance of heat flux among thin films and substrate is a subtle problem,depending on the tungsten film thickness,reflectance, the dielectric film thickness,transmissivity, the filament temperature and so on.

There are,however,some observations suggesting that the tungsten film can be selectively heated up to higher temperature than the silicon substrate or dielectric films. Figures 7 show the evolution of tungsten films deposited onto aluminum via holes in the turbulent flow.

In Fig.7a where the coverage of tungsten film was incomplete, the interface of aluminum layer suffered from no significant deformation. In Fig.7b where the whole surface of PSG was covered with a thick tungsten film ,the underlying aluminum layer melted into the irregular shaped interfaces with silicon and PSG. This drastic change occurred only in a few minutes. It can be concluded from Figs.7 that the temperature of tungsten film became sufficiently high to melt Al-Si alloy ($>660^\circ C$), and not so high to reflow PSG (e.g. $<1000^\circ C$),though the deposition parameter was specified to be $460^\circ C$.

If the large discrepancy in activation energies of 0.38eV and 0.7eV is mainly due to this additional rise of temperature, and if the tungsten film can be selectively heated by an infrared radiation, it is crucially important to protect the W-Si interface of contacts from thermal damages.

REFERENCES

[1]R.H.Wilson,R.W.Stoll and M.A.Calcone:This proceedings,(1985)35
[2]E.K.Broadbent and C.L.Ramiller,J.Electrochem.Soc.:131,(1984)1427
[3]W.A.Bryant, J.Electrochem.Soc.:125,(1978),1534
[4]T.Moriya,K.Yamada,S.Tsunashima,S.Nakata and M.Kashiwagi:Ext.Abst. 15th Conf. on Solid State Devices and Materials,Tokyo(1983)225
[5]C.E.Morosanu and V.Soltuz:Thin Solid Films,52(1978)181

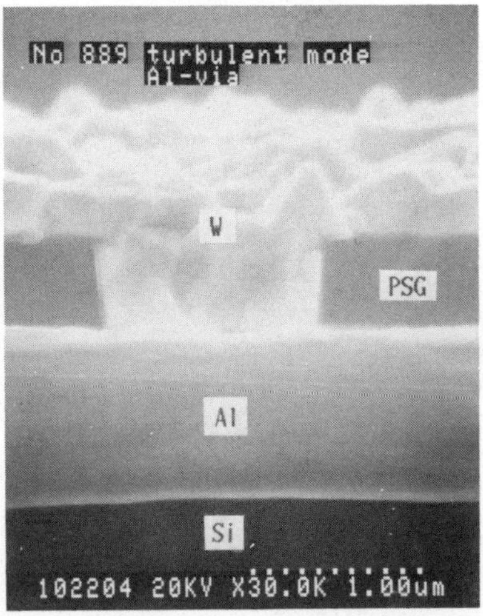

Fig.7a After 15 min. deposition at 460°C,the interfdiffusion occurred at Al-Si interface.

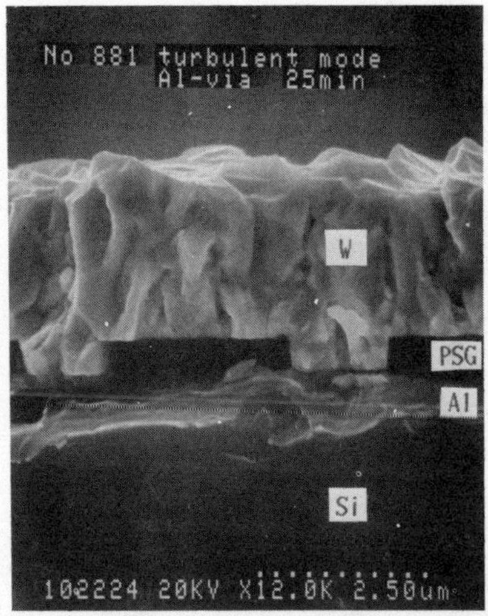

Fig.7b After 25 min. deposition at 460°C,the aluminum layer melted into the irregular shaped interfaces.

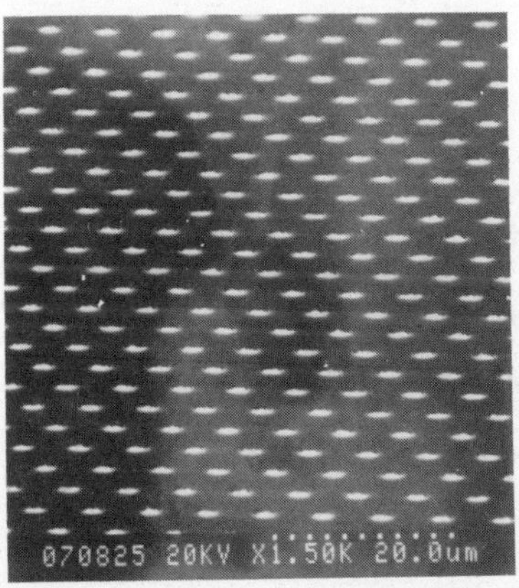

Fig.8 Bird's-eye view of contacts where the total selectivity is preserved.

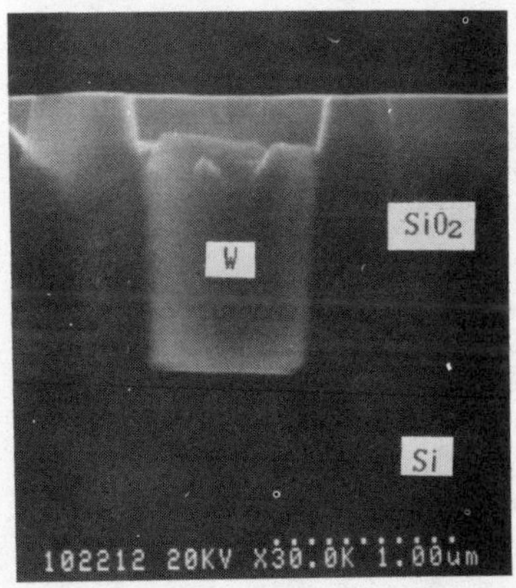

Fig.9 Contact filling free from encroachment.

SELECTIVE AND BLANKET TUNGSTEN BY APCVD

NICHOLAS GRALENSKI AND LAWRENCE BARTHOLOMEW
Watkins-Johnson Company, 440 Kings Village Road, Scotts Valley, CA 95066

ABSTRACT

Atmospheric pressure techniques have generally been considered inadequate for high quality semiconductor coatings. Years of development at Watkins-Johnson, however, have shown this to be untrue. Advanced APCVD techniques can be highly effective for not only oxides but the broad category of non-oxides as well.

Tungsten silicide and tungsten metal have been receiving substantial process attention for the past several years. The status of blanket and selective tungsten, as well as tungsten silicide, by advanced APCVD will be discussed. In general, the atmospheric pressure method is very reliable and cost effective in production, very versatile in process development, and coating results are as good as, if not better than, results by any other deposition method.

* * * * * * *

This paper is about refractory metal coatings, particularly as made by atmospheric pressure CVD. APCVD does not have a good reputation for semiconductor coatings. It is widely believed to be a technique capable, at best, of only second rate coatings and only oxide coatings at that. Our experience is that this is a myth which is simply the result of many disappointing experiences with inadequate equipment.

As our proficiency with advanced APCVD has progressed, it has become increasingly evident that there are inherent advantages, not disadvantages, to an atmospheric process. In general, if the desired result can be obtained at atmospheric pressure, it is not clear that any advantage can be gained by going above or below atmospheric.

The advantages available from advanced APCVD are quite substantial. The inherent simplicity tends to improve cost effectiveness, but there are also productivity and process advantages. Wafers receive individual processing which helps insure that they are all alike; that dopants have been uniformly distributed. A precise relationship between gas source and substrate helps control gas phase reaction and depletion. Particle generation (contrary to what would be expected for a conveyorized system) is the lowest in the industry.[4]

Figure 1 is a diagram of the hardware design. It evolves out of a conveyor furnace product line. CVD technology combined with the furnace technology produces conveyorized APCVD. Substrates on the belt are carried through a horizontal heated metal tunnel, through a coating chamber, and come out done on the other end. The function of the coating chamber and its apparatus is to create and maintain a precision chemical bubble where the coating reaction occurs. Purge and exhaust gas flows are set up to keep the chemicals in the chamber and air out of the tunnel. For some coatings, such as refractory metals, silicides, and silicon, trace oxygen can be a serious contaminant. In our experience, more careful design consideration of purge flows, coating chamber isolation, exhaust metering, and sealing of connections is all that is needed to effectively

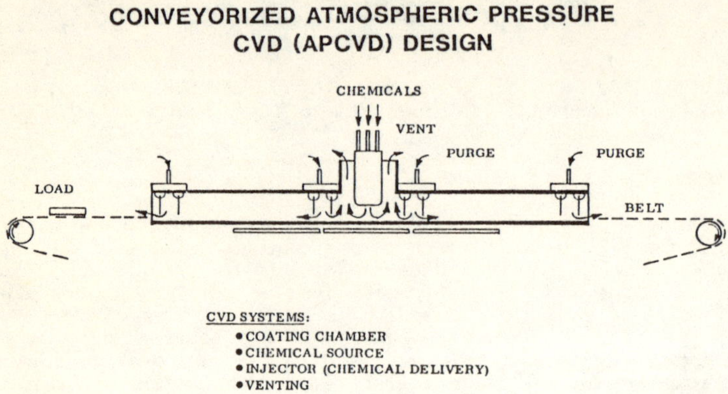

Figure 1

exclude oxygen. (For more details about design, performance and application, see references [1-5]).

Since the introduction of production conveyorized APCVD equipment 12 years ago, the design has advanced so that oxygen levels below one part per million are achievable. The oxygen content of the gas in the deposition zone is especially important. Typical measurements show the whole system to be within .2 ppm of the input liquid nitrogen source gas oxygen level. Auger and ESCA analysis of tungsten silicide films show no detectable oxygen.

A continuous conveyor system excluding oxygen allows in-line multistep processing, enhancing productivity and yield by eliminating intervening exposure to contamination and cleaning steps. A polycide process and a contact/via fill and metallization deposition are two possible multistage applications shown in Figure 2. An in situ surface preparation step, such as gaseous etching of native oxide from silicon contacts, can be accomplished without exposing the device to air prior to selective tungsten growth. The etching can be controlled in the same manner and to the same degree as the deposition. One pass through the furnace can accomplish multilayer deposition without native oxide interfaces. The polycide deposition of silicon plus tungsten silicide has been demonstrated (Figure 3). The annealed structure (Figure 4) can be achieved in-line in a controlled atmosphere with the advantages of a minimal impact on the thermal budget of the device and plastic relief and stabilization of stress in the film prior to strain incurred during cooling due to the mismatch of thermal expansion coefficients.[6]

Uniform delivery of reactants to the whole substrate surface is very important. Vacuum systems rely on long mean free path to improve distribution, but then must coat in batches to meet production demands. Batch coating invites serious depletion problems, especially for complex doping.[7] The problems get worse with larger wafers. At atmospheric pressure, gases can be precisely metered at a controlled temperature at close proximity (3 to 15 mm) to the wafer surface. Our experience is that the hardware required to do this, especially the gas injector, is quite sophisticated. But practical designs have been developed and show consistent excellent production performance.

MULTI-STAGE CONVEYORIZED APCVD

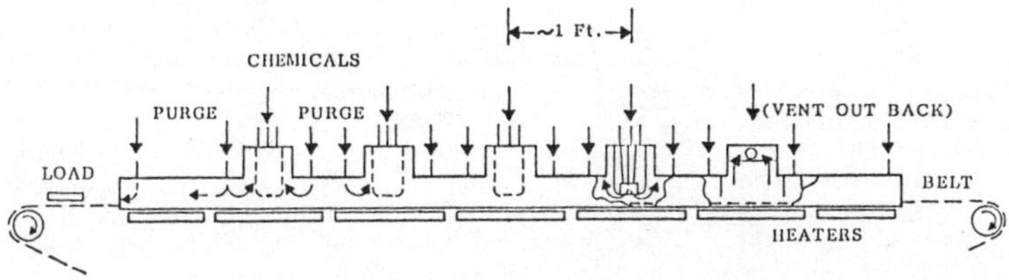

POLYCIDE PROCESS	NATIVE OXIDE ETCH	α-Si(P) 570°C	WSi_2 360°C	WSi_2 360°C	ANNEAL 1000°C
CONTACT/VIA FILL AND METALLIZATION	NATIVE OXIDE ETCH	SELECTIVE W 280°C	SELECTIVE W 320°C	WSi_2 360°C	BLANKET W 450°C

Figure 2

Such an injector is able to deliver a controlled continuous laminar chemical flow directly to the wafer surface across the width of the conveyor belt. Each wafer receives the same presentation of gases regardless of its size. There is a better chance for reaction to occur on the heated substrate rather than in the gas phase. Distant gas sources, premixing and poor gas temperature control encourage the gas phase reaction.[8] At atmospheric pressure, a small deposition zone is sufficient for high throughput because of the high deposition rate. Doped silicon oxide deposition typically occurs at a rate of 13000 A/min., amorphous silicon at 8000 A/min., and tungsten silicide and blanket W at 5000 A/min. We have not seen any penalty in step coverage or stress inherent to atmospheric pressure operation.[2] Figure 5 shows the coverage of amorphous silicon deposited in 5 micron deep oxidized silicon trenches. Complete filling of trenches without voids has been achieved, as deposited, by a borosilicate glass film.

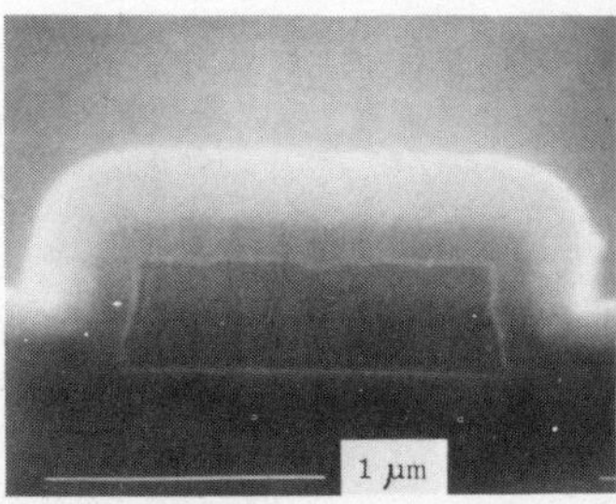

Fig. 3 Polycide as deposited on SiO_2

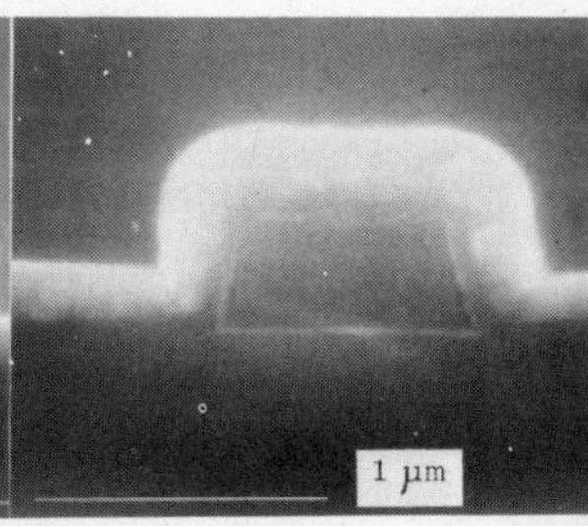

Fig. 4 Annealed polycide on SiO_2

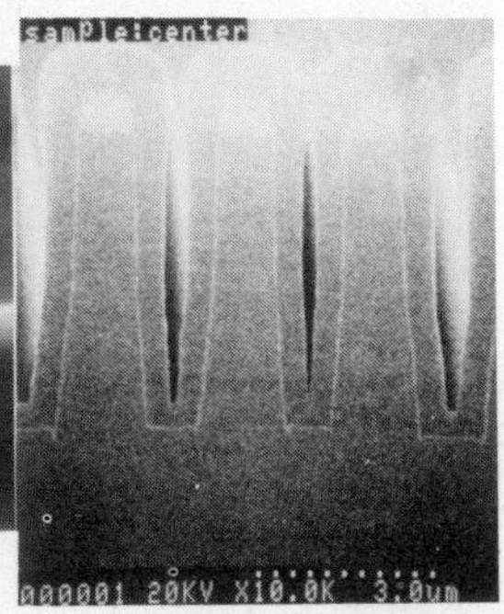

Fig. 5 Amorphous silicon

Another distinction of atmospheric operation is that the presentation of chemicals to the surface can be controlled. There is no restriction on total flow rate to maintain a low pressure; the residence time of the reactant species is controllable. The concentration of chemicals can be optimized for both film quality and safety. Coating chemical concentration is typically about one percent. The reactants can be delivered to the surface in the optimum sequence; the chemicals can either be kept separate or mixed in the gas stream above the substrate. The same ratio of chemicals in the gas input is delivered directly to each wafer surface; film composition shows good correlation to gas composition. For example, tungsten silicide ($WSi_{2.4}$) is deposited using approximately a 3:1 ratio of silane to tungsten hexafluoride at 360 degrees C.

This wide range of control over the deposition is particularly advantageous for tungsten deposition from tungsten hexafluoride. Several unfavorable reactions can occur which damage device integrity if not controlled by the deposition conditions. The silicon reduction reaction can cause excessive consumption of silicon (especially of N+ substrates) and encroachment assisted by stress at an oxide interface.[10] Tungsten hexafluoride can also attack silicon oxide, leading to pinholes and deposition of tungsten oxide which may become volatile during subsequent thermal processing.[11] Saturation with hydrogen enhances the hydrogen reduction reaction, but reduces selectivity of the silicon reaction and generates hydrogen fluoride as well as nonvolatile tungsten subfluorides that may be incorporated at low temperatures.[9] Hydrogen fluoride adsorbed on tungsten, driven by interfacial stress at the perimeter of contacts, can create wormholes into the substrate.[10] The hydrogen fluoride by-product can also diffuse through tungsten to the silicon interface and react with native oxide to form a thin silicon layer and cause loss of adhesion.[12] Also, the exposed insulator material affects the dissociation of hydrogen and availability of nucleation sites.[13]

These complications are not eliminated by coating at atmospheric pressure. However, atmospheric operation does allow the flexibility to adjust the coating conditions at each stage of deposition using in-line multiple chambers while assuring that each substrate is treated the same. For example, selective tungsten deposition may require the following deposition sequence to prevent device damage. First, an in-situ etch of the native oxide is needed for uniform tungsten growth having a smooth interface.[14][15] Second, the initial growth of tungsten just thick enough to cover the exposed silicon contacts may require a low deposition temperature, a high flow rate to remove by-products, and a low ratio of H_2:WF_6 but enough H_2, perhaps incident on the surface first, to prevent excessive silicon consumption without loss of selectivity. Third, with no exposed silicon remaining, a high deposition temperature and high ratio of H_2:WF_6 may favor selectivity and reduce etching of oxide and formation of tungsten oxide and subfluorides.[11] Fourth, a reduction in WF_6 concentration or other change may be necessary to preserve selectivity while building up the thickness to one micron.[9] All of these process conditions can be accomplished successively in-line at atmospheric pressure with continuous treatment of each substrate individually.

APCVD W coating is currently in a developmental mode at Watkins-Johnson. Selective deposition occurs between 280 and 340 degrees C, with spillover nucleation dependent upon the insulator material and percentage area of exposed silicon. The silicon reduction reaction using nitrogen atmosphere without hydrogen produces a self-limiting thickness of about 400 A, but encroachment is severe. Hydrogen minimizes silicon consumption and encroachment, particularly when it impinges upon and saturates

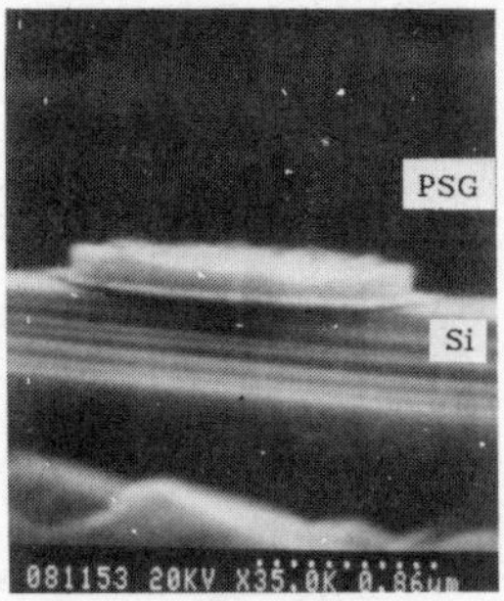

Fig. 6 Selective W in Si contact hole

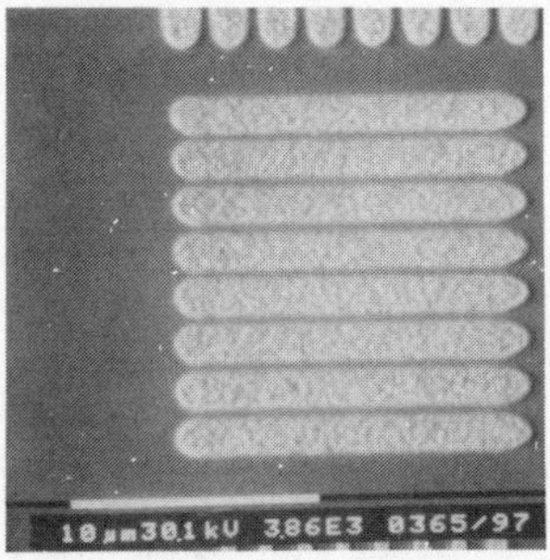

Fig. 7 Selective W on Si through PSG

Fig. 8 Selective W on Si through PSG

the surface first, prior to the tungsten hexafluoride. Figure 6 shows 1300 A W selectively deposited in a 1.5 micron diameter contact hole at a low temperature using 3 sccm tungsten hexafluoride with initial hydrogen saturation. It is believed that the lateral growth is due to etching of the PSG at the interface rather than encroachment since it appears to be above the silicon surface. Verification of the absence of silicon consumption is required. Optimum selectivity and highest deposition temperature are achieved with PSG as the insulator (Figure 7). The thickest film with excellent selectivity deposited at this point is about 3000 A (Figure 8).

Blanket W has been deposited from 350 degrees up to 600 degrees C, with higher surface temperature reaction yielding better adhesion, reduced stress, and enhanced chemical efficiency (up to 50%) (Figure 9). Tungsten silicide as an interface layer promotes adhesion on oxide. The $H_2:WF_6$

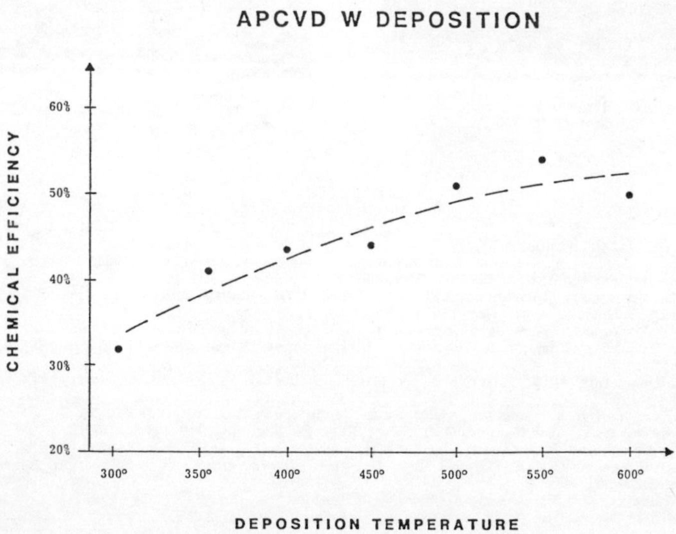

Figure 9

ratio affects the deposition rate and surface morphology. Silicon consumption, etching of silicon dioxide, and formation of tungsten oxide have all been observed under poor process conditions. Typical resistivity varies from 10 to 20 micro ohm-cm, as deposited, depending upon process parameters. Resistivity as low as the bulk value for W (5.3 micro ohm-cm) has been achieved in some depositions or by annealing films at 600 degrees C in N_2. Higher tungsten hexafluoride flows are used than those typical for selective deposition. An 8000 A thick blanket W deposition at 450 degrees C on thermal oxide with contacts to silicon used 21 sccm of tungsten hexafluoride; the resistivity was 12 micro ohm-cm (Figures 10 and 11).

Overall, the deposition mechanism and character by APCVD seems to be the same as reported for low pressure processes. However, at atmospheric pressure, the continuous individual wafer processing permitted by a high deposition rate and the precise relationship between the gas source and substrate provide increased versatility and greater opportunity for an optimum production process. The use of multiple chambers in-line allows sophisticated substrate surface preparation and coating steps to be accomplished without sacrificing productivity. In conclusion, APCVD has far more potential for high quality coating than traditionally believed for both oxides and non-oxides.

Fig. 10 Blanket W over contact in SiO_2

Fig. 11 Blanket W over SiO_2

REFERENCES

1. N.M. Gralenski, "Thin Films by Conveyorized Atmospheric CVD," ISHM-Internepcon, Tokyo, Japan, January 18, 1983
2. L.W. Winkle, C.W. Nelson, Solid State Technology, 24, (10), 123 (1981)
3. N.M. Gralenski, Hybrid Circuit Technology, May 1984
4. N.M. Gralenski, "Advanced APCVD: BPSG Film Quality and Production Reliability Report," Semicon West, May 1986
5. N.M. Gralenski, "New Developments in the Deposition of Tungsten Silicide and Selective Tungsten Films," Semiconductor International, Birmingham, England, September 24, 1986
6. P. Townsend, MRS Symposium on Materials Issues in Silicon Integrated Circuits, Spring 1986
7. W. Kern and R.K. Smeltzer, Solid State Technology, 28, (6), 171 (1985)
8. W. Kern, W.A. Kurylo, and C.J. Tino, RCA Review, 46, 117 (June 1985)
9. M.L. Green and R.A. Levy, J. Electrochem. Soc., 132, (5), 1243 (1985)
10. M.L. Green and R.A. Levy, "Selective LPCVD Tungsten as a Diffusion Barrier," Semicon West, May 1985.
11. J. Carlsson and M. Boman, J. Vac. Sci. Technol. A, 3, (6), 2298 (1985)
12. C.M. Melliar-Smith, A.C. Adams, R.H. Kaiser, and R.A. Kushner, J. Electrochem. Soc., 121, (2), 298 (1974)
13. D.R. Bradbury and T.I. Kamins, J. Electrochem. Soc., 133, (6), 1214 (1986)
14. E.K. Broadbent and W.T. Stacy, Solid State Technology, 28, (12), 51 (1985)
15. H.H. Busta and C.H. Tang, J. Electrochem. Soc., 133, (6), 1195 (1986)

SELECTIVE TUNGSTEN DEPOSITION IN A BATCH COLD WALL CVD SYSTEM

R. CHOW, S. KANG, W. R. HARSHBARGER AND M. SUSOEFF
GENUS, Inc., 515 Ellis Street, Mountain View, CA 94043

INTRODUCTION

Selective deposition of tungsten offers many advantages for VLSI technology. The process can be used as a planarization technique for multilevel interconnect technology, it can be used to fill contacts and to provide a barrier layer between Al and Si materials, and the selective W process might be used as a self-aligned technology to provide low resistance layers on source/drain and gate conductors. Recent publications have indicated that cold wall CVD systems provide advantages for development of selective W process [1]. Genus has investigated selective W deposition processing, and we have developed a selective W deposition process for the Genus 8402 multifilm deposition system. This paper will describe the Genus 8402 system and the selective W process developed in this reactor. To further develop selective W technology, Genus has signed an agreement with General Electric establishing a joint development program. As a part of this program, we have characterized the selective W process for encroachment, Si consumption and degrees of selectivity on various dielectrics. The status of this development activity and process characterization will be reviewed in this paper.

GENUS 8402 MULTIFILM SYSTEM

The selective W process developed at Genus has been implemented in a Genus 8402 batch cold wall CVD reactor. Figure 1 shows a schematic of this system. Previously, this system has been utilized for CVD processing of WSi_2, W and SiO_2 [2,3,4]. With modest modification of the system, selective W deposition can also be performed. System modifications to implement selective W deposition include quartz fixturing to optimize the selective deposition process, new software to provide multiple temperature control and gas manifold modifications to provide the proper reactant gases.

The selective W deposition process developed at Genus consists of a standard 2 step deposition process. In the first step, the WF_6 is reduced by Si to form a thin dense layer of tungsten. H_2 is used to reduce WF_6 to obtain the required W thickness in the second step. Figure 2 illustrates the capability of this process. The figure shows contacts that are 1.2 micron in diameter in a layer of undoped LTO film that is 1.5 micron thick. The contacts have been selectively filled with W. With the Genus selective W process, deposition has been demonstrated at rates from 100 to 500 Å/min.

GENUS - GENERAL ELECTRIC AGREEMENT

In June, 1985, Dr. Ronald Wilson of the General Electric CRD Laboratories described the development of a high rate, highly selective W deposition process [1]. This research established the advantages of a cold wall reactor to selective W deposition. To expedite the development and commercialization of selective W deposition, Genus signed an agreement with General Electric for an exclusive license to the technology developed

Figure 1 - Schematic of Genus 8402 System

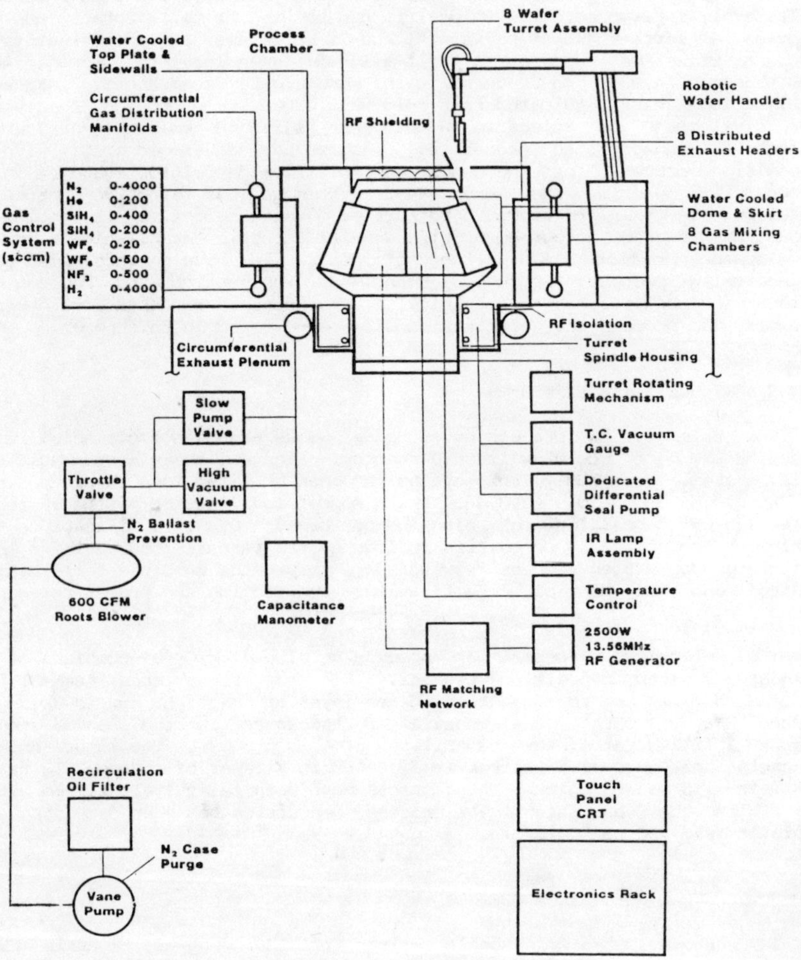

at General Electric CRD. In addition to the technology exchange provided by this license, a joint development program for further development of selective W technology has been established between Genus and General Electric. This joint development program is investigating many aspects of selective W deposition, including wafer preparation, selective deposition on various metals, metal silicides and insulators, encroachment, Si consumption, junction leakage, contact resistance, kinetics and thermodynamics of the process. Much of the information presented in the remainder of this paper was obtained as a part of this joint development program.

Figure 2 - Selective W Deposition in 1.5 Micron Thick LTO Film. Contact Diameters are Approximately 1.2 Micron

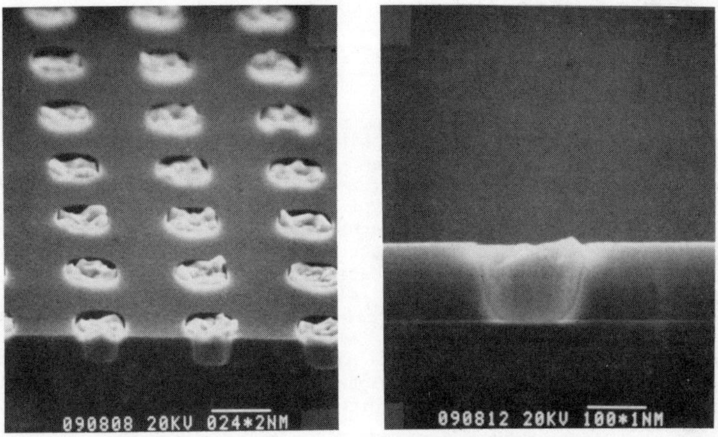

PROCESS CHARACTERIZATION

The Genus selective tungsten process consists of two steps, in which process parameters are varied according to the applications. The silicon reduction step uses a temperature between 350°C and 580°C, a pressure of 100 mtorr, and argon to tungsten hexafluoride ratio between 60 and 600. The hydrogen reduction step uses a temperature betwen 450°C and 580°C, a pressure between 400 mtorr and 600 mtorr, and hydrogen to tungsten hexafluoride ratio between 30 to 60. Typical film resistivity is 9.5 micro-ohm-cm.

A statistical procedure, known as orthogonal design, was used to optimize this process. The experiments were conducted by varying temperature, pressure, hydrogen and tungsten hexafluoride flows at three levels. Deposition rates were measured on the bare silicon wafers by a profilometer, and selectivity was determined on patterned wafers with various sizes of contacts in 1.5 micron of LTO. Wafers were microscopically examined at five positions - top, center, bottom, left and right - to quantify selectivity. The nucleation sites were counted and averaged. The process parameters and results are shown in Table I.

Table I - Orthogonal Design with Four Variables at Three Levels

Turret Temp ^{o}C	Pressure TORR	H_2 SCCM	WF_6 SCCM	Growth Rate Å/min	Sheet Resistance mohms/Square	Selectivity No./sq. mm
520	300	3000	150	947	29.37	8
520	400	3500	200	1167	28.87	59
520	500	4000	250	1403	24.86	28
550	300	4000	200	1220	25.92	20
550	400	3000	250	1433	25.23	59
550	500	3500	150	1187	24.03	17
580	300	3500	250	1440	20.94	28
580	400	4000	150	1260	22.51	8
580	500	3000	200	1870	16.74	70
520	300	3000	150	843	40.50	5

After data analysis, the following conclusions are made. The growth rate increases with temperature and pressure. The growth rate also increases with WF_6 flow, however this rate becomes constant when the WF_6 flow is around 200 sccm. This indicates that so long as WF_6 flow is at least 200 sccm in the Genus reactor, there is no depletion of reactant gas. Sheet resistance decreases with temperature and pressure. Selectivity improves with the increase of hydrogen. The deposition rates in Table I were determined with bare Si wafers. The growth rates on patterned wafers are about 3 to 4 times slower. A deposition rate as high as 500 Å per minute at a wafer temperature of 420°C (corresponding to the turret temperature of 580°C) has been achieved to fill 1.2 micron contacts with good selectivity. Higher deposition rates are possible, if the wafer temperature is increased further.

A study was made to determine the effects of backside coatings on the growth rate. Deposition rates were measured for wafers backside coated with LTO, Mo on LTO, Mo, plasma oxide on Mo and a wafer with backside of exposed silicon. The fronts of these wafers were coated with 1.5 micron of LTO with a contact mask pattern. The wafers were implanted with 5E15 As+ at 50 Kev. A control test wafer, backside coated with 1500 nm of LTO but with no implant, was also included in this test. After deposition, the W thicknesses in the scribe-lines at five areas on each wafer were measured by a profilometer. The measured W film thicknesses on these wafers after a 20 minute deposition are shown in Table II.

The data indicates that wafers with dielectrics in direct contact to the turret face have a higher deposition rate than wafers with metal in contact to the turret face. In addition, when the growth area is implanted, the deposition rate is reduced .

Table II - Backside Coating Effect on Growth Rates

Backside Coatings	Average Thickness in Scribe-lines
1500 nm Of LTO, No Implant Control Wafer	636 nm
1500 nm of LTO	498 nm
500 nm Mo/1500 nm Plasma Oxide	386 nm
No coating, Si exposed	348 nm
1500 nm of LTO/500 nm Mo	286 nm
500 nm of Mo	288 nm

ENCROACHMENT PHENOMENON

Several experiments were performed to study encroachment effects. Initially, wafers without implants were used to obtain a process with minimum encroachment. Then, wafers implanted with As, BF2 or P with an implant dose of 5E15, implant energies between 40 to 50 Kev and activated at 900°C for 120 minutes in N_2 were used to further optimize the process and eliminate encroachment. From these experiments, a process was developed in which there was no discernible encroachment for wafers implanted with BF2 or P. Encroachment was still observed for As implanted wafers. An additional series of experiments was performed to reduce encroachment for wafers with As implantation. In this experiment the As dose was increased to 8E15, and the implant energy was increased to 90 Kev. Eight runs were made for this experiment. The process parameters for these runs are shown in Table III.

Table III - Process Parameters for the Encroachment Study

Run No.	Si Reduction P = 100 mTORR t = 1 min.			H_2 Reduction P = 400 mTORR t = 30 min.		
	Turret Temperature °C	Ar Flow SCCM	WF_6 Flow SCCM	Turret Temperature °C	H_2 Flow SCCM	WF_6 Flow SCCM
1.	580	1000	5	580	6000	200
2.	350	1000	15	580	6000	200
3.	350	1500	15	580	6000	200
4.	350	1000	15	580	7000	200
5.	350	1000	5	580	6000	200
6.	350	3000	15	580	6000	200
7.	580	3000	15	580	6000	200
8.	350	3000	15	580	6000	200

Cross section SEM pictures were taken for these wafers to measure encroachment. Most of the SEM pictures did not show encroachment. However, some of pictures showed encroachment on only one side of the contact hole. This is illustrated in Figure 3, in which encroachment was observed on the right side of the contact but not on the left side.

Figure 3 - Encroachment on One Side Only

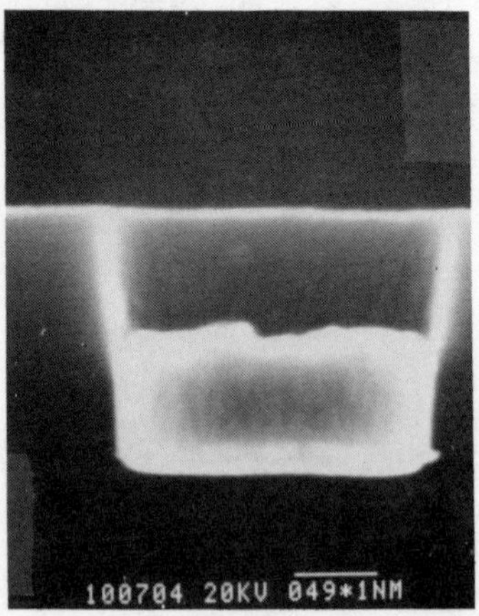

This phenomenon is due to the fact that minor encroachment is not continuous along the edge of the contact. A two dimensional cleaved cross section can easily miss encroachment at certain areas of the contact edge. For the sample shown in Figure 4, both tungsten and LTO were etched from the sample. The silicon consumption mark left behind by the encroachment on the edge of the contact can be seen to be discontinuous. Since a SEM picture with a cross section of one plane of view does not adequately represent encroachment on the entire opening, other analysis techniques, such as electrical measurements of junction leakage, are necessary to efficiently monitor encroachment.

Figure 4 - Encroachment Mark Along the Edge of a Contact

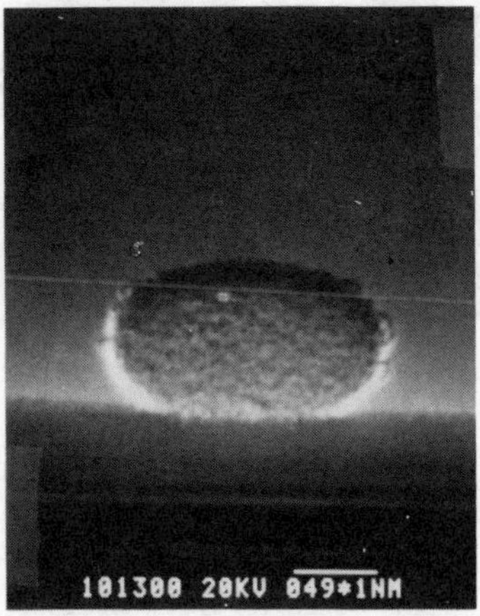

In general, high Ar to WF_6 ratio in the Si reduction step reduces encroachment. Encroachment may be enhanced during a long hydrogen reduction step. In addition, Moriya [5] has reported that high temperature deposition reduces encroachment. The standard Genus selective W deposition process has reduced encroachment on As implanted wafers to less than 600 Å. For wafers implanted with P or BF_2, encroachment was not observed by SEM analysis.

SELECTIVITY ENHANCEMENT

A study has been performed to determine the selectivity order for seven different dielectrics - plasma oxide, LPCVD oxide, low temperature oxide, thermal oxide, PSG, LPCVD nitride and plasma nitride. The test wafers were patterned to be half silicon and half dielectric, and the wafers were the same type used by A. Williams and R. Wilson for their study presented in this workshop. A two step process with a Si reduction step at a temperature of 580°C, pressure at 100 mtorr, WF_6 flow at 5 sccm, Ar flow at 1000 sccm and time for 1 minute, and a hydrogen reduction step at a temperature of 580°C, pressure at 400 mtorr, WF_6 flow at 200 sccm, H_2 flow at 6000 sccm and time for 30 minutes was used for this investigation. After 31 minutes of deposition, wafers with plasma oxide, LPCVD oxide and LTO did not show any tungsten growth on the dielectric adjacent to the silicon growth area. Wafers with thermal oxide and PSG (3%) showed some tungsten on the dielectric adjacent to the silicon. Wafers with nitrides had considerable tungsten on the dielectrics. Figure 5 shows three examples for the dielectric effect on selectivity.

Figure 5 - Dielectric Effect on Selectivity

Plasma Oxide Thermal Oxide Plasma Nitride

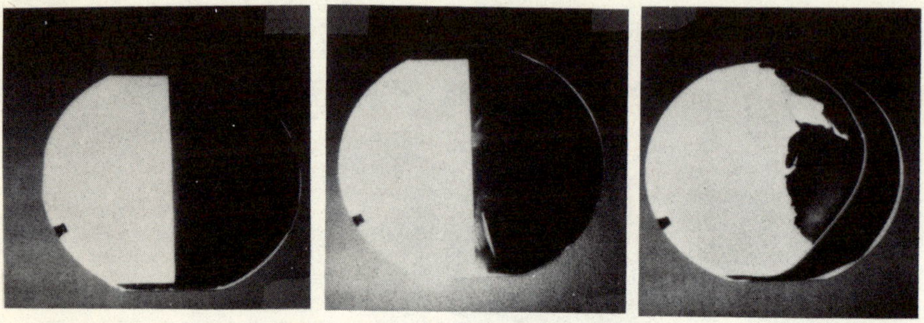

Since there was no tungsten growth on plasma oxide, LPCVD oxide and LTO, the number of nucleation sites were counted for these wafers. Ten areas on the half dielectrics of each wafer were microscopically examined. five locations were examined 1/2" from the center line, and five other locations 1 1/4" from the center line were examined. The results are shown in Table IV.

Table IV - No. of Nucleation Sites on Three Dielectrics (Normalized Units)

	Plasma Oxide		LPCVD Oxide		LTO	
	1/2" Ave.No.	1 1/4" Ave.NO.	1/2" Ave.No.	1 1/4" Ave.No.	1/2" Ave.No.	1 1/4" Ave.No.
Run 1	17	6	45	7	63	9
Run 2	51	2	78	14	102	13

Based on visual inspections, the selectivity order was determined as follows: Plasma Oxide > LPCVD Oxide > Low Temperature Oxide > Thermal Oxide > PSG (3%) > LPCVD Nitride > Plasma Nitride. In general, the results agree with the publication by Broadbent and Stacy [6]. However, we did not observe significant reduction in nucleation sites with the PSG (3%) as reported by Bradbury and Kamins [7]. The number of nucleation sites on plasma oxide at 1-1/4" away from the growth area seems to be less than those on LPCVD oxide and LTO. In addition, we observed that non-selectivity starts near the growth area.

We have also studied the effects of process parameters on selectivity. The experimental results agree with the predictions by Carlsson [8]; low pressure and high H_2 to WF_6 ratio enhance selectivity. Minimization of heated metal surface areas in the reaction chamber also enhances selectivity. In addition, we observe that selectivity is enhanced by deposition at higher temperatures with shorter deposition times.

ADVANTAGES OF COLD WALL BATCH CVD SYSTEM

For selective **W** deposition a cold wall CVD reactor produces higher selectivity than a hot wall CVD reactor. This is primarily due to the reduced amount of hot surface area. There will be less deposition on areas other than the wafer. The total amount of reaction by-products are greatly reduced, and it has been reported that some reaction by-products lead to loss of selectivity. [6,7]

The deposition rate of selective W in the cold wall system is often higher than in a hot wall reactor, because improved selectivity in the cold wall reactor allows for deposition at higher temperatures. In a cold wall system, selectivity is enhanced with high temperatures and short deposition times, whereas in a hot wall system, processes are carried out at lower temperatures and longer times. Green and Levy [9] stated that in a hot wall reactor, at temperatures below 400°C, good selectivity is observed. Above this temperature, W deposits on both Si and glass. In addition, the deposition rate reported by Green seems to be a function of WF_6 flow, implying the growth mechanism is mass transport limited. In the Genus cold wall reactor, the growth rate is not a function of WF_6 flow at WF_6 flow rates greater than 200 sccm. A surface controlled reaction mechanism can be readily achieved in a cold wall reactor. Another advantage for high temperature **W** deposition in the cold wall system is that the fluorine content is reduced in the W film. Shim, Haskell and Byrne [10] stated that fluorine content decreases with increasing temperature.

Experimental results indicate that the Genus batch system does not require a load lock to achieve repeatable processes. Thus, a simpler loading mechanism can be used to obtain high throughput and reliable processing.

CONCLUSION

A selective W deposition process has been developed for use in the Genus 8402 multifilm reactor. The process is enhanced by a simple modification kit for the system. With this process, we have demonstrated selective W deposition to fill contacts as deep as 1.5 micron. Encroachment for the process has been reduced to less than 600 Å. Genus has obtained an exclusive license to selective W deposition technology developed at General Electric CRD, and a joint development program has been established to further optimize the processes. Based on the progress of development of selective W deposition technology, we are confident that a high rate, highly selective **W** CVD process is being developed to incorporate in future VLSI technology.

ACKNOWLEDGMENTS:

The authors gratefully acknowledge valuable discussions with R. H. Wilson and A.G. Williams during the development of the selective W CVD process. We also acknowledge B. Gorowitz for preparation and analysis of samples used in this study.

REFERENCES

1. R. H. Wilson, R. W. Stoll and M. A. Calacone, 1985 V-MIC Conf., June 25, 1985.

   R. W. Stoll and R. H. Wilson, Electrochem. Soc. Meeting, Vol 86-2, Abstract 354, San Diego, October 1986.

2. K. C. Saraswat, D. L. Brors, J. A. Fair, K. A. Monnig and R. Beyers, IEEE. Trans. Electr. Dev. ED-30, 1497 (1983).

3. D. L. Brors, K. A. Monnig, J. A. Fair, W. Coney and K.C. Saraswat, Solid State Technology 27, No 4, 313 (1984).

4. E.J. McInerney, 1986 V-MIC Conf., June 9-10, 1986.

5. T. Moriya, S. Shima, Y. Hazuki, M. Chiba, and M. Kashiwagi, IEDM Technical Digest, 550 (1983).

6. E. K. Broadbent and W. T. Stacy, Solid State Technology 28, No 12, 51 (1985).

7. D. R. Bradbury and T. I. Kamins, J. Electrochem. Soc. 133, 1214 (1986).

8. Jan-Otto Carlsson and Mats Boman, J. Vac. Sci. Technol. A, 3, 2298 (1985).

9. M. L. Green and R. A. Levy, J. Electrochem. Soc. 132, 1243 (1985).

10. H. S. Shim, R. W. Haskell and J. G. Byrne, in Proceedings Of The Third International Conference On Chemical Vapor Deposition, edited by F. A. Glaski (American Nuclear Society, Hinsdale, Il., 1972), P. 715.

SELECTIVE TUNGSTEN DEPOSITION IN A VARIAN/TORREX 5101 COLD WALL CVD REACTOR

R.F. FOSTER, D.L. BRORS and S. TSENG
Varian/Torrex CVD, 2054 Martin Avenue, Santa Clara, CA 95050

ABSTRACT

Selective tungsten deposition was carried out in a Varian/Torrex radiantly heated single-wafer cold wall 5101 CVD reactor. The wafer holder configuration is such that the wafer is heated directly by halogen lamps, allowing the use of a variety of wafer heating rates. A unique feature of this system is that the wafer temperature is measured directly by the use of an infra-red sensor mounted on the bottom plate. This configuration results in maximum process flexibility, which we will show as essential for selective tungsten process.

The deposition of one micron of selective tungsten has been realized in a reproducible basis, and, in a limited set of experiments, selective deposition was carried out for oxide thickness of up to ~1.7 microns. Silicon consumption is kept to a minimum by insuring good nucleation of tungsten on the contact at the beginning of the process. In this manner a one micron deposition results in less than 300 angstroms of silicon loss. Deposition rate versus temperature data will be presented for a limited set of process conditions. These data illustrate the behavior of selective tungsten kinetics for a cold wall chamber in which the wafer temperature is known. Data will also be presented on the temperature uniformity of this system. Wafer cleaning prior to selective tungsten deposition will be discussed in reference to contact size and type of clean used.

Introduction

The successful deposition of one micron of selective tungsten has been demonstrated in several single wafer cold wall reactors [1]. This appears to be a function of the availability of the cold wall deposition system. With such systems a "high rate plug filling process" can be contemplated. Using the non-isothermal environment of a cold wall system presents not only advantages and certain process disadvantages but also challenges for equipment manufacturers. The main requirements are tight temperature and gas flow controls, repeatability in wafer handling and process chamber cleaning in order to utilize fully isolated wafer processing. The evolution of selective tungsten into a "production process" provides an excellent vehicle on which to develop these concepts. In this paper, we report on our initial efforts to deposit selective tungsten in a Varian/Torrex 5101 reactor, with emphasis on some of the problems as well as the successes.

Deposition System

The Varian 5101 (Figure 1) is a single wafer cold wall CVD reactor equipped with a vacuum loadlock. The silicon wafers are clamped to a quartz holder and are heated radiantly with halogen lamps from the backside. Since the back of the wafer is not exposed to process gas, backside deposition is eliminated. Wafer temperature measurement is accomplished with the use of an infra-red sensor which collects 8 - 14μm radiation from a 1 inch diameter spot in the center of the wafer. The signal from the pyrometer is fed into a programmable controller, which in turn regulates the lamp power for maximum process control. The "actual" wafer temperature and the pyrometer reading are correlated by thermocouple measurements at various heater power settings. In this way, meaningful pyrometer temperature measurements can be made over the entire instrument span of 100 - 800°C [2]. A water cooled chuck insures that no metal surfaces heat significantly even at 600°C wafer temperatures. A simple gas distribution system provides for mixing of the cold gases to suppress any unwanted reaction and provides uniform delivery of gas to the wafer.

RESULTS:

The deposition of one micron of selective tungsten on patterned one micron oxide over silicon substrates was carried out reproducibly at 1,000 to 1,500 Å/minute deposition rates without loss of selectivity. Other experiments on ~1.7μ thick oxide films resulted in two microns of selective tungsten with a 0.4 micron overfill. The deposition rate was found to depend exponentially on the temperature as measured on the wafer (Figure 2). While there is substantial error in the accuracy of these temperature measurements, for a given wafer type, the repeatability is ±2°C. Such problems put any activation energy derived from these data in doubt [3]. Nevertheless, no saturation effect on deposition rate versus temperature is observed even at ~3500 Å/minute. Figure 3 illustrates the effect of WF6 flow in the deposition rate at 525°C and the expected saturation effect is observed [4], [5], [6].

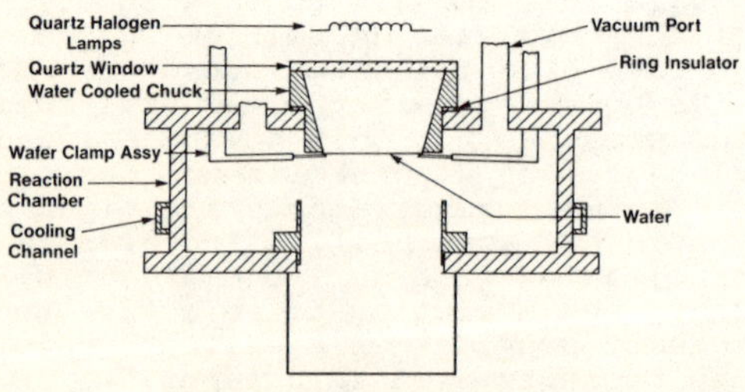

FIGURE 1

RTCVD

RAPID THERMAL CVD

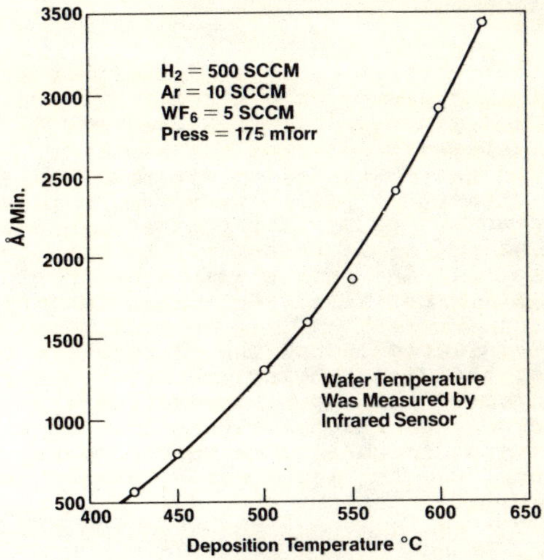

FIGURE 2

SELECTIVE TUNGSTEN

Deposition rate vs temperature

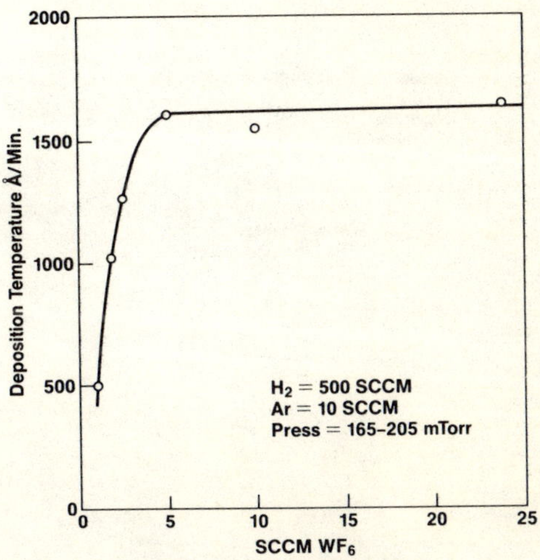

FIGURE 3

SELECTIVE TUNGSTEN

Deposition vs WF6 flow @ 525 C

Uniformity

The deposition uniformity is a direct result of the temperature uniformity across the wafer. Because the wafer resides in a radially non-uniform environment, it is expected that the deposition uniformity would reveal some edge effects. Figure 5 shows the deposition uniformity across a four inch wafer from our initial chuck design. It is evident that a thermal gradient exists for approximately 1/2" from the edge of the wafer. This pattern is altered by directing more radiation to the edge of the wafer. Such an arrangement results in an improvement to ±7% half-range uniformity.

The results in Table 1 show the effect of WF6 flow on the uniformity at 550°C. Changing the flow from 3 to 5 sccm dramatically increases the deposition rate at the edge of the wafer. Since we can not reconcile these two experiments on the basis of temperature change, we conclude that this improvement is due only to flow.

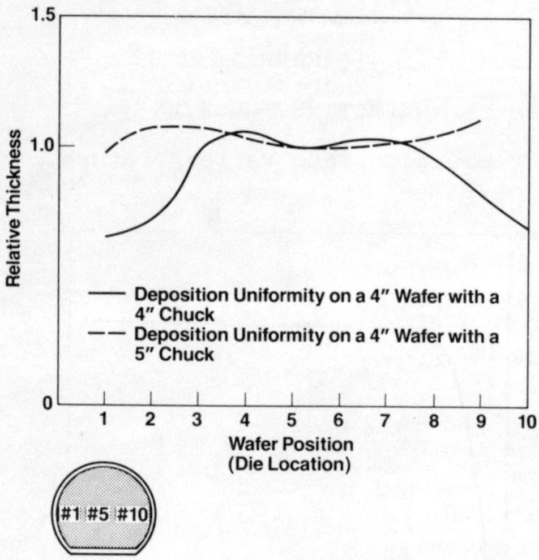

FIGURE 5

DEPOSITION UNIFORMITY

TABLE 1

SELECTIVE TUNGSTEN

Temp (°C)	WF_6 (SCCM)	Dep Rate (Å/Min) Center Thickness Å	Edge Thickness Å
550	3	1050	~0
550	5	1600	1000

[H2] Dependence

The dependence of the deposition rate at 575°C on hydrogen partial pressure is given in Fig. 4. In each deposition the total pressure, P_T, was kept constant as was the WF6 and Ar flows. It is evident that a reasonable correlation is achieved when the data are plotted against [H2] . These results are consistent with those reported previously [5], [6]. Such a fractional power dependence can be rationalized by a variety of surface reaction mechanisms. However, a simple model can be constructed using the hydrogen dissociative adsorption as the rate-determining-step. All other steps, such as WF6 adsorption and subsequent dissociation and HF desorption are considered to be fast. Using this scheme and the steady-state approximation one can arrive at an expression for the rate of reaction:

$$V = k' \, PWF6 \;\; PH2^{1/2} \qquad (1)$$

where V is the reaction velocity, PWF6 and PH2 are the WF6 and H2 partial pressures, respectfully and k' is a complex rate constant. If a step, such as HF desorption, becomes slow so that HF competes with H2 for surface sites, then the kinetic expression becomes more complicated.

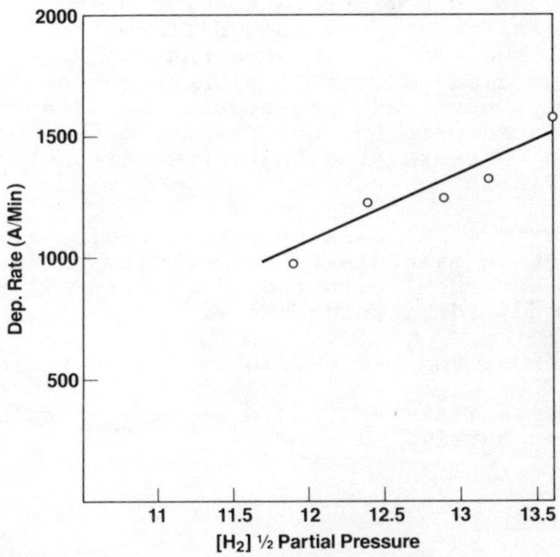

FIGURE 4

DEPOSITION RATE

Dependence on hydrogen partial pressure

Selectivity

For substrates of <30% open area in the oxide, it is possible to grow up to 2μ of tungsten with excellent selectivity (<5 nucleations/view at 200x dark field) at deposition rates of 1,000 - 1,500 Å/minute and temperatures up to 600°C. If the open area is 40 -50%, some loss of selectivity can occur, especially near large silicon areas. At lower wafer temperatures such as for aluminum or $PtSi_2$ substrates, good selectivity (5 - 10 nucleations/view) has been obtained up to 1μ of deposition.

Several qualitative facts can be stated about selectivity. First is that longer deposition times for a given thickness (i.e. lower deposition rate) lead to worse selectivity. These observations are derived from one micron depositions at 450°C in which few wafers have <5 nucleations per view. Second, for a given wafer type (e.g. Torrex patterned thermal SiO_2 with ~20% open area) we observe qualitatively that the selectivity gets worse as the temperature increases. At temperatures >625°C using Torrex wafers the nucleation can be >100/view. It seems that whatever mechanisms exist for tungsten nucleation on the oxide, at least some are temperature sensitive. These observations, of course, are wafer dependent; that is, the wafer history seems to be important. For example, two seemingly identical wafers may show different selectivities under identical conditions. In some cases one can by experience assign a major role to either contact etching or some related step, substrate pre-deposition clean and rinse or a process related problem, but in many cases these variables are so inter-related that only circumstantial evidence is available.

That the surface and bulk consumption of the insulating film is important in maintaining selectivity is indicated by our qualitative agreement with the literature [7] that the order of preference for insulators is:

BPSG > PSG > Thermal and CVD oxides >> Plasma silicon nitride

A more quantitative statement would require an extensive and carefully planned matrix.

Silicon Consumption

Because H_2 does not dissociate appreciably on silicon at these temperatures, tungsten metal must be nucleated in a contact before further deposition can take place. If this nucleation is allowed to take place at elevated temperatures, uncontrolled consumption of silicon will take place. However, if WF_6 and silicon are brought in contact at low temperature, then efficient deposition of tungsten occurs with minimal silicon consumption. Tungsten deposition can then be completed at high temperatures. Using this technique, we have measured <400 Å of silicon consumed for 1u of tungsten grown in wet etched or dry etched contacts. Furthermore, this silicon consumption is independent of deposition time; that is the first seconds of the deposition are the most important. There are important differences in the nature of the contact etch. Figure 5 shows the SEM photographs of different wet and dry etch wafers which were stripped after tungsten deposition. The wet etched contacts with a 45 slope show pitting of the silicon at the edge of the contact (Figure 6a), but these pits are not evident in a similar dry etched wafer with approximately 200 Å over etch (Figure 6b). Both contacts show a relatively smooth silicon surface. By contrast Figure 6c shows a dry etch wafer (over etch unknown) with a somewhat rougher silicon surface and no pitting at the edge. Finally, no pitting or rough interface can be seen for deposition over WSix.

Other Substrates

A >1μ thick selective tungsten film was grown over aluminum substrates at ~430°C at a deposition rate of ~400 - 500 Å/minute (Figure 7). No obvious change in the aluminum substrate could be detected, but the contact resistance to aluminum is unknown. The deposition rate is comparable to that of a silicon substrate at the same temperature, which implies a similar deposition mechanism. Some tungsten nucleations were observed on the oxide especially at the longer deposition times. This fact also indicates that silicon consumption or its by-products (e.g. SiF4) are not solely responsible for selectivity loss.

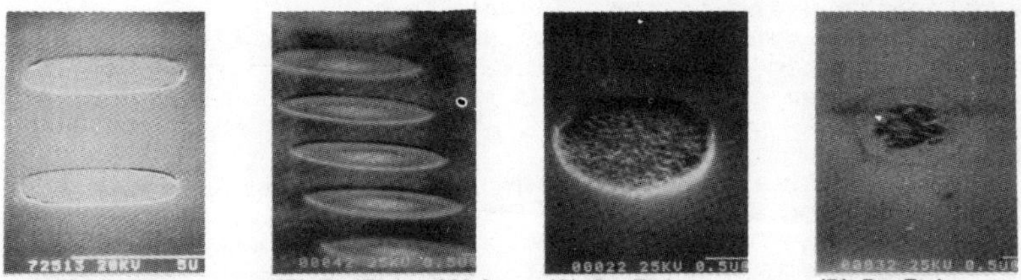

(A) Wet Etched Contact (B) Dry Etch, ~200 A Overetch (C) Dry Etch, Overetch Unknown (D) Dry Etch Contacts to W Six

FIGURE 6

SUBSTRATE SURFACE AFTER DEPOSITION

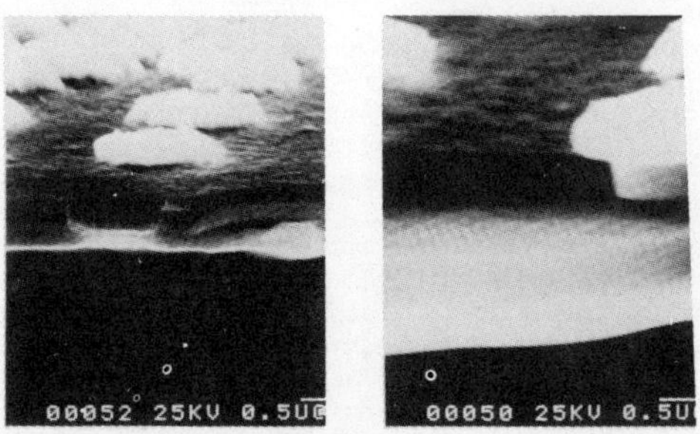

FIGURE 7

1u OF SELECTIVE W OVER AL

At 430 C dep. rate ~400 A/min

Discussion

The deposition of selective tungsten in our cold wall reactor is consistent with simple chemical models already proposed. The silicon reduction of WF_6 proceeds initially (nucleation) and is moderately temperature sensitive. At high temperatures the hydrogen reduction of WF_6 dominates and is more strongly temperature dependent. Because the initial tungsten nucleation, contact silicidation and selectivity are all temperature dependent it is also important that the temperature be known as accurately as possible. IR temperature measurement seems to be well suited for this problem. However, different surfaces have different emission probabilities (emissivity) and these emissivities change with temperature. In addition, as tungsten is deposited the wafer emissivity changes further. Measuring the temperature on the backside would appear to be a better solution, however, different materials and surface finish present similar problems. What is needed is a method of measuring the wafer emissivity in real time during the process. We are exploring one such solution in our laboratories.

Another strong influence on the process is the nature of the contact surface. Any residual native oxide, etching or photoresist residue in the contact will suppress efficient nucleation and cause excessive silicon consumption. In some cases, this spiking can be >0.5μ and causes vertical tungsten growth on the sides of the contact. Another factor is silicon damage caused by dry etching, ion implantation, or other differences due to dopants. Preliminary indications are that silicon consumption in p+ contacts can be up to twice that of n+ contacts.

Argon sputter cleaning can reduce native oxides, but adversely affect selectivity. High temperature H_2 reduction may be possible, but wafer temperature uniformity might be a problem. A variety of plasma cleans are possible, such as H_2, H_2/NF_3 and other etching gasses. In part, one solution is to sputter Ti and silicide the contact to get rid of silicon damage. We have already demonstrated selective 1u deposition over $TiSi_2$ at ~550 - 600°C, but have not yet determined effects on contact resistance.

Control of selectivity is also related to control of the wafer temperature, but is less sensitive than silicon consumption. Equally as important are the oxide surface condition and the process conditions. Fortunately the process variables, WF_6 flow and total pressure, are easily controlled and result in a forgiving process window. We have recently begun to explore special surface pre-treatments and the results are inconsistent at best. We believe that a more efficient surface clean must be instituted after the previous processing, and perhaps this step could be a simple etch. However, with materials other than silicon, such etches may be difficult.

Conclusion

Deposition of >1μ of selective tungsten has been reproducibly achieved at rates of >1,000 Å/minute. The degree of selectivity achieved over silicon and other substrates is the result of using complex variables but can be controlled efficiently. The major process variable in this cold-wall CVD reactor is the wafer temperature and uniformity. Corrections must be made for different wafer emissivities before the process temperatures are well characterized and other deposition variables can be properly investigated. However, the current state of pyrometry affords control so that silicon consumption and encroachment can be minimized.

The next phase of our program will be to integrate this knowledge and measure electrical properties on real devices. These measurements will undoubtedly provide us with new and better information.

References

1. R.H. Wilson, R.W. Stoll and M.A. Calacone, Proceedings 2nd. International IEEE VLSI Multilevel Interconnect Conference, p.343, IEEE Cat. no. 85CH2197-2, Santa Clara, Ca. June 25-26, 1985.
2. 2 pyrometers were used for these experiments:
   a). IRCON - MODLINE 4, SIERES 47, 400 - 1,200°C
   b). Linear Laboratories - TherMonitor 1000L, 100 - 800°C
3. N.E. Miller and I. Beinglass, Solid State Technology, Vol. 23, p.85, 1982.
4. Eliot K. Broadbent and William T. Stacy, Solid State Technology, Vol. 27, p.51, 1985.
5. Y. Pauleau and Ph . Lani, J. Electrochem. Soc., Vol. 132 (11), p. 27, 79.
6. M. Boudart and G. Djéga - Mariadassou, "Kinetics of Heterogeneous Catalytic Reactions", Princeton University Press, Princeton, N.J., 1984.
7. D.R. Bradburry and T.I. Kamins, J. Electrochem. Soc., Vol. 133 (6) 1214, 1986.

Blanket CVD Tungsten: Metal Gate Applications and Adhesion Studies

HIGHLY RELIABLE TUNGSTEN GATE TECHNOLOGY

N. KOBAYASHI, S. IWATA, N. YAMAMOTO AND N. HARA
Central Research Laboratory, Hitachi Ltd., Kokubunji, Tokyo, Japan

ABSTRACT

In this paper, the reliability of W gate process will be discussed from three aspects; thermal stability of gate/SiO_2/Si structure, oxidation resistance of W and high-purity film formation. In terms of thermal stability of refractory metals /SiO_2/Si structure, it is shown that Si/SiO_2 reaction can be a key problem, rather than metal(W or Mo)/SiO_2 reaction in the case of (W or Mo)/ SiO_2/Si structure. A new oxidation method, wet hydrogen oxidation, can overcome the oxidation problem of W gate process, in which Si can be oxidized without oxidizing W. W gate technology using CVD method has been developed to reduce mobile ion contamination and is shown to be very promising for use in VLSIs.

INTRODUCTION

Recently, refractory metals such as W and Mo have been extensively studied for VLSI applications as gate electrodes (1,2), alternative interconnections to Al (3) and barrier metals for contacts (4,5). Refractory metal gate technology is one of the most promising approach in submicron VLSIs to obtain a lower gate resistance (< 0.5 Ω/□) than those of silicides. Historically, the metal gate technology has been studied since 1970's (6, 7, 8), however, it has not been widely used for manufacturing as compared with poly Si or polycide technology. It is mainly due to poor reliability of the metal gate technology in processings as well as in electrical characteristics.

We have selected W as a most promising gate metal among various refractory metals, and have developed a highly reliable W gate technology for MOS VLSI applications (9,10,11). In the W gate technology, new techniques have been developed to improve the process reliability and process compatibility with conventional Si gate technology. In this paper, the reliability of W gate technology, including these techniques, are discussed.

One important problem is the thermal reaction of W with gate SiO_2. It becomes the more important as the gate oxides becomes the thinner by scaling device dimensions. Because, in order to guarantee an excellent MOS characteristics, the gate structure is required to have a metallurgical stability at up to 900 ~1000 °C even for thin gate oxides. In our study, the typical gate oxide was 20nm thick. The thermal stability of the W/SiO_2/Si structure is discussed as compared with other refractory metals (Mo, Ti, and Ta).

The oxidation of W is a serious problem because W can be very easily oxidized even at a low temperature (~300 °C). A new oxidation method, using H_2 with an appropriate amount of H_2O, has been developed to enable Si oxidation without oxidizing W (10, 11). We call this new oxidation method <u>W</u>et <u>H</u>ydrogen <u>O</u>xidation, or in short "WHO". This oxidation method can overcome a most serious restriction of the metal gate technology that high-temperature oxidation process must be avoided after gate metallization or that some passivation film over the gate metal is indispensable to prevent oxidation of the gate metal.

Third, high-purity W film formation has been pursued. In the metal gate, mobile ion contamination by impurities such as Na and K was mainly caused during film formation, making device characteristics unstable. We have developed phosphosilicate glass (PSG)/W gate structuring (9, 11) and a sputtering method using a high-purity W target (12) to reduce the mobile ion contamination. Here, the other approach, non-selective CVD-W, is discussed in terms of high-purity and conformal film formation (13).

THERMAL STABILITY OF W/SiO_2/Si STRUCTURE

1. Experimental results

The thermal reaction of refractory metal (W, Mo,Ti or Ta)with SiO_2 was compared and analyzed by XD, AES, ESCA, and SEM as well as electrical characteristics (sheet resisitance, breakdown voltage of the gate oxide) . These refractory metals were selected because they are familiar as the materials for IC fabrication. The conditions of sample preparation is summarized in Table 1. The refractory metals (~350 nm thick) were deposited by sputtering or electron beam evaporation method onto SiO_2 substrates.The 20 nm thick gate oxide was formed by dry O_2 oxidation at 1000 °C for 20 min on Si (100) p-type substrate with resistivity of 8 ~ 13 Ω.cm. Either dry or wet etching of the metals was done to make the metal gate MOS capacitors. The gate area was 200 x 200 μm^2.

The sheet resistance change after annealing is shown in Fig. 1. The sheet resistance was measured by the four-point probe method. As shown in the figure, the large reduction of sheet resistance was observed in the Ti and Ta samples after annealing in H_2 at 1000 °C for 1 hr. According to XD analysis, $TiSi_2$ or Ta_5Si_3+Ta_2O_5 was detected as reaction products in those samples , respectively. As for the W and Mo, there was not a large change in the sheet resistance except a slight increase of the Mo sample after annealing at 1000 °C, and no reaction product was detected by XD. The change of gate breakdown voltage after annealing is shown in Fig. 2. The gate breakdown voltage was apparently reduced after annealing of the Mo, Ti, and Ta samples, whereas, no large degradation was observed in the W samples.

Table 1. Sample conditions.

	deposition method	thickness (nm)	resistivity (μΩ·cm)	etching method
W	DC magnetron sp.	350	11	dry (SF_6)
Mo	⇑	350	12	wet*
Ti	electron beam ev.	370	62	dry (CF_4+O_2)
Ta	RF sp.	330	191	⇑

* H_3PO_4+HNO_3 +CH_3COOH+H_2O

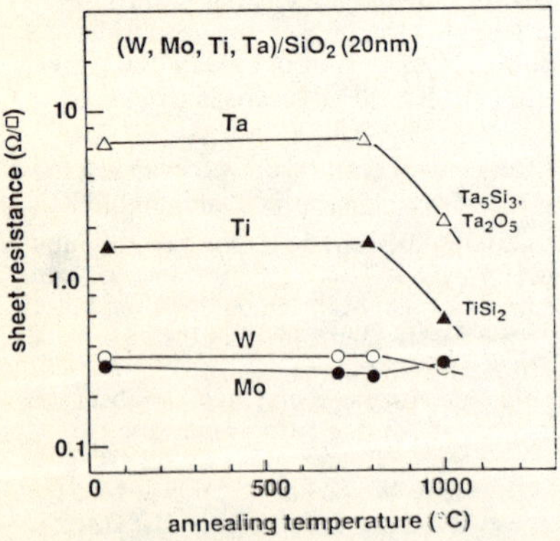

Figure 1. Sheet resistance change after annealing (1 hr in H_2).

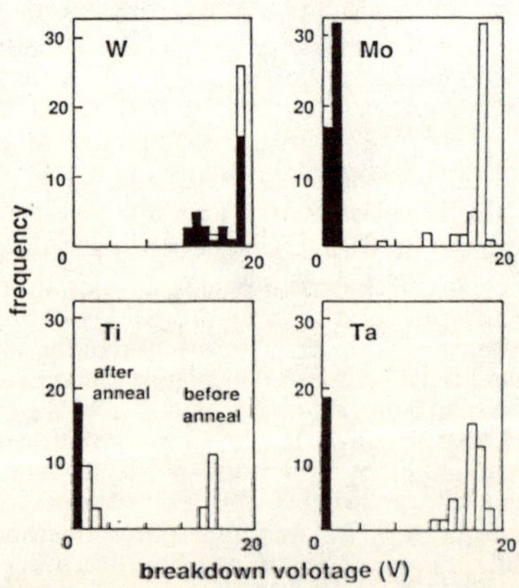

Figure 2. Change of gate breakdown voltage after annealing at 1000 °C for 1 hr in H_2 (SiO_2: 20 nm, I_c= 10^{-8} A).

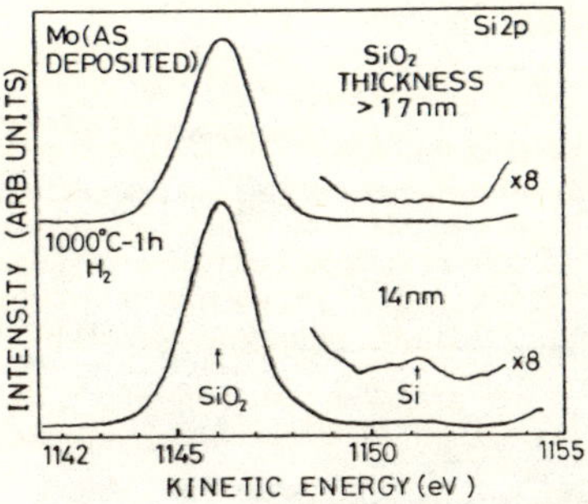

Figure 3. Change of Si_{2p} spectrum of SiO_2 in $Mo/SiO_2/Si$ structure after annealing.

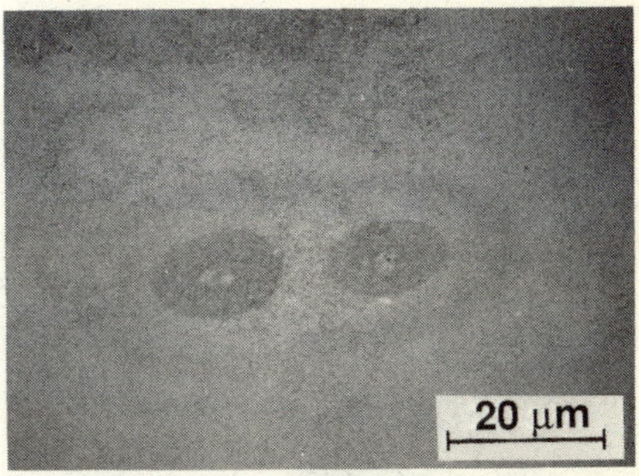

Figure 4. SEM micrograph of reaction area of the Mo sample.

The gate breakdown voltage lowering in the Ti and Ta samples can be understood considering the fact that they reacted with SiO_2, thus forming silicides as shown by XD results. However, the problem is why the degradation was observed in the Mo samples. Figure 3 shows the change of Si_{2p} spectrum of SiO_2 as measured by ESCA (11). The spectrum was measured after chemically removing the Mo. The figure shows an increased peak intensity for unoxidized Si after annealing, which indicates that the SiO_2 was partially reduced. No such changes were observed with the W sample. Figure 4 shows the SEM micrograph of reaction area of the Mo sample. The sample was annealed in vacuum (~ 10^{-5} Pa) at 1200 °C for 40 min. According to the micro-AES measurement, Si signal was detected only at the reaction area. These results indicate that the reaction with SiO_2 was occured in the Mo sample after anneling at more than 1000 °C, but it was locally and scarcely occured as compared with the Ti and Ta samples. And, the W sample was more stable than the Mo sample in the reaction with SiO_2.

2. Mechanism of the reaction

For the Ti and Ta samples, the thermal reaction of Ti or Ta with SiO_2,

$$3Ti + 2SiO_2 = TiSi_2 + 2TiO_2, \quad \text{or} \quad 37Ta + 15\,SiO_2 = 5Ta_5Si_3 + 6Ta_2O_5,$$

can be expected, because the free energy of the reaction is negative (14). When the 20nm thick SiO_2 was reduced by these reactions, the Ti or Ta is considered to react with the Si substrate and form silicides. On the contrary, for the W and Mo samples, W or Mo does not react with SiO_2 because the free energy of the reaction is positive. However, it is very important to notice that the reaction between Si and SiO_2 is apparent at these high temperatures (> 1000 °C)(15,16). Figure 5 shows SEM micrographs of poly $Si/SiO_2/Si$ structure after annealing in vacuum at 1200 °C. The micrographs indicate the interfacial SiO_2 film disappears after annealing. The reaction of poly Si with SiO_2 can be expressed as, $Si + SiO_2 = 2SiO$. Also, the reduction of SiO_2 by H_2, $H_2 + SiO_2 = SiO + H_2O$, will be thermodynamically expected when the sample is annealed in H_2. Thus, as for the thermal stability of (W or Mo)$/SiO_2/Si$ structure, it is concluded that the SiO_2/Si interface reaction is dominant in comparison with (W or Mo)$/SiO_2$ reaction.

The similar implication as well as other important information was obtained by the ESCA measurements of those samples (11, 17). The metal oxides (WO_x, Mo_x) were observed at the metal/SiO_2 interface in as-deposited sample. And, MoO_x could be reduced to Mo when annealed in H_2 at 1000 °C, however, the WO_x could not be reduced after the annealing. These interfacial metal oxides are considered to provide an oxidizing atmosphere for SiO_2, making the SiO_2 more stable by their presence. Therefore, the $W/SiO_2/Si$ structure is more stable because the interfacial metal oxide (WO_x) could not be reduced as much as the MoO_x in the $Mo/SiO_2/Si$ structure.

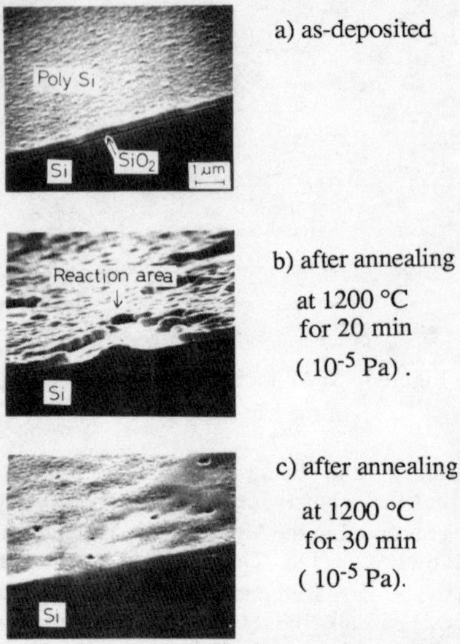

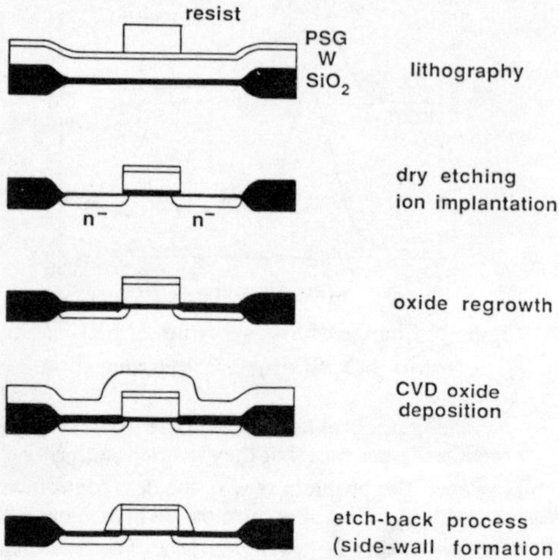

Figure 5. SEM micrographs of poly Si/SiO_2/Si structure before and after annealing in vacuum.

Figure 6. Process flow of W gate technology.

WET HYDROGEN OXIDATION (WHO)

1. The oxidation problem

First, it is explained why the metal oxidation is so serious in VLSI processings. Figure 6 shows a typical flow of W gate process. After conventional lithography, a W gate is formed by dry etching. Source and drain are formed by ion implantation using a W gate as a mask. PSG film is coated on a W gate to prevent an ion channelling as well as to reduce mobile ion contamination (9, 11). And, oxide side-walls are formed by the etch-back process if the LDD (Lightly Doped Drain) structure is used.

The problem is how to regrow the gate oxides around a W gate. The thinned oxides around a W gate has to be regrown thicker to prevent the leakage failure at gate edges. And, the oxide regrowth can prevent an exposure of Si surface during side-wall formation. This oxidation process can be applied to poly Si and polycide gate, however, it can not be applied to refractory metals because of the oxidation.

2. Wet hydrogen oxidation

A new oxidation method, wet hydrogen oxidation, has been developed to overcome the problem (10, 11). Using hydrogen with an appropriate amount of water, Si can be oxidized without oxidizing W. The mechanism is closely related with the stability of the oxides, as previously discussed in the reaction of refractory metal with SiO_2. Figure 7 represents the temperature dependence of the equilibrium partial pressure ratio of H_2/H_2O for the oxidation of W, Mo, Ti or Ta and Si. These values were calculated using the Gibbs free energy for these oxidation reactions (14). For Si, oxidation will proceed in the upper side of the equilibrium curve and SiO_2 reduction will proceed in the under side of the curve. The situation is the similar to the oxidation of refractory metals. Therfore, selective oxidation can been achieved in the region surrounded by two equilibrium curves (for example, between W, Mo and Ti, Si , Ta). Figure 8 shows the cross-section TEM micrograph of W gate MOSFET after WHO. You can easily see that the oxides around a W gate grow thicker after wet hydrogen oxidation (11, 16).

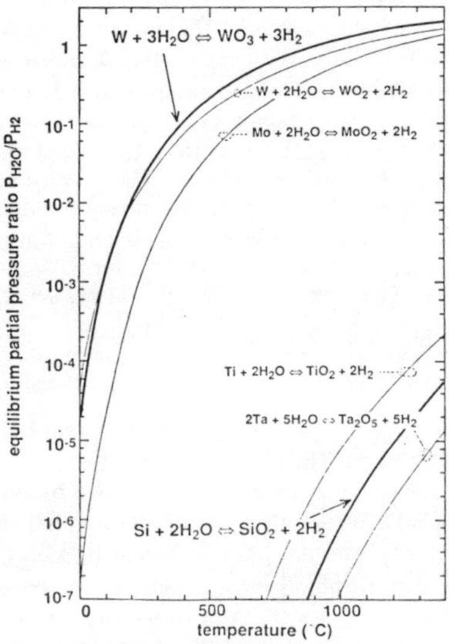

Figure 7. Wet hydrogen oxidation (WHO).

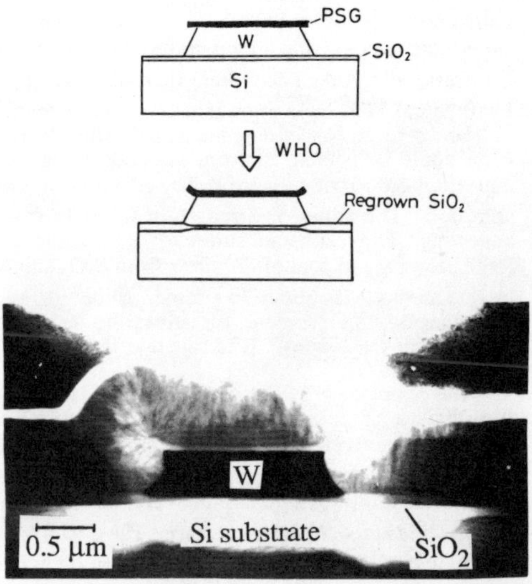

Figure 8. Cross-section TEM micrograph of W gate MOSFET after WHO.

3. Electrical characteristics

Si oxidation characteristics (10) and electrical properties of SiO_2 (16) grown by WHO were investigated. Si oxidation rate was determined by three parameters, that is , the oxidation temperature, time and the partial pressure ratio of H_2O/H_2. The oxidation characteristics can be almost undersood as "diluted" wet oxidation. The oxides (15 ~ 40 nm thick) grown by WHO had slightly higher dielectric strength with less deviation in their values and almost same value of interface charge trap density as compared with conventional dry O_2 oxidation.

W gate technology using WHO was applied to 1.3 µm n-MOS technology as the method for the oxide regrowth around W gates, which have been developed in our laboratory (18). Using WHO, the gate breakdown voltage was improved with no unfavorable effects on other device characteristics (threshold voltage, etc.) (16).

NON-SELECTIVE W CVD TECHNOLOGY

1. Motivation

For applications of gate electrodes and interconnections, W has been usually deposited by the sputtering method mainly because it is a suitable method for the material with a high melting point and the resulting W films show good oxide adhesion. However, the sputtering method can lead to damage and contamination of the thin gate oxides (< 20 nm thick) as well as having poor step coverage over abrupt steps. Thus, non-selective CVD-W onto the thin gate oxides has been pursued to overcome these problems.

Non-selective CVD-W onto the thick oxide (> 100 nm) have been studied , in which other interfacial films, such as WSi_x and Si , between W and SiO_2 are used for adhesion improvement (19, 20). However, non-selective CVD-W technology, which can be used for gate formation on thin gate oxides (< 20 nm thick), has not yet been developed.

2. Experimental results

CVD-W is usually deposited using WF_6 and H_2 at 300 ~ 350 °C. It is well known that selective deposition is achieved by this method and CVD-W films can be scarcely formed on oxides, and that they can have a poor oxide adhesion even if the deposition occurs.

In our experiment, a small amount of SiH_4 was added to the forming gases in order to obtain non-selectivity. CVD-W was deposited in a tube-type hot wall LPCVD reactor. Most films were deposited at a temperature of 350 °C with a WF_6 flow of 20 cm^3/min and WF_6:H_2 flow ratio of 1:100. The typical flow of SiH_4 was 4 cm^3/min. The total pressure was $2.3x10^3$ Pa (0.7 torr). The deposition rate under these conditions was about 7 nm/min.

Under these conditions, the as-deposited W films had poor adhesion to the oxides, and the films could be peeled off from the underlying oxides with an adhesive tape. Then, W films deposited onto oxides were followed up with annealing in a N_2 or WHO to improve oxide adhesion. The adhesion strength of CVD-W films to underlying oxides was evaluated afer annealing. The results are summarized in Table 2. The adhesion strength was measured by the force necessary to peel off W films from SiO_2 substrates. The maximum strength was measured by this method was about 7 kg/mm^2. Although an annealing atmosphere had a slight effect for the adhesion improvemenr, the annealing temperature was larger than 700 °C to ensure good adhesion to the oxides. It is felt that the reduction in the W film stress results in adhesion improvement.

The plain-view TEM micrographs of as-deposited films are compared in Fig. 9. CVD-W films have a larger (~100 nm) grain size than sputtered W films. The CVD-W film has excellent step coverage over SiO_2 steps as compared with the sputtered W film. This is shown by the SEM micrographs in Fig. 10. The AES analysis of the CVD-W film after annealing for adhesion improvement (800 °C, 30 min in WHO) is ahown in Fig. 11. An interfacial film such as WSi_x is not detected at the W/SiO_2 interface. The CVD-W film contains no other high-resistivity material except a small amount of oxidized silicon, which will not increase the film resisrivity. These results indicate that SiH_4 added to the system was not consumed in the WSi_x formation after reacting with WF_6, but in such reaction, $2WF_6 + 3SiH_4 = 2W + 6H_2 + 3SiF_4$, in which Si can not be involved in the W film because of the high vapor pressure of SiF_4.

3. Electrical characteristics

The resistivity of as-deposited CVD-W films (60 ~ 200 nm thick) is approximately 10 $\mu\Omega \cdot cm$ as shown and can be reduced to 8 $\mu\Omega \cdot cm$ by annealing at up to 1000 °C. This is an even lower resistivity value than that of W films deposited with planar DC magnetron sputtering.

The electrical characteristics of CVD-W gate MOSCs (Capacitors) were compared with those of sputtered W gates. The gate SiO_2 was nominally 20 nm thick. The results are summarized in Table 3. The breakdown voltage of the gate oxide is near 22 V, which is as high as that of sputtered W gates. In addition, the CVD-W gate MOSCs have slightly lower interface trap charge densities. C-V measurements of MOSCs show no difference between CVD-W and sputtered W gates. Flatband voltage of the CVD-W gates is almost the same as that of the sputtered W gates. It indicates that the CVD-W films have the same work function as the sputtered films (4.7 eV), thus implying the absence of other species between CVD-W film and gate SiO_2.

Table 2. Change of W film adhesion strength after annealing.

TEMPERATURE* / ATMOSPHERE	500°C	600°C	700°C	800°C	900°C
N_2	poor	poor	good	good	good
H_2+H_2O	poor	poor	fair	good	good

*annealing time : 30 min

poor: easily peeled off by adhesive tape

fair : partially peeled off by adhesive tape

good : adhesion strength > 7 kg/mm^2

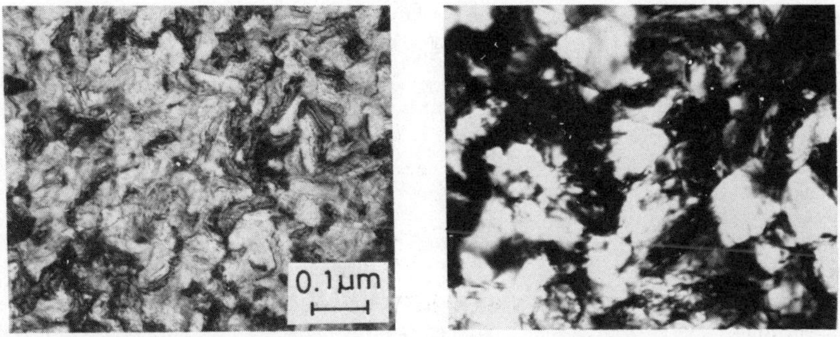

Figure 9. Plain-view TEM micrographs of W films (as-deposited).

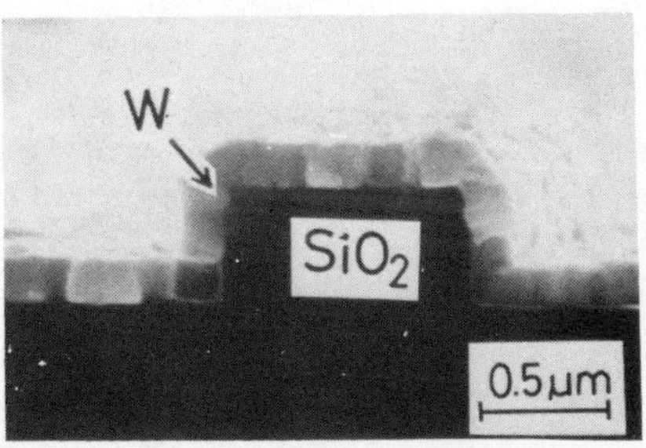

Figure 10. SEM micrograph of CVD-W film over SiO_2 steps.

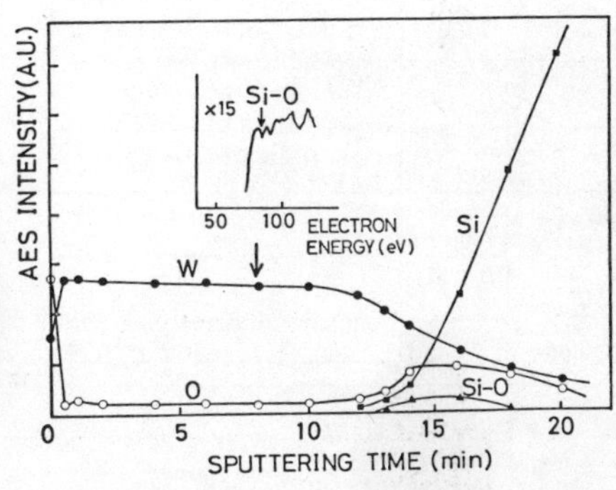

Figure 11. AES analysis of CVD-W film (after annealing in WHO at 800 °C for 30 min.)

The flatband voltage shift is below 0.01 V under a 2 MV/cm stress condition for 30 min at 200 °C, as shown in Fig. 12. This shows that the mobile ion density in CVD-W films is less than 0.01 ppm. This value is as low as that of W films using a high-purity sputtering target (12). It is important to notice that the low content of mobile ion density was obtained in the CVD method using a conventional WF_6 bottle (purity ~ 2-N), however, the low content was not obtained in the sputtering method until a high-purity target (3-N ~ 4-N) has been developed(11,12).

To practical use, the sputtering method is still advantageous over the CVD method at present because of easiness and reliability of the process. However, it is convinced that the CVD method is very promising for damageless and high-purity film formation in the future gate metallization.

CONCLUSION

The reliability of the W gate technology was discussed from three aspects; thermal stability, oxidation resistance and high-purity film formation. Among various refractory metals (W, Mo, Ti and Ta), the W/SiO_2/Si structure is shown to be thermally the most stable. The reaction mechanism is proposed considering that the reaction of SiO_2/Si interfacial reaction is more predominant than the (W or Mo)/SiO_2 interfacial reaction in the metal/SiO_2/Si structure. A new oxidation method, Wet Hydrogen Oxidation (WHO), has been developed to overcome the metal oxidation problem. WHO improved the gate breakdown volotage without unfavorable effects on other device characteristics. Non-selective W CVD method has been applied to a gate process and it can form a conformal and high-purity W film with a lower resistivity (8 μΩ.cm) than the sputtering method. Electrical characteristics of CVD-W gate MOS capacitors are as good as those of the sputtered W gate.

Table 3. Comparison of W film characteristics.

		CVD-W	Sput. -W (High purity)
STEP COVERAGE (SiO_2 : 500nm)		1.0	0.5 ~ 0.6
ADHESION [Kg/mm²]		>7	>7
RESISTIVITY [μΩcm]		8	10
MOS CAPACITOR (GATE SiO_2 20nm)	V_B [V] (200μm□)	22.5	22.6
	V_{FB} [V]	0.1	0.1
	ΔV_{FB} [V] (BT STRESS)	< 0.01	< 0.01

V_B : BREAKDOWN VOLTAGE
V_{FB}: FLATBAND VOLTAGE
ΔV_{FB}: SHIFT OF FLATBAND VOLTAGE UNDER BIAS TEMPERATURE STRESS

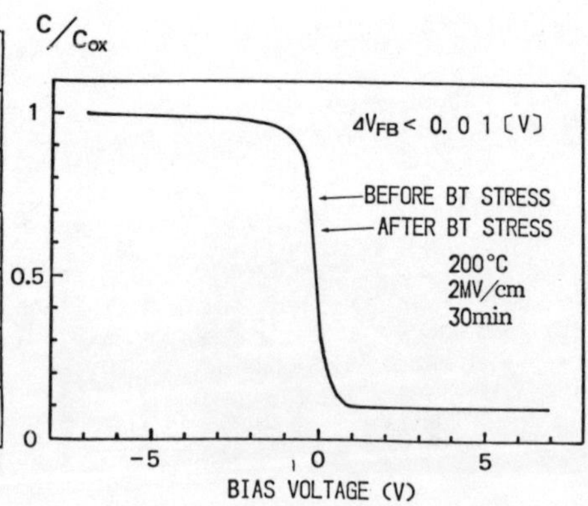

Figure 12. Flatband voltage shift before and after annealing.

ACKNOWLEDGMENT

The authors would like to express their thanks to Dr. S. Asai, Dr. K. Ito, Dr. N. Hashimoto, Dr. K. Yagi, Dr. H. Sunami, Dr. Y. Wada and Dr. T. Terada for discussions and encouragements during the course of this work, and to the processing group in VLSI Center for their contributions to this work. The authors also would like to thank Dr. K. Hinode and S. Aoki for the TEM observations and to Dr. T. Yoshimi of Musashi Works for the discussions and measurements of the adhesion strength of W films.

REFERENCE

(1) T. Amazawa and H. Oikawa, Proc. of 15th Conf. Solid State Devices and Materials, P 229, 1983

(2) R.F. Kwansnick, B.F. Griffing, E.B. Kaminsky, P.A. Frank, and M.J. Kim., Proc. of Symposium on VLSI Technology, P 11, 1986

(3) J. M. Mikkelson I. A. Hall, A. K. Maihotra, S. D. Seccombe and M. S. Wilson , Proc. of IEEE International Solid Circuits Conf., P 106, 1981

(4) K. Saraswat, VLSI Science and Technology, Proc. of Electrochem. Soc., P 409, 1984

(5) T. Moriya, S. Shima, Y. Hazuki, M. Chiba and M. Kashiwagi, Proc. of IEEE International Electron Devices Meeting, P 550, 1984

(6) D. M. Brown, W. E. Engeler, M. Garfinkel and P. V. Gray, Solid-State Electronics, vol. 11, p 1105, 1968

(7) A. K. Sinha, Japan. J. Appl. Phys., Suppl. 2, pt. 1, P 487, 1974

(8) P. L. Shah, IEEE Trans. Electron. Devices, ED-26, p 631, 1971

(9) N. Yamamoto, S. Iwata, N. Kobayashi and T.Terada, Proc. of 15th Conf. Solid State Devices and Materials, P 217, 1983

(10) N. Kobayashi, S. Iwata, N. Yamamoto and T. Terada, Proc. of Symposium on VLSI Technology , P 92, 1983

(11) S. Iwata, N. Yamamoto, N. Kobayashi, T. Terada and T. Mizutani, IEEE Trans. Electron Devices, ED-31, P 1174, 1984

(12) N. Yamamoto, H. Kume, S. Iwata, K. Yagi, N. Kobayashi, N. Mori and H. Miyazaki, J. Electrochem. Soc. vol. 133, p 401, 1986

(13) N. Kobayashi, N. Hara, S. Iwata and N. Yamamoto, Proc. of IEEE VLSI Multilevel Interconnect. Conf., P 436, 1986

(14) O. Kubaschewski and C. B. Alcock, Metallurgical Thermochemistry, 5th ed., Oxford, Pregamon Press, P 267, 1979.

(15) N. Kobayashi, S. Iwata and N. Yamamoto, Abstracts for Fall Meeting, Japan Appl. Phys. Soc., P 389, 1983

(16) N. Kobayashi, S. Iwata, N. Yamamoto, T. Mizutani and K. Yagi, Proc. of IEEE International Electron Devices Meeting, P 122, 1984

(17) S. Iwata, Proc. of JIMIS-4, Grain Boundary Structure and Related Phenomena, Supplement to Trans. Japan Institute of Metals, P 979, 1986

(18) Y. Wada, H. Sunami, N. Yamamoto, Y. Kawamoto, T. Mizutani, K. Yagi, Y. Homma, N. Hashimoto and S.Asai, Proc. of IEEE Electron Devices Meeting, P 323, 1983

(19) G. C. Smith, Proc. of IEEE VLSI Multilevel Interconnect. Conf., P 350, 1984

(20) M. A. Metz and E. A. Beam, Proc. of IEEE Multilevel Interconnect. Conf., P 357, 1984

A REFRACTORY METAL GATE APPROACH FOR MICRONIC CMOS TECHNOLOGY

Véronique LUBOWIECKI* AND Jean-Luc LEDYS*
* MATRA HARRIS SEMICONDUCTEURS, BP 942 - 44075 NANTES CEDEX, FRANCE
AND
C.PLOSSU**, B.BALLAND**
** Laboratoire de Physique de la Matière, INSA Lyon - FRANCE

INTRODUCTION

In the future, devices scaling down, integration density and performance improvements are going to bring a number of conventional circuit design and process techniques to their fundamental limits. To avoid any severe limitations in MOS ULSI (Ultra Large Scale Integration) technologies, interconnection materials and schemes are required to emerge, in order to face the Megabits memory field. Among those, the gate approach will obviously take a keyrole, when the operating speed of ULSI chips will reach the practical upper limits imposed by parasitic resistances and capacitances which stem from the circuit interconnect wiring. Even if fairly suitable for MOS process, doped polycrystalline silicon is being gradually replaced by refractory metal silicide [1] or polycide [2] structures, which match better with low resistivity requirements.
However, as we approach the submicronic IC's, higher conductivity materials will be paid more and more attention. Recently, works have been devoted and published on refractory metal gate technologies [3,4]. Molybdenum or tungsten, deposited either by CVD or PVD methods, are currently reported even if some drawbacks in their process integration still remain.
This paper is willing to present such an approach based on tungsten (more reliable than Molybdenum [5]) deposited by LPCVD (giving more conductive and more stable films than PVD). Deposition process will be first described. Then CMOS process flow will allow us to focus on specific refractory metal gate issues. Finally, electrical and physical properties will be assessed, which will demonstrate the feasibility of such a technology as well as the compatibility of the tungsten with most of the usual techniques.

W CVD DEPOSITION PROCESS

Attempts to use pure CVD tungsten as gate material have been complicated by poor adhesion of tungsten to silicon dioxide. A "glue material" is then required, compatible either with the underlying dielectric as well as with the upper W film.

* Adhesion layers

Polycrystalline silicon and sputtered tungsten have been studied to meet this requirement.

i. Polysilicon

1000 Å thick polysilicon was deposited by pyrolysis of SiH_4 at 620°C in an LPCVD reactor, and doped with $PoCl_3$ leading to a sheet resistance of 100Ω/sq.

ii. Sputtered Tungsten

Using a 4 N purity target, fairly adherent thin W layers have been sputtered in a SGUN 3125 Varian equipment. Powered at 1200 watts, films were deposited at 80 Å / mn. up to an acceptable thickness of 500 Å.

* CVD Tungsten Deposition

Tungsten films were deposited in a LPCVD system using the well-known reduction reactions :

1 $2WF_6 + 3Si \longrightarrow 2W + 3SiF_4$

2 $WF_6 + 3H_2 \longrightarrow W + 6HF$

which have been extensively studied in the literature [6,7].

Depositions were carried out in a commercial hot-wall horizontal LPCVD reactor, described in figure 1, for temperatures ranging from 250 to 325°C and pressures between 0.1 to 0.4 torr, whereas the H_2 to WF_6 flow ratio was kept between 20 to 25.

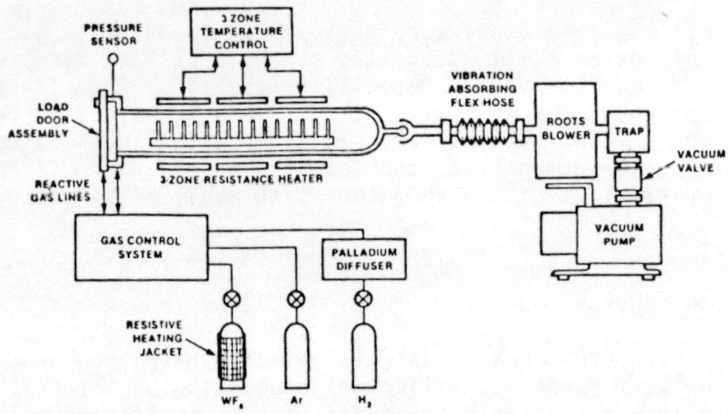

Figure 1: Hot-wall Low Pressure Tungsten System.

Using the best set of process conditions, deposition rate of 40 Å/mn. was achieved, whilst the resistivity was found between 8 to 10 μΩ.cm. For a 25 wafer load, wafer-to-wafer thickness uniformity was less than 10% and on-wafer dispersion ranged toward + 5%.
Typically, the thickness target was 3500 Å, leading to a gate sheet resistance of 0.3 to 0.4Ω/sq.
Adhesion tests were performed, thru the so-called "scotch tape test", on monitor wafers for each run and for each adhesion material.

Films were considered as adherent to "glue layer" (and further, the glue layer to the substrate), if there was no flaking during the test ; indeed very good adhesion was obtained when tungsten was deposited on top of either LPCVD phosphorus doped polysilicon, or sputtered tungsten.

CMOS TUNGSTEN GATE PROCESS

Tungsten, even though generating a large interest, for use as gate material, still exhibits several shortcomings :

- its unability to withstand any high temperature treatments in oxidizing ambient, due to volatile oxide formation.
- its poor ion implant blocking efficiency in case of self-aligned device fabrication.
- its poor chemical stability, especially in HF containing solutions.

To cope with such limitations, several process flow modifications have to be undertaken, which are outlined in table 1.

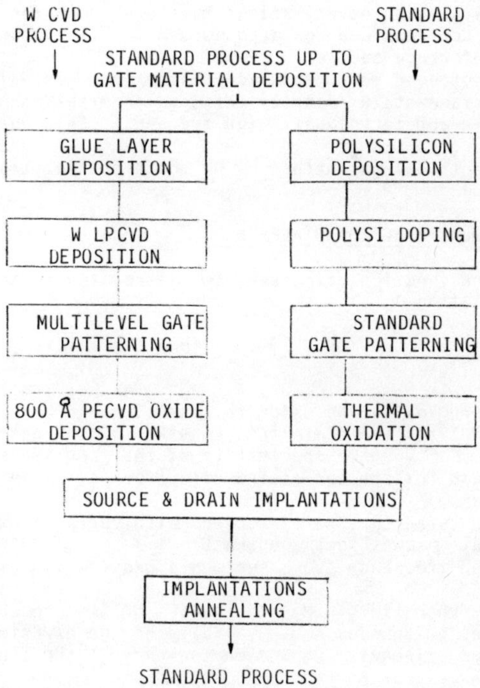

TABLE 1 : PROCESS FLOW COMPARISON

P-type single-crystal silicon wafers were processed through conventional front end steps up to gate oxidation. Then, adhesion layer and W CVD were sequentially deposited.
Gate patterning was achieved by standard photolithographic techniques. These multilevel structures were then etched in a reactive ion etcher using SF_6 chemistry, with a good control of the slope and line width.
Largely textured, W CVD layers are very propitious to ion channeling during source and drain implantations. Since thermal oxidation could not be carried out, like for polysilicon gates, an amorphous scattering layer must be capping the patterned tungsten gates : in our experiments, an 800 Å thick undoped SiO_2 layer was PECVD deposited.
After source and drain implantations, a phosphosilicate glass (PSG) layer, able to supply any gettering effect similar to the ones associated with the phosphorus doped polysilicon, was CVD deposited.
Next the activation of implanted ions as well as the PSG reflow occured thru a high temperature treatment which has to be tuned to prevent any detrimental effects in case of Poly Si/W structures. Contacts are finally patterned before metallization.

ELECTRICAL CHARACTERIZATION RESULTS AND DISCUSSION

The W CVD gate technology has been applied to a 1.5 µm CMOS process. Electrical measurements have been carried out on a specific test vehicle including capacitors and variable channel length N- and P-channel transistors (drawn length between 1 and 25 µm).
All the characterizations have been done for the two above mentioned gate structures, and compared to usual polysilicon gate results, referred to as "reference samples".
Electronic work function difference ϕms, fixed oxide charge Qox, fast surface state density Nit, oxide breakdown voltages were assessed by the relevant techniques (high frequency C(v), quasistatic measurements, I(v), charge pumping).
Electrical parameters of N- and P-channel transistors are presented.

* WORK FUNCTION DIFFERENCE

Work function difference ϕms is related to the flatband voltage Vfb by the relation :

$$\phi ms = Vfb + Qox \,.\, Tox \;/\; \varepsilon o \,.\, Kox$$

where Tox is the oxide thickness.
Kox the dielectric constant of the gate oxide
ε o the permeability of the free space
and Qox the sum of the effective net oxide charge per unit area.

MOS capacitors were prepared with variable oxide thickness in such a way that Qox was kept constant.
High frequency CV measurements provide Vfb as a function of Tox.

As shown in fig.2, the values of ϕms are extracted at Y-axis intercept, i.e. for Tox = 0. Validity of the experiment is afforded by the fairly good linearity shown on the curve, the slope of which gives the oxide charge Q ox.

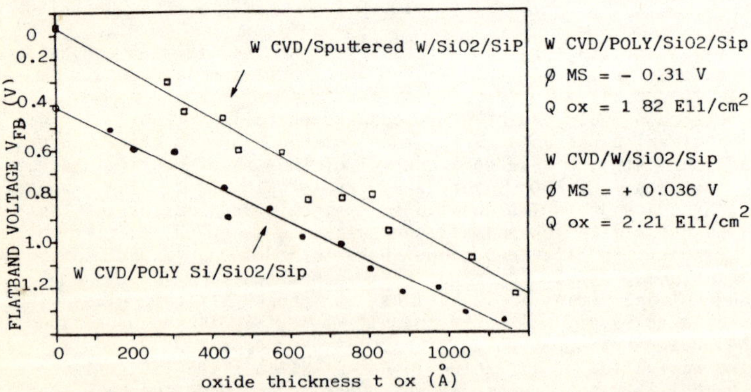

FIGURE 2 : Ø MS DETERMINATION

* SURFACE STATE DENSITY AT THE Si/SiO2 INTERFACE

It is well-known that tungsten can generate electronic traps in the dielectric and induce a modification of interface state density. Distribution of surface state densities Nss (E), in the bandgap of silicon has been determined by comparison of high frequency and quasistatic C-V characteristics.
Slow surface states determined for the various W CVD gate structures are compared to reference results in figure 3.
Even if the surface state distributions are slightly different, the average values are all together in good agreement. Nevertheless, further investigations are now required for a detailed understanding.

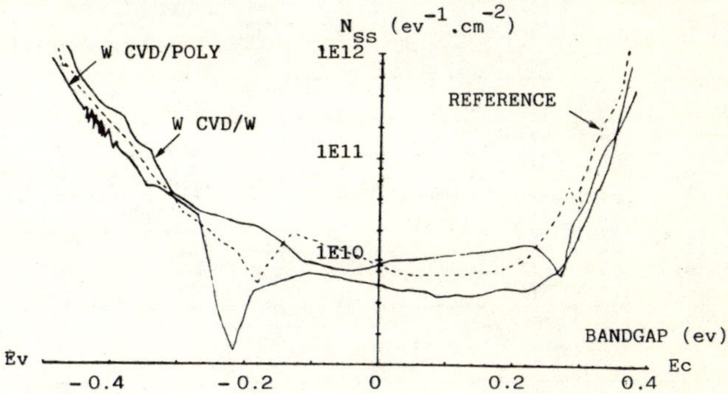

FIGURE 3 : DISTRIBUTION OF INTERFACE STATE DENSITIES FOR THE DIFFERENT GATE STRUCTURES

* OXIDE BREAKDOWN

In order to assess any interaction of gate material with the gate dielectric, breakdown voltages of 300 Å thick SiO_2 have been measured on fully processed wafers. Results presented in figure 4 show that no detrimental effect appears in the oxide, obviously related to the W gate technology.

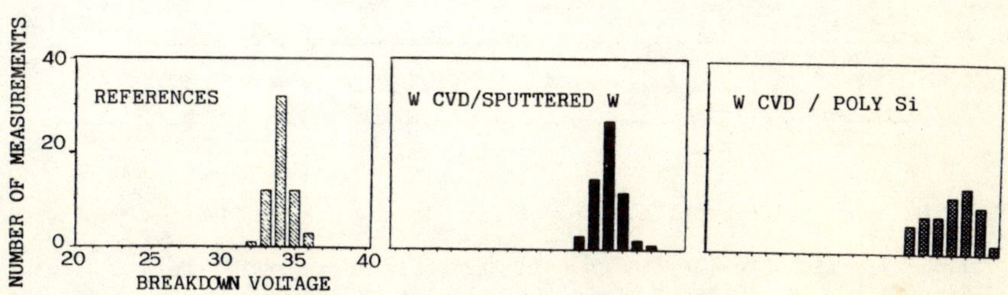

FIGURE 4 : BREADOWN VOLTAGE HISTOGRAMS

* TRANSISTOR BEHAVIOUR ASSESSMENT

Main electrical parameters of N- and P-channel transistors are presented in the table, thereafter.

GATE MATERIAL / PARAMETERS	A REFERENCES	B POLY Si + W CVD	C SPUTTERED W + W CVD
Leff N 1.5 µm	1.28 µm	1.28 µm	1.28 µm
Leff N 1.0 µm	0.80 µm	0.80 µm	0.80 µm
VT N 1.5 µm	+ 0.702 V	- 0.056 V	- 0.302 V
VT N 1.0 µm	+ 0.692 V	- 0.11 V	- 0.280 V
BV N 2.5 µm	14.45 V	15.8 V	17.5 V
BV N 1.5 µm	7.4 V	8.8 V	11.2 V
BV N 1.0 µm	6.4 V	7.7 V	8.4 V
VT P 1.5 µm	- 0.802 V	- 0.350 V	- 0.212 V
VT P 1.0 µm	- 0.788 V	- 0.282 V	- 0.175 V
BV P 2.5 µm	15.7 V	16.4 V	19.2 V
BV P 1.5 µm	8.1 V	9.8 V	12.3 V
BV P 1.0 µm	7.4 V	9.2 V	11.1 V

Despite some out-of-control parameters, fairly working devices have been obtained.
Obviously related to the Φms change, threshold voltages of the W CVD transistors will have to be adjusted by a relevant channel counterdoping ion implantation.
On top of that, some devices of groupe B have been affected by a silicidation phenomenon, occuring during the high temperature reflow cycle, which of course has to be avoided to prevent any physico-chemical damages.

* CHARGE PUMPING MEASUREMENTS

C-V method is classicaly used to characterize the Si/SiO2 interface, however, a more accurate technique, called "Charge Pumping", is now emerging, which is able to assess the transistor channel interface behaviour itself. This method, specifically designed for short channel MOS transistors, has been used on our samples.
As shown in figure 5, when applying voltage pulses on gate electrode, the silicon surface is cycled between accumulation and inversion. Then a pulsed current flows from source and drain toward substrate. This current can be directly related to the channel surface state density [8].

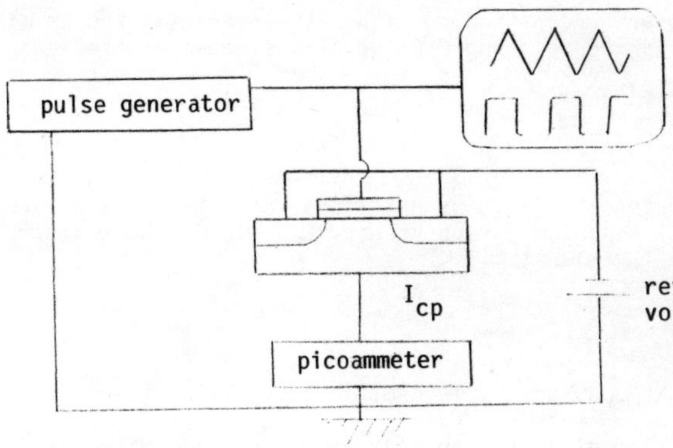

CHARGE PUMPING EXPERIMENTAL SET-UP

FIG.5

Table figures out the average surface state densities and capture cross sections of traps on our different samples.

GATE MATERIAL	Nss ($ev^{-1}.cm^{-2}$)	σ (cm^2)
W CVD / POLY Si	$8.9 . 10^{9}$	$6.2 . 10^{-16}$
W CVD / SPUTTERED	$4.9 . 10^{9}$	$5.5 . 10^{-16}$
REFERENCES	$5.7 . 10^{9}$	$7.2 . 10^{-16}$

CONCLUSIONS

The above-presented work has to be considered as a first approach to use W CVD gates. The electrical characterizations as well as some process parameters indicate that W CVD on top of sputtered W could very well compete with the phosphorus doped polysilicon. However, several process steps are still questionable and will require further detailed experiments.

In order to reach a mature enough feasibility level, W CVD process and equipments have to quickly progress : a single reliable and production-type deposition equipment will be the keypoint to reach the submicronic submicronic MOS technology field.

As last, a better understanding of W CVD interactions with the device behaviour, will drive this approach up to the necessary reliable and manufacturable state.

ACKNOWLEDGMENTS

This work, carried on, only by MHS Technology Development Group in its first step, will be jointly supported by E.E.C., as part of the so-called ESPRIT contract "SPECTRE".

REFERENCES

[1] S.P. MURARKA, J.Vac.Sc. Technol., 17 (4), 775, (1980)

[2] M.Y. Tsai and al, J.Electrochem. Soc., vol.128 N°10, 2207, (1981)

[3] N.YAMAMOTO, H.KUMI, S.IWATA, K.YAGI and N.KOBAYASHI, J.Electrochem. Soc., 133 (2), 401, (1986)

[4] T.P. CHAU and A.J. STECKL, 513, IEDM 83

[5] P.L. SHAH IEE Trans. on Electron. Devices, vol. ED-26, N°4, 631, (1979)

[6] K.Y. TSAO and H.H. BUSTA, J.Electrochem Soc., vol. 131, N°11, 2702 (1984)

[7] E.K. BROADBENT, ECS Meeting Washington, Oct.83

[8] G.GROESENEKEN, H.E.MAES, N.BELTRAN, R.F.de KEERSMAECKER, IEEE Trans. on Electron. Devices, vol. ED-31, N°1, (1984)

ADHESION OF BLANKET TUNGSTEN TO OXIDE

K. C. Ray Chiu* and N. E. Zetterquist**
* Process Industries, Inc., 720 Tabor Drive, Scotts Valley, CA 95066.
** GENUS, Inc., 515 Ellis Street, Mountain View, CA 94043.

ABSTRACT

The adhesion of CVD tungsten to Silicon Dioxide has been a difficult and poorly understood problem. [1] The present study looks at the adhesion issues as they relate to equipment, processing and the condition of the substrate surface layers, (i.e., type and cleanliness.)

The goal of this work is to move to a quantitative description of the components of adhesion (i.e., tungsten film adhesion strength, stress, strain, tensile strength, etc.).

Bulk and interface properties as they relate to the adhesion will be discussed. Types of adhesion failures of tungsten layers on oxide surfaces are described. Finally, we will review the present methods of testing for adhesion and their limitations.

INTRODUCTION

The CVD blanket tungsten film can be used both as an interconnect and plug "filling" of contact or vias. It is one of the most promising metallic materials for VLSI and ULSI applications. In the past, blanket tungsten use meant "glue" layers of "other" metallic materials and irreproducible adhesion results to SiO_2. Until recently the knowledge of the parameters of adhesion was very limited.

In this paper we describe the studies and results of tungsten adhesion to SiO_2 using a blanket tungsten process. The process developed at GENUS uses a WSi_x layer as the adhesion layer between tungsten and SiO_2. It is our intention to continue work toward better understanding of the properties that effect adhesion to obtain better quality blanket tungsten films.

BACKGROUND

The adhesion of blanket CVD tungsten depends on interface properties between layers of W, WSi_x and SiO_2 films and/or the Si substrates, as well as bulk properties of these layers. These properties are adhesion strength between layers, as well as tensile strength, stress and strain within these layers. The material properties are the product of the deposition processes which depends on such things as:

* The substrate surface properties

* The metallurgy of the deposited films

* The process recipe

THE PROBLEM: CRACK AND PEEL

Thick blanket tungsten films fail to adhere to SiO_2 without a "glue" layer at low temperatures (<$500^{\circ}C$). The reasons for this failure mode is tied to the bulk and interface properties of the films (W and SiO_2). Fig. 1 shows an example of crack and peel. Fig. 2 shows a drawing of the types of failures that can be seen within the tungsten layer and between the tungsten and SiO_2 layers.

EXPERIMENT

The work described herein was done on a GENUS 8402 system. Methods of quantitatively checking the performance characteristics of the deposition system to ensure predictable adhesion results for a tungsten film of 1.0 micron thickness have been derived but will not be covered in this paper.

The standard blanket tungsten process uses six sub-processes:

* Flash tungsten - To deposit selective tungsten within the contacts where Silicon is present. The WF_6 also acts as a mild surface etch to the SiO_2 in the presence of water vapor, thus cleaning native oxide for good ohmic contact.

* Ar sputtering - Used to prepare the SiO_2 surface. The tungsten within the contacts is also sputtered somewhat, but protects the bare Silicon surface from sputter damage.

* WSi_x deposition - Used as "glue" layer. It has bonding and mechanical properties which favor its use.

* NF_3 etch plus H_2 Plasma - NF_3 is used to condition the WSi_x surface and the H_2 plasma is used with the idea of reducing the residual Flourine. These steps ensure the adhesion of the tungsten layer to the WSi_x surface.

* W deposition

* WSi_x coating (cap) - To prevent oxidation of the tungsten layer when the wafers are being pulled from the deposition system.

Samples were prepared, and taken out at consecutive steps to check the surface and bulk properties of the deposited film (Table I). Thermally oxidized test wafers were used for most of the experiment. Auger, XPS and SIMS techniques were used to determine the surface and bulk compositions of the films. The tape test was used to determine the adhesion. A stress gauge was used to measure the tensile stress of the deposited film. A computer process modeling technique was used to determine the process conditions and to assist in the quantitative analysis.

ANALYTICAL METHODS

Surface Analysis

Less than 1 atomic percentage of carbon is found at the WSi_x - SiO_2 interface for clean wafers deposited in a GENUS reactor. "Bad" adhesion can be observed despite the use of "clean" conditions. It can then be concluded that the "bad" adhesion can also result from other causes such as film properties.

We did not use any gross contamination methods to test adhesion characteristics of the film because these would be abnormal conditions.

We determined the Silicon content in the film. The phase diagram shows that the Si solid solubility in W films is slightly above 1.0 atomic percent at 450°C and is slightly below 1.0 atomic percent at room temperature.

CVD deposited W can cause the Si incorporated in the W layers to be in excess of solid solubility because of non-equilibrium nucleation at lower than melting point temperatures. Some of this excess Si segregates to the grain boundaries and can precipitate there. For W columnar grains of 500A to 2000A, a grain boundary can accommodate approximately one atomic percent of Si.

Auger analysis of our samples indicate a total of 2-4 atomic percent of Si can be accommodated as well as 0.5 atom percent of oxygen; therefore, SiO_2, which is preferred to WO_y compounds, will form at the grain boundaries.

This oxide formation at the grain boundary is the mechanism for "suicide", which will be discussed later. Almost all wafers with the "suicide" phenomenon can be traced to the oxygen or water vapor contamination of the W films. This points out the importance of depositing W film in clean, dry systems on clean dry substrates.

Stress Measurement

The method we have used to determine the bulk property of a tungsten film to a Silicon or Silicon Dioxide coated substrate is by use of a stress gauge measurement. The commercial IONIC Stress Gauge is used to measure the strain deformation of the Silicon wafer to deduce the stress of the deposited film.

Tape Test

The most widely used method for determining film adhesion to Silicon and Silicon Dioxide coated silicon surface is the well known tape test.

There is no specific way the adhesion test should be performed, essentially the tester cuts a piece of tape from the roll. The tape is then applied to the wafer surface and is pressed back and forth to ensure that part of the tape sticks onto the surface. The tape is pulled off from the surface in any direction, angle, speed and force the tester prefers.

The interpretation of the results is also very arbitrary, essentially a film that peels off with the tape has "bad" adhesion. If a film does not come off with the tape then this is considered "good" adhesion. Whatever is in-between is either good to fair to unacceptable.

There are different types of tapes available from the manufacturers. Table II lists some of the tape properties. They are very different. Furthermore, the adhesion properties vary with time and environmental condition such as humidity and temperature.

DISCUSSION: MECHANISM

o **Adhesion Strength**

For a clean interface, the adhesion strength of the interface between the films used for GENUS tungsten was found to have the following order:

$$Si/SiO_2 > Si/WSi_x > Si/W > W/WSi_x/SiO_2 > W/SiO_2$$

The adhesion strength is the magnitude of the force between the layers to hold them together against the sheering internal and external tensile stresses. When the total internal stress of the composite film plus any additional external stress applied to the film exceed the adhesion strength limit between WSi_x and SiO_2 layers, the film peels off at the SiO_2 interface. We do not observe any peel off between the W and WSi_x; therefore, we need only focus our attention to the system and process functions which affect the interface properties of the WSi_x and SiO_2 layers.

o **Stress and Strain**

The tensile stresses of the as-deposited W and WSi_x films and the thermally grown SiO_2 film as given below: (In unit of 10E9 dyne/cm^2)

$$W(10\text{-}20) > W\ (Si\ doped\ (5\text{-}15) > WSi_x\ (2\text{-}5) > SiO_2\ (0.5\text{-}2) > Si\ (neutral).$$

Stress causes strain, and therefore deformation. The stress and strain are a result of deposition conditions. The total deformation increases with film thickness. In our tungsten process the W layer is the thickest layer (8,500A of W against 1200A WSi_x and 1000A SiO_2) therefore, it is the major contributor to the stress of the composite film.

Within the elastic limit, the strain of each layer is proportional to its elastic modulus and the stress. The bulk modulus of the related materials are given below in units of 10E9 dynes/cm^2:

$$W(3200),\ WSi_x(N/A),\ Si(988),\ and\ Quartz\ (350).$$

The stress force transmits continuously and constantly throughout the composite film. However, the strain changes from layer to layer. The Si substrate curves up to relax the tensile stress and strain of the composite film. The residual stress is balanced by the composite strain of the composite film. At equilibrium the deformation of the individual layers can increase or decrease and change from tension to compression or vice versa.

For Si wafers, the back side is in tension and the front side is compressed thus the net result is that the W film strain is reduced.

Most likely the strain components of the WSi_x and the SiO_2 films are also reduced. All this happens at the expense of the strain increase of the Si wafer as it is forced to curve up in its equilibrium state.

o Tensile Strength

The tensile strength of the pure W filament is 20-40 X10E9 dynes/cm^2. Usually a filament has higher tensile strength than does a film.

Under tensile stress, the composite film maintains its surface intact by its material strength, the tensile strength. When the total stress is less than the tensile strength, the film remains intact. Otherwise the film cracks to relieve the excess stress. It has been our observation that the tensile stress value of the deposited film is very close to the tensile strength; therefore, under improper processing conditions the films can crack and peel.

DISCUSSION: PHENOMENA, CRACK AND PEEL, MECHANISM

As stated before, for an intact film, the stress force is distributed uniformly over the entire surface of the wafer. For a film with defects, either inflicted by scribing (as used for the tape test) or as a result of deposition such as bubbles or pinholes in the film, stress concentrates at the defect.

Failure Modes

"Suicide"

When the internal stress overcomes the maximum binding force of the material tensile strength, cracking of the film occurs. Usually a crack appears at the defect where the stress is concentrated. Once the crack is formed, stress will continue to act at the crack and cause the crack to enlarge and propagate until a ribbon of W film is peeled off spontaneously from the substrate. The film first cracks then peels. This type of adhesion failure is designated as "suicide". It sometimes occurs when the wafer is taken out of the system. The difference in contractions of the film and substrate produces a thermal stress addition to the stressed film. The initial breakline of the crack is perpendicular to the substrate surface. (Figure 2)

"Homicide"

The initial crack of the film is induced by a scribing defect inflicted after the film is grown. The film then crack and peels similar to the "suicide" film. This kind of crack and peel is called "homicide". Apparently either the tensile strength of the film is higher or the stress is lower than the "suicide" case.

Tape Test

As stated before, film lifting or peeling caused by the tape test is called "bad" adhesion. The "bad" adhesion may have a better tensile strength and lower stress. But the internal tensile stress of the film and the additional stress by pulling off of the tape overcomes the bonding force of the adhesion strength between the film and the substrate. The initial break line of the peel is parallel to the substrate. (Figure 2)

DISCUSSION: MODIFICATION OF THE OXIDE SURFACE

Hydrocarbon or water vapor contamination of the oxide surface alters the adhesion strength and usually causes bad adhesion and therefore unpredictable results. The cleanliness and condition of the oxide surface is critical for adhesion. Therefore, in-situ cleaning of the oxide surface is preferred prior to the deposition. Flash W and Ar sputtering cleans are used in the GENUS process. The effects of these steps are discussed in the following sections.

* During the Flash W step, no W is deposited on the SiO_2 due to the selective nature of the H_2 reduced W process. However, the W does cause asperity at the SiO_2 surface thus enhancing cone formation during the Ar sputter step.

* Ar sputter produces etched cones on the SiO_2 surface. Physically the cones increase the surface contact area. From the sputter time and sputter rate, we estimated a cone will have base diameter of 10-50A. The increase in the surface area is therefore 2-5 times. The cone formation also helps to interlock the contact surfaces to resist the shear force movement to withstand higher tensile stress.

* The Ar sputtering also enriches the Si content on the oxide surface. The silicon rich oxide surface provides more Si for the WSi_x to adhere to and slightly shift the adhesion strength toward the stronger Si/WSi_x adhesion strength away from the adhesion strength of the WSi_x to SiO_2.

* For standardized experiments, thermal oxide was used. However, in real application, different types of oxides are used for different applications. The wide variation of the properties of the different oxides make the adhesion properties significantly different. It is known that thermal SiO_2 is the most difficult oxide to be adhered to, as compared to CVD and PECVD oxides. In real device processing the oxide surfaces are modified and have Si exposed, and also have steps and edges. All these conditions provide better adhesion than untreated SiO_2.

IMPROVEMENTS

Although our understanding of the blanket tungsten adhesion is still somewhat qualitative, we have identified areas of improvement for producing good quality, adhesive films. They are: 1) Reduce stress, 2) Reduce strain, 3) Improve tensile strength and 4) Improve adhesion strength.

1) <u>Reduce Stress:</u>

The following method could further reduce the stress of the W film:

- Deposit at higher temperature than standard Genus process of $450^{\circ}C$ but maintain the same deposition rate.

- Deposit at lower total pressure and lower partial pressures of inert gases.

- Use PECVD W deposition. The ion bombardment of the PECVD process will reduce the tensile stress of the deposited film.

2) Reduce Strain (of the W film):

- Use a thicker silicide pad. The contact resistance will not change and the interconnect resistance of the composite film (which is in parallel) will only increase slightly.

- Deposit on the topology wafer, such as a device wafer. The morphology of the device wafer surface allow the stress relieves on steps at a micron dimensions to reduce stain build up.

3) Improve Tensile Strength:

- The GENUS system is set up and run to specification for W deposition. These baseline specifications are available from GENUS. This will reduce the oxygen and water vapor contamination levels to an acceptable levels. This will give films of higher tensile strength. These films will be more resistant to crack and peel.

- Reduce pinholes, bubbles and other type defects on the substrate to allow for better film adherence.

- Look at addition of more Si or other desired "impurity" incorporation into the W film to increase the tensile strength.

4) Improve Adhesion Strength:

- Use oxides other than thermal oxide. As stated earlier, oxides other than thermal oxide have better surface properties.

- Sputter etch the oxide surface more. The Ar sputter etch cleans the surface, increases the Si content on the oxide surface, forms etched cones to increase contact area, and roughens the surface to increase mechanical interlocking and cementing.

Overall, film quality usually improves with lower stress. Reducing the stress is the best approach to better film adhesion.

CONCLUSION

This report is a semi-quantitative approach to a better understanding of the adhesion mechanisms. Existing data analyzed and new experiments are designed to confirm the speculation of the adhesion mechanisms.

The different nature of the failures can be used to identify the process problems and to pin down the most sensitive processing steps or system character that caused the failure. For example: Oxygen or water vapor contamination in the process gas or in the system will cause suicide; surface contamination will cause "bad" adhesion.

The Improvement Section addresses the subject of improved film adhesion and methods of changing the bulk properties of the film. Added improvement by using device wafers can be obtained due to better adhesion of the substrate surface and better stress relaxation due to the topology of the wafer surface. With these improvements the blanket tungsten film adhesion to oxide surfaces should be enhanced to increase its application for future devices.

ACKNOWLEDGMENT

The authors wish to thank Laura Watson and Jack McInerney for providing most of the processing information and for accumulating information on adhesion and other Silicide problems during the course of this study. We also wish to thank Alvin Mitchell for his assistance in carrying out most of the experiment.

REFERENCE

(1) D. W. Woodruff, R. H. Wilson and R. A. Sanchez-Martinez, Workshop on Tungsten and Other Refractory Metals for VLSI Applications, Editor Robert S. Blewer (MRS) p. 173.

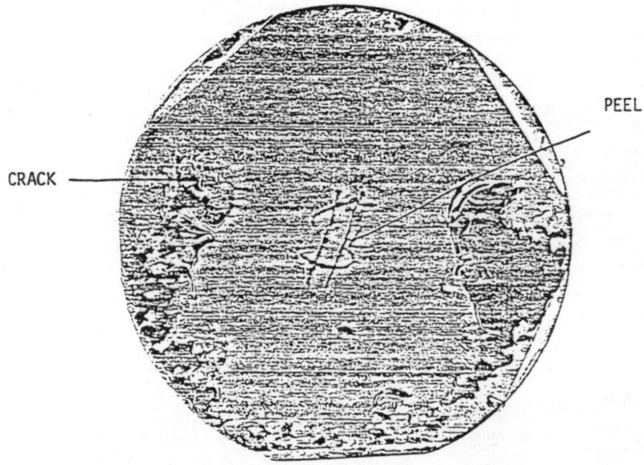

FIGURE 1: EXAMPLE OF CRACK AND PEEL

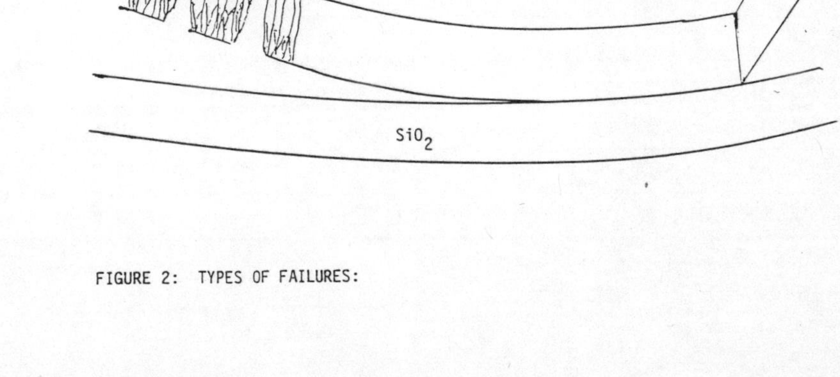

FIGURE 2: TYPES OF FAILURES:

TABLE I

BULK PROPERTIES

LAYER	GRAIN SIZE MICRON	BULK MODULUS 10E9 DYNE/CM^2	TENSILE STRESS 10E9 DYNE/CM^2	TENSILE STRENGTH 10E9 DYNE/CM^2
$WSi_{2.6}$	FINE < 0.05			
W		3200	™ 10-20	™ 20-40
W (2-5% Si)	COLUMNAR 0.05-0.2		™ 5-15	
$WSi_{2.6}$	FINE < 0.05		™ 2-5	
SiO_y y<=2	AMORPHOUS	350 (QUARTZ)	™ 0.5-2	
Si	S. XTAL	988	0	

TABLE II

TAPE PROPERTIES

TAPE TYPE	TENSILE STRENGTH lbs/in	ADHESION TO STEEL ozs/in	THICKNESS OF TAPE mils	ELONGATN AT BREAK %
3M 600	15	30	2.5	40
3M 610	30	45	3.0	15
3M 810	15	25	2.5	40
3M 845	--	45	3.4	120
LEPAGE 2000	25	32	2.2	20
LEPAGE 3000	20	45	2.4	18
PROSPECTOR 3001	20	45	2.4	18

THIN LAYERS OF TiN AND Al AS GLUE LAYERS FOR BLANKET TUNGSTEN DEPOSITION

V. V. S. RANA, J. A. TAYLOR, L. H. HOLSCHWANDNER, AND N. S. TSAI
AT&T Bell Laboratories, Allentown, PA 18103

ABSTRACT

The sputtered TiN and Al have been found to be very effective glues for blanket tungsten deposition over oxide. Thicknesses as low as 5 nm for Al and 10 nm for TiN provided excellent adhesion. The adhesion is related to the presence or absence of compounds at the W/glue layer interface.

On metal substrates such as Al and Ti, the W film initially forms due to the reduction of WF_6 by the substrate itself. This leads to the formation of fluorides that stay at the interface. On TiN, no such fluorides are formed. The film deposits because of the nucleation and growth of W brought on by the reduction of WF_6 with H_2, exclusively. The absence of O or F at the W/TiN interface makes it ideally suited for use as the glue layer under W, in contacts.

INTRODUCTION

For the use of tungsten either as an interconnect or an etch-back plug material, a blanket deposition of tungsten over an oxide surface is required. However, LPCVD tungsten does not adhere to oxide. To prevent the peeling of tungsten from oxide surfaces, an adhesion promoter (glue layer) is employed between the tungsten film and the oxide surface. We have found titanium nitride and aluminum to be very effective glues for blanket tungsten deposition. Although sputter metal such as Ti [1], and Mo [2] and LPCVD WSi_2 [3] have been used as glue layers for the blanket W deposition, Al and TiN have not yet been proposed. A thick layer of Al under W, in contacts, would cause junction spiking, while the conductive nitrides have not been investigated as glue layers, so far. We report here the results of our analysis of TiN and Al as glue layer materials for blanket W deposition and also suggest their possible application in metallization schemes. Titanium layers were investigated for comparison with these two materials.

EXPERIMENTAL

The starting material was 125 mm dia., lightly doped, high resistivity (5-50 Ω cm), p-type < 100> Si, covered with 100 nm or more of thermal oxide. The various glue layers were sputter deposited on these wafers using a Varian 3180 sputter gun. The aluminum film had a composition of 0.75 at % Si, 0.5 at % Cu, and the rest Al. It was sputter deposited at 7 mtorr argon pressure with a deposition rate of 2.5 nm/min.. Titanium was sputter deposited under the same conditions at a rate of 160 nm/min. TiN was formed by reactively sputtering Ti with N_2 in argon. The deposition rate was 40 nm/min at 7 mtorr.

The W deposition was carried out in the ANICON deposition system. It is a hot-wall, vertical flor reactor [4]. The operating conditions were

Gas flows:	H_2 = 1000 sccm, WF_6 = 200 sccm
Pressure:	0.35 to 0.4 torr
Temperature:	450° C

A number of different analytical techniques were used to characterize the thin films. X-ray photoelectron spectroscopy (XPS or ESCA) and Auger depth profiling were performed using a Perkin-Elmer model 555 ESCA and Auger system. This instrument uses a double-pass cylindrical mirror analyzer, which can be operated in either the AES or XPS mode. Typical settings for the electron beam were 5 kV and 10 μA. A 4 kV ion beam with a rastered crater of 3 mm^2 was used for the ion milling, giving an approximate sputtering rate of 10 nm/ min. For XPS data, Mg Kα was used for x-ray excitation, and the binding energies were calibrated to Au $4f_{7/2}$ peak taken at 84.0 eV. The various interfaces were also analyzed with a Philips model 400T transmission electron microscope. The thickness of the W films was obtained from SEM cross sections, by using a Hitachi field-emission secondary electron microscope (model S-800).

RESULTS

Adhesion

To check on adhesion, typically 1.0 to 1.5 μm thick W films were deposited. The adhesion of these W films to the oxide wafers was tested using the scotch-tape peel test. The adhesion of the glue-layer films to the oxide film was never a problem. Whenever the W film peeled, it always did so at the W-film to glue layer interface.

The blanket W film was deposited on the glue layers with and without prior wet cleaning. The adhesion of W films to uncleaned Ti was poor. The adhesion improved when the Ti layers were cleaned before the W film deposition. Even then there were instances of the delamination of the W films. In contrast, there was strong adhesion of the W film to both the Al and TiN layers, even if no clean was given before the W deposition.

The different thicknesses of Ti, Al, and TiN layers were investigated. The adhesion, when the as deposited thickness of the Ti films was less than 50 nm, was bad. The films of different thicknesses, down to 5 nm for Al and 10 nm for TiN, were tried. There was strong adhesion of W films to these materials even at these thicknesses.

Native Oxides

XPS was used to determine the nature and thickness of the native oxides, typically found on the as deposited glue layer films before tungsten deposition. The photoelectron lines for the outer surfaces of very thin layers of these materials are depicted in Figs. 1 to 3. For Al and TiN, Figs. 1 and 2 respectively, the oxide and the underlying unoxidized Al and TiN layers are clearly seen. After light Ar milling, the oxides were completely removed, indicating that the native oxides were thin. The binding energies observed indicated that these layers were Al_2O_3 and TiO_2, respectively. The thickness of the native oxides were calculated from the area ratios of the oxide and metallic peaks, yielding 2.3 nm for Al_2O_3 on Al and 1.5 nm for TiO_2 on TiN. However, in case of as deposited Ti

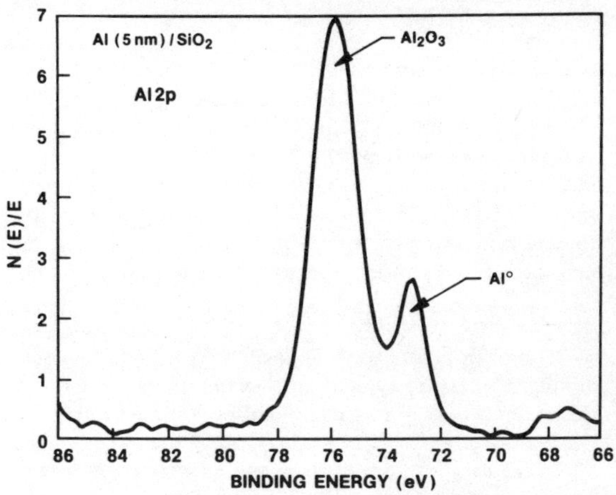

Figure 1: The ESCA spectrum for as deposited 5 nm thick Al layer on SiO_2.

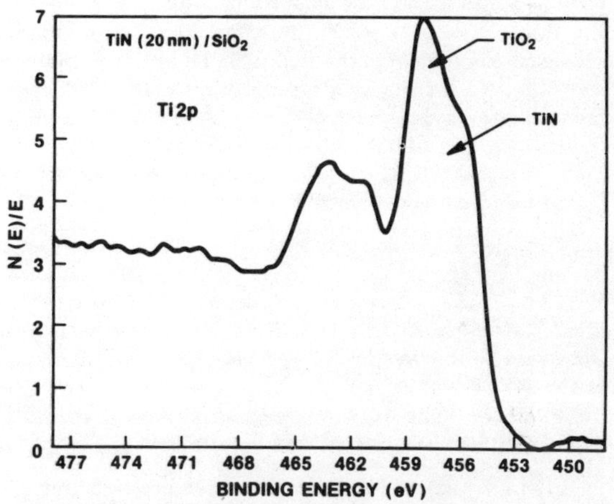

Figure 2: The ESCA spectrum for as deposited 20 nm thick TiN film on SiO_2.

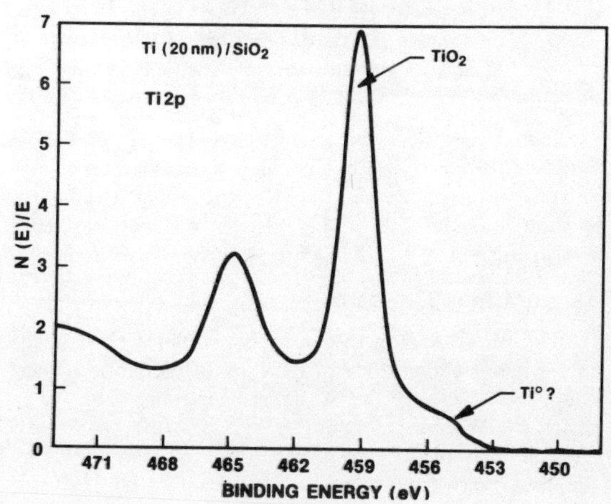

Figure 3: The ESCA spectrum for as deposited 20 nm thick Ti film on SiO_2.

films thinner than 20 nm, the line for the metallic Ti was just barely detected (Fig. 3), and results after Ar ion milling indicated that there was little or no metallic Ti, confirming that these Ti films were nearly completely oxidized. For Ti films with an as deposited thickness of 50 nm or greater, the outer oxide layer (TiO_2) could be removed by Ar milling, to obtain a metallic Ti layer. No oxygen was found in the metallic Ti layer. Also, for thick Ti films the Ti $2p_{3/2}$ line for the metallic Ti on the outer surface was more intense, which allowed the thickness of the native oxide to be calculated from the area ratios, indicating 3.7 nm of TiO_2. This thickness of the native oxide is much less than the thickness of the mostly oxidized thin Ti films. The higher oxidation of the thin Ti films may be related to the structure of these films.

W/(Ti, Al, TiN) Interface

Tungsten layers were deposited on the glue-layers, and Auger depth profiles were obtained. Several typical profiles are shown in Figs. 4 to 6. For thin Ti films ($\leq$ 20 nm), the entire Ti layer was completely consumed during the tungsten deposition, and only a trace amount of fluorine was found at the W/oxide interface. These films adhered poorly and were easily removed from the substrate. No Ti was observed on either of the two freshly exposed surfaces by XPS or Auger, and depth profiles showed only W and SiO_2. For thicker ($\geq$ 50 nm) Ti films, metallic Ti was observed at the interface with some dissolved oxygen throughout this Ti layer. It is to be noted that no such dissolved oxygen was observed for the as-sputtered films before W deposition. Significant amounts of fluorine and oxygen were observed at the W/Ti interface. Analysis of the delaminated film (for 50 nm of Ti) indicated that the break occured at the W/Ti interface. For Ti films thicker than 50 nm, the quality of the adhesion varied, but similar amounts of fluorine and oxygen were always observed at the W/Ti interface.

Tungsten films deposited on Al and TiN could not be delaminated. Auger profiles, obtained for the Al films, showed considerable amounts of fluorine but little oxygen at the W/Al interface. The Al layer seemed to be intact at the W interface even for the 5 nm film. Relatively slightly less fluorine was observed at the interface of the 5 nm film as compared to the 20 nm one. In contrast, no fluorine and very little oxygen was observed at the W/TiN interface. In fact, more oxygen was found in the TiN film itself than at the interface. The oxygen was incorporated in the TiN structure during sputter deposition; its composition was determined as $TiN_{0.95}O_{0.10}$ by using electron microprobe.

Figures 7 (a) to (c) show the TEM cross-sections of the W interface with Al, TiN, and Ti, respectively. It can be seen that the W/Al and W/TiN interfaces are sharp, while the W/Ti interface is broad. The broad interface in the later case is because of the presence of significant amounts of O and F in this region. The surface of TiN is smooth as compared to Al. This is probably because of the larger grain size for Al.

The microdiffraction patterns from W/Ti and W/Al interfaces, taken using the TEM, contained some spots that could not be ascribed to either W, Ti or Al. In the diffraction pattern from the W/Ti interface, there was a spot at d = 2.33Å, which can be attributed to rhombohedral TiF_3 (210). In the diffraction pattern from the W/Al interface, a spot was present at d $\simeq$ 5.5Å which can be attributed to tetragonal AlF_3 (001).

Tungsten Film Growth

The tungsten film thickness dependence on deposition time for Si, Al, and TiN substrates is shown in Fig. 8. The difference in growth on TiN as compared to Si and Al is apparent. On Si and Al substrates there is an initial rapid growth of W, following which the W film thickness increases at a rate of nearly 21 nm/min. However, there is an incubation time for film growth on TiN. On TiN, the W crystallites first nucleate on the surface. With time these crystals grow in number and size. Eventually, the crystal

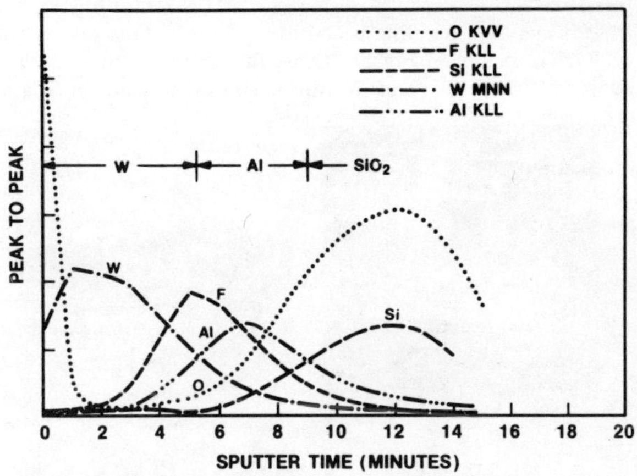

Figure 4: The Auger depth profile of $W/Al/SiO_2$ layers.

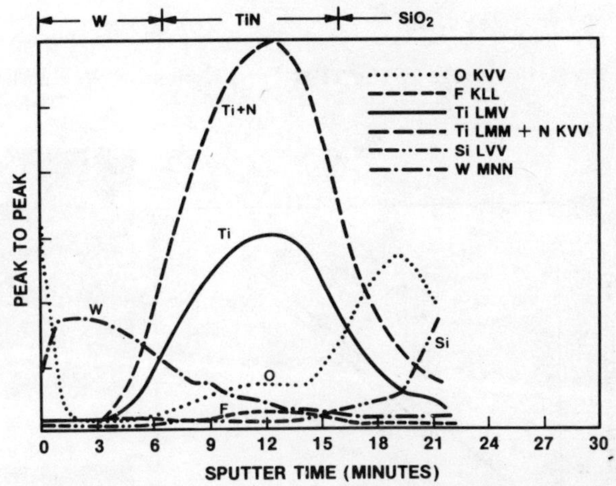

Figure 5: The Auger depth profile of $W/TiN/SiO_2$ layers.

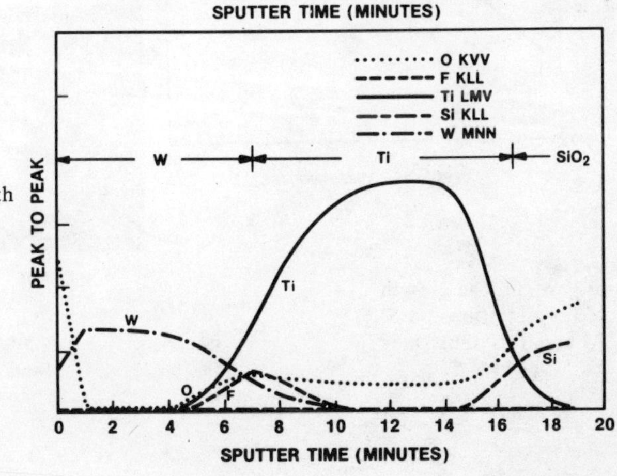

Figure 6: The Auger depth profile of $W/Ti/SiO_2$ layers.

islands coalesce forming a continuous film. This process of W nucleation and film growth on TiN is shown in Fig. 9. Once the complete film coverage is obtained on the TiN substrate, the growth of W film is similar to that on Si and Al.

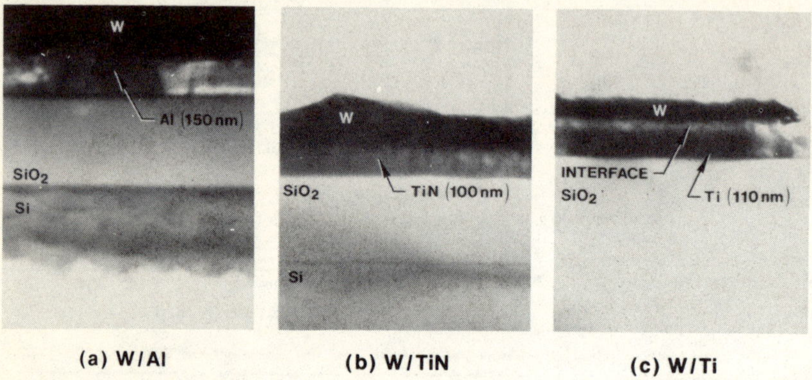

Figure 7. The TEM cross sections of W/glue layer interfaces.

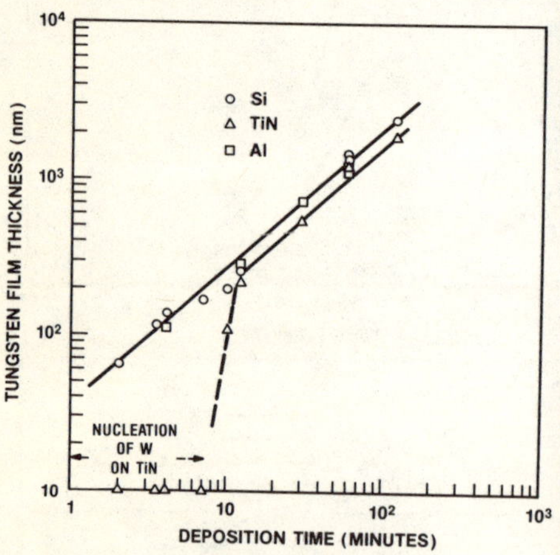

Figure 8. The growth of tungsten films on Si, Al and TiN substrates at 450° C.

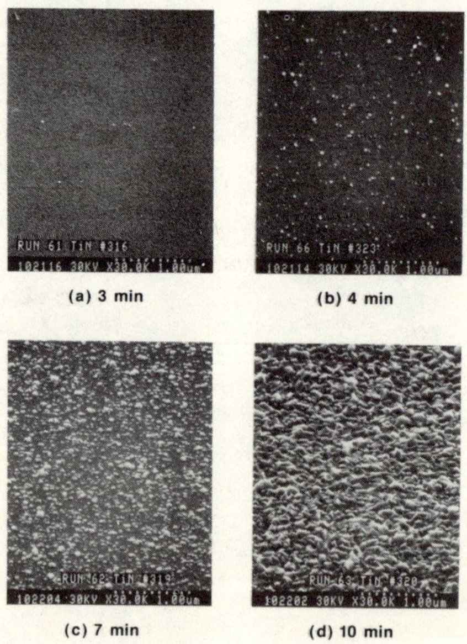

Figure 9. The nucleation and growth of W on TiN at 450° C, for different deposition times.

DISCUSSION

Adhesion at the glue layer/oxide interface

The studies of glass to metal adhesion [5] have shown that in order to obtain strong adhesion, a strong chemical bond has to develop at the interface. For strong adhesion the metal atoms have to bond to the oxygen ions at the surface of the oxide. In such a case a transition layer of metal oxide is formed at the metal/oxide interface resulting in a gradual change of structure rather than a sharp boundary at the interface. The oxygen atoms serve as a bridge between the metal and the oxide film. The bonding at the interface can be represented by the reaction

$$\frac{y}{2}\,SiO_2 + xM = M_xO_y + \frac{y}{2}\,Si \qquad \Delta G^\circ < 0 \qquad (1)$$

where ΔG^o is the free energy for the reduction of SiO_2 by the metal, thereby forming the metal oxide. In order to be able to form a transition layer of metal oxide at the metal/oxide interface, ΔG^o should be negative. In other words, the metal oxide has to be more stable than SiO_2. Since the metals such as Ti and Al have free energies for oxide formation that are more negative than SiO_2 [6], the formation of a strong bond with oxygen is favored. The strong adhesion of Ti to SiO_2 probably plays a key role in the adhesion of TiN films to oxide. The adhesion is also improved by the use of sputtering for the deposition of the glue layer films. The sputter deposited atoms would penetrate the oxide surface and be mechanically interlocked with the oxide. Such interlocking may enhance chemical bonding also. Since films as thin as 5 nm of Al or 10 nm of TiN adhered strongly to the oxide, a strong chemical bond may be formed between these materials and the oxide substrate.

Adhesion at the W/glue layer interface

As noted earlier, all the adhesion failures in our work occurred at the W/glue layer interface. From the results of the cleaning studies, it is clear that the presence of oxide at the W/glue layer interface is detrimental to adhesion. Figure 1 to 3 show that thicker native oxide grows on Ti than on Al and TiN, while thinner films of Ti (20 nm or less) show more oxidation than thicker films. During the W deposition some oxide is removed from the substrate surface by WF_6. The Auger spectra of layers after the W deposition (Figs. 4-6) shows that little or no oxygen is left at W/Al or W/TiN interfaces, while greater amounts were seen at the W/Ti interface. Although the native oxide on thicker Ti layer (3.7 nm) is not significantly greater than on Al (2.3 nm), it may be just above the critical thickness that may be removable by WF_6.

The loss of Ti during the tungsten deposition occurs because of the interaction of WF_6 with Ti to form volatile titanium fluorides or oxy-fluorides. The formation of several compounds such as TiOF, $TiOF_2$, $TiOF_4$, and TiF_3 is thermodynamically favorable. The volatile TiOF, $TiOF_2$ and TiF_4 (sublimes at 285° C) would be lost during the tungsten deposition at 450° C. The interfacial fluoride has been identified as TiF_3, which is quite stable at 450° C (sublimation point 930° C). The adhesion of W films was bad for Ti films thinner than 50 nm, as the Ti was completely consumed during the W deposition. For thicker films, the poor adhesion can be attributed to the presence of oxides and fluorides at the interface.

For Al glue-layers, as far as we can tell, little or no Al is lost during the tungsten deposition. This allows very thin Al layers (5 nm) to be successfully used as glue-layers. Even though no volatile products seem to be formed, the WF_6 reacts initially to form a stable AlF_3 layer (which sublimes at 1291° C) just at the W/Al interface. The sharpness of the W/Al interface indicates that the AlF_3 layer is extremely thin. The discontinuities in the layer would allow for the strong metallic bonds to form between W and Al, leading to good adhesion.

For TiN no fluorine was seen at the interface indicating that the presence of Ti-N bonds prevent the interaction of Ti with WF_6, preventing the formation of fluorides. As seen in Fig. 7 (b) the W/TiN interface is sharp, and no evidence of any fluorides or oxides are seen in the Auger profiles. The strong adhesion of W to TiN is probably because of direct bonding between W and Ti.

Tungsten Film Growth

For metals and silicides, it is known that, even in the presence of H_2, the W film initially forms due to the reduction of WF_6 with the substrate [7,8]. These so-called displacement reactions can be written as

$$2\ WF_6 + 3\ Si\ (Ti,\ Al) = 2\ W + 3\ SiF_4\ (TiF_3,\ AlF_3) \qquad (2)$$

These reactions provide the W nuclei for further film growth and are the reasons for the selective deposition of W obtained on metals and silicides. Once the substrate surface is completely covered with a tungsten film, further growth is considered to take place by the diffusion of the substrate species (e.g. Si) through the W film and its consumption due to the displacement reaction [8]. However, a limiting thickness of the W film is soon reached, when the contact between the Si and WF_6 is suppressed at greater thicknesses. Then, the growth of thicker W films takes place by the reduction of WF_6 with H_2 according to the reaction

$$WF_6 + 3H_2 = W + 6\ HF \qquad (3)$$

At 450° C, a self-limiting thickness of about 30 nm has been reported for lightly implanted wafers. Also, the self-limiting layer grows rapidly, reaching its final thickness in typically less than 1 min. [7,8]. In our case the film growth data for Si (Fig. 8) has an intercept of nearly 35 nm, indicating this to be the thickness of the W film deposited because of the displacement reaction. The W-film growth data in Fig. 8 is for the film growth due to the reduction of WF_6 with H_2 following this initial film deposition. For Al and Ti, similar thicknesses to that for Si were observed.

Since TiN does not appear to react with WF_6 the nucleation and growth of W on TiN is expected to be similar to SiO_2. The rate controlling step for the growth of W films is considered to be the dissociation of adsorbed H_2 on the surface [9]. On continued exposure to WF_6 and H_2, H and F atoms would react, nucleating W on the surface. The advantage of not using any kind of displacement reaction is that no fluorides are formed at the interface.

CONCLUSIONS

It has been shown that the presence of compounds such as oxides and fluorides at the W/glue layer interface is detrimental to adhesion. For glue layers with thicker native oxides, the quality of adhesion may vary with the effectiveness of the cleaning operation needed to reduce the thickness of this native oxide. The strong adhesion of W to TiN and Al, independent of any cleaning operation, makes these materials suitable candidates for glue layer applications in production environments.

Since the glue layer will be present under W, in contacts, the presence of interfacial fluorides is a cause for concern. The presence of AlF_3 has been reported to cause an increase in contact resistance of two orders of magnitude in Al/W/Al multilevel interconnect structure [10]. We have not seen such changes in contact resistance in our work involving encapsulation of Al runners with selective W [11]. It may be because of a much larger contact area between W and Al in our case. If it is so, then a continuous Al layer under W interconnects (runners) should provide low resistance contacts. The use of Al is particularly suited for upper levels, in multilevel metallization schemes where junction spiking is not a consideration. The absence of O and F at the W/TiN interface makes TiN ideally suited for use as glue layer especially for etch-back plugs, where contact area between W and the glue layer will be small.

ACKNOWLEDGEMENTS

The authors thank Fred Degler for making the tungsten deposition runs, Bruce Snyder for SEM and V. C. Kannan for the TEM work. The authors also thank A. J. Manocha, J. B. Bindell and A. K. Sinha for their support of this work.

REFERENCES

1. G. C. Smith and R. Barry Jucha, Third Int. IEEE VLSI Multilevel Interconnect Conf., 403 (1986).
2. D. W. Woodruff, R. H. Wilson and R. A. Sanchez-Martinez, Tungsten and Other Refractory Metal for VLSI Applications (MRS, Pittsburgh 1986), Proc. of 1984 and 1985 Workshops, ed R. S. Blewer, 173.
3. G. Higelin, C. Wieczorek and V. Grewal, Proc. Third Int. IEEE VLSI Multilevel Interconnect Conf., 443 (1986).
4. A. J. Learn, J. Electrochem. Soc., 132, 390 (1985).
5. J. A. Pask and R. M. Fulrath, J. Am. Ceram. Soc., 45 (2), 592 (1962).
6. R. W. Swalin, Thermodynamics of Solids (John Wiley and Sons, 1972), Ch. 5.
7. K. C. Saraswat, S. Swirhun and J. P. McVittie, VLSI Science and Technology 1984, Eds. K. E. Bean and G. A. Rozgonyi, Proc. Electrochem. Soc. Vol, 84-7, p. 409.
8. M. L. Green, Y. S. Ali, B. A. Davidson, L. C. Feldman and S. Nakahara, Proc. Mat. Res. Soc. Symposium (MRS, Pittsburgh), vol 59, 1986, p. 723. 723 (1986).
9. E. K. Broadbent and C. L. Ramiller, J. Electrochem. Soc., 131 (6) 1427 (1984).
10. T. Moriya, S. Shima, Y. Hazuki, M. Chiba and M. Kashiwagi, IEDM Tech. Dig., CH 1973-7/83, 550 (1983).
11. H. P. W. Hey, A. K. Sinha, S. D. Steenwyk, V. V. S. Rana and J. L. Yeh, to be presented at IEDM, Los Angeles, CA (1986).

Chemical Vapor Deposition of Molybdenum from MoF_6

LOW PRESSURE CHEMICAL VAPOR DEPOSITION OF MOLYBDENUM: 1. KINETICS AND REACTION MECHANISMS

TURGUT SAHIN, EDWARD J. FLANIGAN, AND JOHN T. SEARS
Department of Chemical Engineering, Montana State University, Bozeman, Montana 59717

ABSTRACT

Chemical kinetics of low pressure chemical vapor deposition (LPCVD) of molybdenum on single crystal p-type (100) oriented silicon substrates were studied by hydrogen reduction of MoF_6. The pressure and temperature ranges covered were 0.9-10 torr and 250°C to 400°C, respectively. The rate of molybdenum deposition showed a square root dependence on the hydrogen partial pressure and no dependence on the MoF_6 partial pressure. The molybdenum film characteristics were dependent on the deposition conditions. Higher temperatures and pressures appeared to favor the formation of molybdenum silicide films near the molybdenum-silicon interface.

INTRODUCTION

Refractory metals such as Mo, W, and Ta have several potential integrated circuit applications [1]. The application of Mo films as Schottky barriers and gate metals for high speed MOS devices have been successfully demonstrated [2-6]. There has been a significant number of studies on W deposition on Si. However, Mo deposition has not received as much attention as W. Much of the Mo CVD work was done at high pressures and temperatures for solar cell and metallurgical applications [7].

Mo can be deposited on Si by H_2 reduction of MoF_6 or $MoCl_5$ according to the following reactions:

$$MoF_{6(g)} + 3H_{2(g)} \rightarrow Mo_{(s)} + 6HF_{(g)} \quad (1)$$
$$2MoCl_{5(g)} + 5H_{2(g)} \rightarrow 2Mo_{(s)} + 10HCl_{(g)} \quad (2)$$

The H_2 reduction of $MoCl_5$ at temperatures between 600 and 1300°C and the properties of the resultant Mo films formed have been studied [8-10]. A majority of these studies were at high pressures.

There is a need for more data at low pressures and temperatures. To the authors' knowledge, there has not been any significant published work covering the low pressure H_2 reduction of MoF_6 on a single crystal Si. In this work MoF_6 was chosen instead of $MoCl_5$ because of its high vapor pressure and lack of significant kinetic data. In addition, MoF_6 was relatively easier to handle for our experimental set up, and it may be possible to compare results of MoF_6 with the WF_6 results reported in the literature [11,12]. The objective of this work was to study reaction kinetics and mechanism of the H_2 reduction of MoF_6 on Si at low pressures and temperatures. In this article, the kinetics and reaction mechanism implications will be presented. Also, some film characteristics will be discussed.

EXPERIMENTAL

The molybdenum film depositions were carried out in a 6-way-cross stainless steel prep-chamber of a ultra high vacuum system. The sample heater was a 4cm x 5cm x 1.8cm stainless steel block mounted on a conflat flange. Inside the stainless steel block were four Ni-Cr (Nichrome) wires encased in ceramic tubes. The Si wafers were cut to 3cm x 4cm pieces to fit on the heating block. The chamber was evacuated by two mechanical pumps. Gases were introduced from the top and pumped out from the bottom so that they would be introduced directly on the Si surface. Gas flow rates were measured and controlled by MKS gas flow controllers. The chamber pressure was controlled by a throttle valve from MKS Instruments. The pressure measurement was achieved by a barotron type pressure transducer.

The gases used were molybdenum hexafluoride (99.8% MoF_6), hydrogen (99.9997%) and helium (99.9997%). The MoF_6 bottle and lines were heated to approximately 35-40°C to keep MoF_6 in gaseous form. Temperature on the heating block surface (between the heater and the sample) was monitored by a spot-welded type K thermocouple. The thermocouple wires were isolated from the gases by a ceramic tube.

The exhaust gases were passed through a heated (300°C) reactor filled with stainless steel shavings to reduce excess MoF_6 and through a gas absorber filled with high surface area (250m^2/gm) Al_2O_3 to capture the HF generated. Al_2O_3 also captures any unreacted MoF_6 as a result of reaction of MoF_6 with the surface hydroxyl groups. The exhaust gases continued through the mechanical pump into the hood. A dilute NaOH solution with an acid-base indicator was placed at the exhaust of the pump to detect any escaping HF or MoF_6 gases.

The wafers used were P-type (100) oriented single crystal Si. The Si samples were etched in a dilute HF solution and rinsed by deionized water, followed by an acetone and methanol rinse and air dried. The thickness of the deposited Mo films was measured by weight difference after acid dissolution.

Prior to the deposition of Mo, the cleaned sample was loaded to the reactor and evacuated to a base pressure of less than 0.010 torr before heating the substrate. When the substrate temperature reached 80°C, H_2 was introduced and the flow rate was gradually raised to the operating flow rate to avoid any thermal shock to the substrate. When the desired operating temperature was reached, the chamber was brought to the operating pressure with H_2 and the metered MoF_6 was allowed into the chamber. Constant pressure and temperature was maintained during the deposition period. At the end of deposition, MoF_6 flow was stopped and H_2 continued to flow for an additional 5 minutes before the chamber was evacuated. Then the chamber was filled with He or H_2 while the sample cooled down.

The Mo deposition was carried out in a pressure range of 0.9 - 10 torr for a temperature range of 250° - 400°C. The gas flow rates were between 2 - 5 and 2 - 65 cm^3/min for MoF_6 and H_2, respectively. Deposition times ranged from 5 to 20 minutes.

RESULTS

Reaction Kinetics

The substrate temperature dependence of the molybdenum deposition rate by hydrogen reduction of MoF_6 was studied between 523K and 623K. Arrhenius plots made by plotting the rate versus temperature data at 0.9 torr and 5 torr are shown in Figure 1. From the slopes of Arrhenius plots the activation energy of the reaction was calculated as 65 + 5 kJ/mol. As can be seen on the Arrhenius plots, a strong pressure dependence of the deposition rate is evident. Higher pressures favor higher deposition rates.

To determine the order of reaction with respect to the hydrogen partial pressure, the hydrogen partial pressure was varied while keeping the MoF_6 partial pressure and temperature constant. The data were collected at the substrate temperatures of 523K and 573K while keeping the MoF_6 partial pressure at 0.3 torr. From the slopes of the lines

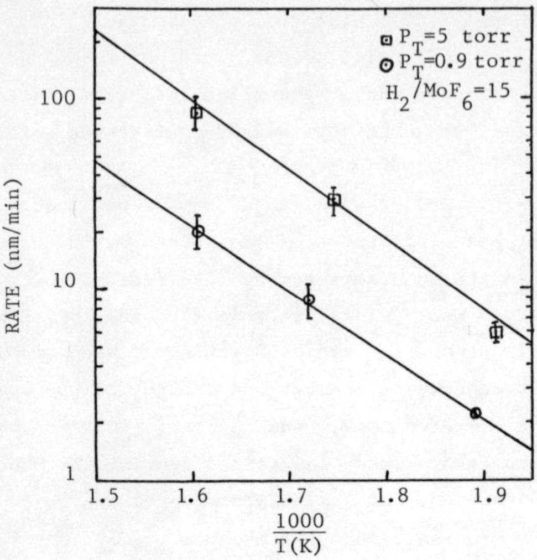

Figure 1. Arrhenius plot; temperature dependence of deposition rate.

(Figure 2), the order of reaction with respect to the hydrogen partial pressure was found to be near 1/2.

The MoF_6 partial pressure dependence of the deposition rate was also studied while keeping both temperature and hydrogen partial pressure constant. The slopes of the lines generated by plotting rate versus

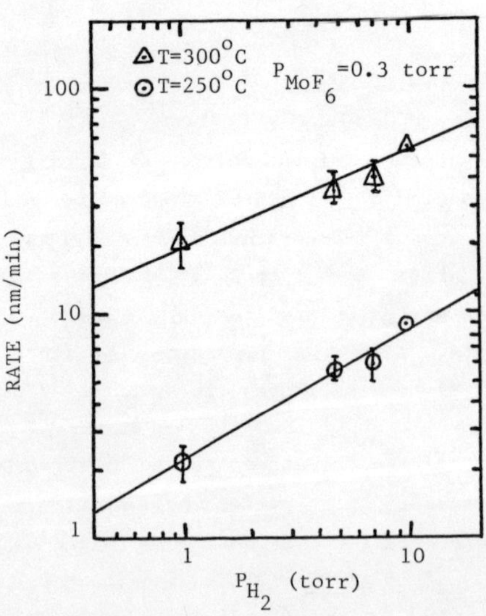

Figure 2. Deposition rate as a function of H_2 patial pressure.

MoF_6 partial pressure (Figure 3) were approximately zero. Therefore, the rate of molybdenum deposition is independent of the MoF_6 partial pressure.

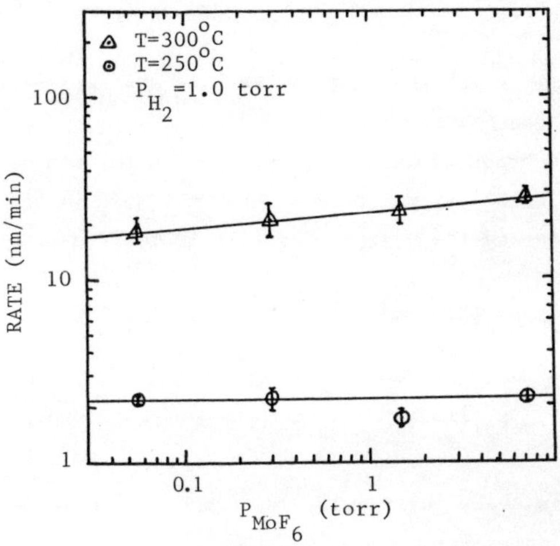

Figure 3. Deposition rate as a function of MoF_6 partial pressure.

The growth rate of molybdenum films in nm/min is given by the equation:

$$r = 1.07 \times 10^{7}\ e^{\frac{(-7800)}{T}}\ P_{H_2}^{\ 1/2}\ P_{MoF_6}^{\ 0} \qquad (3)$$

The deposition of molybdenum by H_2 reduction of MoF_6 follows very similar kinetics to that of tungsten by H_2 reduction of WF_6 [11-13]. Similar to McConica's [12] mechanism for hydrogen reduction of WF_6, we can write the following elementary reaction steps:

$$MoF_6 + * \xrightarrow{k_1} MoF_6* \qquad (4)$$

$$MoF_6* + H* \xrightarrow{k_2} MoF_5* + HF* \qquad (5)$$

$$MoF_5* + H* \xrightarrow{k_3} MoF_4* + HF* \qquad (6)$$

$$MoF_4* + H* \xrightarrow{k_4} MoF_3* + HF* \qquad (7)$$

$$MoF_3* + H* \xrightarrow{k_5} MoF_2* + HF* \qquad (8)$$

$$MoF_2* + H* \xrightarrow{k_6} MoF* + HF* \qquad (9)$$

$$MoF* + H* \xrightarrow{k_7} Mo* + HF* \qquad (10)$$

$$Mo^* = * \qquad (11)$$

$$HF^* \underset{k_9}{\overset{k_8}{\rightleftharpoons}} HF + * \qquad (12)$$

$$H_2 + 2^* \rightleftharpoons 2H^* \qquad (13)$$

Where * represents an empty site and MoF_x*, H* and HF* represent the adsorbed species of MoF_x, H and HF.

Any one of the MoF_x* reduction reactions (equations 5 - 10) can be rate limiting, and satisfy equation 3 if we assume that the MoF_x is the most abundant surface intermediate and the number of available surface sites are constant.

Desorption of HF* (equation 12) can also be rate limiting and would yield,

$$r = k\ (H_2)^{1/2}\ (MoF_6)^{1/6}$$

Since measuring 1/6 order is very difficult, we can conclude zero order with respect to MoF_6 partial pressure.

The adsorption of the hydrogen can also be rate limiting and satisfy equation 3 if we assume gas phase dissociation of H_2 and then adsorption on the surface sites. The corresponding two elementary reactions will be

$$H_2 = 2H \qquad (15)$$

$$H + * = H^* \qquad (16)$$

Any one of these elementary reactions can be rate limiting. Further information is needed to be able to tell the true reaction mechanism.

Film Characteristics

The deposited molybdenum films were characterized by scanning Auger microprobe (SAM) and RBS. There was a significant amount of oxygen, fluorine and carbon in the molybdenum films. The O/Mo ratio was between 0.15 and 0.30. The F/Mo ratio was always below 0.1. SAM studies have indicated the formation of some molybdenum silicide near Mo-Si interface at high temperature and pressure deposition conditions.

ACKNOWLEDGMENTS

Many thanks go to Richard J. Smith of Physics Department at Montana State University for doing the RBS work for us. We also thank J.R. Anderson and M.C. Jaehnig of CRISS (Center for Research in Surface Science) at Montana State University for making the SAM available for us.

REFERENCES

1. S.K. Ghandhi, VLSI Fabrication Principles, 1st ed. (Wiley, New York, 1983), p.437.

2. S.S. Simeonev, E.I. Kafedjiiska, and A.L. Guerassimou, Thin Solid Films 115, 291 (1984).

3. D.M. Brown, W.E. Engeler, M. Garfinkel, and P.V. Gray, J. Electrochem. Soc. 115, 874 (1968).

4. F. Yanagawa, K. Kiuchi, T. Husoya, T. Tsuchiya, T. Amazawa, and T. Mano, Trans. IEEE Elect. Device Conf., Washington, D.C., 362 (1979).

5. M. Fukumato, K. Inoue, S. Ogawa, S. Okada, and Kugimiya, Symposium on VLSI Technology, Digest of Technical Papers, Maui, HI, USA, 98 (1983).

6. G.G. Pinneo in The Third Int. Conf. Chem. Vapor Deposition, edited by F.A. Galaski (The Am. Nucl. Sac., Hinsdale, IL, (1981), p. 462.

7. A.M. Shroff and G. Delval, High Temp. - High Pres. 3, 695 (1971).

8. J.J. Casey, R.R. Verderber, and R.R. Garnache, J. Electro-chem. Soc. 114, 201(1967).

9. A.H. El-hoshy, J. Electrochem. Soc. 118, 2029(1971).

10. D.K. Seto, V.Y. Aop, and S. Dash, in The Second Int. Conf. on Chem. Vapor Deposition, edited by J.M. Bloches, Jr. (Electrochem. Soc., New York, N.Y., (1970), p. 659.

11. E.K. Broadbent and C.L. Ramiller, J. Electrochem. Soc. 131, 1427(1984).

12. C.M. McConica and K. Krishnamani, Tungsten and Other Refractory Metals For VLSI Applications, R.S. Blewer, ed, Materials Research Society (Pittsburgh, PA), p. 433 (1986).

13. H. Cheung, in The Third International Conference on Chemical Vapor Deposition, edited by F.A. Glaski (The American Nuclear Society, Hinsdale, IL, 1981), p. 136.

CVD MOLYBDENUM FROM MoF_6

DAVID W. WOODRUFF AND RONY A. SANCHEZ-MARTINEZ
General Electric Corporate Research and Development,
1 River Road, K1/CEB 402, Schenectady, NY 12301

ABSTRACT

Molybdenum films were deposited from the hydrogen reduction of MoF_6. The reaction was completely selective against deposition on silicon dioxide. The reaction with silicon is very high rate, and is not self-limiting, in contrast to the analogous reaction between WF_6 and silicon. Films of selective CVD tungsten, and sputtered TiW and Mo were ineffective as barrier films to the MoF_6-Si reaction. The TiW and Mo films, when deposited over SiO_2, were etched clean. MoF_6 also etched the oxide film. The films deposited over silicon were high in oxygen content and porous.

INTRODUCTION

Tungsten deposition from the hydrogen reduction of tungsten hexafluoride for use in microelectronics is well known [1]. Both selective and non-selective modes of deposition are known, and a number of potential applications have been identified. Even though the technology has seen rapid development, there remain some unresolved technical problems. The technology of selective deposition has advanced considerably [2], but additional advances in terms of controlling selectivity would be welcome. Adhesion to oxides for blanket deposition can also be an issue.

This work was undertaken to explore chemical vapor deposition of molybdenum from the hydrogen reduction of molybdenum hexafluoride. Molybdenum and tungsten chemistry is similar enough that one might expect similar and possibly advantageous CVD behavior. Molybdenum CVD has been reported previously from MoF_6 at atmospheric pressure [3] and at low pressures on molybdenum substrates [4]. There was also a brief report of CVD Mo on silicon substrates by LPCVD [5], though few details were given. Plasma CVD of Mo has also been reported [6] from MoF_6 and H_2. These latter two were not very successful. Much more work has been reported for CVD of Mo from the chloride [3,7-9], including gate formation [10], and from the carbonyl [11-14].

Chemically the tungsten and molybdenum compounds are quite similar. The vapor pressure of MoF_6 is slightly lower than that of WF_6 making it more difficult to introduce to a CVD system. WF_6 freezes at 2.5^{o}C and boils at 17.5^{o}C, where MoF_6 freezes at 17.5^{o}C and boils at 35°C. Thermodynamically, MoF_6 is more favored to react than WF_6 with both hydrogen and silicon, and so both molybdenum reactions are expected to proceed.

EXPERIMENTAL

Experiments for this study were carried out in a hot wall, low pressure CVD reactor, shown in **Figure 1**. The chamber was a 2.5 inch diameter quartz tube inside a 3-zone furnace. The MoF_6 was introduced by bubbling hydrogen through MoF_6 liquid held at 20°C. The flowrates of hydrogen to both the bubbler and the

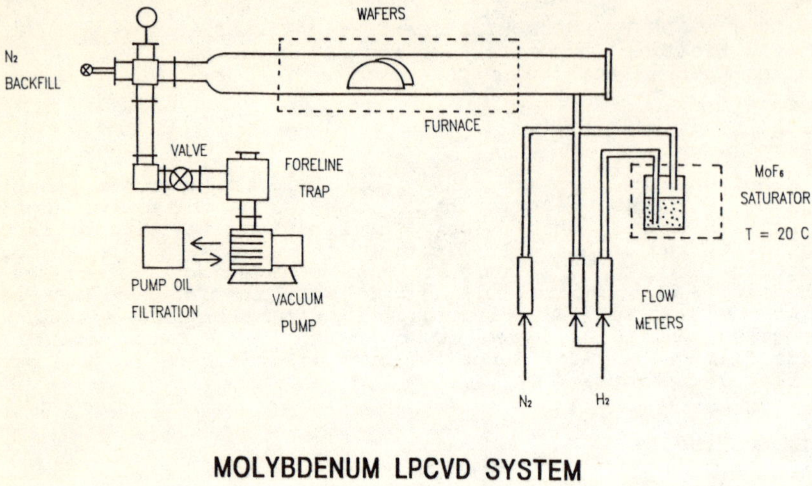

Figure 1. Hotwall LPCVD System for CVD Molybdenum

reactor were adjusted to give the desired H_2/MoF_6 ratio. Table I gives the range and prefered conditions for deposition.

TABLE I. Deposition Conditions

CONDITION	RANGE	PREFERRED
TEMPERATURE	$20^oC - 500^oC$	$200^oC - 250^oC$
H_2 FLOW	100 sccm	100 sccm
MoF_6 FLOW	5 - 25 sccm	5 sccm
PRESSURE	2 - 5 torr	2 torr

RESULTS

Self-Limiting Behavior

The reaction of MoF_6 with silicon proceeds in this system with **no self-limiting behavior.** The deposit thickness increases with deposition time with or without hydrogen present, and generally at a very high rate. For example, at 300°C with 5% MoF_6 in H_2, 11 microns of molybdenum was deposited, etching 17.5 microns of silicon (see **Figure 5**). For just the MoF_6/Si reaction, one would expect 9.2 microns of molybdenum. The observed film thickness is greater both because of the reaction with hydrogen, and because the deposit is not completely dense. This lack of self-limit was repeated many times at different conditions, and is not believed to be the result of excess residual native oxide on the surface.

Film Quality

The films deposited over silicon were porous, with the films separated in some cases at the grain boundaries, and had a high oxygen content. **Figure 2** is an SEM of a crossection of a 10 micron thick film deposited over silicon. The voids are visible at the grain boundaries. The film stress has also resulted in a loss of adhesion due to failure of the silicon. Auger analysis of a 1500 A film showed very high oxygen content distributed uniformly through the film, resulting in a high sheet resistance. Mass spectrometry showed no oxygen in the source MoF_6, and the system was tested for leaks with a He leak detector. Other possible sources of oxygen were the reactor walls (oxide etching has been observed) or oxidation of the film on removal from the reactor. A TEM of the film at a point near the silicon interface is shown in **Figure 3**. The film grain size is 300-500 A.

Figure 4 shows different regimes of deposition on silicon observed under various conditions. The first regime is smooth deposit, also seen in **Figure 2**. The second is a rough deposit with obvious large etch pits extending under the film. These regions tended to have thinner deposits. The third is a regime where silicon etches **without** the expected deposition of molybdenum. The silicon etch rate here is higher than in areas of deposition. Many times any or all of these regimes would be present on a single wafer.

Figure 2. SEM Crossection of an 8 micron Mo film deposited on a silicon substrate. Voids are visible in the Mo film.

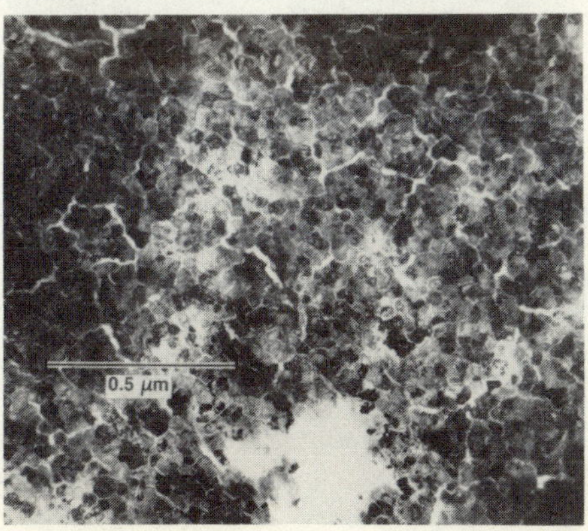

Figure 3. STEM of Mo film, taken near the Mo/Si interface The grain size ranges from 300 to 500 Angstroms.

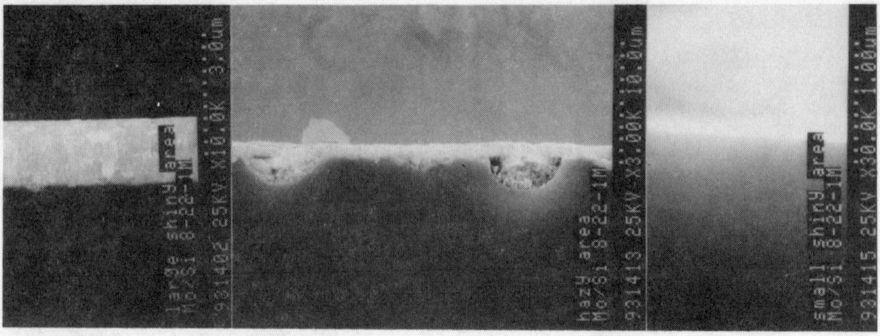

Figure 4. SEM Crossection of deposition regimes on the same wafer. Left-to-right: Thick, smooth deposit; Thinner, rough, pitted film; Severe etching with no deposition.

Selectivity and Encroachment

For all conditions run, the deposition of molybdenum was 100% selective. In no case was there any nucleation on the oxide surfaces present, including reactor walls and wafer carriers. This is in contrast to tungsten deposition, where above 300°C in a hotwall reactor, there is, with time, considerable tungsten nucleation on an oxide surface.

Figure 5 shows the boundary between oxide-covered and bare silicon, exposed to deposition conditions of 300°C for 30 minutes. The deposit was 11 microns thick, consuming 17.3 microns of silicon, and resulted in severe encroachment under the oxide. **Figure 6** shows the same behavior for deposition in a 1.3 micron diameter via, deposited at 220°C for 30 minutes. For all deposition conditions above room temperature, the encroachment was unacceptable.

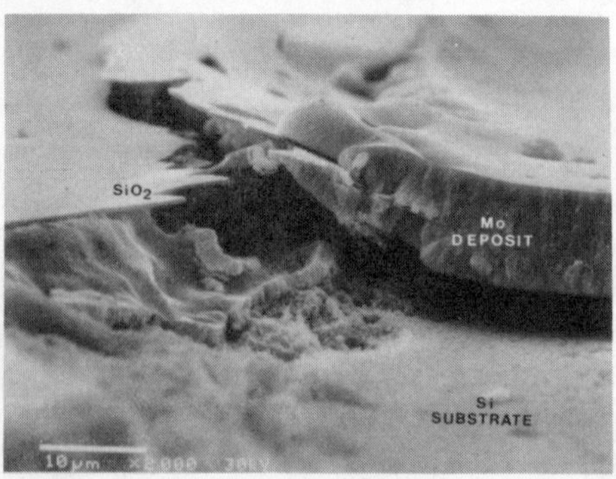

Figure 5. Mo deposition at an oxide/silicon boundary.

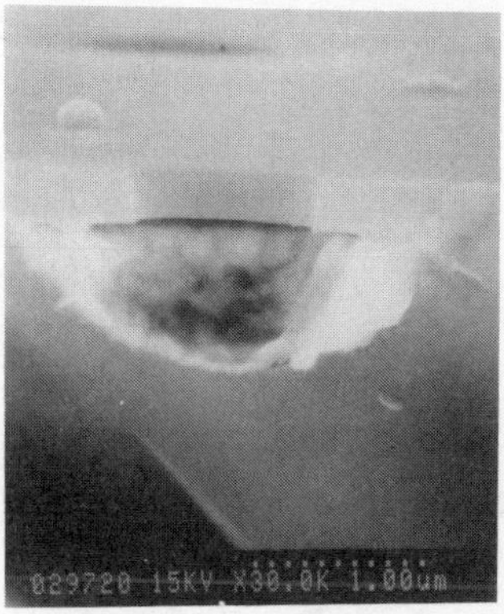

Figure 6. Deposition in a 1.3 micron diameter via through SiO_2 to silicon. Severe encroachment and some deposition are visible.

Capping Layers

After observation of the severe etching of silicon, attempts were made to put a capping layer between the silicon and the depositing molybdenum to stop this reaction. The obvious choice was selectively deposited tungsten since it is such an effective barrier to the WF_6/Si reaction. Sputtered TiW and Mo were also attempted. None of these materials were effective at stopping the MoF_6/Si reaction.

Capping experiments were done on wafers processed so that there were four areas tested on each wafer. One area was bare silicon, one was bare oxide, one was the capping layer over silicon, the last was capping layer over oxide. In all cases the sputtered metal over the oxide was etched away, leaving bare oxide. From this it is apparent that MoF_6 is a strong fluorinating agent for many metals, including molybdenum. It was also seen that the unprotected oxide had been etched. Etch rates were indeterminant because the time required to clear the metal-covered oxide was not known.

Figure 7 shows the effect of sputtered Mo over silicon. The deposit was not completely removed, but there was severe pitting of the film, similar to one of the deposition regimes over bare silicon. **Figure 8** shows the effect of sputtered TiW over silicon. The area with a deposit was bare silicon, where the heavily etched region without a deposit originally had the TiW cap. It apparently catalysed the silicon etching reaction. In this region 5.9 microns of silicon was etched, whereas over bare silicon 3 microns was etched to give a 1.5 micron deposit. These capping experiments were run at 250°C for 30 minutes.

CONCLUSIONS

Molybdenum hexafluoride has been ineffective as a source for CVD of molybdenum for microelectronic applications. Due to the lack of a self-limitation for the MoF_6/Si reaction, it encroaches severely in both the vertical and lateral directions. The reason for the difference between this and tungsten behavior is unclear. Molybdenum hexafluoride is a more powerful fluorinating agent than WF_6, as evidenced by it etching of tungsten, titanium-tungsten, silicon dioxide, and even molybdenum metal. This is

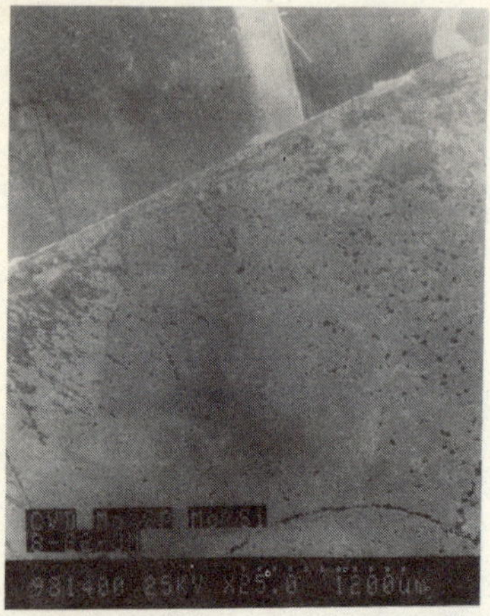

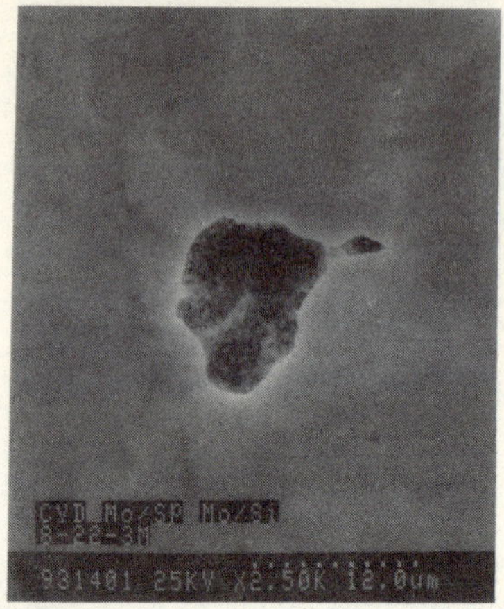

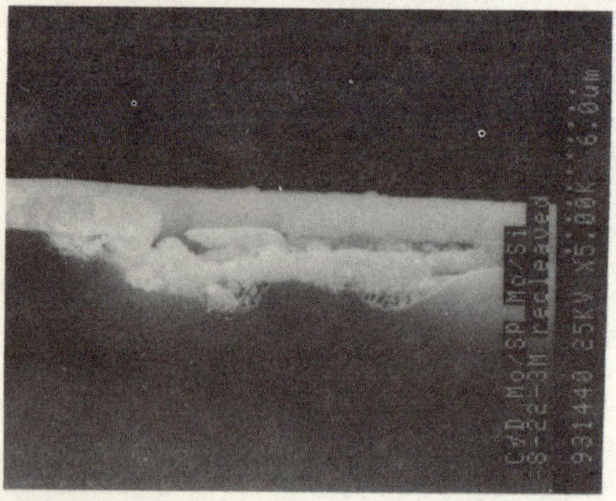

Figure 7. Top and crossection views of Mo deposition over sputtered Mo on silicon substrate. Pitting of the film is visible. Sputtered Mo was etched off an SiO_2 surface.

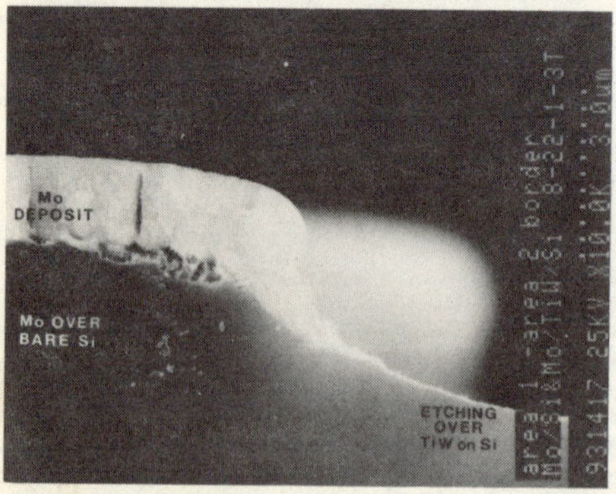

Figure 8. Mo deposition at a boundary where there was bare silicon (left) and sputtered TiW on Si (right). The TiW coated Si showed severe etching with no Mo deposition. TiW was also etched off an SiO_2 surface.

due to a higher free energy of formation (per gm-atom of fluorine) than that of the fluorides of these other metals, including MoF_5. As a result, capping metals were not found to prevent the reaction between MoF_6 and Si.

As a result of this experimental program, and in light of the continued interest in molybdenum metallization for VLSI applications, we conclude that a source other than MoF_6 will be necessary for CVD of molybdenum. The two sources most often discussed are $MoCl_5$ and $Mo(CO)_6$.

ACKNOWLEDGEMENTS

The authors would like to acknowledge the work of D.K. Dietrich on the molybdenum CVD system, C.D. Robertson and N. Lewis for electron microscopy, and M.D. McConnell for the Auger work.

REFERENCES

[1] R.S. Blewer, editor, Tungsten and Other Refractory Metals for VLSI Applications, (Materials Research Society, Pittsburgh, 1986).

[2] R.H. Wilson, R.W. Stoll, and M.A. Callacone, VLSI Multilevel Interconnect Conference, 343, (1985).

[3] R.R. Jaeger and S.T. Cohen, Third International Conference on Chemical Vapor Deposition, 500, (1972).

[4] R. Faron and M Barques, Third International Conference on Chemical Vapor Deposition, 439, (1972).

[5] M. Ayukawa, K. Yamagishi, K. Matsuda, and H. Shimizu, report 14p-C-14, 1984 Fall Meeting of Japan Society of Applied Physics,[JST News,3,(6),(1984)].

[6] C.C. Tang, J.K. Chu, and D.W. Hess, Solid State Technology, 125,(1983).

[7] T. Sugano, H. Chou, M. Yoshida, and T. Nishi, Japanese Journal of Applied Physics 7 (9), 1028,(1968).

[8] D.K. Seto, V.Y. Doo, and S. Dash, Second International Conference on Chemical Vapor Deposition, 659, (1970).

[9] K. Hieber and M. Stoltz, Fifth International Conference on Chemical Vapor Deposition, 437, (1975).

[10] D.M. Brown, W.R. Cady, J.W. Sprague, and P.J. Salvagni, IEEE Transactions, ED-18, 931, (1971).

[11] L.H. Kaplan, and F.M d'Heurle, J. Electrochem. Soc.,117,(5),693,(1970).

[12] W.J. McCreary, Fifth International Conference on Chemical Vapor Deposition, 714, (1975).

[13] G.E. Carver and B.O. Seraphin, Appl. Phys. Lett, 34 (4), (1979).

[14] G.E. Carver, Thin Solid Films,(63), 169,(1979).

MICROSTRUCTURAL CHARACTERIZATION OF MOLYBDENUM FILMS DEPOSITED BY LPCVD.

N. LIFSHITZ, J. M. BROWN AND D. S. WILLIAMS
AT&T Bell Laboratories, 600 Mountain Avenue Murray Hill, New Jersey 07974

ABSTRACT

Molybdenum films have been deposited by Low Pressure Chemical Vapor Deposition (LPCVD) on silicon substrates by the reduction of molybdenum hexafluoride in hydrogen and argon atmospheres. The deposition is extremely selective, with no Mo observed on silicon dioxide surfaces over the temperature range 200-400°C. Reduction by both hydrogen and silicon contribute to the deposition, with approximately equal, extremely high deposition rates; no self-limiting thickness was observed.

Extensive TEM and SEM studies were conducted in order to characterize the microstructure of the deposited films. The main feature of the films is their extreme porosity — about 30%, which can explain many specific properties of the process and the films. On the basis of our experiments we propose a model to explain the porosity of deposited films.

INTRODUCTION

LPCVD of W employing the fluoride chemistry has been a subject of study because of potential application of W to the needs of VLSI. Recently we reported about deposition of Mo using similar chemistry [1]. The deposition of both metals is thought to evolve around two reactions:

$$2MF_6(g) + 3Si\ (s) \rightarrow 2M\ (s) + 3SiF_4(g) \qquad (1)$$

$$MF_6(g) + 3H_2(g) \rightarrow M\ (S) + 6HF(g)\ , \qquad (2)$$

where M stands for W or Mo, and s and g represent the solid and gas phases respectively. The first reaction is the reduction of the metal fluoride by silicon. Obviously, during formation of the metal layer some Si is consumed (about twice the volume of deposited metal [2]). The second reaction is the reduction of metal fluoride by hydrogen. The free energy changes for these reactions are tabulated below [3]:

TABLE I

T = 400°C		Reduction by Si	Reduction by H_2
ΔG	W	−176	−10
(kcal/mol)	Mo	−213	−50

These calculations indicate that reactions 1 and 2 are thermodynamically favorable and, therefore, a selective deposition process of Mo is possible in analogy to tungsten deposition. Also these calculations show that the driving force for the silicon and hydrogen reductions are greater for MoF_6 than for WF_6 and that reduction by silicon has a greater driving force that reduction by hydrogen. In this study, the selective Mo LPCVD process was investigated in the 200-400°C temperature range. The deposited films were studied by Scanning and Transmission Electron Microscopy and by Auger, Rutherford Backscattering and X-ray Photoelectron Spectroscopies.

RESULTS

The molybdenum films were deposited in a hot wall tubular reactor schematically shown in Fig. 1. The detailed description of the reactor and experimental set up can be found elsewhere [1]. The distinctive features specific to the LPCVD of Mo are described below.

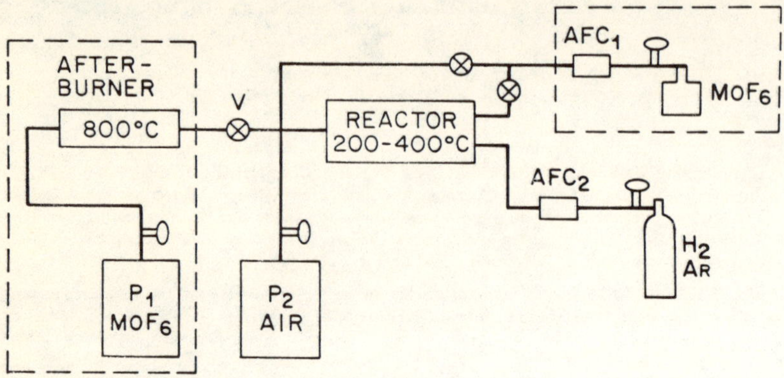

Fig. 1 Schematic diagram of Mo LPCVD station.

a. Deposition rates

In Fig. 2 we plot the film thickness vs. time for different process conditions. The film growth is extremely fast and linear with time. An interesting feature is the incubation time that precedes deposition. This incubation time is shorter for higher deposition temperatures and for hydrogen containing reactant gas streams. Clearly the deposition rates strongly depend on temperature — the films formed at 250°C are an order of magnitude thicker than the films deposited at 200°C. It is also evident that the deposition rate depends on the ambient: in hydrogen the deposition rate is almost a factor of 2 higher than in the inert Ar.

b. Impurity Content

1 XPS and Auger spectroscopies revealed that all LPCVD Mo films contain large amounts of oxygen, though this amount depends on the temperature at which the reactor was open to air. The oxygen content varied from 4 at. % in the films expoosed to air at room temperature to 20 at. % at 250°C. Apparently, the exposure of the films to air at higher temperatures causes the retention of oxygen to be greater. Oxygen was distributed uniformly through the thickness of the layer. This indicates that oxygen can diffuse freely through several microns of the film material. Even films exposed to air at 25°C contained unusually high amounts of oxygen (4%). No other impurities were found in the films.

c. Resistivity

Resistivity of molybdenum films was obtained from sheet resistance and thickness measurements. The sheet resistance of Mo films was measured by the four point probe method and the thickness of the samples was obtained from SEM cross-sectioning. Usually the films were cooled to 150°C before exposing them to air, and their resistivity was measured to be $\sim 90\ \mu\Omega \cdot cm$, a factor of 15 higher than in bulk molybdenum. If the reactor was cooled to room temperature, the resistivity was found to be $\sim 60\ \mu\Omega \cdot cm$, still an order of magnitude higher than in bulk Mo.

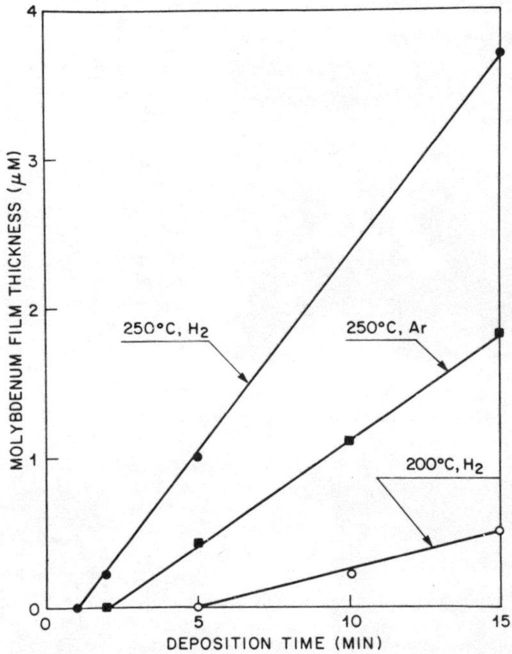

Fig. 2 Dependence of deposited Mo film thickness on time for different process conditions.

FILM MORPHOLOGY

Scanning electron micrographs (SEM) of typical LPCVD Mo films are shown in Fig. 3 and 4. The approximately 2 μm film in Fig. 3 was deposited on Si substrate. The Mo-Si interface is rough, indicating that Si reduction takes place. The film has an open, loose columnar morphology. Figures 4a and 4b show the structure of Mo films at the edge of a 1000Å thick SiO_2 pattern. The film in Fig. 4a was deposited in Ar atmosphere, so that no hydrogen reduction reaction could occur. The film in Fig. 4b was formed in H_2 atmosphere, so that both Si and H_2 reduction reactions contributed to the film growth. The most striking feature of these films is the extensive voids at the SiO_2 edge. These voids are a manifestation of the Si reduction reaction, in which the volume of deposited Mo is only one half of the volume of displaced Si. As seen in Fig. 4a, away from the oxide pattern edge the Mo film surface is below the level of the original interface. The structure in Fig. 4b is produced by a deposition with both the hydrogen and the silicon reduction acting: hence the SiO_2 "marker" is embedded in the deposited film at the original surface position.

The deposited structures shown in Fig. 5a,b also support the conclusion that two reactions take part in the deposition process. The contact window in Fig. 5a was exposed to the MoF_6/Ar atmosphere. A large volume of Si in the substrate has been removed during the deposition of the mounds of Mo. The window in Fig. 5b was exposed to MoF_6/H_2 atmosphere. The substrate also shows voids similar to Fig. 5a, but the volume of deposited Mo is much greater, and the window is filled. The difference in total Mo volume is clearly due to H_2 reduction, while the similarity in the substrate erosion is the result of the Si displacement reaction.

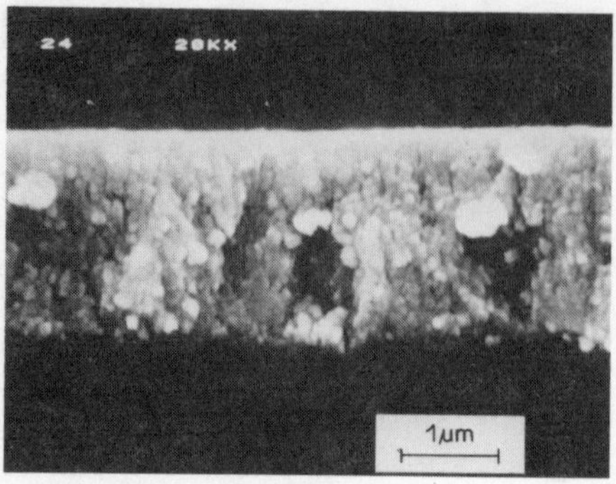

Fig. 3 Cross-sectional SEM micrograph of a Mo film deposited on a flat Si substrate.

REDUCTION BY SILICON

REDUCTION BY HYDROGEN

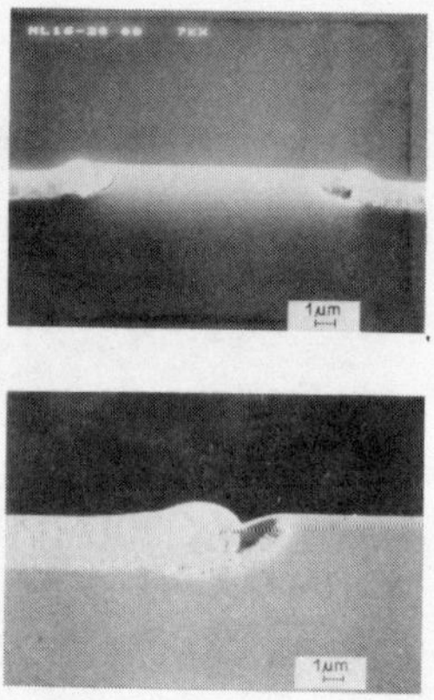

Fig. 4 Cross-sectional SEM micrographs of Mo films deposited on the SiO patterned Si substrate.

a. deposition of MoF_6/Ar atmosphere.

b. deposition in MoF_6/H_2 atmosphere.

REDUCTION BY SILICON

REDUCTION BY HYDROGEN

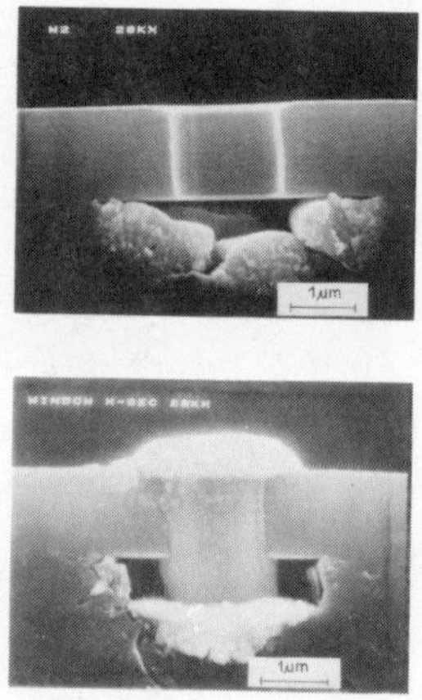

Fig. 5 Cross-sectional SEM micrographs of 1 μm windows in PTEOS, subjected to Mo LPCVD process.

a. in MoF_6/Ar atmosphere.

b. in MoF_6/H_2 atmosphere.

In Fig. 6 a cross-sectional TEM micrograph depicting the structure of the Mo-Si interface is shown. One can see microscopic protrusions densely covering the interface. These protrusions contain small particles which are either metallic or metal containing. Evidence for this comes from a high resolution TEM of the Mo-Si interface (Fig. 7) which clearly shows periodic lattice and Moire-fringe patterns. Similar morphology has been reported to exist at the W-Si interface [4]. The analysis of the electron diffraction pattern identified the particles as either Mo or $MoSi_2$. An even more remarkable similarity between Mo and W LPCVD processes is that the wormholes usually observed at the W-Si interface [2,5] were also observed at the Mo-Si interface (Fig. 8).

POROSITY OF Mo FILMS

We have discussed some specific features of Mo films: extremely high deposition rates, uninterrupted Si reduction, uniform oxidation and high resistivity of the films. All these properties can be readily understood by invoking the porosity of the films.

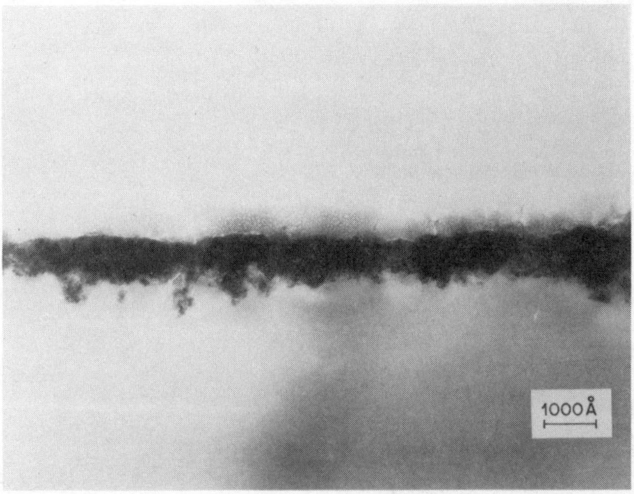

Fig. 6 Cross-sectonal TEM micrograph of Mo-Si interface.

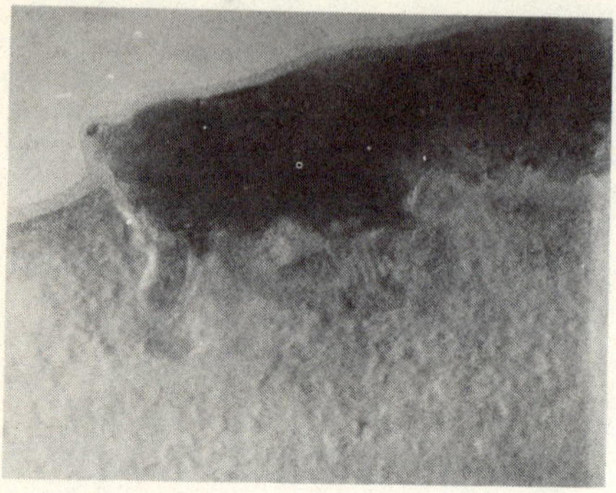

Fig. 7 High magnification cross-sectional TEM micrograph of Mo-Si interface showing periodic lattice and Moire-fringe patterns in the protrusions.

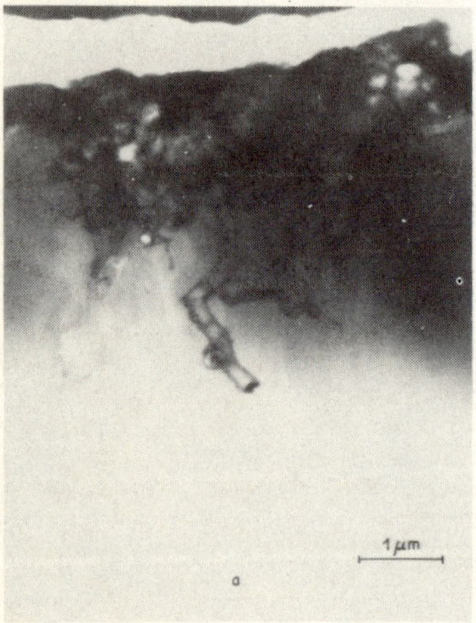

Fig. 8 a. Cross-sectional TEM micrograph of Mo-Si interface showing the
wormholes at the interface.

Quantitatively the porosity was determined by a comparison of the geometrical thickness with the projected thickness of theoretically dense material. The geometric thickness was determined by SEM cross-sectioning, and the thickness of theoretically dense material was calculated from an RBS spectrum of the same sample, taking into account the oxygen content of the films. The ratio of bulk to geometric thickness was found to be 0.7, meaning that the density of the deposited films is only 0.7 of the density of bulk Mo and hence extremely porous. Such porosity of the films allows reactants and products to move freely to and from the silicon/metal interface, and can explain the unusual properties of the deposition process and the deposited film structure. In a random three-dimensional two-phase structure the percolation of one phase occurs when it occupies more than approximately 25% of the volume [6] (percolation indicates a path through a continuous two-phase body lying entirely within one phase). This means that both phases can be mutually percolating if one phase occupies between 25 and 75% of the volume, a condition fulfilled in LPCVD Mo films in which bulk metal occupies 70% and voids occupy 30% of the volume. Moreover, the incompletely randomized structure of the voids with vertical channels between the grains, as shown in Fig. 3, further facilitate the percolation. This model can account for the uninterrupted reaction of Si with MoF_6 — both reactant and product gases can easily diffuse through the open porous structure. The porosity is also responsible for the uniform oxidation across the film thickness and the high resistivity of the films.

Strong evidence of the extreme porosity of Mo films is given in Figs. 9a,b. Here we show a TEM planar view of a 2000Å thick film. These micrographs were made using a 300 keV electron beam. The Si substrate under the Mo layer was thinned but remained intact. The Mo film was also protected from the top by sample mounting material so that the observed voids are not artifacts of sample preparation. In Fig. 9a one can distinguish small crystallites which collectively form into large loose clusters. These clusters are only locally connected and form large voids. The voids appear as long fissures in the microstructure of dimensions ~ 0.5 microns wide and several microns long. Higher resolution TEM (Fig. 9b) suggests that porosity of smaller dimensions (<1000Å) is also present. One can distinguish in the figure small clusters less than 100Å in size which are loosely connected into larger clusters. In fact the film forms almost a fractal like structure with porosity occuring at all scales. However, the presence of diffuse rings on the electron diffraction patterns (Fig. 9a, insert) indicates that the film is polycrystalline. Most interestingly, similar porous texture was recently discovered in W films deposited by silicon reduction [7]. Obviously the two processes have much in common.

The morphology of the films evolves during film growth, as illustrated by the SEM micrographs in Fig. 10. The surface of a very thin (≈ 100Å) film grown in H_2 atmosphere is shown in Fig. 10a. One can distinctly see a dense network of fissures, similar to that shown in the TEM micrograph in Fig. 9. The density of the fissures decreases and their scale increases as the film grows thicker (Fig. 10b, 0.2 μm). The surface of very thick films (Fig. 10c, 2 μm) looks absolutely smooth — all fissures have disappeared. However, this does not slow down the further growth, indicating that the main transport of the reactant material occurs not through the fissures but through the smaller pores.

EFFECT OF SILICON TERTRAFLUORIDE SiF_4 ON THE DEPOSITION PROCESS

The addition of SiF_4 to the reactant gases in W deposition was attempted in study [2] with the intent of minimizing Si erosion by shifting the chemical equilibrium of reaction (1). The authors observed that the sheet resistance of deposited material increased with SiF_4 partial pressure and concluded that Si consumption was indeed suppressed. We carried out similar experiments in Mo deposition and observed a similar increase in sheet resistance. However, we observed that while the SiF_4 curtails the Mo growth it does nothing to decrease the Si consumption. In Fig. 11 we show SEM cross-sections of patterned SiO_2 silicon wafers exposed to the reactant gases with the addition of SiF_4. It is clear the that addition of SiF_4 does not prevent Si erosion, but it effectively prevents metal deposition. Interestingly, for complete Mo removal the flow of SiF_4 should be a factor of 2 higher in the hydrogen/MoF_6 atmosphere than in the Ar/MoF_6 atmosphere. This is clearly due to the difference in the deposition rates in hydrogen and argon atmospheres (see Fig. 2).

From our evidence, the mechanism of this effect is not clear. However, we can speculate that both W and Mo LPCVD are more complex than indicated by reactions (1) and (2). Instead of Eq. 1 a generalized equation may be written:

$$MF_n + SiF_m \rightarrow MF_{n-1} + SiF_{m+1} ,$$

where n runs from 6 to 0 and m from 0 to 4. The intermediate reaction products may have very short lifetimes (only SiF_2, SiF_3, MoF_4, MoF_5 and WF_5 are known to be stable and volatile) but it may be

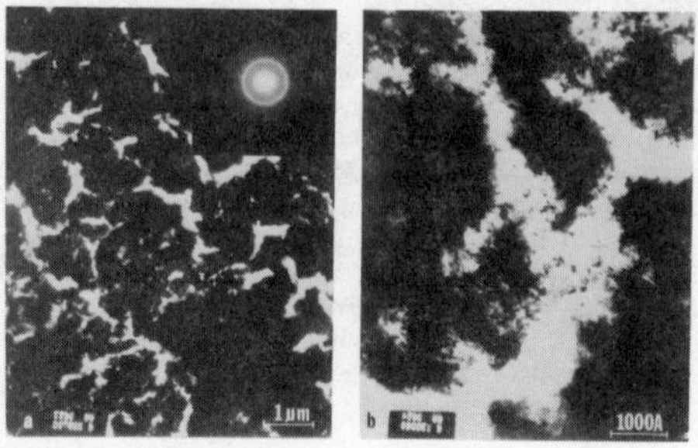

Fig. 9 Planar TEM micrograph of 2000Å thick Mo film.

a. 25,000 magnification.

b. 350,000 magnification.

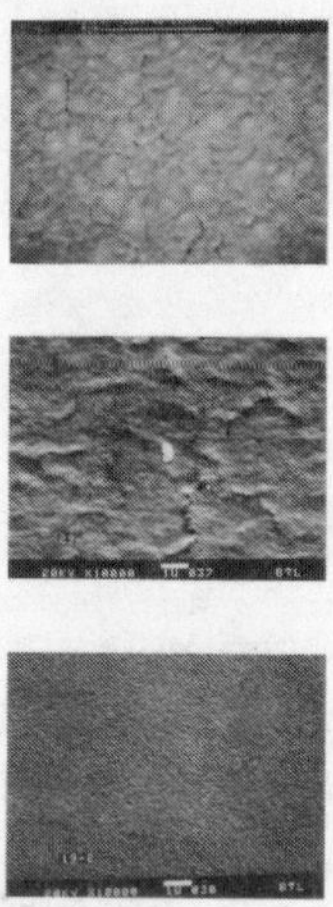

Fig. 10 Surface morphology of Mo films of different thicknesses.

a. <100Å

b. 2000Å

c. 4 μm

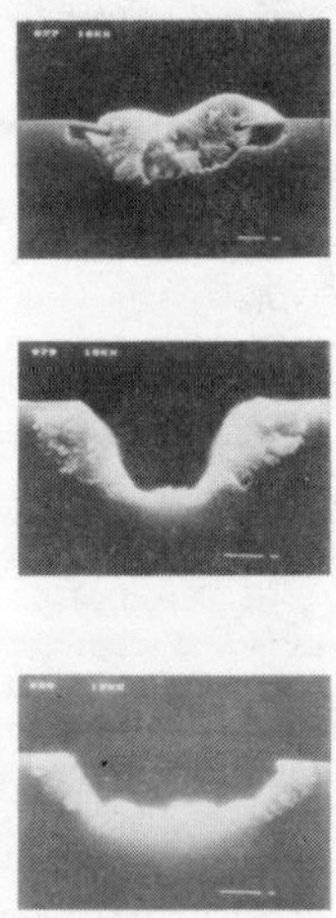

Fig. 11 Morphology of Mo films deposited with SiF_4 in the reactive atmosphere.

sufficient to spacially and temporally separate the event of Si removal at M/Si interface (m changes from 0 to 1) from the event of M formation (n changes from 1 to 0). The addition of SiF_4 may affect one of the intermediate reactions leading to metal atom formation and hence decrease the M deposition rate, but still have no impact on the reaction of Si sublimation. If one accepts this assumption then the next step is to speculate that since SiF_4 is a final product of Si reduction reaction, it is always present in the film during the process and may *locally* impede metal deposition causing porosity in both Mo and W. However there is a significant difference between the two materials: whereas Mo has the same porous structure whether deposited in hydrogen or argon, W exhibits porosity only when deposited in an Ar atmosphere (note that studies conducted on deposition of W at atmospheric pressure in N_2 atmosphere also resulted in extremely porous selective deposits [8,9]). This difference still requires further investigation although some factors can be pointed out: Mo has two lower fluorides, MoF_5 and MoF_4, which are both volatile and stable, while for W only WF_5 is known to have these properties; also, MoF_6 in a H_2 atmosphere is known to etch metal Mo [10] which may result in fractional removal of Mo deposited by hydrogen reduction.

CONCLUSIONS

We conducted an extensive study of LPCVD of Mo using reduction of molybdenum hexafluoride MoF_6 by silicon and by hydrogen. The process is extremely selective, with no Mo observed on silicon dioxide surfaces in the 200-400°C temperature range. Reduction by both silicon and hydrogen contribute to the deposition, with approximately equal, extremely high deposition rates; no self-limiting thickness was observed. The main feature of the deposited films is their extreme porosity — about 30%, which can explain such specific properties of the process as the high deposition rates, ever continuing reaction of Si reduction, high resistivity and uniform oxygen content of the films. SEM and TEM studies indicate that porosity of the films exists on all scales: from tiny pores smaller than 100Å to fissures several microns in size. Inspite of the obvious difference between Mo and W LPCVD processes some similarities are apparent. They are: the similar texture of the metal/silicon interface, the presence of the "wormholes" containing metal particles, and porosity of the (in case of silicon reduced W) deposited material. The results of the experiments with silicon tetrafluoride in the reactive atmosphere offered a glimpse of the complexity of this process and illuminate the origin of porosity in the films.

ACKNOWLEDGEMENT

We would like to thank H. S. Luftman for many illuminating discussions.

REFERENCES

[1] N. Lifshitz, C. D. Capio, D. S. Williams, J. M. Brown, to be published.

[2] M. L. Green and R. A. Levy, J. Electrochem. Soc. *132*, 1243 (1985).

[3] JANAF Thermochemical Tables, 2nd Edition, D. R. Stull and H. Prophet, Eds., National Bureau of Standards, Washington, DC, 1971.

[4] D. C. Paine, J. C. Bravman and K. C. Saraswat, Tungsten and Other Refractory Metals for VLSI Applications. Robert S. Blewer, ed MRS, Pittsburg, Pa (1985) p. 117.

[5] E. K. Broadbent, W. T. Stacy, Solid State Technology, Dec. 1985, p. 51-59.

[6] B. I. Shklovskii and A. L. Efros, "*Electronic Properties of Doped Semiconductors*," Springer-Verlag, Berlin, 1984, p. 104.

[7] M. L. Green, Y. S. Ali, T. Boone, B. A. Davidson, L. C. Feldman and S. Nakahara. To be published.

[8] C.-E. Morosanu and V. Soltuz, Thin Solid Films, *52*, 181 (1978).

[9] R. J. Miankovski, K. Y. Tsao and H. A. Waggener, Tungsten and Other Refractory Metals for VLSI Applications. Robert S. Blewer, ed. MRS, Pittsburg, PA (1985) p. 145.

[10] Yu. M. Korolev, A. V. Rychagov, Fiz Khim. Obrab. Mater. *2*, 128 (1978).

Characteristics of CVD Tungsten Films in Device Structures

The Influence of Selective Tungsten Deposition on Shallow Junction Leakage

G. E. Georgiou, J. M. Brown, M. L. Green, R. Liu, and D. S. Williams
AT&T Bell Laboratories, Murray Hill, New Jersey 07974

R. S. Blewer
Sandia National Laboratories, Albuquerque, NM 87185

ABSTRACT

We report on the onset of reverse bias diode leakage induced by CVD selective tungsten deposition on (a) shallow n^+(As)/p diodes made by implantation into silicon and (b) n^+(As or P)/p or p^+(BF_2)/n diodes made by implantation into $CoSi_2$. Tungsten, nominally 1000Å thick, was selectively deposited in a hot wall tubular reactor using either a one-step or a two-step deposition sequence at 290°C. The n^+(As)/p junctions show an increase in leakage population after tungsten deposition, as the junction depth decreases. The two-step process was found to produce greater leakage in shallow n^+(As)/p junctions than the one-step process.

Arsenic implanted junctions show severe degradation in the leakage distribution of shallow diodes beneath $CoSi_2$. In contrast to arsenic, shallow junctions made by outdiffusion of phosphorus or boron implanted into $CoSi_2$ do not show significant changes in leakage population after selective tungsten deposition. Junction leakage measurements, and their temperature dependence, will be correlated with transmission electron micrographs.

1. INTRODUCTION

The LPCVD of selective tungsten has been studied by many groups as a barrier material to the Aℓ-Si interaction. Depositions have been reported using both hot wall tube systems [1,2,3] with ≈30Å/min deposition rate and a cold wall single wafer system [4] with ≥500-1000Å/min deposition rate. The initial stages of deposition are known to involve an interaction of the WF_6 source gas with the silicon substrate resulting in a net consumption of silicon. However, a systematic study of the effect of selective tungsten deposition on diode leakage in shallow junctions (≤2500Å) has not been reported.

After a description of the experimental procedure, we will present reverse bias diode leakage data obtained after thin selective tungsten is deposited on arsenic-doped silicon n^+/p diodes with progressively shallow junction depths. We will also describe data obtained after selective tungsten deposition on n^+/p and p^+/n junctions formed by outdiffusing As, P or B from $CoSi_2$. The barrier properties of selective tungsten are studied by measuring the junction leakage before and after aluminum deposition and after hydrogen anneal. This leakage data is interpreted using (1) the temperature dependence of reverse bias leakage to determine the activation energy of the leakage mechanism and (2) transmission electron microscopy to identify microstructural features that may cause leakage.

2. Experimental Procedure

The test vehicle used in this work is the p-n diode made using the lithography masks which make up the M7118 MOS monitor system [5]. As is shown in Fig. 1, 130 diode sets of various areas and configurations are arranged in a "cross" pattern covering the central region of the wafer.

Bulk <100> p-type (6-8 Ω-cm) and n-type (10-20 Ω-cm) wafers are used.

Diode processing begins with the growth of a 3500Å thick field oxide in dry oxygen followed by a steam ambient. To avoid surface inversion, p-type wafers are given a 2E12 cm^{-2} boron "chanstop" implant at 160keV. The field oxide is then patterned by a reactive ion etch. The processing then diverges for the various junctions types.

The n^+/p diodes without $CoSi_2$ received 7E15 cm^{-2}, 30-60keV arsenic implantation through a 100Å thick deposited oxide and are annealed for different temperatures and times to achieve the desired junction depths. A ≈1μm thick SiO_2 film is then deposited.

The n^+/p and p^+/n diodes with $CoSi_2$ [6] received 200Å cobalt which is sputter deposited, and then annealed in hydrogen to form CoSi. An etch to remove the unreacted metal from the field oxide is followed by the final anneal to form $CoSi_2$ at 700°C. After the appropriate implantation into $CoSi_2$, a ≈1μm thick SiO_2 is deposited. A heat cycle at 800°C is used to outdiffuse the As, P or B from the $CoSi_2$ to produce the appropriate junction depths.

Windows are then reactive ion etched into the deposited oxide. A hydrogen anneal at 500°C, 60 min. is done to remove any process-related damage that may affect reverse bias junction leakage. This leakage current is then measured at several reverse bias voltages. A map is recorded for each substrate and is compared to that after selective W deposition, aluminum deposition and hydrogen anneal.

Tungsten is then selectively deposited in a hot wall tube reactor using either a one-step or a two-step process. The one-step process introduces WF_6 and H_2 (1:400 flow rate ratio) simultaneously at 290°C. The total deposition time is ≈40 min. at 290°C. The two-step process involves an initial tungsten displacement reaction with silicon using WF_6in argon for 2 min. at 350°C followed by hydrogen reduction of WF_6 (10:4000 sccm WF_6:H_2) at 290°C. The deposited tungsten thickness is ≈1000Å. In order to ascertain whether the electrical results vary with reactor operation, depositions were performed at AT&T Bell Laboratories and Sandia National Laboratories with subtle differences in operating procedures noticeable. In each case, the silicon or $CoSi_2$ surface was cleaned in dilute 100:1HF immediately before loading into the reactor to minimize the presence of any oxide.

The completed diode cross-section (for the $CoSi_2$ case) is shown at the bottom of Fig. 1. The lateral distance between the field oxide and the window edge is 10μm which should eliminate the concern of perimeter leakage from these experiments.

The aluminum + 0.5% copper metallization is 1μm thick and is sputter deposited with the substrate preheated to 300°C after an in-situ argon backsputter clean. The Al deposition is followed by a 330°C or 450°C H_2 anneal for 45 min.

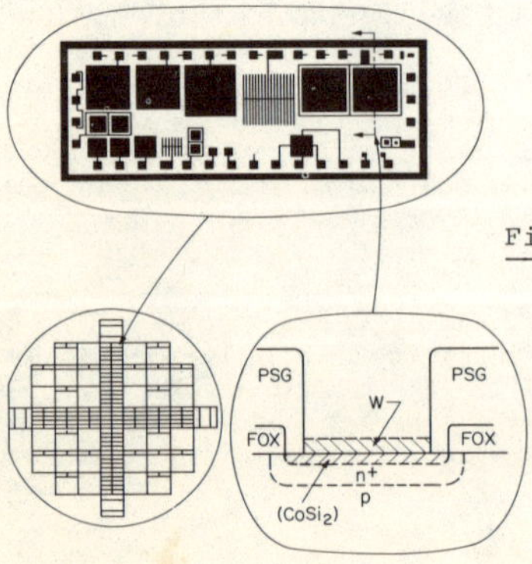

Fig. 1: Diode array layout and n^+/p diode cross section below $CoSi_2$

3. Dopant Profiles

For implants into silicon, the dopant profiles are calculated [7] and shown in Fig. 2 for the various anneal conditions used to obtain 1600, 2100, 2600 and 3400Å deep junctions with ≈2×10^{20} cm^{-3} As surface concentration. The implant dose is a constant 7E15 cm^{-2} with the implant energy, the anneal temperature and the anneal time chosen to obtain low leakage diodes with the desired junction depth.

The dopant profiles for $CoSi_2$ junctions have been determined by SIMS and RBS profiling. The junction depths resulting from an 800°C, 120 min. outdiffusion of 5E15 cm^{-2} P, 5E15 cm^{-2}As or 2E15 cm^{-2} BF_2 implanted into $CoSi_2$ are respectively, 1000Å, 500-800Å or 400Å below the $CoSi_2$/Si interface.

4.0 Experimental Results:

The reverse leakage current from 130 diodes on each wafer were obtained by automatic probing at room temperature. Arrhenius plots of leakage current as a function of temperature were obtained by hand probing select diodes on a thermal chuck. Here, we present a summary of the leakage behavior. The data expressed as current density is obtained from large area diodes with the metallization area being approximately the diode area. We also present TEM microstructure results.

4.1 Arsenic n^+/p Junctions in Silicon

A histogram showing the diode leakage resulting from one-step W deposition, and subsequent aluminum deposition and 330°C H_2 anneal, on 1600Å deep arsenic junctions formed in silicon is shown in Fig. 3. Before W deposition, 90% of the diodes have leakage ≈1nA/cm^2 with only 0.5% having leakage greater than 1µA/cm^2. After a one-step W deposition, the high leakage (>1µA/cm^2) population includes 20% of the diodes. Therefore, the degradation caused by selective W deposition is not a uniform increase in mean leakage, but is an increase in the number of diodes having extremely high leakage. After aluminum deposition and 330°C H_2 anneal, the number of low leakage diodes is substantially reduced with less than 15% having leakage less than 10nA/cm^2. About 70% of the population has leakage between 1 and 100µA/cm^2.

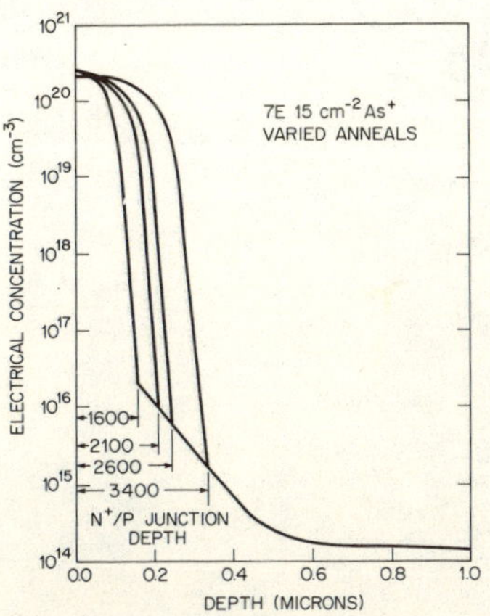

Fig. 2: Calculated As profiles for various n^+/p junction depth diodes in Si.

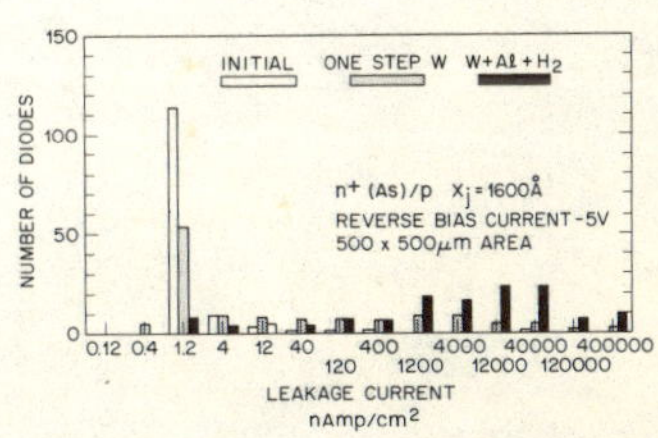

Fig. 3: Histogram of 1600Å deep n^+(As)/p in Si diode leakage for unmetallized diodes (initial) and those after 1-step W and Al+330°C H_2.

Reverse bias diode mean leakage is shown in Fig. 4 as a function of junction depth for $n^+(As)/p$ diodes in Si. The figure is a compilation of the leakage before tungsten deposition, and the leakage after i) two step W CVD, ii) one step W CVD in two different reactors, iii) and after Aℓ deposition and 330°C, 45 min. anneal in H_2. The two step tungsten deposition results in a mean leakage increase of 4 orders of magnitude for junction depths less than ≈2000Å. The one step tungsten deposition process increases the mean leakage current by more than a factor of 10 for junction depths of ≈1600Å. Aluminum deposition and 330°C H_2 anneal further degrades the leakage of ≈1600Å deep junctions by another 2-3 orders of magnitude.

The sport population, defined as the percentage of diodes with leakage greater than >1μA/cm^2, is shown in Fig. 5 for the same diodes. These results parallel those of mean leakage and show an increase in the number of high leakage diodes after two step W deposition for x_j≈2500Å, after one step W deposition for x_j≈2000Å. There is also a dramatic increase in the sport population after aluminum deposition with ≈20% of the ≈2500Å deep junctions having leakage >1μA/cm^2.

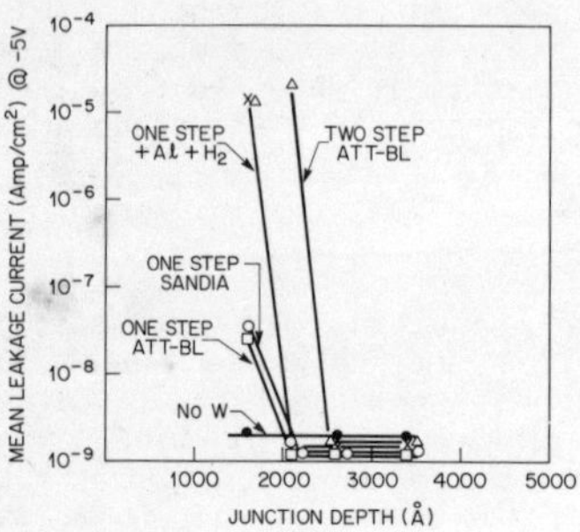

Fig. 4: Mean leakage current density vs. $n^+(As)/p$ junction depth in Si for unmetallized diodes (no W), after 1-step or 2-step W CVD, and after Al+330°C, 45 min. H_2.

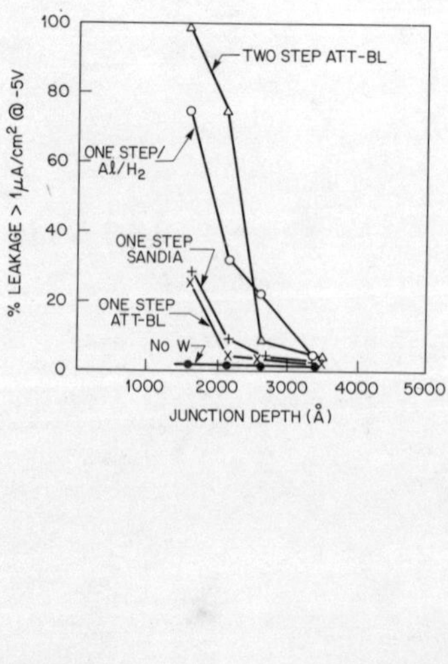

Fig. 5: Sport population vs. $n^+(As)/p$ junction depth in Si for unmetallized diodes(no W), after 1- or 2-step W and after Al+330°C H_2.

4.2 Junctions Below Cobalt Silicide

The influence of tungsten deposition on $CoSi_2$ shallow junctions is shown in Fig. 6. Here, the mean diode leakage at -5V is shown as a function of junction depth below the $CoSi_2$/Si interface for As, P and B doping before and after one step selective W deposition. The data obtained from one-step W deposition on the diodes discussed in the previous section (Fig. 4) is replotted in Fig. 6 for comparison. W deposition on arsenic doped junctions results in a four order of magnitude increase in leakage when x_j<1500Å below the $CoSi_2$/Si interface. The result is expected if the $CoSi_2$ does not prevent WF_6-Si interaction and the results for unsilicided n^+(As)/p junctions are extrapolated to shallower junction depths. In contrast, phosphorus doped junctions do not exhibit as severe an increase in leakage for comparable junction depths.

Ninety percent of the arsenic n^+/p junctions under $CoSi_2$ show a >3-4 order of magnitude increase in junction leakage. In contrast the phosphorous n^+/p junctions show a much smaller leakage degradation. The leakage current distribution shows that for x_j≈1000Å below the $CoSi_2$/Si interface, one step W deposition does not affect the majority of the diodes but appears to increase leakage in primarily the diodes that showed higher initial leakage current. Aluminum deposition and anneal does not affect the majority of the diodes but appears to severely degrade the diode leakage of ≈10% of the population. In complete contrast to n^+/p diodes, p^+/n junctions do not show any change in either the mean value or distribution of leakage current after W deposition or after aluminum deposition and 330°C anneal.

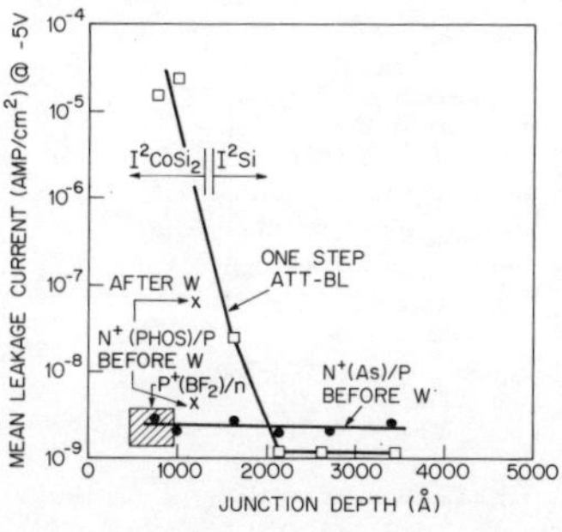

Fig. 6: Mean leakage current vs. As or P n^+/p or B p^+/n junction depth below $CoSi_2$ (x_j<1500Å).

4.3 Temperature Dependence of Diode Leakage

The junction leakage results presented in the previous section have been investigated by Arrhenius measurements of the reverse bias current. Low levels (≤$10^{20}cm^{-3}$) of contamination which can cause leakage if the contamination sites are within the depletion region can be observed by activation energy determinations.

The temperature (T) dependence of the reverse bias leakage current I_R, is given by [8]

$$I_R \propto T^3 e^{-Q/kT} \tag{1}$$

where Q is the activation energy of the leakage current generation mechanism. Figure 7a shows the I-V characteristics of an unmetallized n^+(As)/p junction ≈500Å below $CoSi_2$ at various temperatures. A plot of log (I_R/T^3) vs. 1/T at 5V reverse bias is shown in Fig. 7b. The slope of these data gives an activation energy of Q≈1.1ev corresponding to the band gap of silicon, which indicates that the leakage current of the $CoSi_2$ diode before tungsten deposition is dominated by diffusion current from outside the depletion region. The effect of the one step W deposition process is also shown in Fig. 7b where the leakage mechanism has an activation energy of ≈0.4-0.5 eV. Here, the deposition of W introduces generation centers which are close to the mid-band gap of silicon. Although, the interpretation of this activation energy is not unique, it should be noted that a tungsten impurity in silicon has several ionization potentials between 0.2 and 0.4 eV [9].

Arrhenius plots for two step W deposition on a 2400Å n^+ (As)/p junction in silicon for different degrees of diode leakage degradation are shown in Fig. 8. Diodes which show no increase in leakage yield the expected activation energy of Q≈1.1eV (Fig. 8a). Diodes which show a moderate increase (10-100 times) in leakage indicate two activation energies. At low temperatures, mid-band gap generation centers with Q≈0.4eV are responsible for reverse bias current. At high temperatures Q≈1.1eV indicating that the reverse bias current is dominated by diffusion current outside the depletion region. Diodes exhibiting high leakage degradation are controlled by mid-band gap generation centers with Q≈0.35eV. This activation energy is similar to the value reported earlier for shallow junction n^+(As)/p diodes below $CoSi_2$.

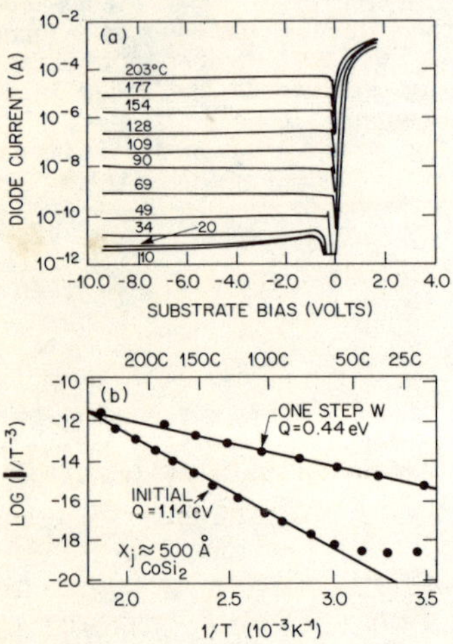

Fig. 7: a. I-V characteristic vs. temperature for As n^+/p diode below $CoSi_2$. b. Leakage current vs. temperature before and after 1-step W CVD.

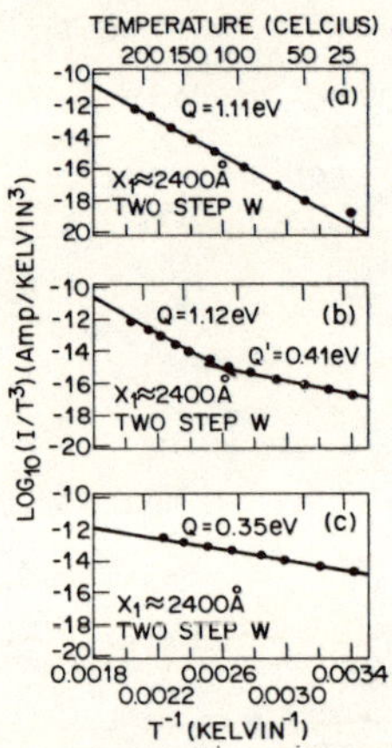

Fig. 8: Leakage current vs. temperature for 2400Å deep As n^+/p junctions in Si with (a) low, (b) medium and (c) high leakage current after 2-step W CVD.

4.4 Transmission Electron Microscopy

The catastrophic degradation of diodes made by implanting arsenic into $CoSi_2$ is readily explained by observing the microstructures. The penetration of tungsten through $CoSi_2$ is clearly seen in the TEM micrographs of Fig. 9. The W penetration occurs primarily at grain boundaries, pinholes or at points where the silicon substrate is exposed. Pinholes observed in $CoSi_2$ are attributed to excessive reactive ion overetch.

The penetration of tungsten through $CoSi_2$ on phosphorus or boron doped silicon is not severe. This is seen in the TEM micrographs of Fig. 10 for one step W on $CoSi_2$ on (a) phosphorus doped Si and (b) boron doped Si.

A TEM micrograph of a high leakage diode made with aluminum on one-step W deposited on a 1600Å deep n^+(As)/p diode in silicon, is shown in Fig. 11 after (a) 330°C, 45 min. and (b) 450°C, 45 min. H_2 anneals. The window edge area, which has been observed [10] to be most susceptible to problems such as wormholes or encroachment, is shown. No major tungsten penetration into the doped region is observed within the limits of TEM sampling statistics. It should be noted that low concentrations of tungsten or point defects would not be detectable by TEM.

The mechanism producing further increase in the leakage of ≤2000Å deep n^+(As)/p diodes after aluminum deposition and 330°C anneal is subtle. The TEM micrograph of Fig. 11(a) does not indicate any microstructural explanation for increased leakage. However, the TEM micrograph of Fig. 11(b) shows aluminum spiking through W after 450°C H_2 anneal. This is not observed at every aluminum grain boundary but about once every 10 aluminum grains or about 50μm.

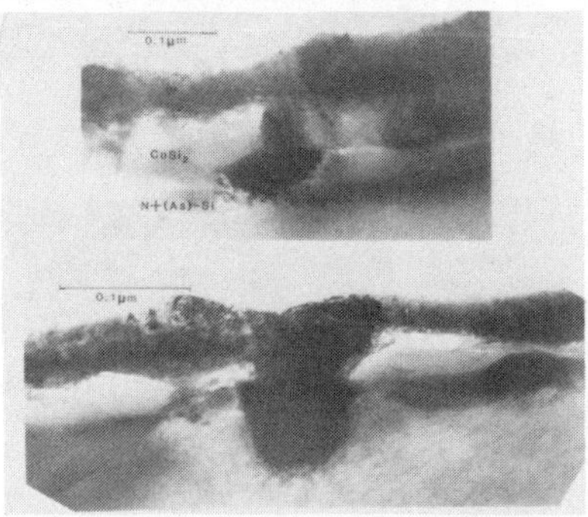

Fig. 9: TEM micrographs of 1-step W CVD on As n^+/p diodes below $CoSi_2$.

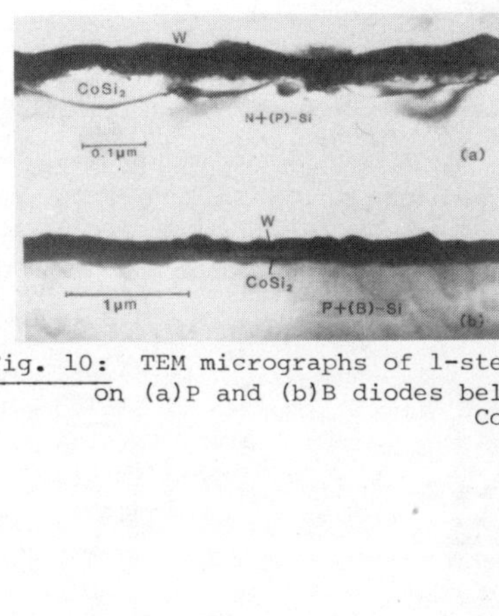

Fig. 10: TEM micrographs of 1-step W on (a)P and (b)B diodes below $CoSi_2$.

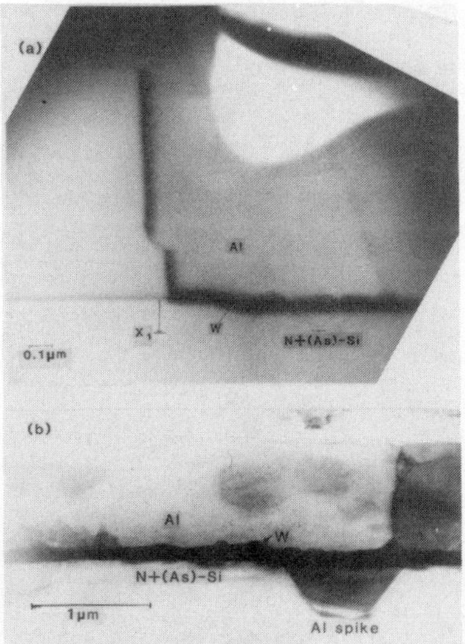

Fig. 11: TEM of Al(.5%Cu) on 1-step W on As junction in Si after (a)330°C and (b)450°C, 45 min. H_2 anneal.

5. Conclusions

Selective tungsten deposition causes an increase in the diode leakage of shallow arsenic n+/p junctions in silicon. The magnitude of diode leakage degradation increases as the junction depth decreases. A microstructural cause of leakage was not observed. However, the temperature dependence of diode leakage indicates impurity or point defect generation centers with an activation energy of about 0.4eV. This activation energy is observed in high leakage arsenic junctions in both silicon for $X_j \geq 2500Å$ and in junctions $\lesssim 1000Å$ below $CoSi_2$.

$CoSi_2$ with the associated processing does not seal the junction against WF_6 penetration into arsenic doped silicon. Tungsten penetration occurs through the $CoSi_2$ grain boundaries or at points where small $CoSi_2$ grains are removed because of reactive ion etching. In contrast to arsenic doped silicon, the WF_6-Si interaction through $CoSi_2$ is not severe for phosphorus or boron doped silicon. The n^+(Phos)/p diodes indicate some degradation after W deposition. The p^+/n diodes, which are shallower than any of the n^+/p diodes fabricated, are not degraded by selective tungsten deposition or subsequent aluminum deposition and 330°C H_2 anneal.

Aluminum deposition and anneal further degrades arsenic n^+/p diode leakage. Although aluminum spiking through nominally 1000Å thick CVD selective W is not observed after a 330°C H_2 anneal, low density aluminum diffusion cannot be excluded. Aluminum spiking is observed after a 450°C H_2anneal.

REFERENCES

[1.] M. L. Green and R. A. Levy, Electrochem. Soc. Fall Meeting Abstracts, 1984, 599.

[2.] R. S. Blewer and V. A. Wells, ibid:, 1984, 601.

[3.] T. Moriya and H. Itoh, Proc. Workshop on Tungsten, 1985, 21

[4.] R. H. Wilson, e.al., Proc. IEEE VLSI Multilevel Metal Interconnection Conf., 1985, 343.

[5.] J. M. Andrews, BSTJ, 62, 1107 (1983).

[6.] S. J. Hillenius, et.al., To be presented at the 1986 IEEE Int. Electron Device Meeting, San Francisco, CA, 1986 (To be published).

[7.] B. R. Penumalli, IEEE Trans. Elecron Devices, ED-30, 986 (1983).

[8.] S. M. Sze, Physics of Semiconductor Devices, 2nd ed. (Wiley-Interscience, New York, 1981), p.88.

[9.] S. M. Sze, ibid., p.21.

[10.] M. L. Green and R. A. Levy, J. Electrochem. Soc., 132, 1243 (1985).

Detrimental Effects of Residual Silicon Oxides on LPCVD Tungsten Depositions in Shallow Junction Devices

ROBERT S. BLEWER AND MAREN E. TRACY
Sandia National Laboratories, P.O. Box 5800, Albuquerque, NM 87185

ABSTRACT

Integrity of shallow junction devices is dependent on maintaining smooth films of constant thickness where metal contacts doped silicon. Localized silicon oxide residue can enhance the WF_6/Si reaction used in LPCVD tungsten depositions resulting in unacceptably large amounts of silicon consumption. We have investigated phenomena which cause excess WF_6/Si interaction that can lead to incursions of tungsten into silicon structures. In layers between 1.0 nm and 2.0 nm thick, silicon oxide appears to catalyze the WF_6/Si reaction and prevent or delay the self-limiting condition of LPCVD W film formation. Consequences for shallow junction devices are discussed, and methods to avoid this problem are presented.

INTRODUCTION

Decreasing feature size, driven by the desire to achieve higher performance ICs at lower cost, affects contact and interconnect technology in several ways: increased current density and poorer step coverage over smaller contact holes stimulates electromigration failures, decreased contact area increases contact resistance, plasma etched cuts through silicon dioxide to silicon are more difficult to clear completely without severely overetching increasingly shallow junctions, and shallower junctions require smaller time-temperature budgets for post-activation process steps.

Self-aligned refractory metals and metal silicides have been used with success to address many of these problems. However, obtaining reproducible self-aligned silicide films or selectively deposited pure metal films requires a clean silicon surface, free of significant oxides or other foreign layers, to achieve optimal results.

The purpose of this paper is to highlight largely unrecognized effects of residual silicon oxides, plasma etch residues and other foreign layers on the deposition chemistry of low pressure chemical vapor deposition (LPCVD) tungsten films and to discuss the influence of these contaminants on Si/W interfacial characteristics as a result of selective tungsten deposition using the Si/WF_6 reaction.

*This work performed at Sandia National Laboratories was supported by the U. S. Department of Energy under contract DE-AC04-76DP00789.

THIN SELF-LIMITING FILM FORMATION

Tungsten and molybdenum LPCVD film depositions on silicon are unusual in that the film grows not only upward from the substrate, as is the usual case in thin film depositions, but also downward, consuming underlying silicon, as well. For CVD tungsten film depositions, the two reactions which cause this behavior are given below.

(1) $2WF_6 + 3Si \rightarrow 2W + 3SiF_4\uparrow$ (silicon reduction reaction)

(2) $WF_6 + 3H_2 \rightarrow W + 6HF\uparrow$ (hydrogen reduction reaction)

Reaction (2) proceeds at a relatively slow pace and results in a dense film which exhibits near bulk conductivity and good diffusion barrier qualities. Since the hydrogen reduction of WF_6 is not a self limiting reaction, the resultant metal film can be built up to any desired thickness. This ability to selectively fill and thus planarize contacts has stimulated much of the recent interest in LPCVD tungsten. However, it is the silicon reduction reaction (Reaction (1)) that determines the critical interfacial properties of a contact plug (e.g., contact resistance, adhesion, and junction proximity effects).

The silicon reduction of WF_6 is, fortunately, a self-limiting reaction. If the silicon surface is properly cleaned prior to exposure to WF_6, silicon consumption and tungsten film formation cease at a small value after only a few seconds, an important consideration for shallow junction devices. If, however, a significant amount of silicon oxide, polymer residue, or other foreign layers are present at the silicon surface, the reaction self-limits only after much more silicon has been consumed [1], and the quality of the resultant tungsten film is very poor [2,3]. Thus, the most critical objective in LPCVD tungsten film depositions on silicon (e.g., for contact barriers, source/drain covers, etc.) is to limit the thickness of silicon consumed to a small fraction of the junction depth to avoid possible leakage problems and other undesirable effects. It has previously been shown that somewhat less than twice as much silicon is consumed in the silicon reduction of WF_6 as is replaced by the tungsten metal that is formed [2,7]. The formation of a 10 nm thick tungsten film indicates that ~20 nm of silicon has been consumed.

At 300°C, the most commonly used temperature for depositing CVD tungsten, no difficulty has been reported in achieving 10 nm - 20 nm self-limiting films when a dilute HF pre-tungsten etch is used to remove native silicon oxide [4-6]. Likewise, electrical data indicates no increase in junction leakage as junction depth is decreased from 300 nm to as little as 100 nm [7]. However, as contact diameters shrink below 1 um, it will be important to assure that plasma etch processes are adequately developed to remove peripheral oxide (which may well go undetected) at the base of contacts and to ash polymer residues that may have been generated.

Several factors have been identified that can result in excessive consumption of silicon in Reaction (1):

(1) Silicon oxide in excess of 1 nm [1].
(2) Plasma polymerization products [8].
(3) Deposition temperatures greater than 320°C [4,9].

Of these, only Item (3) is a purely deposition related concern. Factors (1) and (2) can best be avoided by careful development of plasma etch processes, and so may be more appropriately considered as process integration issues.

DETRIMENTAL EFFECTS OF EXCESS SILICON OXIDE

H. H. Busta et.al. [1], have measured the increase in tungsten film thickness when the self-limiting condition is reached, as a function of native oxide thickness at the beginning of the deposition. As shown in Figure 1 (Zone 1), native oxide layers of less than 1 nm thickness result in <20 nm thick self-limiting films and are thus innocuous insofar as tungsten deposition by Reaction (1) is concerned. In contrast, native oxide layers between 1 nm and 2 nm in thickness (Zone 2 in Figure 1) were found to result in undesirably thick tungsten films. Finally, for native oxide exceeding 3 nm in thickness, the WF_6/Si reaction was effectively suppressed, except at isolated nucleation sites, presumably because the source gas could not penetrate the oxide layer to react with the underlying silicon. Thus, it is critically important that the silicon surfaces that are to be coated with tungsten be cleared of excess oxide before exposure to WF_6 gas.

Fortunately, hydrofluoric acid is normally effective in removing silicon oxides, and a one or two minute 50:1 HF etch is adequate to remove most of the native silicon dioxide from a wafer. Native oxide regrowth measurements have shown that several hours can pass before the film thickness is again sufficiently thick to be detrimental to obtaining a thin silicon reduction reaction tungsten film [10], an acceptably relaxed timing specification between precleaning and tungsten deposition.

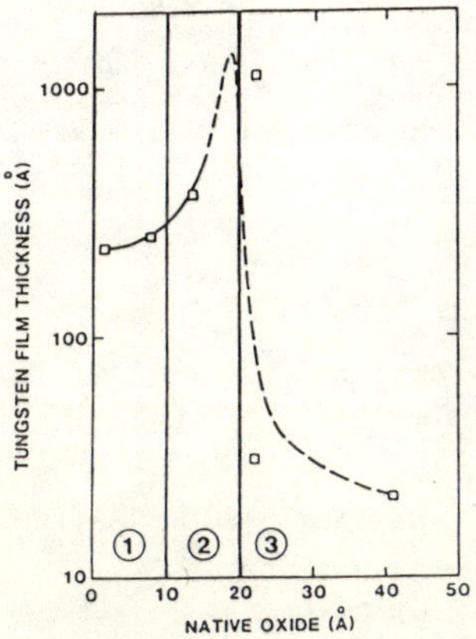

Fig.1 Self-limiting tungsten film thickness from the silicon reduction of WF_6 as a function of native silicon oxide thickness (Ref. 1).

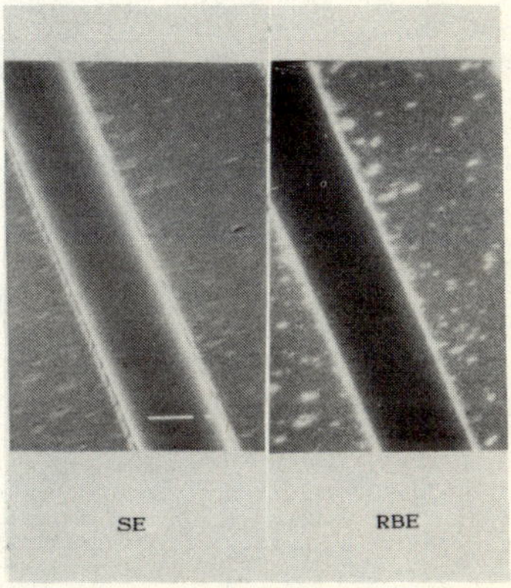

Fig.2 Influence of localized residual oxide on self-limiting film thickness. The magnification marker represents 1um.

EFFECTS OF SILICON OXIDE DISTRIBUTION AND STOICHIOMETRY

Though a dilute HF dip alone is most often sufficient to assure removal of native silicon oxide, thickness, stoichiometry [11], and impurity content may well differ from point to point, complicating complete and uniform removal of native oxides [12]. Thus, HF etch rates developed from tests with thermal SiO_2 should be re-evaluated when using HF to clear native oxide. Patches of unetched oxide can result in localized excess reaction between WF_6 and the silicon substrate, as illustrated in Figure 2. In the secondary electron (SE) scanning electron micrographs (Figure 2a) note the small flat depressions in the silicon areas between SiO_2 lines. These are actually areas of thicker tungsten, as indicated by their brighter appearance on the Robinson backscattered electron (RBE) micrograph, Figure 2b. The tungsten also appears to be thicker at the periphery of the patterned SiO_2 lines, perhaps due to a meniscus remaining from the initial plasma etch, when the cuts through the deposited oxide layers were made. Of the two types of localized excess reaction, the isolated patches are far less common than the wedge shaped peripheral oxide residues at the base of the contact cuts, like those illustrated in Figure 3. Deeply penetrating films stimulated by these residuals should be avoided to minimize concern with respect to potential leakage problems.

Removal of native oxide (including SiO, which may not be effectively etched away by HF) is a wet chemical etching issue. Peripheral oxide, on the other hand, may be too thick to be completely removed by wet etching techniques without causing undesirable lateral etching ("blowout") as well. Thus, peripheral oxide, as well as plasma induced polymer residues, must be removed during plasma etching of the contact cut. Though these residuals have consequences for subsequent metallization process steps, such as tungsten deposition, they are not primarily a metallization issue. Indeed, residual oxide in contacts adversely affects even conventional contacts, but in those cases the evidence of their presence does not present itself until the lot is completed and tested. In contrast, problems of this type can often be detected immediately when a CVD tungsten deposition is involved.

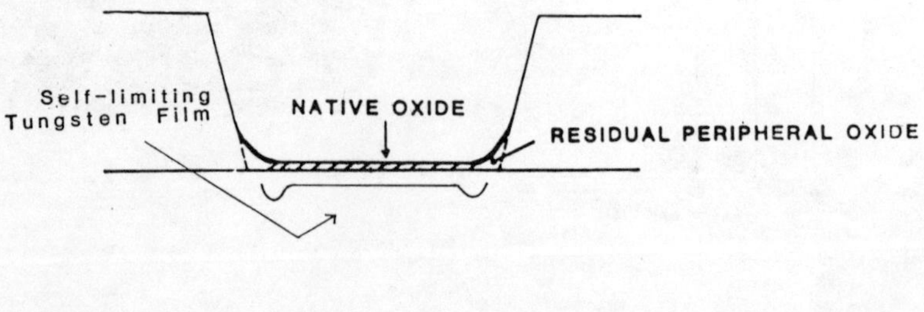

Fig.3 Different forms and types of residual oxide to be considered in clearing contacts for LPCVD tungsten film deposition.

DETECTION OF OXIDE RESIDUES

The presence and quantity of undesirable surface contaminants on silicon can be determined analytically by several techniques, such as X-ray photoelectron spectroscopy, ellipsometry, Auger spectroscopy, or ultraviolet reflectance spectroscopy [12]. However, there are very few nondestructive methods which have both sufficient sensitivity and spatial resolution to detect 2 nm of SiO_2 around the base of a 1 um deep contact hole which may be 1 um or less in diameter. In any case, it is uncommon to have such instrumentation available on-line in an IC fabrication facility. Thus, detection of residual contamination must be inferred in most cases by the effect it has on the silicon reduction reaction through correlations with data such as that shown in Figure 1, and through evaluation of film quality using scanning electron microscopy (SEM) or transmission electron microscopy (TEM). Tungsten films produced by the WF_6/Si reaction are thin (<20 nm), continuous, smooth and >90% dense as in Figure 4 and Figure 5 (lower part) if the silicon is effectively free of residues and oxides. In contrast, those tungsten films for which the silicon native oxide was >10 nm or which had an extraneous plasma-induced layer beneath them, are thick, discontinuous, rough and porous [2, 3, 14, 15], as indicated in Figure 5 (upper part).

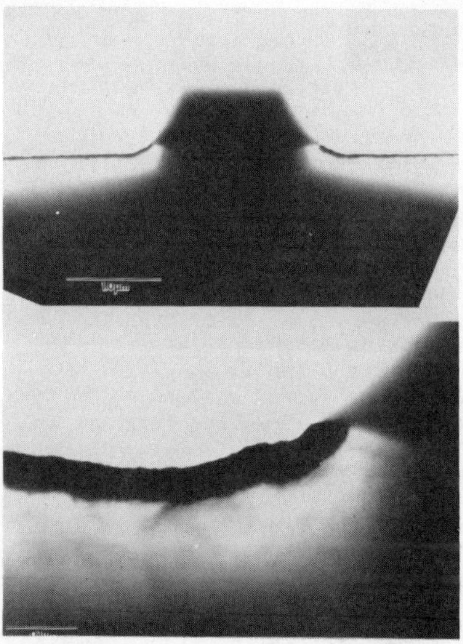

Fig.4 Cross-sectional transmission electron micrograph of a thin self-limiting silicon reduction reaction tungsten film on a patterned silicon chip after a 1 minute 50:1 HF pre-tungsten dip to remove native oxide. The film is dense, smooth, continuous, and shows no evidence of tunnels or encroachment under the silicon dioxide line (Ref. 13).

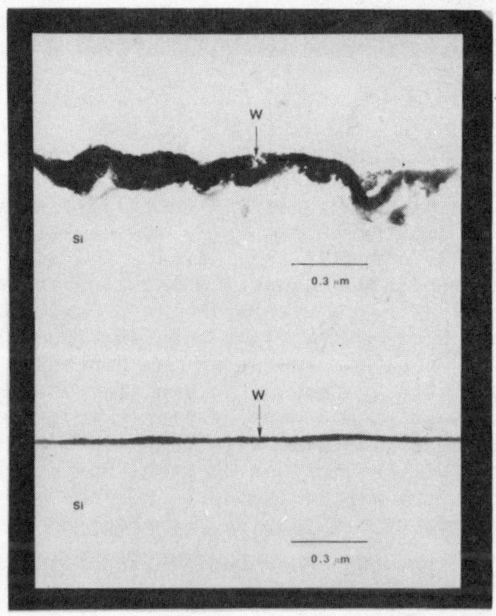

Fig.5 Silicon reduction reaction tungsten films deposited by exposing WF_6 gas to plasma cleaned silicon (upper micrograph) and HF wet cleaned silicon (lower micrograph)(Ref. 2).

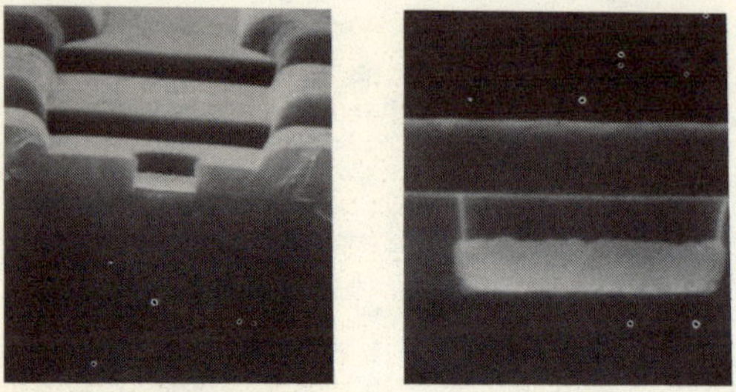

Fig.6 WF_6/H_2 selectively deposited film in 1.5um diameter CMOS contact window. The W/Si interface is smooth with no evidence of excess WF_6/Si reaction or encroachment.

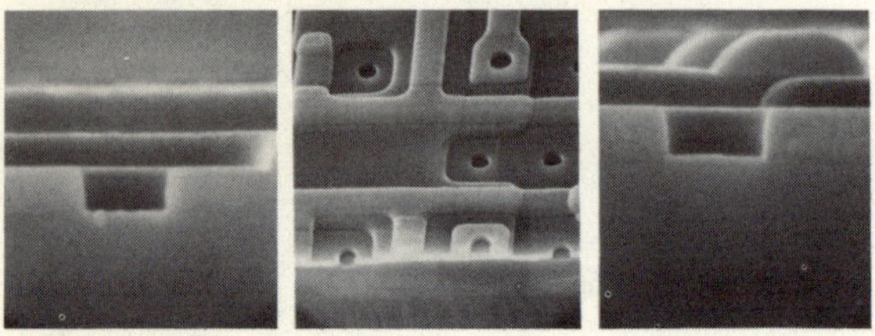

P+ Contact N+ Contact

Fig.7 Cross-sectional SEM micrographs of CMOS contacts after selective tungsten films have been removed by etching in hydrogen peroxide. There is no evidence of excess or non-uniform Si consumption or of encroachment at the Si/SiO_2 interface.

For blank wafers, a beta backscatter areal density measurement [2] is a convenient, inexpensive gauge to determine the efficacy of pre-tungsten cleaning techniques for native oxide removal. For patterened wafers, SEM or TEM sampling is required. Cross sectional SEM views, such as those shown in Figures 6 and 7, can indicate the W/Si interfacial characteristics in the "as-deposited" condition and also after the tungsten deposit has been removed (e.g., with hydrogen peroxide). In Figure 7 it is evident that no peripheral oxide (like that in Figure 3) or oxide patches (like those in Figure 2) were present in the contact windows just prior to tungsten deposition. Cross sectional TEMs, such as Figure 4, can also be definitive, but samples are more difficult to prepare. Surface view SEMs (such as in Figure 8), or plan view TEMs (such as in Figure 9) give information about the full area of the contact, but for the TEM case, require backside thinning up to the W/Si interface. For the samples in these two figures, it is evident that excess WF_6/Si reaction occurred at the periphery and at some isolated sites near the center of the contact, at least for P+ (arsenic doped) silicon contacts.

A method that is potentially more powerful and convenient for evaluating the W/Si interfacial profile is the taper section technique, as illustrated in Figure 10. Plan view SEM can be used on the taper sections to evaluate the characteristics of selective tungsten deposits, as shown in Figure 11. Because the atomic number of tungsten differs so widely from that of silicon and oxygen, tungsten areas appear in high contrast to other areas of the wafer when using the backscattered electron mode of the SEM. Particular sites on the chip can be easily selected for analysis since the taper sectioning can be continued until a specific site of interest has been reached. Moreover, this technique yields quantitative information on the tungsten penetration depth as well as lateral contact dimensions and shapes. Finally, the lateral and depth positions of the tungsten film with respect to other features are also obvious, as can be seen in Figure 11. Here, the tungsten has penetrated deeper into the silicon at the periphery of the contact than at the center, which itself lies 60 nm below the

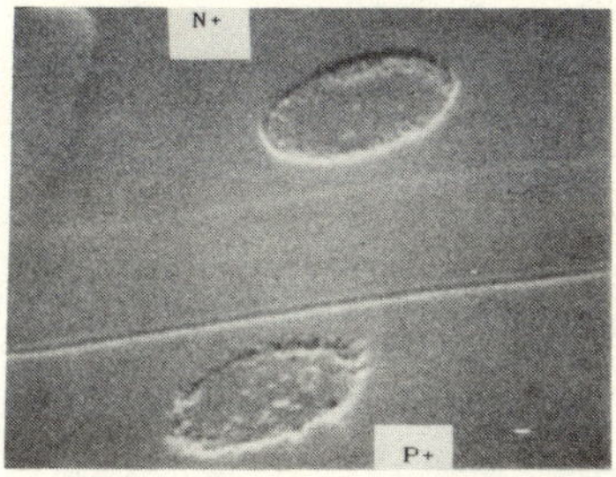

Effect of Peripheral Silicon Oxide

on Silicon Consumption

From the Si/WF6 Reaction

Fig.8 Scanning electron micrograph of 1.5um diameter contacts to N+ and P+ (100) silicon after LPCVD tungsten films were selectively deposited on a wafer that received no pre-tungsten HF dip. The effect of residual peripheral oxide and native oxide patches on silicon consumption from the Si/WF_6 reaction are clearly evident.

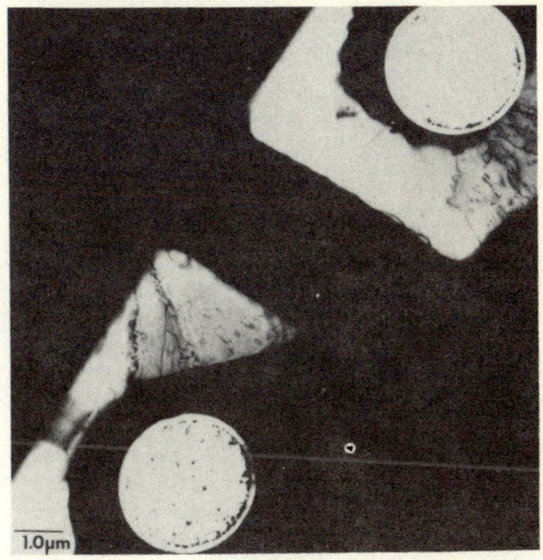

Fig.9 Plan view transmission electron micrograph of CMOS 2.5um diameter contacts after front and back side thinning to reveal the W/Si interface. Because a polymer contamination layer covered the contacts as a result of the plasma etch, the pre-tungsten HF dip was ineffective in removing peripheral oxide, resulting in a ring of excess WF_6/Si reaction.

original silicon surface. For a 5 degree taper on the polishing jig, one micron of horizontal distance implies 87 nm of depth, as measured parallel to the direction of the taper (e.g., from the top of Figure 11 to the bottom).

CONSEQUENCES OF RESIDUAL LAYERS AT THE TUNGSTEN/SILICON INTERFACE

Because the WF_6/Si reaction is sensitive to contaimination on the silicon surface, the LPCVD selective tungsten step in the process flow can be a good indicator of pre-deposition surface purity. Adhesion loss, uneven or delayed film growth, high contact resistance, or the occurence of beta phase tungsten suggest that pre-tungsten cleaning of the silicon surfaces was inadequate. Dilute hydrofluoric acid is normally sufficient to remove silicon oxides but is ineffective against polymer residues which may result from previous plasma etching steps. For example, in Figure 12 are displayed two large area diodes (1 mm^2) and capacitors that were coated with selective tungsten. Polymeric contamination (a fluorocarbon compound) was present on the wafer in Figure 12(b). There is an obvious difference in the adhesion achieved and in the thickness of the deposits. Adhesion tests on blank control wafers can also often be used to give forewarning at the plasma etch step that residual polymers have not been completely ashed.

X-ray diffraction can be used to detect nondestructively the presence of beta phase tungsten (W_3O), which is an indication that silicon oxide

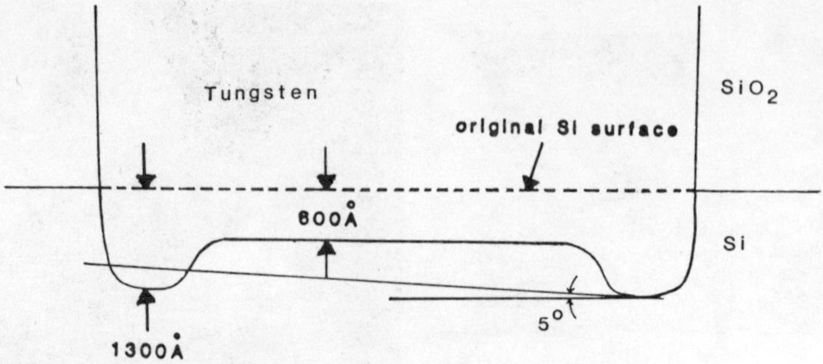

Fig.10 Taper section method of evaluating W/Si interaction from selective LPCVD films in contacts. A five degree tapered jig was used to hold the die for polishing down to the W/Si interface.

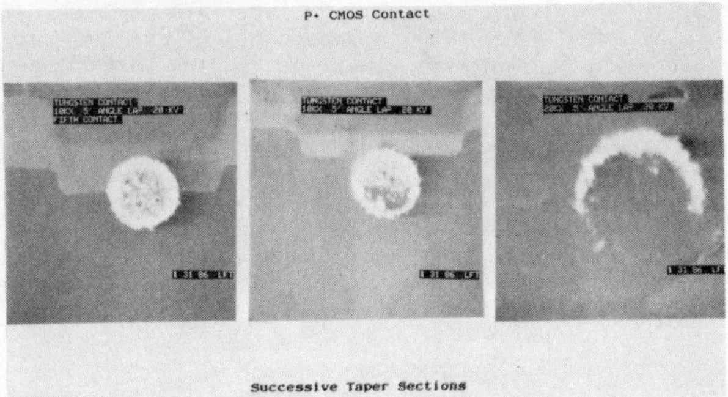

Fig.11 Plan view SEMs of selective tungsten films in 1.5um diameter P+ contacts on a wafer that received no pre-tungsten HF dip show that the tungsten film is still continous at the depth of the SiO_2/Si interface. In the center micrograph the horizontal diameter of the contact is about 600Å beneath the original SiO_2/Si interface, just at the W/Si interface in the central part of the contact.

residues were still present when the tungsten deposition was performed. Surface roughness, which can easily be seen at 200X to 500X magnification using an optical microscope, is also correlated with excess native oxide and beta phase tungsten. Of course, if residual silicon oxide thickness exceeds 2.0 nm, tungsten will fail to nucleate, an indication of gross contamination.

Contact resistance measurements are a very sensitive indicator of interfacial contamination regardless of the metallization being used at the metal/silicon interface. As shown in Table 1, Kelvin contact resistance measurements can be used to gauge the effectiveness of pre-tungsten cleaning procedures. Both the mean contact resistance and the standard deviation can be indicators of undesirable residuals at the interface. Note that P+-to-tungsten contact resistance is more sensitive to adequate pre-tungsten

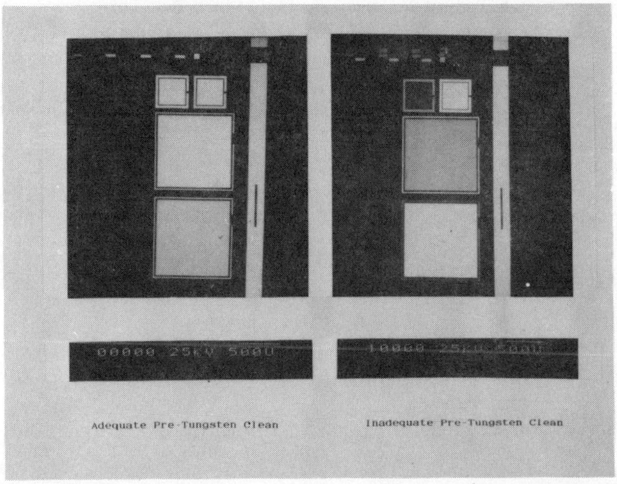

Fig.12 Comparison of selective tungsten film formation and adhesion characteristics for diodes and capacitors which had (a) clean silicon surfaces and (b) polymer contaminated silicon surfaces prior to selective tungsten deposition.

cleaning than N+-to-metal or polysilicon-to-metal contacts. Contact resistance on properly prepared wafers is significantly lower for N+-to-tungsten and polysilicon-to-tungsten contacts and about equal for P+-to-tungsten contacts, as compared to wafers without tungsten barrier contacts.

Functional failures during burn-in testing have also been linked to residual oxide at the W/Si interface. As shown in Figure 13, contacts were examined from die that passed (upper row) and failed (lower row) burn-in testing. The aluminum-silicon and tungsten metallization was removed to expose the silicon contacts. The upper wafer was given a standard 1 minute buffered oxide etch treatment prior to tungsten deposition while the lower wafer received no pre-tungsten etch. Excessive silicon consumption has occurred at the periphery of the uncleaned contacts, as indicated by the dark spots. The properly cleaned contacts exhibited no spikes or trenches.

PREVENTION OF CONTAMINATION AT THE W/Si INTERFACE

Contamination layers on the silicon surface which disrupt a well-behaved reaction between WF_6 and silicon may stem from a variety of sources. A listing of possible problems and suggested solutions is given in Table 2. Contacts that arrive at the LPCVD tungsten reactor with anything more than a thin layer of native silicon oxide on their surface are likely the result of a deficiency in prior plasma etch procedures. Wet chemical etching is not likely to be effective in addressing these problems; thus attention must be given to qualifying plasma etch procedures to obtain pure unobstructed silicon contacts before the metallization is to be applied.

Lot H2204A Contact Resistance Results

	Contact Resistance (Ω) (1.6 μm contacts)		
	Poly-to-Metal Mean (± 1σ)	N+-to-metal Mean (± 1σ)	P+-to-metal Mean (± 1σ)
Tungsten Contact Wafers			
0.5 min. 41:1 BOE Etch	3.2 (± 1.9)	21.0 (± 13)	290 (± 105)
1.0 min. 41:1 BOE Etch	2.6 (± 1.6)	17.0 (± 6.3)	66.5 (± 45)
2.0 min. 41:1 BOE Etch	1.4 (± 1.8)	12.2 (± 5.9)	38.2 (± 12)
Al-Si Contact Wafers	8.4 (± 3.3)	41.2 (± 12.4)	35.0 (± 8.0)

CMOS P+ Contact After Burn-in

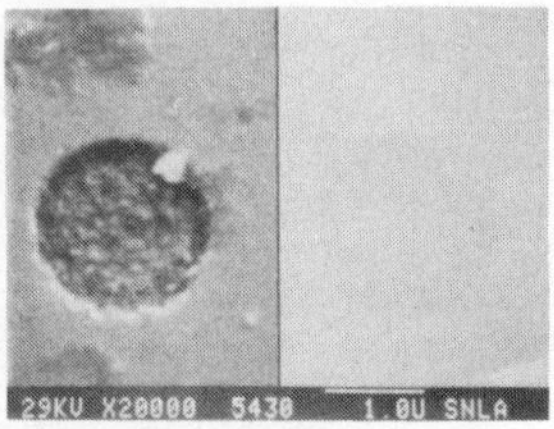

Passed

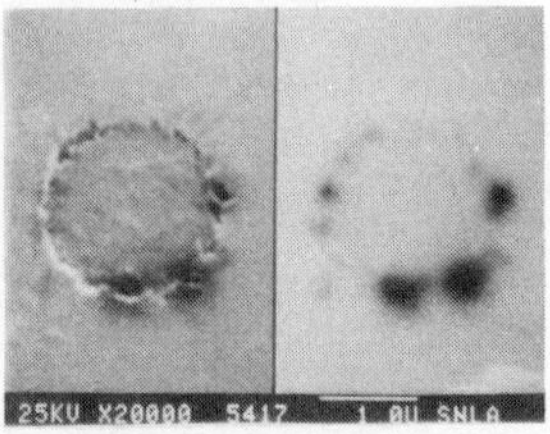

Failed

CMOS N+ Contact After Burn-in

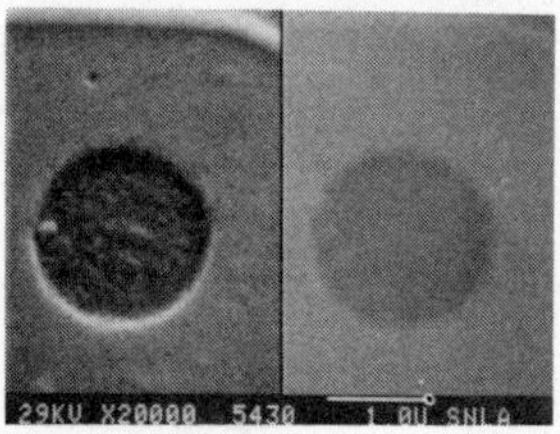

Passed

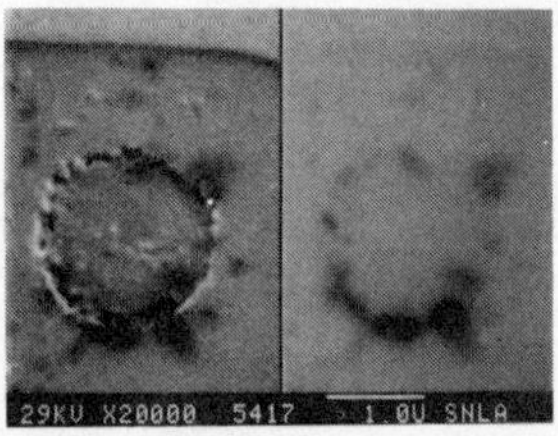

Failed

Fig.13 Scanning electron micrographs of contacts from wafers which had selectively deposited contact barriers. The upper row of contacts had a dilute HF dip prior to tungsten deposition, the lower row did not. Each micrograph displays a secondary electron image (left) and a backscattered electron image (right).

TABLE 2

TYPE OF PROBLEM		HF CLEAN EFFECTIVE?	WHY/WHY NOT?	SOLUTION
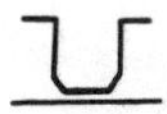	INCOMPLETELY CUT CONTACT	NO	BLOWOUT	DEVELOP BETTER PLASMA ETCH UNIFORMITY
	OXIDE AT CONTACT PERIPHERY	NO	TOO THICK; UNKNOWN AMOUNT	USE CONTROLLED OVERETCH OR DIRECTIONAL ETCH
	NON-SiO2 RESIDUE	NO	"TEFLON" POLYMERS (INERT TO HF)	USE PLASMA ASH OR NON CF PLASMAS; REMOVE UNDERLYING Si
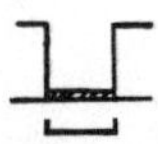	NATIVE OXIDE REGROWTH	YES	UNIFORM OXIDE; THIN LAYERS	USE DILUTE HF ETCH OR REACTIVE INTERMEDIATE CONTACT METAL; IMPLANT THROUGH INTERFACE

CONCLUSIONS

1. Adverse barrier or electrical behavior in LPCVD tungsten contacts can be the result of unrecognized residual silicon oxide or other residues at the base or periphery of the contact cuts.

2. The effects are more serious for small diameter contacts and shallow junction devices.

3. To assure a robust tungsten contact process, pre-tungsten cleans must be developed which address the relevent residue that may be characteristic of a given fab line or device technology.

4. With adequately cleaned silicon, superior results which have been demonstrated for tungsten contacts can be achieved on a repeatable basis.

ACKNOWLEDGEMENTS

It is a pleasure to acknowledge L. F. Tafoya, N. A. Creager and F. P. Ganyard for SEM analysis, P. S. Hund for taper section preparation, R. F. Pinizzotto for TEM analysis, M. B. Chamberlain for Auger Analysis, T. A. Hill for H2204A lot supervision, R. A. Pastorek and K. L. Hughes for electrical measurements, G. M. Gutierrez for drawings and L. M. Tindell for manuscript preparation.

References

1. H. H. Busta and C. H. Tang, J. Electrochem. Soc. 131, 1195 (1986).

2. W. T. Stacy, E. K. Broadbent and M.H. Norcott, J. Electrochem. Soc., 132, 144 (1985). See also E. K. Broadbent, Proc. First and Second Workshops on Tungsten and Other Refractory Metals for VLSI Applications, R. S. Blewer. ed. (Materials Research Society, Pittsburgh, PA) 365 (1986).

3. R. S. Blewer, M. E. Tracy and V. A. Wells, Proc. First and Second Workshops on Tungsten and Other Refractory Metals for VLSI Applications, R. S. Blewer, ed. (Materials Research Society, Pittsburgh, PA) 407, (1986).

4. M. E. Tracy, Ibid, p. 24.

5. For example, see data and references in E. K. Broadbent and W. T. Stacy, Solid State Technol., 28 (12), 51 (Dec. 1985).

6. See references and data in R. S. Blewer, Solid State Technol., 29 (11), 117 (1986).

7. T. Moriya and H. Itoh, Proc. First and Second Workshop on Tungsten and Other Refractory Metals for VLSI Applications, R. S. Blewer, ed., (Materials Research Society: Pittsburgh, PA) 21, (1986).

8. R. Pinozotto and R. S. Blewer, unpublished data, and G. S. Oehrlein, G. J. Coyle, J. G. Clabes and Y. H. Lee, Surface and Interface Anal., 9, 275 (1986).

9. M. L. Green, Y. S. Ali, B. A. Davidson, L. C. Feldman and S. Nakahara, Proc. Mats. Rsch. Soc. Fall Meeting, Abstract E 1.7, p. 156 (1985).

10. G. Mende, J. Finster, D. Flamm and D. Schulze, Surface Sci. 128, 169 (1983).

11. C.R.M. Grovenor, A. Cerezo and G.D.W. Smith, Mats. Research Soc. Proc., 37, 199 (1985).

12. B. E. Deal and D. B. Kao, this volume, p. 27.

13. Micrographs Courtesy of R. Pinizzotto, Ultrastructure, Inc.

14. M. L. Green, Y. S. Ali, T. Boone, B. A. Davidson, L. C. Feldman, and S. Nakahara, this volume, p. 91.

15. M. Wong, N. Kobayashi, R. Browning, D. Paine and K. C. Saraswat, J. Electrochem. Soc. (to be published).

ELECTROMIGRATION-INDUCED SHORT CIRCUIT FAILURE IN ALUMINUM/TUNGSTEN (CVD) CONDUCTORS

ELIOT K. BROADBENT AND JANET M. TOWNER
Philips Research Laboratories Sunnyvale, Signetics Corporation
Sunnyvale, CA 94088-3409

ABSTRACT

Electromigration was studied in Al conductors underlaid with chemically vapor deposited W, and the results compared to the more conventional Al/TiW metallization. In Al/TiW conductors electromigration results in single crystal Al whisker growth, causing metal short rather than open circuit failure. Contrary to expectations, the Al/W metallization was also found particularly prone to the formation of whiskers during current stress. Al/W lifetimes were reduced by about one half compared with Al/TiW; the short circuit failure mode was unchanged. The temperature and current density dependencies were not significantly effected by the composition of the underlayer.

INTRODUCTION

In aluminum thin films, electromigration causes a transport of material in the direction of electron flow. These polycrystalline films contain an abundance of points for flux divergence. Depending on the polarity of the conductor, these points can become the location of either an open or a short circuit. In the past, accelerated lifetesting has relied on the open circuit failure criterion. However, as the use of multiple levels of interconnect has become widespread, reliability studies have also incorporated test patterns to detect electromigration-induced shorts between two levels of metal [1-3]. Using these so-called extrusion monitors, researchers have found that short circuits often form prior to opens.

In homogeneous Al or Al-Cu conductors, mass accumulation and mass depletion are both manifestations of grain boundary diffusion [2], and are equally likely to occur. In this case, the dominant failure mode is determined by the circuit design rather than by the metal properties. In layered conductors, however, electromigration can also occur along the conductive Al interface [4]. This interfacial electromigration can reduce mass flux divergence by balancing grain boundary effects.

Layered conductors, such as Al deposited on a layer of Ti or titanium-tungsten (TiW), are particularly prone to the formation of whiskers and protrusions [5-7]. Figure 1 shows an example of aluminum whisker growth in Al/TiW conductors. The scanning electron migrograph (SEM) is from a representative portion of an unpassivated conductor after open circuit failure. TiW has been used extensively in bipolar IC devices as a diffusion barrier between Al and silicide or silicon contact regions. In a recent work [3], Towner found the dominant failure mode in these conductors to be interlayer metal shorts.

A number of potential applications in microelectronic circuit fabrication are being considered for tungsten films deposited by chemical vapor deposition [8]. Several of these (e.g., contact/via fill for planarity and diffusion barrier) involve the use of tungsten in direct combination with multilevel aluminum interconnects. As a result, a preliminary assessment of the Al/W (CVD) conductor system with respect to electromigration was thought to be important. Experiments were performed to determine the dominant failure mode in layered Al/W conductors, and to evaluate the temperature and current density dependencies for failures.

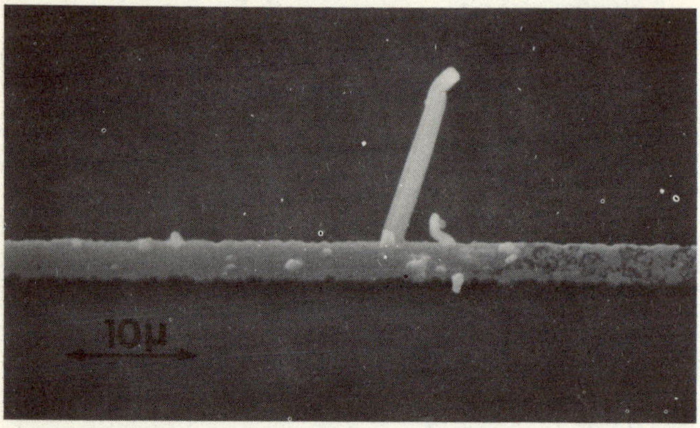

Figure 1: SEM micrograph of Al whisker growth in an unpassivated Al/TiW conductor after current stress.

EXPERIMENTAL

The test vehicle used to measure electromigration-induced shorts and opens is shown in Figure 2. This structure consists of two levels of metal and an intermediate insulating layer. The first metal layer is a current-carrying conductor 2000 μm long and nominally 4 μm wide. On each side of the test stripe, at 4 μm spaces, is a monitor stripe to detect intralayer shorts. The second metal layer, used to detect interlayer shorts, is a 200 μm wide conductor superimposed over the test stripe.

Accelerated life testing was performed on the wafer level. The test procedure has been described in a previous publication [9]. Between eight and sixteen structures were stressed at each test condition. Constant current is passed through the test stripe and the voltage across the stripe, V_o, is monitored for a large increase (>100%) denoting formation of an open circuit. During the test the voltage between the upper monitor stripe and ground (V_1) is continuously recorded. When the voltage acquires a non-zero value, a short has formed between monitor and test stripes. Time to failure is determined by the first event to occur. Due to the test structure geometry, intralayer short circuit formation was not a significant failure mode. To simplify the measurement, electrical isolation between first layer metal conductors was verified only at the conclusion of the test.

Three groups of samples were prepared for this study. Table I summarizes the composite structure of each group. The substrates consisted of 10 cm diameter, <100> Si wafers with a 30 nm layer of thermally grown SiO_2 to minimize Joule heating during testing. First layer Al was sputter deposited from a planar magnetron source onto either the oxidized substrates (Group A) or an intermediate layer consisting of sputtered TiW (Group B) or chemically vapor deposited W (Group C). The Ti content in the TiW was 15 wt.%. Prior to blanket W deposition, the substrates in Group C received a sputtered adhesion bilayer consisting of a thin layer of W on a thin layer of Ti. Blanket tungsten deposition was performed by low pressure CVD using an Anicon hot wall, vertical flow reactor. The deposit conditions were 300°C, 0.55 torr, H_2:WF_6 flow of 44:1, 45 minutes.

Although not reported here in detail, choice of the sputtered bilayer was not arbitrary. Four key requirements for blanket CVD tungsten processing are fulfilled. The lower Ti layer: (1) provides excellent adhesion to dielectric surfaces and (2) assures a high yield, low resistance interface within contact structures when used in combination with a sputter etch. The upper W layer: (3) protects against unwanted WF_6-Ti reactions [10] and (4) readily nucleates CVD tungsten.

The interlayer isolation consisted of 0.9 μm thick CVD phosphosilicate glass (PSG) with a 12 wt.% P_2O_5 content. For the second metal layer, 1.8 μm thick pure Al was deposited using heated substrate electron beam evaporation. All pattern definition was accomplished using conventional photolithographic and wet chemical etching techniques. Wafers were annealed at 450°C in N_2-10%H_2 for 35 minutes. All three sample groups were processed together through common deposition steps. Conductor cross-sections were calculated from electrical resistance, stylus profilometry and linewidth measurements.

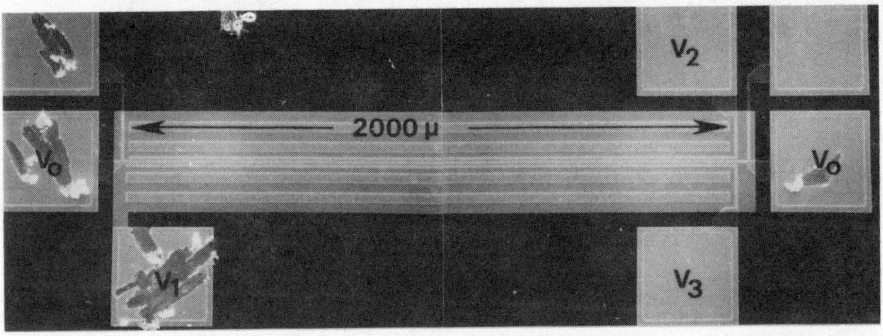

Figure 2: Top view of the two layer metal open/short electromigration test structure.

TABLE I - COMPOSITION OF SHORT CIRCUIT TEST STRUCTURE

Group	TiW Thickness	Adhesion Bilayer	CVD Tungsten Thickness	1st Layer Aluminum Thickness	PSG Thickness	2nd Layer Aluminum Thickness
A	- -	- -	- -	0.45 μm	- -	- -
B	0.15 μm	- -	- -	0.45 μm	0.9 μm	1.8 μm
C	- -	15 nm Ti + 15 nm W	0.15 μm	0.45 μm	0.9 μm	1.8 μm

RESULTS AND DISCUSSION

Failure Mode and Distribution

Examples of the failure distributions in the Al/TiW (Group B) and Al/W (Group C) conductors are shown in Figure 3. This figure is a lognormal probability plot of lifetime for a temperature of 200°C and a current density of 2×10^6 A/cm^2. This data shows clearly that electromigration-induced shorts in Al/W, like shorts in Al/TiW [3] and opens in Al, are well represented by a lognormal distribution of failure times. The coefficients of linear regression were in excess of 0.97. The single failure mode for the Al/TiW metallization was found to be interlayer shorts, with a median-time-to-failure (MTF) of 17 hours. The short circuit failure mode was preserved when the composition of the underlayer was changed from TiW to W (CVD). However, this change in composition did effect electromigration failure times. The MTF for the Al/W metallization was reduced to 8.9 hours.

For comparison, the open circuit failure times for the unpassivated Al conductors (Group A) at these same stress conditions has also been plotted in Figure 3. Coincidentally, the distribution overlies that of the Al/W conductors. With passivation the failure times (mix of open and short circuits) of the Al metallization would be increased by about an order of magnitude [3], making Al more reliable than either the Al/TiW or Al/W metallizations.

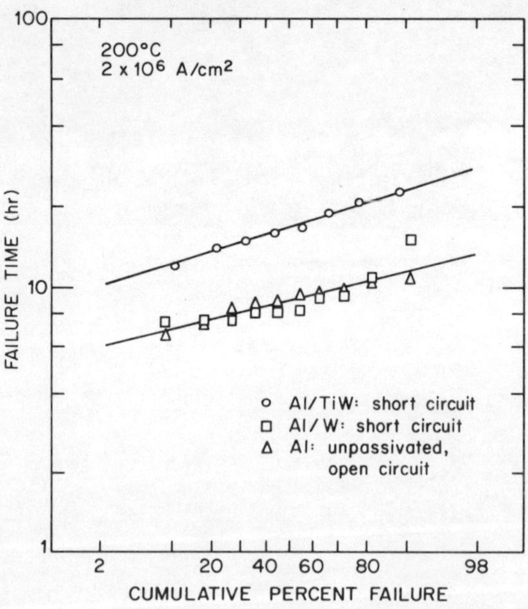

Figure 3: Lognormal probability plot of interlayer short circuit failure times in Al/TiW and Al/W conductors, and open circuit failure times in unpassivated Al conductors.

Temperature Dependence (Activation Energy)

The activation energy (E_a) for interlayer short circuit formation was determined in the temperature range of 125 to 300°C. Al/TiW and Al/W conductors were stressed at a current density of 2x10^6 A/cm^2. Figure 4 shows the relationship between MTF (short) and stress temperature. The E_a for short circuit failure was found to be 0.54±0.05 eV for the Al/TiW, and 0.52±0.04 eV for the Al/W. The dotted lines in Figure 4 represent the 95% confidence level for each group. Al/W lifetimes were reduced by one half compared with the Al/TiW. Since the activation energies for the two groups are comparable, this difference in MTF is consistent over the entire temperature range.

The values of 0.5 to 0.6 eV are typical of electromigration-induced failure resulting from Al grain boundary diffusion. This similarity in E_a for short and open circuit formation suggests that both could result from the same mass transport mechanism. However in Al/TiN conductors, Schreiber found the same E_a for both grain boundary and interfacial diffusion [4]. In either case, mass accumulation contributing to whisker growth is highly directional and occurs at select sites, being fed from relatively large amounts of Al material being swept "downstream". As a result, the likelihood that short circuit failure will occur before a sector of local conductor length can be severed is high.

Current Density Dependence

The influence of current density on short circuit lifetime was measured at a temperature of 250°C. The current densities ranged from 7x10^5 to 2x10^6 A/cm^2. As a power law relationship was assumed, lifetimes were plotted as a function of current density on log-log scales; the results appear in Figure 5. The current density exponent (n) for each type of metallization was calculated from the slope of the plotted data: -2.5±0.4 for Al/TiW and -2.0±0.2 for Al/W. Thus, the MTF (short circuit) for Al/W conductors is inversely proportional to the square of the current density for these test conditions. In a previous study by Towner [3], the current density exponent for short circuit failure in Al/TiW conductors was between 2 and 4 for several layered groups employing different TiW and Al thicknesses.

Failure Morphology

Following the results derived from the two-layer test structures, the existence of crystal whiskers was confirmed by SEM inspection of several unpassivated Al/W conductors after current stress. Figures 6a & 6b show an advanced case where a long single crystal Al whisker has formed near the cathode end of the stressed Al/W line (250°C, 10-40 mA, 20 hours). The length of the whisker exceeds 60 μm with a diameter of 0.5 μm. Several similar whiskers were observed along the conductor length, primarily in the immediate vicinity of denuded portions of conductor (Figure 7). Also evident at numerous locations along the stripe length were local neighboring regions of mass depletion/accumulation where large protrusions formed (Figure 8). This type of mass accumulation may be a precursor of whisker formation, and is most probably as severe a problem as whisker growth.

From comparative observations on failure morphology, we suggest that the layered Al/W conductors fail earlier because the interface is a better diffusion pathway than the Al/TiW interface. This idea is supported by the very extensive voiding witnessed in stressed, unpassivated Al/W conductors.

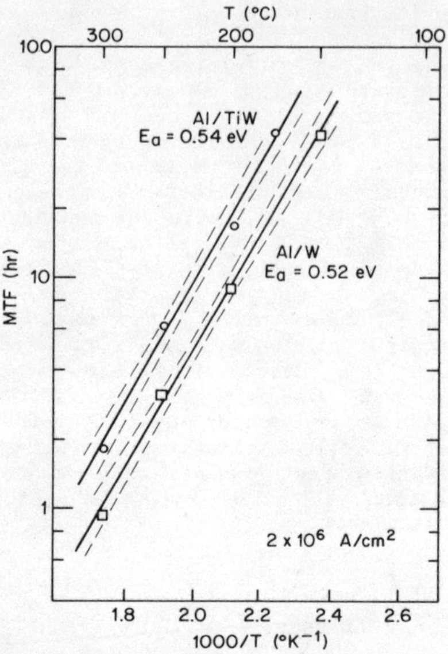

Figure 4: Arrhenius plot of short circuit MTF as a function of stress temperature for Al/TiW and Al/W conductors. (Dotted lines represent the 95% confidence level).

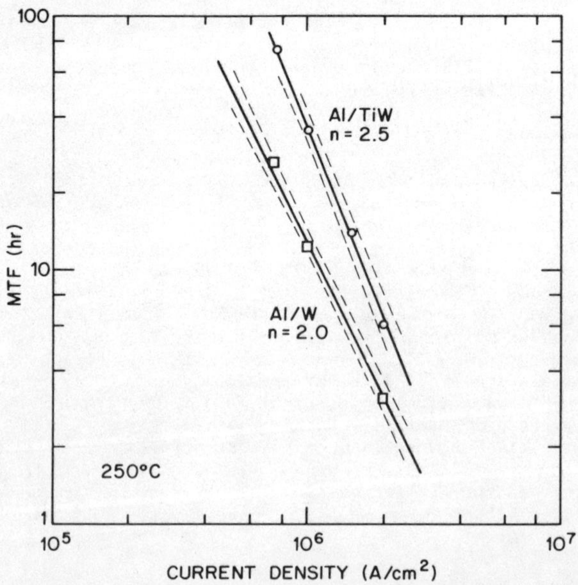

Figure 5: Influence of current density on MTF (short circuit) for Al/TiW and Al/W conductors. (N is the current density exponent, dotted lines represent the 95% confidence level.)

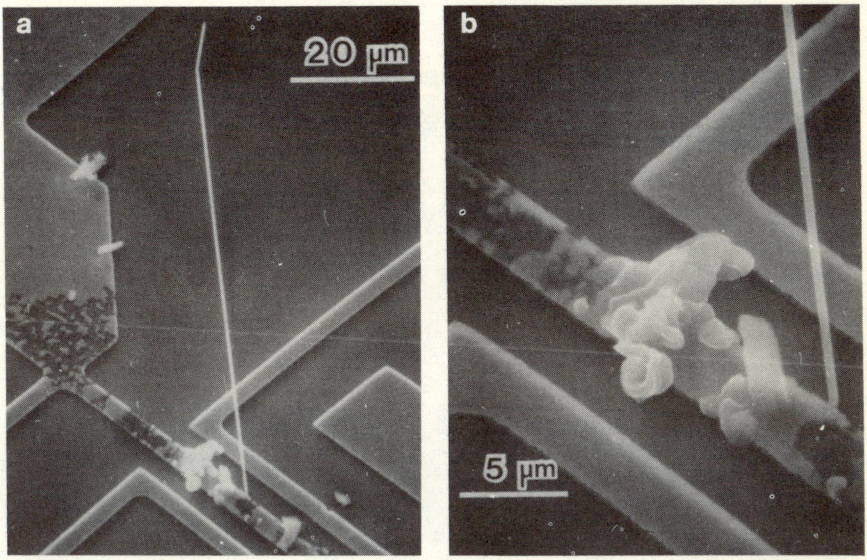

Figure 6: SEM micrographs showing an advanced case of whisker growth in an Al/W conductor after current stress: a) perspective view near cathode end of test stripe, b) higher magnification image of base of whisker.

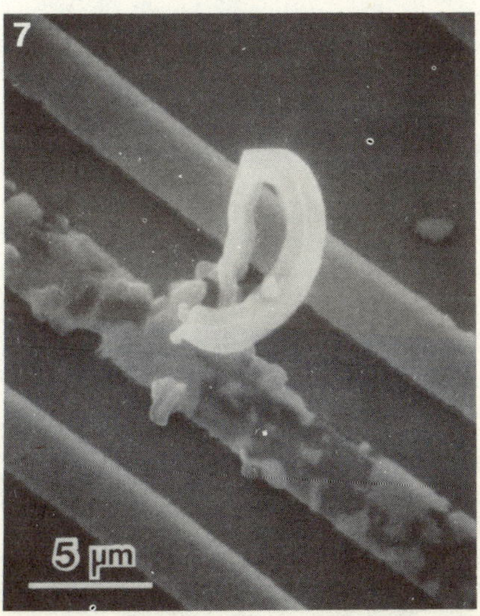

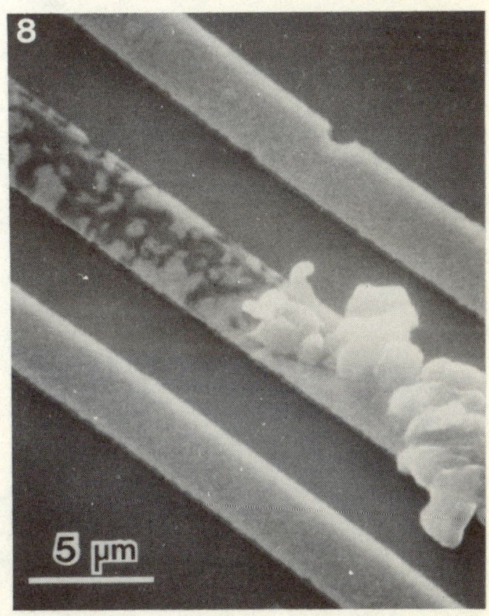

Figure 7: SEM micrograph of a curved growth in an unpassivated Al/W conductor after current stress.

Figure 8: SEM micrograph showing local region of mass depletion/accumulation in a stressed Al/W conductor. Less regular geometries have formed in area of accumulation.

Mechanisms for Whisker Formation in Layered Conductors

Due to the higher melting points of both TiW and W, the conductor rarely opens catastrophically when the Al is depleted from a portion of the line. The layer resistivities of TiW and W (CVD) are 70-120 and 7-10 μohm-cm, respectively, as compared with 3 μohm-cm for Al. Consequently, although TiW and W maintain circuit continuity, both underlayers would cause local increases in line resistance and temperature until Al rebridges the denuded conductor. As illustrated in Figures 1, 6 and 7, single crystal Al whiskers seem to be associated with these depleted regions.

If whiskers form as a result of enhanced current crowding and higher temperature gradients, one would expect the Al/W metallization to be less prone to whisker formation since the W (CVD) underlayer is an order of magnitude less resistive to current flow than TiW (for the same thickness). As discussed in the preceeding sections, however, this hypothesis is clearly not correct.

Another possible cause for whisker generation is the presence of Ti in the Al conductor or at the interface, which might serve to catalyze the single crystal growth perhaps by reducing the surface aluminum oxide. Although this reaction is thermodynamically unfavorable, Al with no underlayer has been reported to form whiskers when stressed in vacuum [11]. This idea is also supported from the results of several studies which report Al whisker growth in both electrically stressed Al conductors underlaid with Ti [5] and TiW [1,3], and in annealed films of Al overlaid with TiN [12]. It was though that the absence of Ti from the Al and the Al-W interface in a layered conductor might prove significant in reducing or eliminating whisker formation. As with the previous resistivity argument, this was obviously not the case.

Compressive stress in an underlayer has also been proposed to play an important role in whisker growth, especially where two different materials are combined to form a layered conductor [12]. Measurements of the as-deposited thin film stress were made prior to patterning using an optically levered laser apparatus described in a previous publication [13]. The TiW was found to be in strong compression with a stress value of -1.71 GPa. In direct contrast the CVD W layer was found to be in strong tension with a stress value of +1.35 GPa. Thus, we might have expected the Al/TiW conductors to be more prone to whisker formation, resulting in earlier MTF than the Al/W. Once again, just the opposite was observed. An evaluation of stress response in these layers as a function of thermal cycling/annealing is under study and will be reported in a later publication [14].

CONCLUSIONS

Electromigration was studied in Al conductors underlaid with CVD W and the results compared to the more conventional Al/TiW metallization. In summary:

- Al/W metallization, as with Al/TiW, was found particularly prone to whisker formation and short circuit failure during current stress.
- Interlayer short circuit failure in Al/W and Al/TiW conductors is well characterized by a lognormal distribution of lifetimes.
- Al/W lifetimes were reduced by one half compared with Al/TiW conductors.
- Short circuit failures displayed an E_a of 0.5-0.6 eV.
- The current density exponent for short circuit failure in Al/W conductors was 2.0.

From this study it is not clear whether short circuit failure is significant in multilayer aluminum metallization where tungsten is used only locally (in contacts and vias), and not as a blanket underlayer. However, the results of this preliminary study indicate this question should be addressed carefully.

Acknowledgements

The authors express appreciation to J. De Blasi for depositing the blanket tungsten films by CVD; and to G. de Groot, P. Bittencurt and S. Kelley for their assistance in wafer processing. We are grateful to P. Townsend and H. Vander Plas (Xerox Palo Alto Research Center) for providing the stress measurements.

References

1.) C. Canali, F. Fantini, E. Zanoni, A. Giovanneti, and P. Brambilla, Microelectron. Reliab. 24, 77 (1984).
2.) J. Lloyd and P. Smith, J. Vac. Sci. Technol. A1, 455 (1983).
3.) J.M. Towner, Proc. 23rd IEEE Rel. Phys. Symp., p. 81, IEEE cat. no. CH2113-9/85, Orlando, Florida (1985).
4.) H.-U. Schreiber, Solid State Electron. 28, 1153 (1985).
5.) B. Grabe and H. Schreiber, Solid State Electron. 26, 1023 (1983).
6.) D. Bartels, A. Capobianco, K. Lasch, and K. Lynch, Proc. 15th IEEE Rel. Phys. Symp., p. 196, IEEE cat. no. 77CH1195-7PHY, Las Vegas, Nevada (1977).
7.) J. Ramsey, J. Vac. Sci. Technol. A1, 721 (1983).
8.) See, for example, this proceedings and those papers in "Tungsten and Other Refractory Metals For VLSI Application", R.S. Blewer, Ed., Materials Research Society, Pittsburgh, Pennsylvania (1986).
9.) J.M. Towner, Solid State Technol. 27, 197 (1984).
10.) E.K. Broadbent, A.E. Morgan, J.M. De Blasi, P. van der Putte, B.J. Burrow, B. Coulman, D.K. Sadana, and A. Reader, J. Electrochem. Soc. 133, 1715 (1986).
11.) R.W. Thomas and D.W. Calabrese, 21st IEEE Rel. Phys. Symp., p. 1, IEEE cat. no. 83CH1846-5, Phoenix, Arizona (1983).
12.) I.A. Blech, P.M. Petroff, K.L. Tai and V. Kumar, J. Crystal Growth 32, 161 (1975).
13.) J.T. Pan and I.A. Blech, J. Appl. Phys. 55, 2874 (1984).
14.) E. Broadbent, J. Towner, P. Townsend and H. Vander Plas, to be published.

THE SELECTIVE CVD TUNGSTEN FOR BIPOLER MULTILEVEL INTERCONNECTION

N.TSUZUKI, M.ICHIKAWA, K.KURITA, K.WATANABE AND K.INAYOSHI
Fujitsu Ltd. Bipolar Process Engineering Dept., 1015, Kamikodanaka Nakahara-Ku, Kawasaki, 211, JAPAN

ABSTRACT

The selective LPCVD tungsten method is widely studied for filling the contact vias of multilayer metallization. However, it is not applicable to Bipolar devices for its high contact resistance. In order to gain low contact resistance, we tried thin buffer layer of refractory metals and refractory metal silicides on top of under metal layer. The depth profiles of elements are observed by AES. Contact resistance is measured by Kelvin method and with via arrays. Step coverage is observed by cross- sectional SEM.

The sputtered tungsten is the best buffer metal for stable contact resistance because of no fluorin peak at the interface between buffer metal and selective tungsten. 500 Å buffer tungsten is enough to stabilize the selective tungsten process. Thus, the stable submicron process is established.

INTRODUCTION

With the progress of multilevel VLSI technology, the via size must be shrinked to submicron region. The accurate via size requires anisotropic etching, where the via wall is vertical. But the DC sputtering, which is widely used for metallization, does not always guarantee good step coverage. The larger the aspect ratio of via becomes, the less reliable step coverage you have.

The selective LPCVD tungsten process is one of the techniques to overcome this problem. [1] [2] However, when we apply this method to the vias of two metal layers the contact resistance is not as low as the one we obtain from the usual process; sputter-etch and the subsequent sputter coating.

To solve the above mentioned problem, we tried various kinds of buffer metal on top of the first metal layer. And as for the buffer metal, sputtered tungsten gives the best result.

SELECTIVE TUNGSTEN

In this work, selective CVD tungsten was deposited by hot wall and vertical type of furnace. Table 1 shows the depositing conditions for selective LPCVD tungsten. There are 2 steps in the depositing conditions. In the first step, nucleation of tungsten on the surface of the first layer is expected. [3] The reaction is descrived as follows;

$$WF_6 + M \text{ ----> } W + MF_x$$

In the second step, tungsten deposition on the tungsten nuclei is expected. The reaction is descrived as follows;

$$WF_6 + 3H_2 \text{ ----> } W + 6HF$$

For the first step, the using gases are WF_6 and argon, and for the second step, WF_6 and hydrogen gas. The reaction temperature is 300 degree Centigrade for both steps. The deposition time is 1 minute for the first step and 40 minutes for the second step respectively. And the thickness of tungsten is 3000 Å. Photo 1 shows anisotropically etched via, and selective tungsten in the via.

ANALYSIS BY AES

These figures are the depth profiles of elements by AES after selective tungsten deposition. Figure 1 is Auger depth profiling of selective tungsten on aluminum without buffer metal. In this case, the peak of fluorine signals exists at the interface between aluminum and selective tungsten.

Figure 2 is for the case of selective tungsten on sputtered tungsten as buffer layer. It shows there is no peaks of fluorine signal at the interface.

Figure 3 is for the case of sputtered molybdenum as buffer layer. In this case, selective tungsten could not be deposited because of molybdenum oxide at the surface .

Figure 4 is for the case of selective tungsten on sputtered titanium as buffer layer. Here, the peak of fluorine signals exists at the interface between sputtered titanium and selective tungsten.

The next three figures are about sputtered refractory metal silicides. These silicides are WSi_x , $TiSi_x$,and $MoSi_x$. In these cases the peak of fluorine signals exist at the interface between sputtered silicide layer and selective tungsten. (Fig. 5, 6, and 7)

Therefore, only sputtered tungsten has no peak of fluorin signal at the interface. In order to obtain a good contact, there should be no peak of fluorine signal at the interface. Then sputtered tungsten as buffer metal should be the best one to make stable electric contact for selective LPCVD tungsten process.

PROCESS FLOW

The first metal layer is Al-2%Cu followed by buffer metal deposited by continuous DC magnetron sputtering, and is etched by reactive ion etch .

The CVD PSG is deposited as dielectric and plannarized by etch back process. The via is also etched by RIE. And the via is filled with selective LPCVD tungsten. Then the second metal layer is deposited by sputtering.

Photo 2 is the cross-sectional SEM photo after this process.

ETCH CONDITION

Table. 2 is etching conditions for the first metal layer and via. For the first metal layer, we use reactive ion etch, and the reaction gases are $SiCl_4$ and Cl_2 mixture . The pressure is 0.03 TORR. The etch rate of Al-2%Cu and W is 4000 Å/min ,400Å/min respectively.

Photo 3 shows the first metallization etched by this etching condition, and it shows a good shape without any under cut.
For via etching, we also use reactive ion etch and the reaction gases are CF_4 and CHF_3 mixture. The etch rate of PSG and W is 2500 Å/min, 80 to 100Å/min respectively. With this etching condition anisotropic etch of PSG is possible.

Table. 3 is ashing conditions of photo resist mask and oxidation of buffer layers. The first sample is sputtered W without photo resist

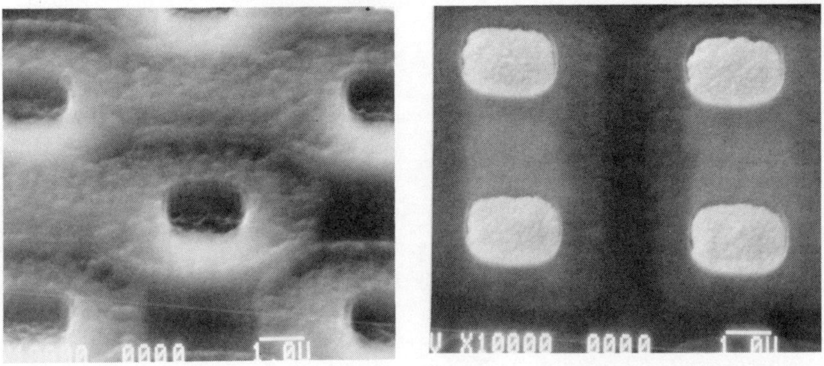

(A) (B)

Photo 1 Anisotropically etched via (A),
and via filled by selective tungsten (B).

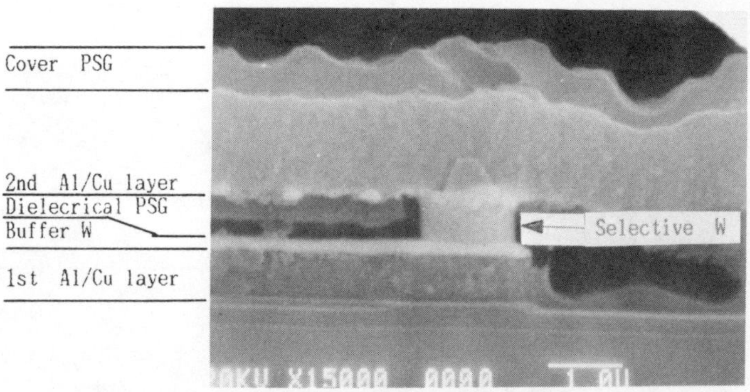

Photo 2 Cross-sectiona SEM of multilevel interconnection using selective tungsten.

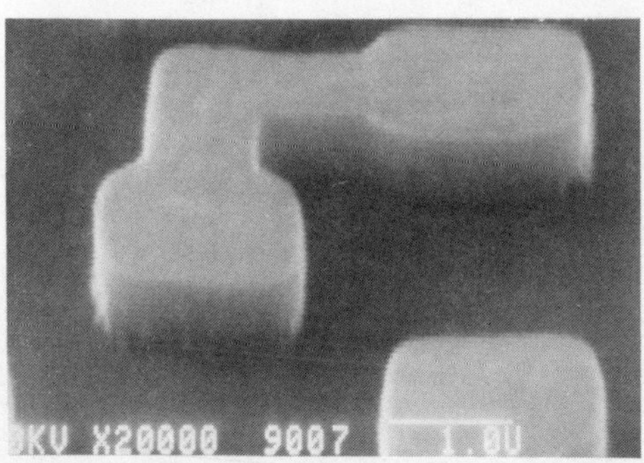

Photo 3 The first metallization of Al-2%Cu with buffer tungsten.

Fig. 3

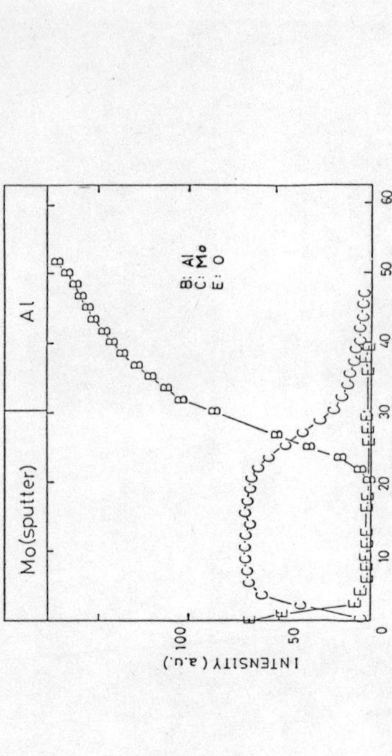

Fig. 4

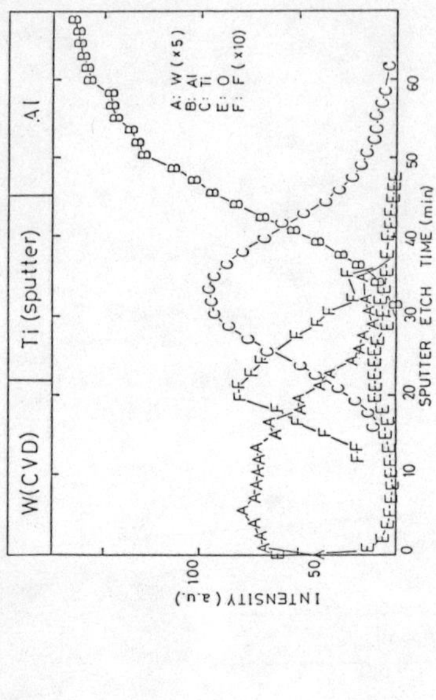

Fig. 1

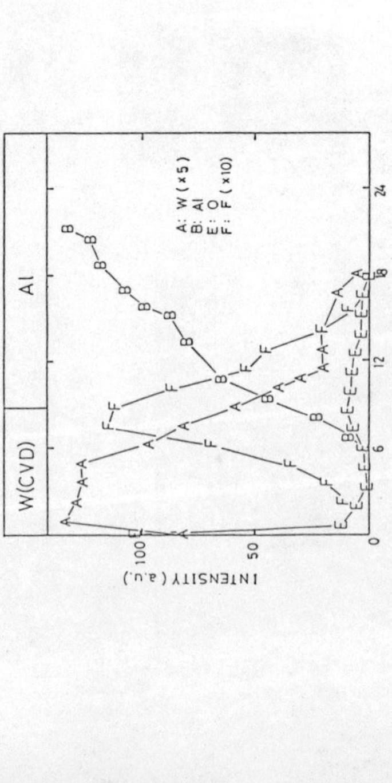

Fig. 2

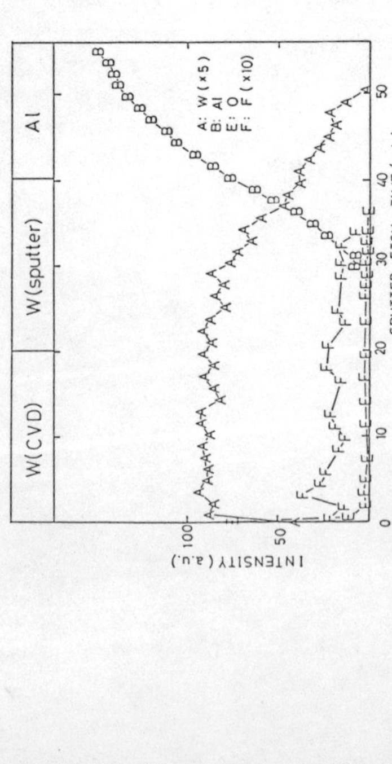

Fig. 5

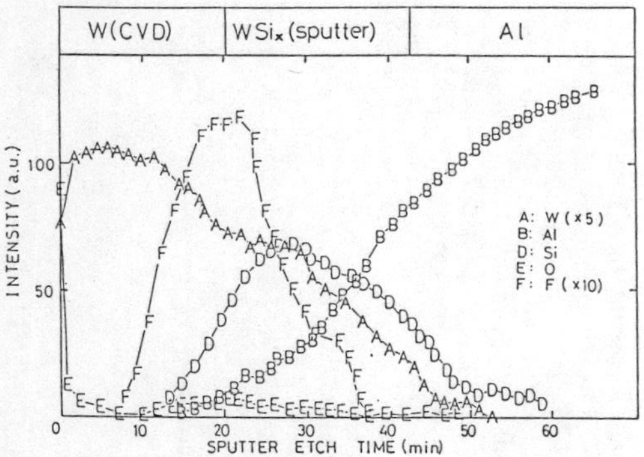

Fig. 6

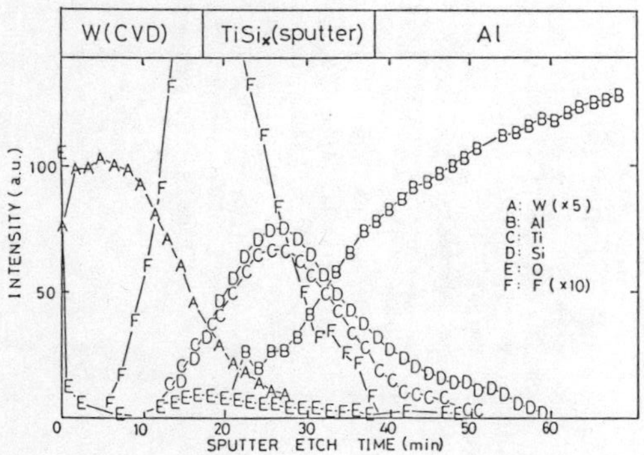

Fig. 7

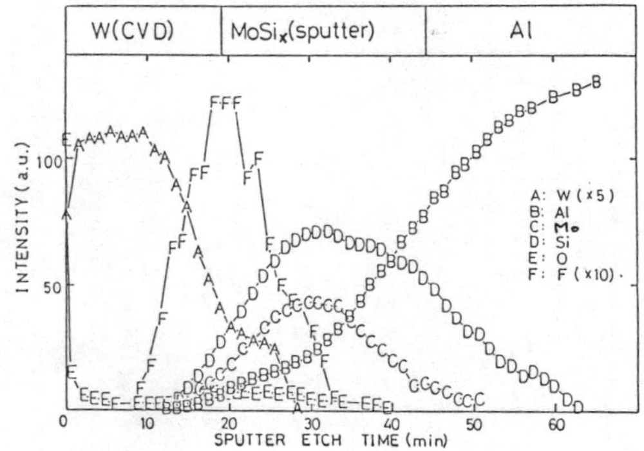

AES Depth Profile			
Fig.	1st Metal	Buffer Metal	Selective W
1	Al-2%Cu	non	500 Å
2	Al-2%Cu	Sputtered W 1000 Å	500 Å
3	Al-2%Cu	Sputtered Mo 1000 Å	500 Å
4	Al-2%Cu	Sputtered Ti 1000 Å	500 Å
5	Al-2%Cu	Sputtered W Si$_x$ 1000 Å	500 Å
6	Al-2%Cu	Sputtered TiSi$_x$ 1000 Å	500 Å
7	Al-2%Cu	Sputtered MoSi$_x$ 1000 Å	500 Å

pattern. The second sample is sputtered W with photo resist pattern. To examine the degree of surface oxidation of buffer metal , break down voltage and color is observed. To complete resist ashing two conditions are tried. In both ashing conditions the color did not change but the break down voltage became slightly higher with increasing plasma power. In regard to tungsten with photo resist mask, the results are the same. In case of molybdenum, surface oxidation takes place much farther than tungsten.

O_2 and CF_4 mixing gases could not be used for resist ashing because of high etch rate of tungsten.

As for contact resistance, we found that break down voltage of less than 6 V showed no problem.

CONTACT RESISTANCE

The contact resistance was measured by Kelvin method. The via size is normally 2 microns. Fig. 8 is the contact resistance of various metal structures of the first metal. Sample 1 shows Al-Al contact of usual process. In the case of sample 2, there is no buffer metal over the first metal layer (Al-Cu) and via is filled by selective tungsten. The contact resistance of sample 2 is much higher than sample 1. This phenomenon would be caused by fluorin at the interface between aluminum and selective tungsten.

Sample 3 to 5 are the case of refractory metals as buffer layer. The thickness of buffer metals is 2500Å. Tungsten for buffer metal shows lower contact resistance than the others. Molybdenum for buffer metal shows higher contact resistance because of surface oxidation of molybdenum. Using titanium, we got higher contact resistance because of fluorin at the interface between buffer metal and selective tungsten.

Sample 6 is for the case of silicides as buffer layer. Molybdenum silicide is selected for the representative of metal silicides. In this case contact resistivity is high as anticipated because of fluorin at the interface.

As a result, tungsten as buffer metal gives the lowest contact resistance. So, tungsten is required as the buffer metal layer of the first metal with selective tungsten method.

Fig. 9 is contact resistance VS thickness of buffer tungsten layer. Thickness of buffer tungsten is 500, 1000, and 2500Å. For these cases contact resistance is almost the same. 500 Å buffer tungsten is enough to stabilize the selective tungsten process.

Fig. 10 shows contact resistance of 330 via array. The via size is 2 microns. This figure also shows low contact resistance, which is consistent with the result of Kelvin method.

Fig. 11 is the specific contact resistivity VS via size. Via size is 0.8 and 2.0 microns. Because of poor step coverage by high aspect ratio, the electric contact could not be made by 0.8 micron via without selective tungsten. In the case of selective tungsten with buffer tungsten, the contact resistance is proportional to area of via. We found that this process is effective to submicron multilayer process.

SUMMERY

1. Using the selective LPCVD tungsten for filling the via, step coverage of 2nd metal layer is improved remarkably.

2. Only sputtered tungsten as buffer layer shows no peak of

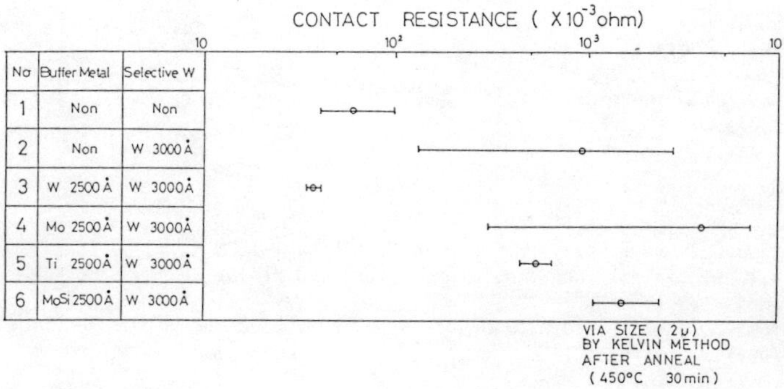

Fig. 8 Contact Resistance by Kelvin Method VS 1st Metal Structure

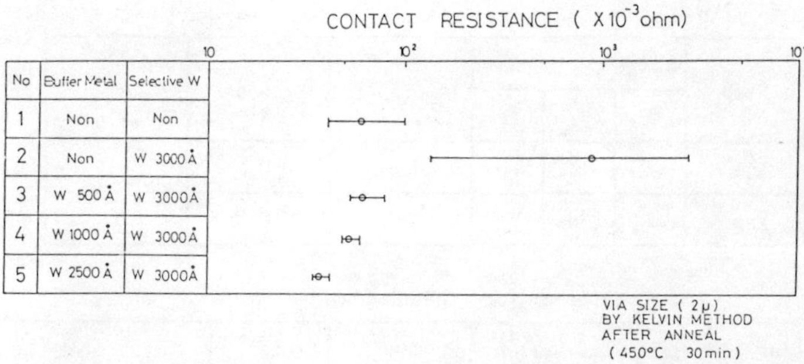

Fig. 9 Contact Resistance by Kelvin Method VS 1st Metal Structure

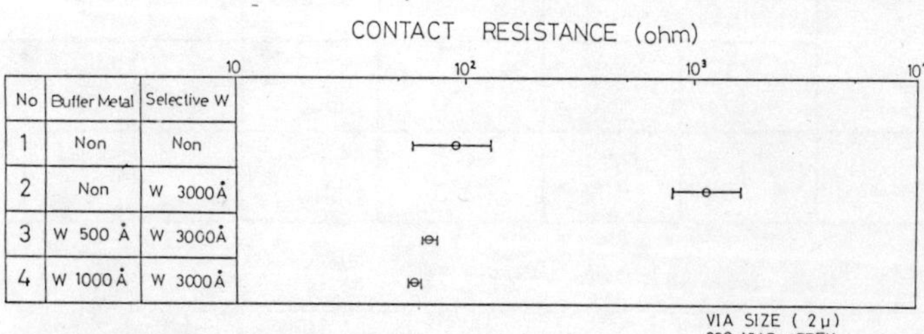

Fig.10 Contact Resistance of Via array

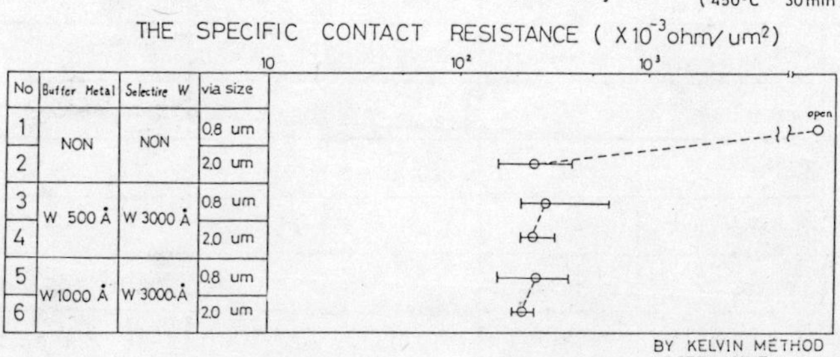

Fig.11 Contact Resistance vs Via. size

fluorin at the interface in the selective tungsten process.

3. The sputtered tungsten is the best buffer metal for low and stable contact resistance.

4. Thus, the submicron multilayer process is successfully established.

REFERENCES

1) T.Moriya, S.Shima, Y.Hazuki, M.Ciba, and M.Kashiwagi, IEDM 83, P.550, 1983 IEEE.
2) R.H.Wilson, R.W.Stoll, and M.A.Calacone, Proceedings of the 1985 Workshop Oct.1984 P.35, 1985 MRS.
3) Y.Ohyama, S.Inoue, Y.Shioya, M.Maeda, and M.Takagi, Electrochemical Soc., 167th Meeting, Vol.85-1, P 282, 1985-May.

Table. 1 The depositing condition for selective LPCVD tungsten

Step	gas flow (SCCM)			Temp.(℃)	Pressure (TORR)	Depo. Time (min.)
	WF_6	H_2	Ar			
1st	30	—	500	300	0.5	1.0
2nd	30	800	—	300	0.3	40

Table. 2 Etch condition for the 1st layer metals and via

Etching process	Gas flow rates (SCCM)		Prasma power (W)	Pressure (TORR)	Etch rates(Å /min)		
					AL /Cu	W	PSG
1st metal layer	$SiCl_4$ 150	Cl_2 30	350	0.03	4000	400	—
Via	CF_4 100	CHF_3 100	450	0.3	—	80 ~100	2500

Table. 3 Oxidation of Buffer Metal by Photo Resist Ashing Process

Metal	methods	As Depo	O_2 Ashing (450W) by Rf Plasma(40min)	O_2 Ashing (1200W) by Rf Plasma(20min)
W	Color	metalic	metalic	metalic
	Breakdown V	1~1.5V	3. 5~4.5 V	5~6 V
W with resist mask pattern	Color	metalic	metalic	metalic
	Breakdown V	1~1.5V	3. 5~4.5 V	5~6 V
Mo	Color	metalic	——	violet or brown
	Breakdown V	1~1.5V	——	2V up to 200V

THERMAL STABILITY OF Al/W/p-n Si CONTACTS DURING POST-METALLIZATION ANNEALING

L. GUTAI, M. DELFINO, and J.M. DE BLASI
Philips Research Laboratories Sunnyvale, Signetics Corporation, Sunnyvale, California 94088-3409

ABSTRACT

Tungsten as a contact metal is selectively deposited at 300°C to a thickness of 430Å onto p-n junctions formed in silicon by implanting B^+, BF_2^+ and P^+ over the dose range of 8×10^{14} to 5×10^{15} ions/cm^2. Electrical activation with minimum dopant diffusion is done by rapid thermal annealing for 10 sec at 950°C. The specific contact resistance and implanted layer sheet resistance are measured using Kelvin cross bridge, sheet end, and transmission line model structures. The effect of 30 min post-metallization annealing in N_2 over the temperature range of 450 to 550°C on these measurements is studied. Data shows that the specific contact resistance is not degraded up to at least 480°C, is lower in p^+- than n^+-Si at low doses, and is lower with BF_2^+ than B^+ at the same dose. These results are not influenced by the choice of test structure and the dimensions of the contacts. Similarly, the junction leakage is not affected up to at least 480°C implying the effectiveness of the tungsten as a diffusion barrier.

Introduction

The properties of selectively deposited LPCVD tungsten films, from their microstructural properties to their deposition kinetics have been extensively studied in recent years [1-3]. Some of the important characteristics of selectively deposited tungsten layers used as a metallization scheme in integrated circuit technology are as follows [4]: (i) It is a diffusion barrier between silicon and aluminum. (ii) It provides low specific contact resistances to appropriately doped n- and p-type silicon. (iii) It has low enough electrical resistivity to serve as a current shunt over poly-silicon gates and over source and drain regions. (iv) It can be used as an etch stop for aluminum interconnect patterning.

The specific contact resistance of selectively deposited tungsten films on diffused n^+-p and p^+-n silicon junctions were measured by Swirhun, Saraswat, and Swanson [5] after a 30 min post-metallization annealing at 450°C by a "four-channel" transmission line model test structure. The electrical properties of selectively deposited (LPCVD) tungsten films on shallow n^+-p and p^+-n junctions (implanted with As^+ and BF_2^+, respectively) were studied by others [4] after a 30 min annealing at 330°C and an additional 30 min at 450°C in forming gas (10% H_2/N_2). Since the non-conventional test structures (four point probe) used in their work yielded end contact resistances rather than contact resistivities, a comparison to our results is difficult. Contact resistivities calculated from end contact resistances can be an order of magnitude smaller than the true specific contact resistance values, depending on the geometry and the transfer length of the contact [6]. A comparison with measurements of the change in the contact resistivity of Al/W/n^+-Si Kelvin structures with extremely wide diffusion taps [7] after annealing at 500°C is difficult also.

In the present work, the specific contact resistance values and areal junction leakage current densities of Al/W contacts onto p^+-n (B^+ and

BF_2^+) and n^+-p (P^+) implanted junctions are measured over a range of implantation doses and post-metallization temperatures. The effect of the measurement method and the characteristic dimensions of the test patterns on the extracted specific contact resistance values are also evaluated.

Sample Preparation

Silicon substrates of p- and n-type doped with boron and phosphorus respectively to approximately 8 x 10^{15}/cm^3 with a (100) orientation were used in these experiments. In order to model the implantation conditions, where normal incidence might be necessary due to the extremely low aspect ratio of small contact openings, a 600Å thick screen oxide was grown in the contact area prior to implantation and a 7° tilt angle and 30° rotation angle away from the <110> was used. Boron was implanted as $^{11}B^+$ and $^{49}(BF_2^+)$ at equivalent energies of 30 keV per atom, whereas $^{31}P^+$ was implanted at 70 keV. Doses of 0.8, 1, 3, and 5 x 10^{15} ions/cm^2 were used except for BF_2^+ which was implanted at a dose of 1 x 10^{15} ions/cm^2 only. After the implantation, a 950°C, 10 sec rapid thermal annealing (RTA) was carried out in N_2 using an A.G. Associates Heatpulse model 610 lamp annealer. The RTA conditions were chosen to minimize dopant diffusion. The temperature was measured with a thermocouple bonded to a reference piece of silicon of similar resistivity.

Prior to the tungsten deposition, contact holes were opened by dry etching into 1000Å thick low temperature oxide (LTO) deposited at 400°C. The tungsten was selectively deposited in an Anicon hot wall LPCVD reactor using a H_2 and WF_6 gas mixture with a gas flow ratio of 2640 sccm to 60 sccm. The reactor pressure was 0.55 torr and the temperature of the chamber was 300°C. The thickness of the tungsten was measured using a beta-backscattering technique [8,9] and found to be 430Å with a uniformity of typically ± 10% across a wafer. The inter-connect pattern was formed by dry etching a 7000Å thick pure aluminum layer deposited by electron beam evaporation. The etching gas used was a mixture of BCl_3 and Cl_2. Post-metallization annealing was done in a furnace for exactly 30 min at temperatures of 450, 480, 500, and 550°C in a dry N_2 ambient. These temperatures are accurate to 0.5°C.

Electrical Measurements

W/p^+-Si/n-Si junction diode characteristics were measured on large area (0.032 cm^2) circular diodes. Leakage currents at 5 V reverse-bias were measured on these diode arrays both prior to and after the 450°C annealing.

The forward- and reverse-bias current-voltage characteristics of a representative p^+-n junction diode are shown in Figs. 1a and 1b respectively. The forward ideality factor is 1.00 over 8 orders of magnitude in current. The areal junction leakage at room temperature at 5 V reverse-bias is less than 0.3 nA/cm^2. These values are typical of both B^+ and BF_2^+ implanted junctions and of all doses used. Considerably higher leakage current densities, i.e. on the order of μA/cm^2 (as will be shown later), were measured on n^+-p diodes in good agreement with data reported in the literature [4]. The leakage currents of both the p^+-n and n^+-p diodes show a slight improvement after the 450°C annealing.

The specific contact resistance data were extracted from measurements on Kelvin cross bridge resistor (KCBR), transmission line model (TLM), and sheet end test structures prior and after annealing at each temperature. Schematic views of the KCBR and TLM structures are given in Figs. 2a and 2b respectively. Nominal contact dimensions of the KCBR structures varied from 1 x 1 μm^2 to 10 x 10 μm^2 with a 1 μm diffusion overlap, while those

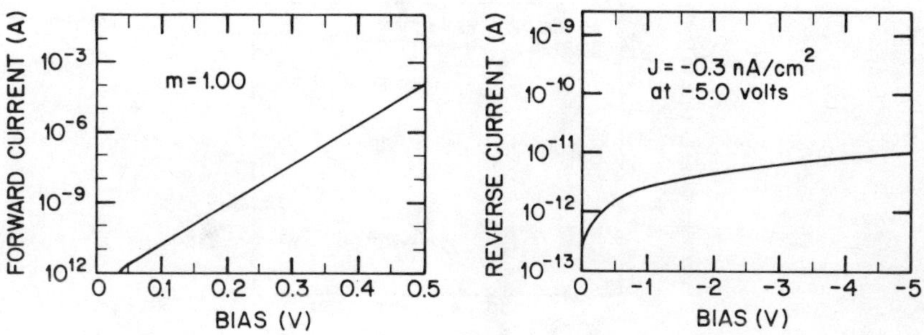

Figure 1. (left) Forward current-voltage characteristic of a typical W/p^{+}-n Si diode. (right) Reverse current-voltage characteristic. The approximate junction depth is 0.3 μm and the diode area is 0.032 cm^2.

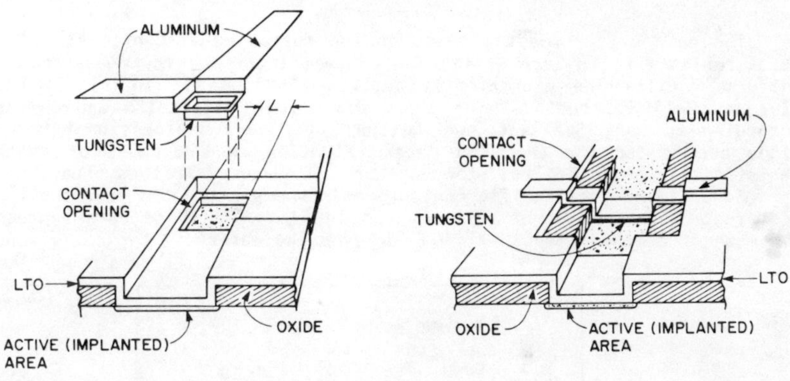

Figure 2. (left) The schematic extended view of the Kelvin cross bridge resistor structure. (right) The schematic view of the transmission line model test structure. One contact is shown only.

of the TLM structures varied from 1.2 x 1.2 μm^2 to 5 x 5 μm^2. The three metal voltage taps were situated 3 μm and 30 μm from each other. The sheetend structure is a 500 x 500 μm^2 implanted area with two 150 x 500 μm^2 tungsten contacts at the opposite ends of the implanted area. Also, there are ten small 3 x 3 μm^2 voltage taps placed along a straight line at equidistant points between the two large area contacts. This structure yielded very accurate sheet resistance values of the implanted layers after activation. Their values were not affected by the annealings. Specific contact resistances can be derived also, on the assumption that the Si sheet resistances are not affected by the tungsten deposition. Therefore, no correction is made for the amount of silicon consumed (as much as 600 Å)[10] during tungsten deposition. Junction leakage is obtained from the sheet end structures also. The effect of annealing temperature on these values is shown in Fig. 3.

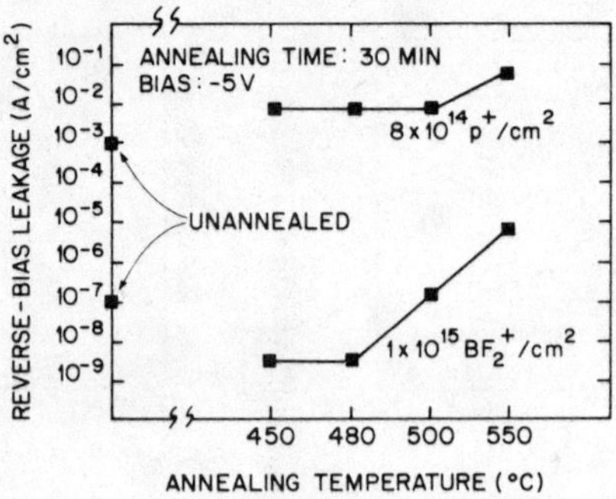

Figure 3. Effect of annealing temperature on the reverse-bias junction leakage obtained from the sheet end structure.

The data extraction was based on the following assumptions: (i) The metal resistivity is zero. (ii) The contact interface is homogeneous and uniform. (iii) The electrically activated silicon is infinitely thin. (iv) In the TLM structure, the third dimension is infinite and uniform. According to Loh, Saraswat, and Dutton [11], no two-dimensional corrections are needed for the KCBR data evaluation in the range of contact resistivities and sheet resistances that are measured in this study.

Some of the the specific contact resistances extracted from the three different test structures are compared in Fig. 4. These data represent the mean of 9 measurements carried out over the wafer. A typical standard

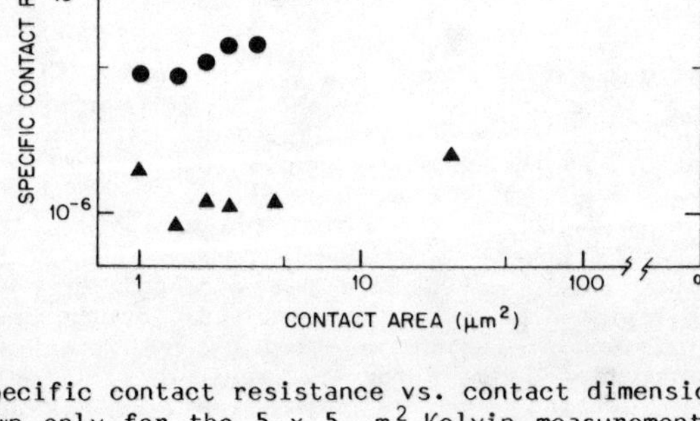

Figure 4. Specific contact resistance vs. contact dimensions. The error bars are shown only for the 5 x 5 μm^2 Kelvin measurement. The implantation is 1 x 10^{15} B^+/cm^2 and the post-metallization annealing is 450°C.

deviation of these measurements is shown for the 5 x 5 μm^2 Kelvin structure. The size dependence of the data shown in Fig. 4 is representative of all measurements made. The sheet end structure gave an accurate value for the semiconductor sheet resistance with small standard deviations (typically 0.5% to 5.0% depending on the value of sheet resistance and the annealing conditions). However, the sheet end structure yielded contact resistivity values with very high standard deviations due to uncertainity in the distance between the large area contact and the voltage taps. Nonetheless, the mean value of the specific contact resistance for each experimental condition was within the error limits of the Kelvin data. The TLM data displayed a smaller size dependence than the Kelvin data.

The size effect, i.e. an increase in specific contact resistance with contact area, in the KCBR structures is opposite to that expected on the basis of the two-dimensional corrections suggested in the literature [11]. The effect, while it is not understood, is not due to the presence of tungsten since a similar effect was found with Al/n^+-Si and Al/p^+-Si both before and after a 30 min annealing at 450°C in N_2. In spite of this unexplained behavior, the large area Kelvin structure was used for the following reasons: (i) This structure yields the specific contact resistance value without any correction. (ii) The size effect saturates with contact area. Furthermore, the general conclusions and trends of the data are independent of the choice of test structure as well as of the dimensions. This is shown in Fig. 5, where specific contact resistance vs. sheet resistance data is plotted with the annealing temperature as a parameter using data obtained from 10 x 10 μm^2 and 2 x 2 μm^2 Kelvin structures.

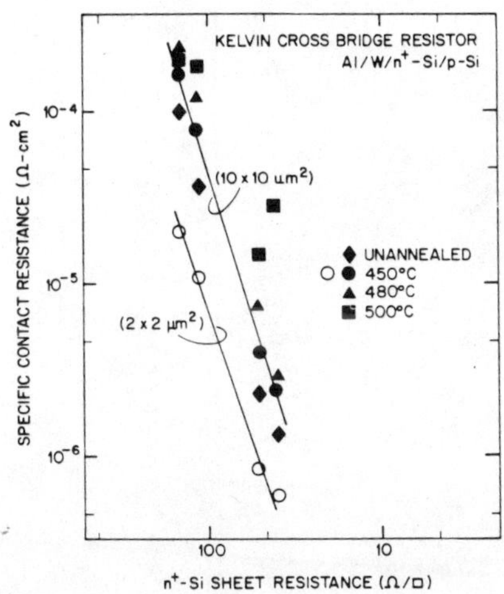

Figure 5. Specific contact resistance vs. silicon sheet resistance with the annealing temperature as a parameter. The implantation ion is $^{31}P^+$ at 70 keV.

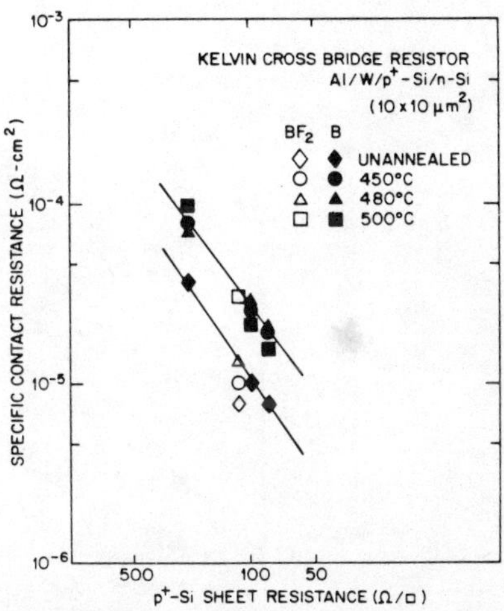

Figure 6. Specific contact resistance vs. silicon sheet resistance with the annealing temperature as a parameter. The implantation ions are $^{11}B^+$ at 30 keV and $^{49}(BF_2^+)$ at 134 keV.

Results and Discussion

The specific contact resistance values for the different implantation doses and post-metallization temperatures are plotted as a function of the sheet resistances of the implanted semiconductor layers derived from measurements on the sheet end structures in Figs. 5 and 6. In order to compare those graphs with graphs plotted as contact resistivity vs. surface concentration the following has to be taken into account. To a first approximation, the peak concentration (essentially the surface concentration after RTA) of a Gaussian distribution is linearly proportional to the implantation dose at a constant energy. Since the sheet resistance, assuming constant bulk mobility, is inversely proportional to the implant dose, the peak concentration is inversely proportional to the sheet resistance. Using the measured sheet resistance data rather than calculated [4] surface concentration data, a direct experimental relationship results between the specific contact resistance and the surface concentration of the electrically activated implanted atoms provided that the bulk mobility is independent of the implantation dose in the range of the experiment. The linear relationship in Figs. 5 and 6 suggests that, in our case, the sheet resistance is inversely proportional to the surface concentration of the electrically activated implanted atoms. This justifies the assumption of constant bulk mobility. Consequently the saturation effect in conductivity at high dose, i.e. $5 \times 10^{15}/cm^2$ shown in Fig 7. suggests that this saturation is possibly due to incomplete electrical activation and dopant precipitation, rather than a lowering of the carrier mobility. Another consequence of Figs. 5, 6 and 7 is that the log contact resistivity vs. log implant dose can not be linear at high doses.

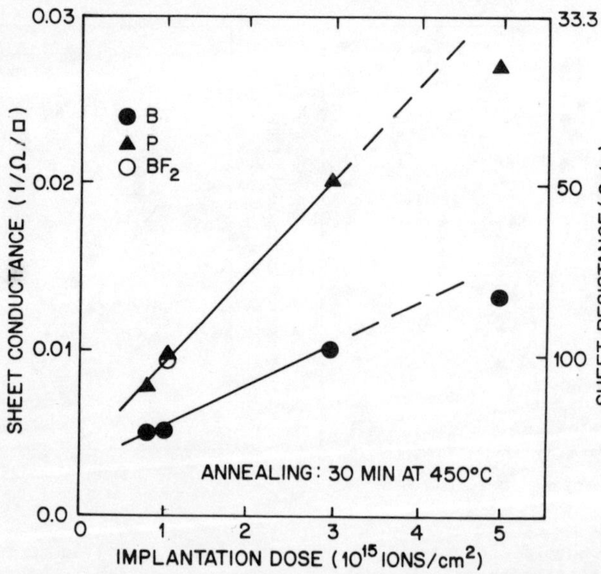

Figure 7. Dependence of the silicon sheet resistance on the ion implantation dose. These data are not affected by the post-metallization annealings. The correlation coefficients are 0.99.

The data points corresponding to the Al/W/n^+-Si/p-Si structure (Fig. 5) are basically on the same straight line up to a post-metallization annealing temperature of 480°C, indicating that the annealing had no effect on the contact properties. Even though the contact resistivity values are not increasing considerably above 500°C, the departure from the linear relationship indicates an adverse effect of this annealing on the contact properties. Similarly, the plot (Fig. 6) of specific contact resistance vs. sheet resistance of Al/W/p^+-Si/n-Si is linear in a log-log scale. However, in the latter case, there is a definite increase in contact resistivity after the 450°C annealing with no apparent change in slope. These values remain constant up to at least 500°C.

Another observation is that the contacts on p^+-Si implanted with BF_2^+ yield a contact resistivity that is three times lower than the respective B^+ implanted contact at the same sheet resistance value. However, it should be noted that the B^+ dose was three times as high as the BF_2^+ dose to achieve the same sheet resistance. Interestingly, after annealing at 500°C, the contact resistivity of the BF_2^+ implanted contact increases and approaches that of the B^+ implanted contact.

The slopes (change in specific contact resistance with silicon sheet resistance) of both the Al/W/n^+-Si/p-Si (Fig. 5) and the Al/W/p^+-Si/n-Si (Fig. 6) contact resistivities are identical to slopes (change in specific contact resistance with surface concentration) published for deeply diffused junctions [5]. It is noted that our spreading resistance measurement of a 1×10^{15} B^+/cm^2 implanted contact showed a surface concentration of approximately 5×10^{19} carriers/cm^3. Using this value, reasonable agreement of the absolute values of our specific contact resistance data with this previous work [5] is found.

Conclusions

As a result of this study, the following conclusions can be drawn:

(i) The leakage current of Al/W/p-n Si junctions is not affected by annealing for at least 30 min up to at least 480°C, however the leakage is almost 6 orders of magnitude lower in p^+- than in n^+-Si.

(ii) The specific contact resistance of Al/W/p^+-n Si junctions after 30 min annealing up to at least 500°C remains the same as after 30 min annealing at 450°C. The specific contact resistance of Al/W/n^+-p Si junctions are not degraded by annealing for at least 30 min up to at least 480°C.

(iii) The specific contact resistance is lower in p^+- than in n^+-Si after 30 min annealing at 450°C for sheet resistances higher than 80 Ω/sq.

(iv) Implanting and activating $^{49}(BF_2^+)$ yields lower sheet resistance at the same dose ($1 \times 10^{15}/cm^2$) and lower contact resistivity at the same sheet resistance than does $^{11}B^+$.

(v) The log of the measured contact resistivity vs. the log of the measured sheet resistance is linear, whereas the log of the measured contact resistivity vs. the log of ion implantation dose is not.

(vi) The foregoing results are independent of the three different test structures (KCBR, TLM, and sheet end) and of the linear dimensions (1 μm to 500 μm) used.

Acknowledgements

The authors thank G. de Groot for help with the wafer processing and for providing useful suggestions. We also thank our colleagues in Philips Research Laboratories in Eindhoven, The Netherlands for supplying the VIAPEM-2 mask set and measurement program.

References

1. W.T. Stacy, E.K. Broadbent and N.H. Norcott, J. Electrochem. Soc., 132, 444 (1985).
2. P.A. Gargini and I. Beinglass, IEEE IEDM Tech. Dig., p. 54 (1981).
3. E.K. Broadbent and C.L. Ramiller, J. Electrochem. Soc., 131, 1427 (1984).
4. R.A. Levy, M.L. Green, P.K. Gallagher and Y.S. Ali, J. Electrochem. Soc., 133, 1905 (1986).
5. S. Swirhun, K.C. Saraswat, and R.M. Swanson, IEEE Trans. Electron Device Lett., EDL-5, 209 (1984).
6. H.H. Berger, Solid-State Electron., 15, 145 (1972).
7. Y. Shioya, M. Maeda and K. Yanagida, J. Vac. Sci. Technol., B4, 1175 (1986).
8. A. Ott, Chemie-Ingenieur-Technik, 46, 805 (1974).
9. R.H. Muller, Analy. Chem., 29, 969 (1975).
10. J.M. De Blasi, D.K. Sadana and M.H. Norcott, Mat. Res. Soc. Symp. Proc., 71, 303 (1986).
11. W.M. Loh, K. Saraswat, and R.W. Dutton, IEEE Trans. Electron Device Lett., EDL-6, 105 (1985).

Properties of Refractory Metal Deposited by Sputtering or Evaporation

THE CHARACTERIZATION OF TITANIUM-NITRIDE THIN FILM AS DIFFUSION BARRIER FOR VLSI METALLIZATION

Kyung-ho PARK, S. MIHARA, Y. SATO*, H. TSUCHIKAWA* and M. YOSHIDA
Fujitsu Ltd. Advanced Technology Division, * Process Development Division
1015 Kamikodanaka, Nakahara, Kawasaki, 211, JAPAN

ABSTRACT

Titanium-Nitride (TiN) films prepared by reactive sputtering have received considerable attention in recent years because of their applicability as a diffusion barrier for semiconductor metallization. However, the ability of these films to serve as diffusion barriers is not yet fully understood. Moreover, none of the previous studies were performed on patterned samples which would simulate the application of these barrier films to VLSI circuitry. In this work we have used Transmission Electron Microscopy (TEM) to study the thermal stability of a TiN barrier layer between polycrystalline Silicon and an Aluminum overlayer on patterned wafers. It was found that when the TiN is annealed that the film which is deposited on the side walls of the pattern behaves differently than the film which is deposited on the flat areas on the top of the pattern and in the trenche.

INTRODUCTION

In recent years, the complexity and scale of integrated circuits has increased, this trend has resulted in greater device densities and decreased feature sizes. Because of this, the requirements for VLSI contact metallurgy have become more and more strict due to the need for shallow junctions and sub-micron contacts. Conventional Al-Si metallization has serious problems making reliable contacts and surviving all of the thermal cycles associated with complex VLSI processes. The use of diffusion barriers in contact structures has become the most common approach for achieving stable and reliable contacts.

To date, several kinds of barrier layers have been proposed with particular emphasis on the application to silicon devices which employ aluminum metallization [1-8]. However, an ideal barrier which has perfect thermal, chemical and mechanical stability, good electrical conductivity, and inhabits interdiffusion between its bordering media does not exist, as has been pointed out by Ting and Wittmer [3]. In reality, a barrier layer may interact with neighboring layers to a limited extent, if after subsequent thermal cycling the material has not been totally consumed. Diffusion across the barrier layer should be small enough so that there is no noticeable effect on the device properties.

One of the most attractive materials for a barrier layer is TiN, since it has a lower bulk resistivity than pure titanium and it is a very stable compound with a high melting point. Although the use of TiN thin film as a diffusion barrier have been extensively studied [9-16], there is still very little knowledge of the film properties and the mechanisms governing this particular diffusion barrier.

It is recognized that the barrier performance of thin films are influenced by both structural and chemical aspects of the film. It is also well known that both of these properties are strongly dependent on the method of film preparation.

In the present study we describe an investigation of TiN films which were prepared by reactive sputtering. These films were used as diffusion barrier between polycrystalline Si and an Al overlayer. The purpose of this paper is to consider the structural properties of sputtered TiN barrier layer observing the microscopic phenomenon of diffusion using cross-sectional TEM. Prior work in this field has been performed using RBS, SIMS and Auger(AES). These types of analytical tools give imformation from relatively large sampling area and tend to give an average value for the analytical region. For the fabrication of VLSI circuitry, it is necessary to investigate the film performance on patterned surfaces as well as planar surfaces. This is particularly relevant to VLSI geometries because it is well known that sputtered films can exhibit poor step coverage.

EXPERIMENTAL PROCEDURE

Two different types of samples were prepared for this investigation. Both of these sample structures had a multilayer thin film structure with a reactively sputtered TiN barrier layer interposed between a top Al layer and an underlying polycrystalline Si layer. The first type of samples was an unpatterned SiO layer on Si substrate as schematically shown in Fig. 1(a). The second sample type was patterned with a line and space pattern etched into a 700 nm thick PSG film, again on a Si substrate. This second sample type is shown schematically in Fig. 1(b). For both sample types, the top Al layer was deposited by sputtering after the TiN deposition without breaking vacuum in between the deposition.

The nominal TiN deposition parameters are listed in Table I. Samples which did not receive heat treatment are referred to as as-deposited, those samples which were annealed three times at 723 K for 30 minutes and once at 753 K for 30 minutes in nitrogen atmosphere are refered to as annealed.

Elemental analysis and depth profiles of these elements in the unpatterned samples were determined using SIMS and AES. Cross-sectional TEM specimens were prepared for both types of samples. The structure and morphology were examined in STEM at 400 kv(JEOL 4000EX equipped with EELS and EDX). The formation of the different microstructures was also investigated by electron diffraction, EDX and EELS. Nano-probe microdiffraction studies were carried out in order to determine different microstructures in nanometer scale areas.

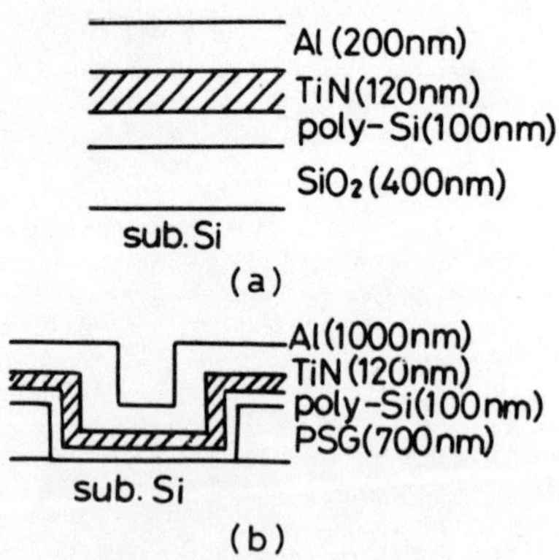

Fig. 1 Schematic representations of the cross-sectional view for samples (a) unpatterned and (b) patterned.

Table I.
Conditions for TiN deposition

Method	DC Magnetron Sputtering
Gas	Ar/N_2 (=1/2)
Flow Rate	50 sccm (total)
Power	8 kw
Pressure	5 mtorr
Deposition Rate	200 nm/min

EXPERIMENTAL RESULTS and DISCUSSION

The depth profiles obtained from the SIMS and AES analysis for the annealed, unpatterned samples were essentially the same. For this reason only the SIMS results are presented here. In Fig. 2, the depth profiles of Al, Ti, N and Si are shown. The shape of the Al and Ti depth profiles are significantly different in the annealed versus as-deposited samples. The depth profiles for each of the elements in the as-deposited samples are sharply defined at the interface of each layer. In contrast, in annealed samples the Al and Ti profiles are considerably broader at the interfaces.

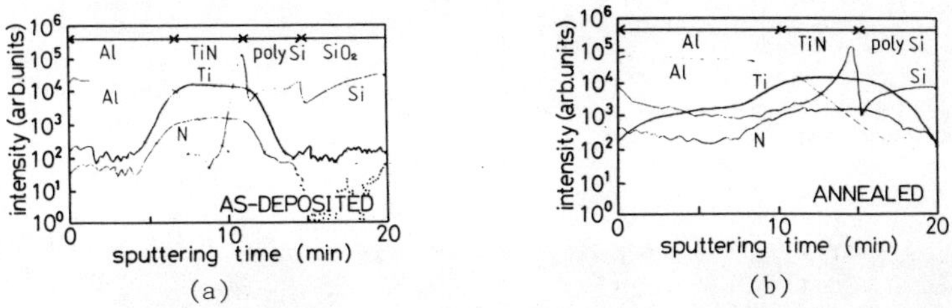

Fig. 2 SIMS depth profiles of unpatterned samples for (a) as-deposited and (b) annealed.

Cross-sectional TEM micrographs are shown in Fig. 3. The micrograph taken from the as-deposited sample shows that each layers are clearly identifiable and have sharp interfaces. Selected area diffraction (SAD) patterns from each layer were analyzed for the elements of interest and are shown in Fig. 4. EDX spectra were obtained from each layer using a 7 nm beam diameter spot. These EDX spectra show only a single element in each layer and there is no evidence of any interdiffusion. These results for unpatterned, as-deposited samples are in very good agreement with the data obtained by SIMS and AES analysis.

A cross-sectional TEM micrograph for the unpatterned, annealed samples is presented in Fig. 5. Many precipitates have appeared in Al layer near the TiN layer interface. EDX spectra of these precipitates and the surrounding Al matrix are shown in Fig. 6. These spectra clearly demonstrate that the precipitates are titanum rich, suggesting that they may be a TiAl compound. Obtaining a quantitative element ratio for the precipitates is not feasible because the exact thickness of TEM specimen is unknown and the precipitates are surrounded by an Al matrix. This TEM observation of heterogeneous precipitation at the interface of Al and TiN layers explains why the Al and Ti SIMS and AES depth profiles broaden for the annealed samples.

The TEM observation for the patterned, as-deposited samples are similar to those obtained from the unpatterned, as-deposited samples, as is shown in Fig. 7. For the annealed patterned samples, the film on the top of PSG lines and in the trenches between the lines reacts in a fashion similar to the unpatterned samples. In contrast, the film on the side walls exhibits a different phenomenon after annealing. A typical TEM micrograph is shown in Fig. 8, note that the contrast due to polycrystalline Si has disappeared. EDX spectra taken from each of the layers on the side walls indicated that the degree of interdiffusion was greater than that in the trenches or on the top of the PSG lines. The EDX also indicates that the disappearance of polycrystalline Si contrast is due to a substitution of Al into the Si as shown in Fig. 9. This data is evidence that the TiN layer at side wall of pattern step is not adequate for use as diffusion barrier.

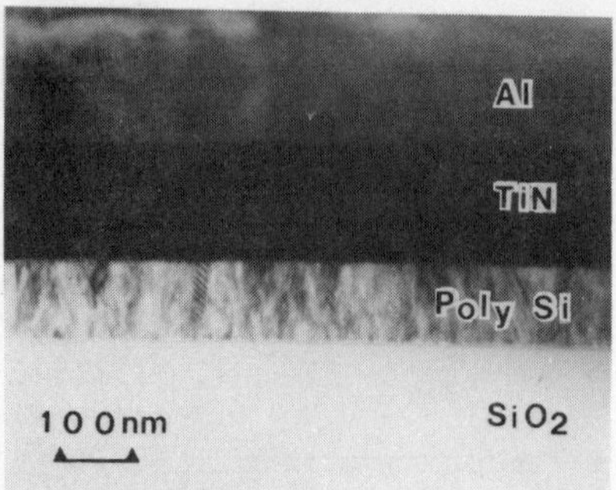

Fig. 3 The cross-sectional TEM micrograph of unpatterned, as-deposited samples.

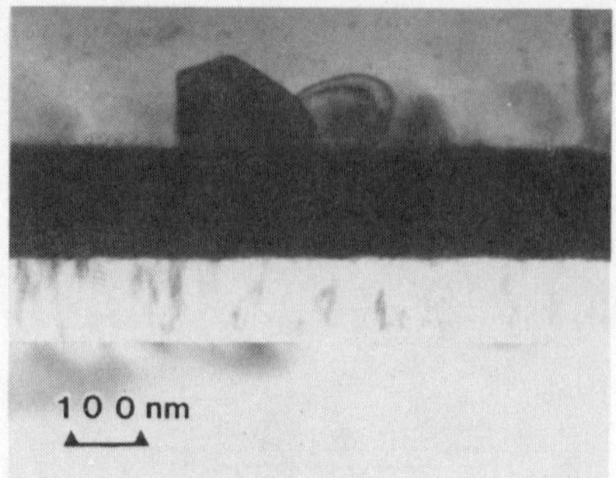

Fig. 5 The cross-sectional TEM micrograph of unpatterned, annealed samples.

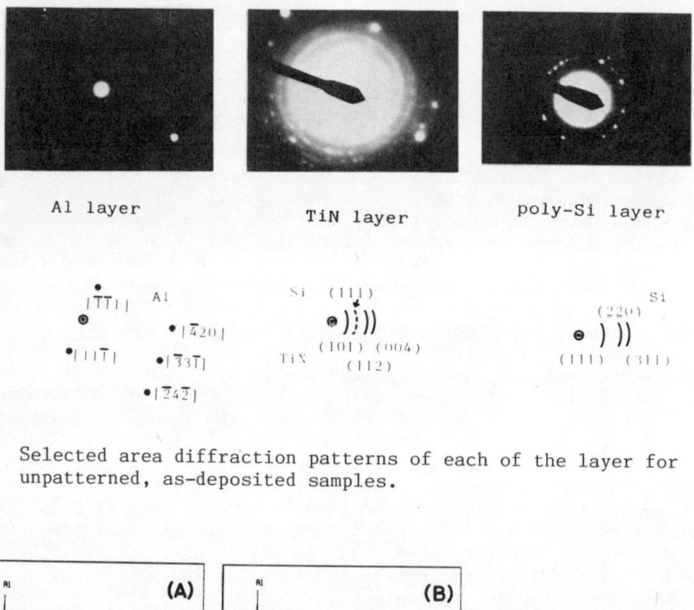

Fig. 4 Selected area diffraction patterns of each of the layer for unpatterned, as-deposited samples.

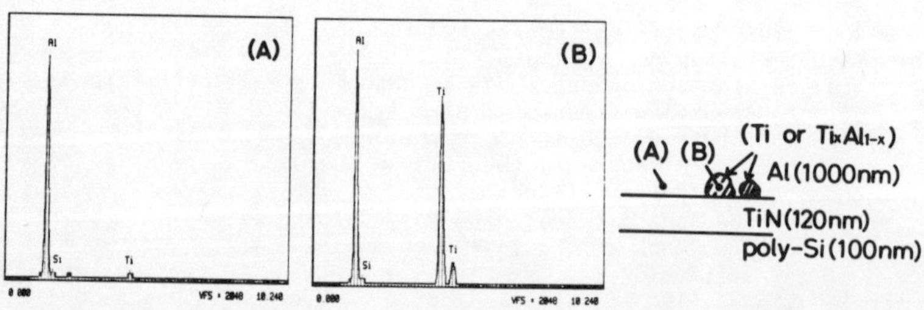

Fig. 6 EDX spectra of (A) Al matrix and (B) precipitates.

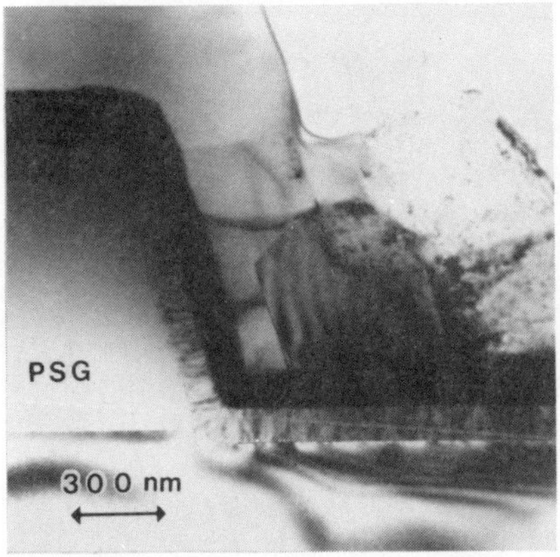

Fig. 7 The cross-sectional TEM micrograph of patterned, as-deposited samples.

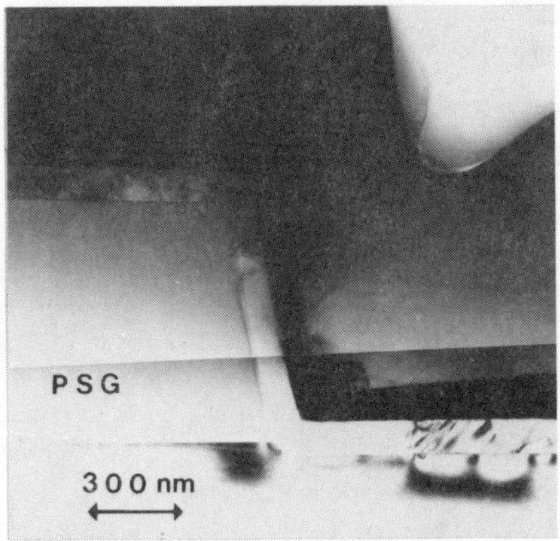

Fig. 8 The cross-sectional TEM micrograph of patterned, annealed samples.

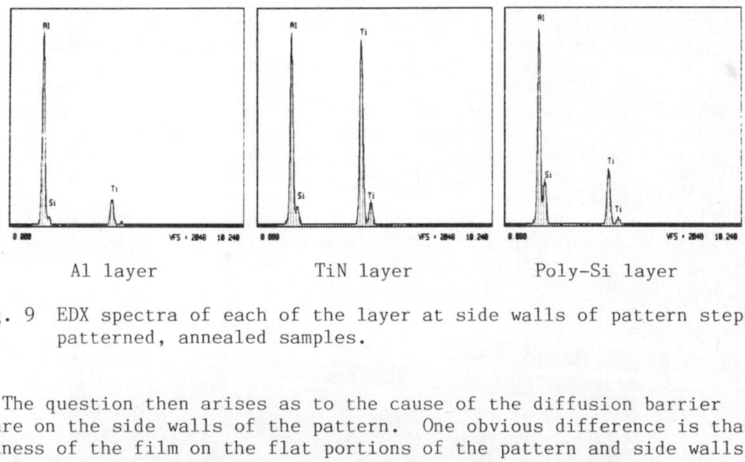

Al layer TiN layer Poly-Si layer

Fig. 9 EDX spectra of each of the layer at side walls of pattern steps for patterned, annealed samples.

The question then arises as to the cause of the diffusion barrier failure on the side walls of the pattern. One obvious difference is that the thickness of the film on the flat portions of the pattern and side walls is different. However, this does not appear to be the reason for the failure since the phenomenon was not observed in any of the flat areas when the samples recieved additional annealing. Moreover, in the EDX spectra obtained from the patterned, as-deposited samples there is already evidence of diffusion in the side walls as shown in Fig. 10. In these spectra the Ti peak in Al layer of side wall is clearly distinguishable while it is not in spectra taken from non-side wall areas of the same sample.

In answering this question there are many factors which should be considered, including grain size, grain structure, chemical composition and impurities. Fig. 11 is a high magnification TEM micrograph of the TiN layer in one of the flat areas (trenches) between the pattern lines. This micrograph indicates that the grain size is less than 10 nm. It also indicates that the grain structure is composed of ordinary fine grains and is not columnar. Chemical composition and inpurity analysis was obtained using

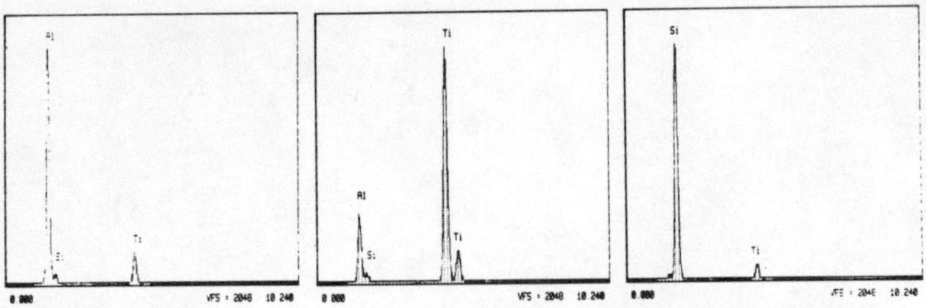

Al layer TiN layer Poly-Si layer

Fig. 10 EDX spectra of each of the layer on the side walls of pattern staps for patterned, as-deposited samples.

EELS. Spectra were obtained using STEM, 12 nm beam diameter spot mode, from both areas of the patterned, as-deposited samples. No impurity elements, such as oxigen or carbon were detected in either areas. However, a difference was detected in the peak ratio of Ti(L-edge, 455eV) and N(K-edge, 399eV) between the spectra obtained from the side walls and flat areas as shown in Fig. 12. Again a quantitative value for the nitrogen-to-titanium ratio cannot be obtained because the exact specimen thickness is unknown and varied for each layer. It was observed that the nitrogen peak for spectra taken from the side walls was always smaller than that obtained from the flat areas. As before, SAD showed only the TiN lattice spacing. The SAD signal was obtained from relatively large area compared to TiN grain size. To obtain additional information, nano-probe microdiffraction on both areas was performed. Typical results are shown in Fig. 13. Note that some of the diffraction spots can be identified as resulting from a Ti_2N phase. The amount of material in the Ti_2N phase is apparently greater in side walls when compared to that of flat areas. From these observations, it can be stated that the TiN thin film consists of both TiN phase and Ti_2N phase material. In addition the TiN layer on patterned substrates have different ratios of TiN and Ti_2N when the side walls are compared to the flat areas of the pattern.

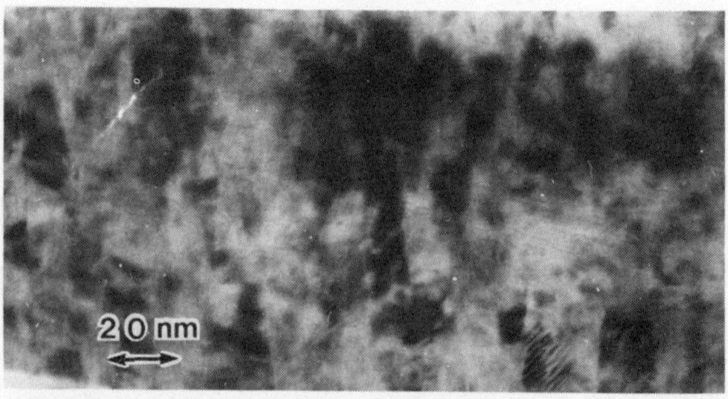

Fig. 11 The cross-sectional TEM micrograph of the TiN layer for patterned, annealed samples.

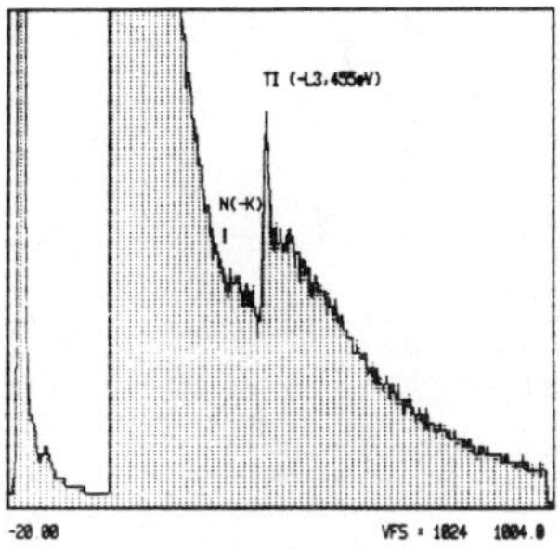

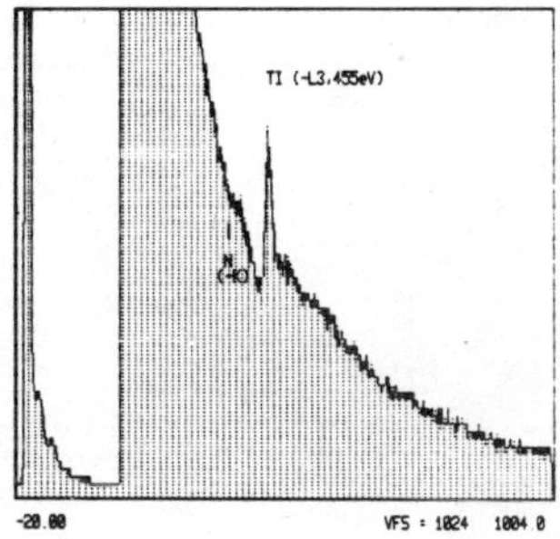

(a) (b)

Fig. 12 EELS spectra of (a) side wall areas and (b) flat areas of pattern stapes for patterned, as-deposited samples.

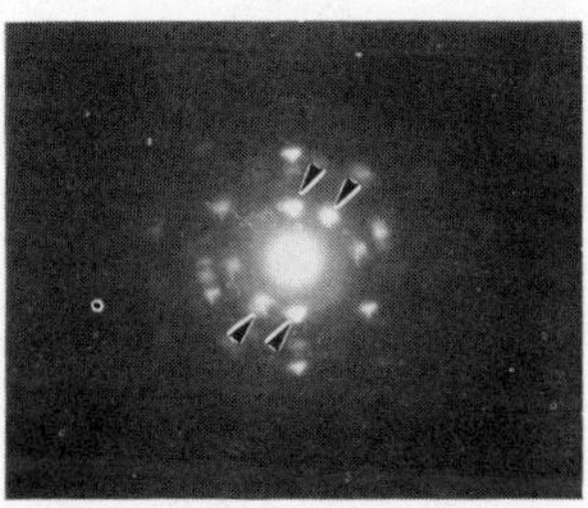

(a) (b)

Fig. 13 Nano-probe microdiffraction patterns of (a) side wall areas and (b) flat areas of pattern steps for patterned, as-deposited samples. Arrow marked spots were identified the Ti_2N phase.

SUMMARY

1) The TiN thin film deposited by reactive dc magnetron sputterig is a small grain structure with an average grain size of less than 10 nm.

2) After annealing, heterogeneous Ti rich precipitates were found in Al layer near the TiN layer interface.

3) The TiN film deposited on patterned surface had two phases, TiN and Ti_2N, the ratio of these phases are varied with respect to the position of the film on the pattern.

Acknowledgements

The authers would like to thank the people working in the process development facility for their efforts in sample fabrication. The authers are grateful to Drs. Y. Ban, M. Nakano, K. Fujino and H. Yagi for their helpful discussions and to Mr. Y. Kataoka for his contibution in SIMS analysis.

REFERENCES

1. M.-A. Nicolet, Thin Solid Films 52, 415 (1978).
2. M. Wittmer, J. Appl. Phy. 53, 1007 (1982).
3. C.Y. Ting and M. Wittmer, Thin Solid Films 96, 327 (1982).
4. I. Suni, M. Maenpaa, M.-A. Nicolet and M. Luomajarvi, J. Electochem. Soc. 130, 1217 (1983).
5. M.F. Zhu, F.C.T. So and M.-A. Nicolet, Thin Solid Films 130, 245 (1983).
6. M. Wittmer, J. Vac. Sci. Tech. A3, 1797 (1985).
7. H.P. Kattelus, E. Kolawa, K. Affolter and M.-A. Nicolet, J. Vac. Sci. Tech. A3, 2246 (1985).
8. R.A.M. Wollters and A.J.M. Nellissen, Solid State Tech. 29(2), 131 (1986).
9. C.Y. Ting, J. Vac. Sci. Tech. 21(1), 14 (1982).
10. M. Wittmer and H. Melchior, Thin Solid Films 93, 397 (1982).
11. K.Y. Ahn, M. Wittmer and C.Y. Ting, Thin Solid Films 107, 45 (1983).
12. I. Suni, M. Blomberg and J. Saarilahti, J. Vac. Sci. Tech. A3, 2233 (1985).
13. A. Armigliato, M. Finetti, J. Garrido, S. Guerri, P. Ostoja and A. Scorzoni, J. Vac. Sci. Tech. A3, 2237 (1985).
14. W. Sinka, G.P.A. Frijilink and F.W. Saris, Appl. Phys. Lett. 47, 471 (1985).
15. S. Kanamori, Thin Solid Films 136, 195 (1986).
16. S. Iwabuchi, S. Shima, T. Maeda and T. Moriya, VLSI Inter. Symp. Proceeding 1986 55 (1986).

THE DEPOSITION, CONTACT AND ELECTROMIGRATION PROPERTIES OF REACTIVELY SPUTTERED TITANIUM NITRIDE FILMS

R.C. ELLWANGER* AND J.M. TOWNER
Philips Research Laboratories Sunnyvale, Signetics Corporation, Sunnyvale, CA. 94088

Titanium nitride (TiN) has been reported to be a good diffusion barrier between silicon or silicides and aluminum [1,2], also providing a low resistance contact in such a structure [3,4]. It also has been reported that TiN does not promote Al whisker growth when sandwiched between two Al layers [5].

The purpose of this work was first to characterize the deposition process and film properties of reactively sputtered titanium nitride. A specific focus then was the comparison of TiN to the TiW films used within IC processing. The results of Towner [6] show the presence of an underlayer to be of major importance in the electromigration behavior of Al. Accordingly, emphasis has been placed on determining the effect of a TiN underlayer on Al electromigration performance, as well as its performance as a diffusion barrier and contact material.

Films of TiN were deposited onto 10 cm diam. <111>, p-type Si substrates using a Balzers BAS 450 dc magnetron sputter system. This system uses a so-called 'drum within a drum' design. The cathode is located on the outer cylinder facing inward, while the substrates are mounted on the rotating inner cylinder. The target-to-substrate distance was 8 cm. The target material was 99.999% purity Ti. The working gas for reactive sputtering was a mixture of Ar (99.995%) and N_2 (99.99%).

The system was evacuated to a pressure less than 2×10^{-7} mbar, and then backfilled with Ar to a pressure of 2×10^{-3} mbar, followed by a target sputter clean of 5 min at 2 kW. Subsequently, metered flows of N_2 and Ar gases were introduced and allowed to stabilize to the desired pressure for 5 min each. With both Ar and N_2 now present, the dc plasma was then re-ignited and the target was sputtered for 5 min just prior to opening the shutter. Unless otherwise noted, all experiments were performed at a total chamber pressure of 2×10^{-3} mbar and a target power of 2 kW.

Rutherford backscattering spectrometry (RBS) was used to obtain the Ti/N compositional ratio in the as-deposited films. Both channeled and glancing angle spectra were analysed. Film depth homogeneity and impurity levels were determined using Auger electron spectroscopy (AES) depth profiling. Specific layer resistivity was calculated from four-point probe (electrical) and stylus profilometry (thickness) measurements.

Diffusion barrier properties were evaluated using an extended guardring PtSi Schottky diode structure. A cross-sectional schematic and device structure specifics are given in Fig. 1. After receiving the appropriate diffusion barrier material (as outlined in Section III B), the substrates received a 0.7 μm layer of evaporated Al. 0.1 μm thick TiW (10 wt.% Ti) layers, used for comparison, were deposited in a Varian 3125 sputter system from a conical magnetron source.

Contact resistance was measured on both p^+ and n^+ Si regions using Kelvin structures with a 4x4 μm^2 contact area. All samples received a sputtered 1.0 μm thick layer of Al-1%Si and were annealed at 450°C for 30 minutes in flowing N_2 (10% H_2) prior to testing.

* Present Address: Philips Research Laboratories Eindhoven
5600 JA Eindhoven, The Netherlands

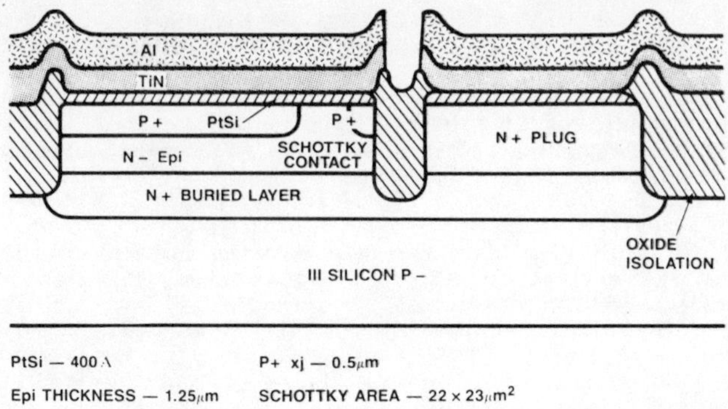

Figure 1: Cross-sectional schematic of extended guard-ring PtSi Schottky diode. Device structure specifics are listed.

Electromigration testing was performed on patterned whole wafers according to a previously described procedure [6]. The test structure consisted of two levels of aluminum with an intermediate 1.0 μm thick dielectric layer of phosphosilicate glass (PSG). A conductor stripe defined in the first layer of metal was stressed at constant current. Failure occurred when either an open formed in the first layer test stripe or a short formed between first and second layer metal.

The effect of N_2 partial pressure on the chamber total pressure is shown in Fig. 2 for both the presence and absence of a dc plasma. The Ar partial pressure was a constant 1.25×10^{-3} mbar. Without a plasma, each unitary increase in N_2 partial pressure corresponds to a unitary increase in chamber pressure. After plasma ignition substantial gettering occurs. For the 2 kW plasma two distinct slopes are observed. For N_2 partial pressures up to 0.9×10^{-3} mbar, the amount of gettered nitrogen increases with increasing pressure. Beyond this pressure no further gettering occurs and the slope returns to unity. We will refer to this point of discontinuity as the critical pressure of the system.

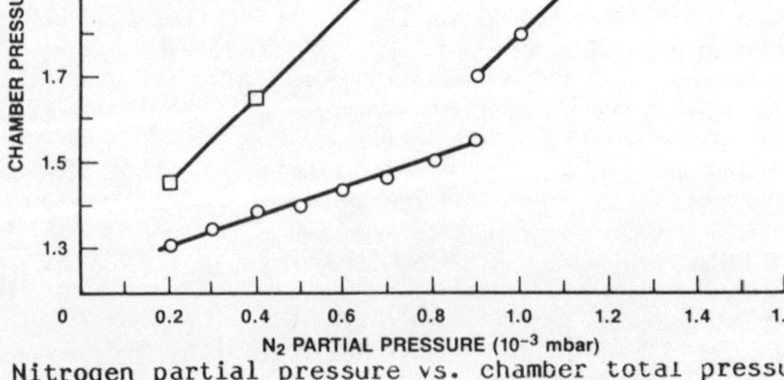

Figure 2: Nitrogen partial pressure vs. chamber total pressure for both the absence of a plasma and the presence of a 2kW plasma.

Fig. 3 shows a plot of the film composition and deposition rate as a function of increasing N_2 concentration (% total working pressure). For these experiments the total pressure was held constant at 2×10^{-3} mbar. A sharp decrease in deposition rate is seen at the point of critical N_2 partial pressure. This sharp reduction in deposition rate has been commonly reported in the literature, and has been correlated to the range of nitrogen partial pressures at which the deposited film changes from an hcp (α-Ti) to an fcc (TiN) microstructure [7-9]. This reduction in the deposition rate is best explained by changes occurring at the target surface. The difference in sputter efficiency between N_2, N, and Ar would not account for such a sharp decrease in deposition rate for such a small change in partial pressure. Also, neither gas phase nor surface interactions can account for the sharp retardation in nitrogen gettering observed once the critical pressure has been achieved. The discontinuity and retardation in nitrogen gettering can be attributed to the sudden decrease in sputter yield as the target surface becomes coated with a continuous layer of titanium nitride. Having reached the critical pressure, the target nitridation rate now remains constant.

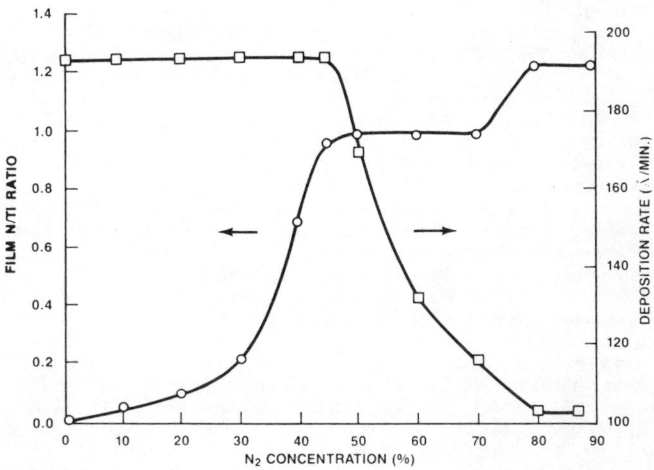

Figure 3: Film N/Ti ratio and deposition rate vs. N_2 concentration in N_2 + Ar working gas.

Several systems of metal nitride reactive sputtering have previously been studied by Shinoki and Itoh [10]. Using mass spectroscopy they have shown that even above the critical pressure, the concentration of metal, N, and N_2 ions are an order of magnitude higher than those of the metal nitride ions. These results suggest that titanium nitride is removed from the target as individual titanium and nitrogen atoms rather than as titanium nitride molecules. The reactivity of atomic nitrogen is known to be very high [11] and as such explains the sudden increase in substrate film nitrogen content upon reaching the critical pressure.

Further examination of Fig. 3 shows that the substrate film composition remains stoichiometric from the point of critical pressure, N_2:Ar $\approx$1:1, to a ratio of greater than 2.3:1. Since the mean free path is greater than the target-to-substrate distance, titanium nitride gas phase recombination (Ti + N) will be minimal. Thus it appears that substrate film nitridation is caused by surface recombination of titanium and highly reactive atomic nitrogen.

The film N/Ti ratio ranged from 0 to 1.25. Work by both Wittmer, et al [12] and Sundgren, et al [13] indicates that for films with a N/Ti ratio <0.15, none of the reported phases of titanium nitride were formed. It is believed that the nitrogen in these films occupies interstitial sites in the α-Ti lattice. For compositional ratios ranging from 0.15 to near 1.0, the films are a mixture of α-Ti, Ti_2N and TiN. Combined phases of α-Ti and Ti_2N shift to that of Ti_2N and TiN as stoichiometry is approached. At the point of film stoichiometry only the TiN phase is seen. In agreement with the literature, electron diffraction performed from cross-sections shows films which are of stoichiometric composition to be single phase fcc TiN. Cross-sectional TEM shows this film to be polycrystalline, containing grains which are vertically elongated at an average diameter of about 0.02 μm.

For the range of N_2/Ar partial pressures studied, Auger depth profiles showed the film N/Ti ratio to be homogeneous with respect to depth. The oxygen content in the film bulk ranged from below the Auger detection limit (<1 at.%) for sub-stoichiometric to a high of about 4 at.% for hyper-stoichiometric films. The value for stoichiometric films was about 1-2 at %.

Guard-ring Schottky diodes (Fig. 1) were used to evaluate the effectiveness of TiN layers as a barrier to Al/Si interdiffusion. These diode structures allowed the assessment of Al movement at three distinct junction depths. A decrease in barrier height indicates interaction at the 400 Å PtSi/Si interface. Slight increases in diode ideality and reverse current leakage indicate an increasing interfacial roughening as the interaction approaches the 0.5 μm p^+/n junction. Large decreases in breakdown voltage evidence that the reaction was extended below the 0.5 μm p^+ junction, while a short circuit indicates penetration to the 1.25 μm epi-to-buried layer junction.

The sample groupings, anneal conditions, and diode measurements are listed in Table I. All anneals were performed sequentially, with diode measurements preceding and following each anneal cycle.

The initial diode characteristics were not effected by any of the five depositions. After the 450°C anneal the aluminum has already moved to the PtSi/Si interface where no barrier layer was present; the reduction of barrier height to 0.75 eV is typical of an aluminum Schottky diode to n-Si. This data is in good agreement with Hosack [14] who reports a value 0.72 eV for a 450°C anneal of the Al/PtSi/Si contact system. A slight increase in the diode ideality factor and reverse current leakage is also seen at this point in the Al/PtSi system. After the 450°C anneal all barrier systems studied demonstrated absolute electrical integrity.

With the exception of the stoichiometric TiN/PtSi system a slight but significant decrease in the barrier height, accompanied by an increase in the reverse leakage current characteristics, was observed for all barrier systems after the 500°C anneal. One explanation for this small decrease in barrier height is the presence of a small amount of either Ti or Al at the silicide-Si interface. In this case a low and high barrier Schottky diode in parallel could cause this effect.

Table I: Diode characteristics for several contact systems

	Barrier Height (eV)	Ideality Factor	Breakdown (V)	Leakage at 15V (nA)
450°C ANNEAL, 30 MIN.				
TiN	0.795	1.04	17.5	2
TiN/Ti	"	"	"	"
$Ti_{1.02}N_{.98}$	"	"	"	"
TiW	"	"	"	"
Al	0.715	1.06	"	6
500°C ANNEAL, 30 MIN.				
TiN	0.795	1.04	17.5	2
TiN/Ti	0.78	"	"	4
$Ti_{1.02}N_{.98}$	"	"	"	4
TiW		"	"	4
Al	0.715	1.06	16.5	6-50
550°C ANNEAL, 30 MIN.				
TiN	0.71	1.05	10-14	-
TiN/Ti	0.71	1.05	13-16	-
$Ti_{1.02}N_{.98}$	0.62	1.44	4	-
TiW	shorted	shorted	shorted	-
Al	shorted	shorted	shorted	-

After the 550°C anneal substantial interdiffusion was seen for all groups. The TiW and Al groups are completely shorted. The titanium-rich ($Ti_{1.02}N_{.98}$) barrier film group shows a large amount of interdiffusion as witnessed by the high (1.44) ideality factor and low breakdown voltage. The barrier height of 0.62 eV is not indicative of that material which comprises the metal-to-silicon interface. Rather, it has been artificially lowered due to the high leakage (non-ideality) of the junction. Although the stoichiometric titanium nitride barrier film group clearly exhibits an Al Schottky behavior, these films have prevented any large amounts of interdiffusion and the associated deep spiking. The diodes still exhibit near-ideal behavior; the breakdown, although lowered is still abrupt (low leakage levels at reverse bias until avalanche breakdown).

Table II: Normalized n^+ and p^+ contact resistance for various layerings. (All samples annealed at 450°C, 30 min, N_2/H_2).

	Normalized Contact Resistance	
Material	n^+ ($\approx 4\times10^{19}$ As/cm^3)	p^+ ($\approx 1\times10^{19}$ B/cm^3)
Al/TiN/60ÅTi/Si	1.0	2.4
Al/TiN/$TiSi_2$/Si	5.5	1.0
Al/TiN/Si	190	46
Al/$Ti_{1.02}N_{.98}$/Si	195	48
Al-1% Si/Si	230	1.0

In Table II we list those metal/Si contact layerings studied, the dopant concentration, and the relative n^+ and p^+ contact resistance for the nominal 4 x 4 μm^2 contacts. The Ti interfacial layer was observed to reduce contact resistance by a factor of 5 over that of $TiSi_2$, and by greater than 2 orders of magnitude over that of Al-1%Si contact to n^+ Si. For contact to p^+ Si areas the TiN/Ti structure yielded values slightly higher than the Al-1%Si and $TiN/TiSi_2$ systems. The composition of the nitride layer appears to have no influence on those p^+ and n^+ regions contacted directly by TiN. The $TiN/TiSi_2/Si$ system also yields acceptably low values of contact resistance.

To evaluate the effect of the TiN underlayer on both open and short circuit failure, five groups of substrates were prepared. As shown in Table III, two compositions of TiN were used: stoichiometric TiN with a resistivity of 165 μΩ-cm and Ti-rich TiN with a resistivity of 75 μΩ-cm. For purposes of comparison, Al was also deposited directly on oxide and on a layer of sputter deposited TiW (10 wt.% Ti). In the final group, Al was deposited onto a layer of unreacted Ti. The underlayer thicknesses were 0.1 μm. All samples were metallized with 0.7 μm of Al in the same deposition cycle. Wafers were then annealed at 450°C for 35 minutes.

Table III illustrates the effect of various underlayers on the electromigration failure mode. Whenever an underlayer was present, first-to-second layer metal shorts were the predominant failure mode, accounting for 85% to 100% of all failures. In contrast, only 30% of the failures were due to interlayer shorts when these layers were omitted.

Table III: Materials used in open/short circuit electromigration study.

Conductor Material	Aluminum Cross-section	Failure Mode
Al	3.0 μm^2	30% short
Al/TiW	1.9 μm^2	90% short
$Al/Ti_{1.02}N_{.98}$	1.8 μm^2	85% short
Al/TiN	2.8 μm^2	90% short
Al/Ti	2.5 μm^2	100% short

Preliminary tests with single level, unpassivated Al/TiN test structures revealed a substantial reduction in the density of Al whiskers as compared to unpassivated Al/TiW conductors. The Al did, however, accumulate in large, nondescript mounds. Apparently, both mound formation and whisker growth preferentially accelerates short circuit failure.

Fig. 4 provides a comparison of the effects of the TiN and TiW barrier layers. This figure is a lognormal probability plot of lifetimes at a temperature of 300°C and current densities near 2×10^6 A/cm^2. At this test conditions, the Al conductors exhibit the longest lifetimes. When Al is underlaid with TiW, MTF is reduced by a factor of five. Even greater reductions in lifetime occur with either of the two TiN compositions.

As a final experiment, the Al/Ti conductors were tested. It has previously been reported that layered Al/Ti alloy conductors exhibited substantial improvements in electromigration resistance in terms of the open circuit failure mode [15,16]. These reports also suggest that shorts, not opens, will ultimately limit the reliability of the conductors.

The sheet resistance of the Al/Ti conductors increased following the 450°C anneal, from 4.9×10^{-2} to 6.2×10^{-2} Ω/square. Fig. 5 is a lognormal probability plot of failure times for Al, Al/TiW, and Al/Ti conductors. The stage temperature was 300°C and the current density was 3×10^6 A/cm^2. Although this current density is considered to be excessive due to the large amount of joule heating (10-20°C), it was chosen to minimize test times for the Al/Ti conductors. Fig. 5 shows that the reliability of Al/Ti conductors is determined by the time to

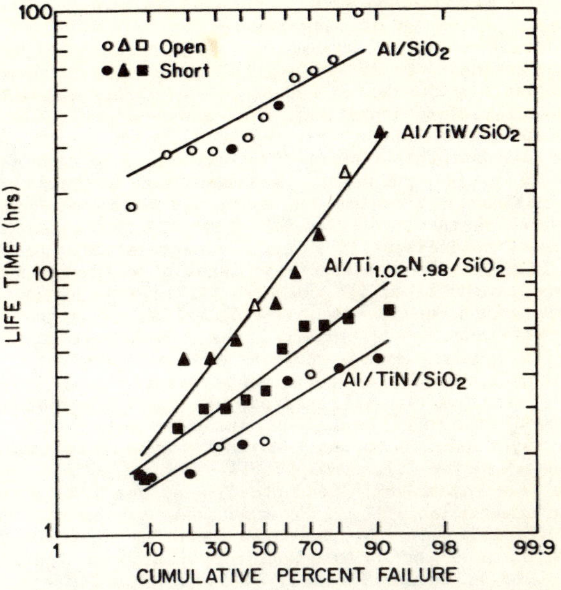

Figure 4: Logonormal probability plot of interlayer short circuit failure times in the Al, Al/TiW, $Al/Ti_{1.02}N_{.98}$ and Al/TiN conductor systems. Current densities were between 2.0 and 2.2 x 10^6 A/cm^2.

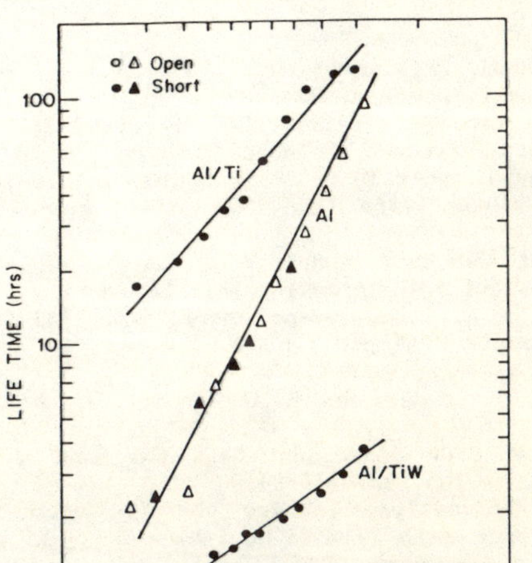

Figure 5: Lognormal probability plot of failure times for Al, Al/TiW, and Al/Ti conductors. The stage temperature was 300°C and the current density was 3 x 10^6 A/cm^2.

short circuit failure. However, lifetimes were increased substantially compared to the Al or Al/TiW conductors.

In conclusion, the mechanisms of titanium nitride film formation by reactive sputtering in an Ar + N_2 ambient have been discussed. The primary determinant is the degree and stability of target nitride coverage. This coverage is a function of target atom removal rate and system nitrogen partial pressure. For a given removal rate, a critical N_2 partial pressure can be defined, at which the target surface becomes fully covered with a TiN layer. The primary mechanism for substrate nitridation is the recombination of titanium and monatomic nitrogen, both originating from the target surface. Ti-rich TiN films can be deposited by maintaining the N_2 partial pressure at a level below this critical pressure.

Stoichiometric TiN films possess superior diffusion barrier properties compared to sputtered TiW (10 wt.% Ti) layers. A layered TiN/60Å Ti structure demonstrated the same superior diffusion barrier properties as the TiN film alone. Additionally, the TiN/60Å Ti structure yielded acceptably low contact resistance to n^+ and p^+ Si regions.

The single area of poor performance observed for TiN films is one which has not previously been reported, namely, its effect upon the electromigration of the overlying Al. The electromigration failure mode shifts from open circuit to short circuit formation when Al stripes are underlaid by a TiN (or Ti) layer. For TiN, this shift in failure mode is accompanied by a reduction in conductor MTF of one order of magnitude.

Helpful discussion with E. Broadbent and W. van den Hoek is gratefully acknowledged. The authors also thank G. de Groot for substrate preparation, T. Bril and B. Westlund for etching of the structures, A. Morgan for AES, T. Kuiper for RBS, D. Sadana for TEM and electron diffraction, J. DeBlasi for contact resistance structures, H.-P. Vu for TiW depositions, and D. Stadtler for sputter deposition support.

REFERENCES

1. M. Wittmer, Appl. Phys. Lett. 37, 540 (1980).
2. H. Norstom and T. Donchev, Physica Scripta 28, 633 (1983).
3. C.Y. Ting, J. Vac. Sci. Technol. 21, 14 (1982).
4. C.Y. Ting and M. Wittmer, Thin Solid Films 96, 327 (1982).
5. B.B. Grabe and H. Schreiber, Solid State Electronics 26, 1023 (1983).
6. J.M. Towner, Proc. 23rd IEEE Reliability Physics Symposium, p. 81 (1985).
7. M. Wittmer, J. Vac. Sci. and Technol. A3(4), 1797 (1985).
8. G. Lemperiere and J.M. Poitevin, Thin Solid Films 111, 339 (1984).
9. F. Schinoki and A. Itoh, J. Appl. Phys. 46, 3381 (1975).
10. H.F. Winters, D.L. Raimondi and D.E. Horne, J. Appl. Phys. 40, 2996 (1969).
11. M. Wittmer, B. Studer and H. Melchoir, J. Appl. Phys. 52, 5722 (1981).
12. J.E. Sundgren, B.O. Johansson, S.E. Karlsson and H.T.G. Hentzell, Thin Solid Films 105, 367 (1983).
13. Independent RBS analyses showed the 70 μΩ-cm film to be $Ti_{1.02}N_{.98}$ and the 165 μΩ-cm film to be TiN.
14. S.S. Iyer and C.Y. Ting, Proc. 22nd IEEE Reliability Physics Symp. p. 273 (1984).
15. R.E. Jones, Ir. and L.D. Smith, Proc. 2nd VLSI Multilevel Interconnect Conf. p. 194 (1985).
16. S. Swirhun, K.C. Saraswat and R.M. Swanson, IEEE Electron Device Lett., Vol. EDL-5, no. 6, 209 (1984).

FORMATION KINETICS AND PROPERTIES OF TITANIUM NITRIDE FORMED BY RAPID THERMAL ANNEALING.

RAMA K. SHUKLA* AND STEPHANIE A. YOSHIKAWA**

* TECHNOLOGY DEVELOPMENT,INTEL CORPORATION, 3601 JULIETTE LANE, SANTA CLARA, CA 95051
**DEPT. OF MATERIALS SCIENCE AND ENG., UNIV. OF CALIFORNIA, BERKELEY, CA 94720

ABSTRACT

Titanium Nitride thin films formed by nitrogen ambient rapid thermal annealing (RTA) of 300-1000 Å titanium sputtered on silicon dioxide and silicon nitride substrates have been characterized through electrical resistivity measurements, etch rate studies, Auger spectroscopy and transmission electron microscopy. Under the reaction conditions used, polycrystalline, cubic TiN formation is suggested at or above 800 oC. The specific resistivity of the film is formation temperature dependent, decreasing with increasing temperature due to grain growth. Etch rate data suggests that nitridation of titanium over silicon nitride is incomplete due to diffusion limitations resulting in a TiN/Ti bilayer structure while nitridation of titanium over silicon dioxide under similar conditions results in complete nitridation of the titanium film to cubic titanium oxynitride, the formation of which is aided by the oxygen provided from the underlying silicon dioxide due to an interfacial redox reaction. A thermodynamic model to account for the substrate effects on the reaction products is presented.

INTRODUCTION

Titanium nitride (TiN) thin films have been widely investigated for use as a diffusion barrier between silicon and contact metals, as etch stop and as local interconnect shunts in the VLSIC technologies[1,2]. The low resistivity and high chemical stability of this material make it an excellent candidate for these applications. As silicon device dimensions continue to be scaled down, Rapid Thermal Anealing (RTA) is being considered as an alternate to the standard furnace annealing techniques because of numerous advantages it offers. Thermal nitridation of titanium thin films in conventional furnaces is extremely difficult due to the high oxidation potential of titanium. It has been found that the properties of titanium nitride(such as composition, resistivity, hardness and microstructure) made by conventional furnace annealing in nitrogen can vary substantially with the processing parameters used[3].

While previous studies[4,5] have shown that rapid thermal nitridation of Ti over SiO_2 and Si_3N_4 substrates leads to the formation of TiN, the various properties of these films have not been well characterized. In this paper, titanium nitride thin films produced by rapid thermal annealing of titanium in a nitrogen ambient have been characterized as a function of the formation temperature and times. Sheet resistance measurements, Auger and TEM analyses and etch rate results are presented. The influence of the underlying substrates, silicon oxide vs. silicon nitride has been studied and a model to account for the behavior observed is presented.

EXPERIMENTAL DETAILS

Test specimens were prepared on p(100) single crystal silicon wafers of 100 mm diameter with resistivity > 5 ohm-cm. Following substrates were examined: 1000 Å and 300-400 Å Ti on 800 Å thermally grown SiO_2 and on 500 Å of LPCVD Si_3N_4. Prior to the titanium deposition, the wafers received a buffered HF clean equivalent to the removal of approximately 100 Å of thermal oxide. The titanium was sputter deposited in a Varian 3180 sputtering system utilizing heated substrates (~200 ^oC) with a base pressure of ~$2X10^{-7}$ torr. Utilizing a tungsten-halogen rapid thermal annealing system, the samples were annealed in a 99.999% pure nitrogen ambient at the temperatures of 700 to 1100 ^oC for 15 to 60 seconds. The sheet resistance of the films was measured using a standard four-point probe method. Etch studies in titanium selective etch ($H_2O/NH_4OH/2H_2O_2$) and buffered HF were carried out by etching the films repeatedly for various fixed times and measuring the changes in sheet resistance. The reaction products were analyzed by Auger analyses and by horizontal and cross sectional TEM methods.

RESULTS AND DISCUSSION

<u>Sheet Resistivity Changes</u>

Figure 1 shows the specific resistivity of the films as a function of the RTA anneal temperatures (all anneals for 20 seconds). The specific resistivities have been calculated by assuming full nitridation of the deposited Ti layers (the maximum theoretical TiN thickness is 1.08 x Ti thickness). In fig. 1, the initial increase in the resistivity of as deposited Ti upon annealing at the lowest temperature (700 ^oC) is most likely due to the dissolution of nitrogen (and oxygen as an impurity) in the metal and has been observed by several researchers in the past. As the temperature of anneal increases further, there is a continuous decrease in the specific resistivity from a high value of ~ 160 μ ohm-cm for 700 ^oC anneal to 35 μ ohm-cm for 1100 ^oC anneal. This decrease in resistivity implies that there is a structural and/or compositional difference of the films formed at each anneal temperature.

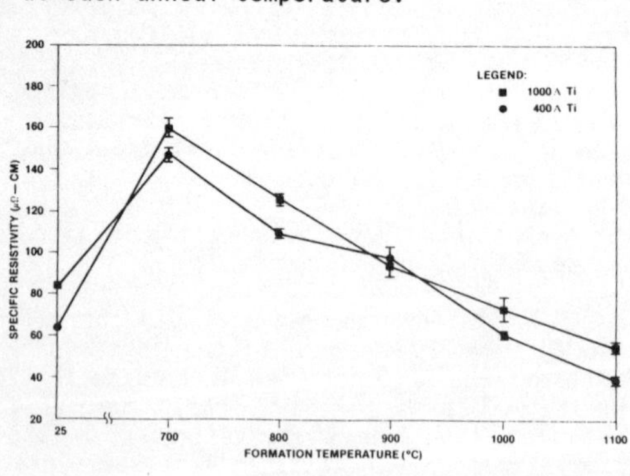

Figure 1: Specific resistivity vs. annealing temp. by RTA of Ti on SiO_2 (all anneals for 20 seconds).

From Fig. 1, it can also be observed that the films of different initial thicknesses have approximately the same values of specific resistivity for the processing temperatures greater than or equal to 800 ^{o}C. From this information, it can be concluded that the assumption of full nitridation of titanium is essentially correct for anneals at temperatures greater than 700 ^{o}C and that the reaction product is independent of the initial Ti thickness. The behavior of Ti on Si_3N_4 (Fig. 2) is essentially similar, except that the specific resistivity is lower than the corresponding values for Ti on SiO_2 in Fig. 1. The specific resistivity values for both silicon oxide and nitride substrates appear to converge for the 1100 ^{o}C anneals.

The color changes associated with anneals also provide an insight into the nature of the reaction products. At the low annealing temperature of 700 ^{o}C, there is no visible change in the color of the film. At higher temperatures, the film turned to a golden yellow color, characteristic of cubic titanium nitride[4]. This again suggests incomplete or no TiN formation at 700 ^{o}C.

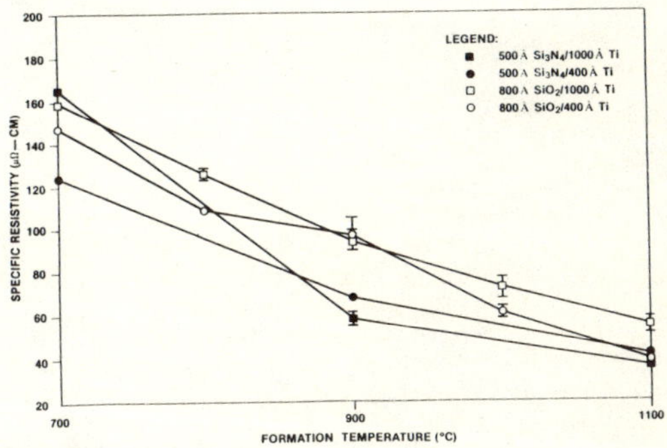

Figure 2: Specific resistivity vs. annealing temperature of TiN formed by RTA of Ti films on Si_3N_4, assuming complete conversion of Ti to TiN (all anneals for 20 seconds).

Reaction Kinetics

Figure 3 shows the sheet resistance of 1000Å Ti on SiO_2 annealed at 900 ^{o}C for various times. The relative change in resistivity after the initial 20 seconds of anneal is relatively small, changing less than 10% as the anneal time increased. This result implies that the reaction kinetics are fast enough for this case that the structural/compositional changes in the film are essentially complete in the anneals of 20 seconds or less. Thus, it appears that thermal nitridation under these conditions is essentially complete in the first 20 seconds of anneal. Based upon the similarity of the calculated specific resistivities of thick and thin Ti films on SiO_2 in the previous discussion, the conclusion of complete nitridation in 20 seconds anneals is substantiated for anneals at temperatures equal to or greater than 800 ^{o}C.

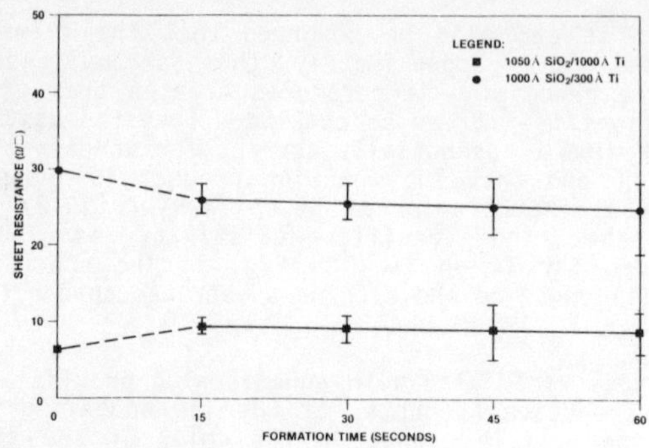

Figure 3: Sheet resistivity vs. anneal time of Ti thin films on SiO_2 (anneal temp. = 900 ^{o}C).

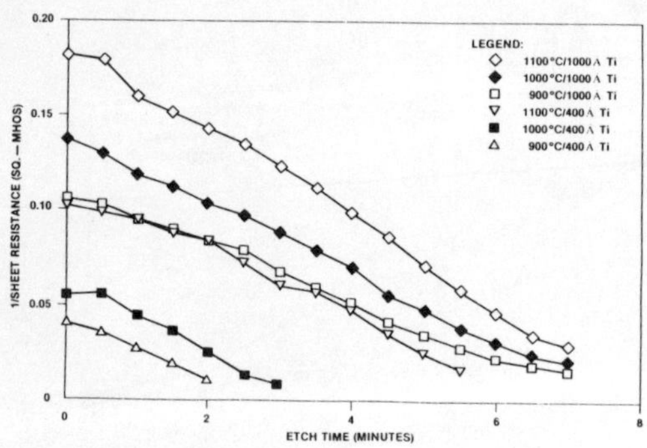

Figure 4: Inverse sheet resistivity vs. etch time for TiN formed on SiO_2 substrates by RTA for 20 seconds. (See text for etch formulation).

Etch Rate Studies

Figure 4 shows the inverse sheet resistance of TiN formed on SiO_2 films as a function of the etch time in the titanium selective etch. The relatively linear nature of the data suggests that the nitrided wafers (Ti on SiO_2), regardless of the processing temperatures, etched at a constant rate indicating that the films are chemically uniform in composition. In order to calculate the etch rates from these data, a normalization procedure was done using the assumptions that the etch rate 'R' and the specific resistivity are constant through the bulk of the films; which leads to the following equation:

$$1 - S_i/S_t = R\, t/L_i \tag{1}$$

where L_i is the initial film thickness (for complete conversion of Ti to TiN, the resulting film thickness will be approximately equal to the initial Ti thickness), S_i is the initial sheet resistance and S_t is the sheet resistance after etching for time, t. Using this normalization method, the etch data of Figure 4 has been replotted in Figure 5.

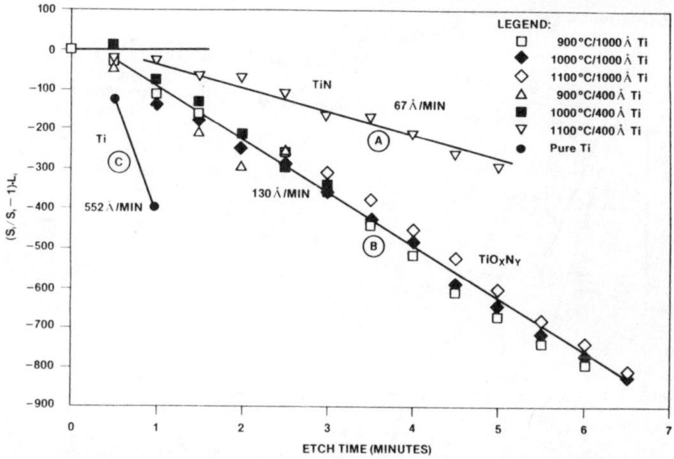

Figure 5: Normalized etch rate data (per equation (1) in text) for TiN formed on SiO_2 substrates by RTA for 20 seconds.

According to the equation (1), the slope of the lines in Figure 5 represents the etch rate of the films. A non-linear behavior in this normalized representation of the data would indicate that the film has varying etch rate and hence varying composition through the depth of the film. Figures 4 and 5 show that the thin Ti film (400 Å) on SiO_2 processed at 1100 ^{0}C forms a film (type A in Fig. 5) which etches at a very low value of ~67 Å/min. as compared to an etch rate of about 130 Å/min. for all the other films (grouped together as film type 'B' in Fig. 5). For comparison, the etch rate of pure Ti, as shown in Fig. 5, is ~550 Å/min. in the selective etch. Since the plots in Fig. 5 are highly linear and all the data for type B films fall on the same straight line, a single product layer through the bulk of the film (having uniform etch characteristics) is suggested and is consistent with the previous discussion.

TiN formed over Si_3N_4 has considerably different etch characteristics as shown in the Figure 6. This figure indicates that the effort to unify the etch rate data in a normalized manner as before does not yield a linear behavior for films annealed at 900 and 1100 ^{0}C. The two limiting slopes from the non linear normalized etch data in this figure have been used to calculate two limiting etch rates and imply that there are at least two distinct layers with different etch characteristics. The etch rate data for the 700 ^{0}C annealed film in this figure do show a linear behavior with an etch rate of 320-430 Å/min (depending upon the initial Ti thickness). These values are closer to the etch rate of as deposited Ti and imply that these films may be a solid solution of Ti with nitrogen (confirmed by TEM microdiffraction). A comparison of the two etch rates derived from the two limiting slopes in Figure 6 for the higher temperature anneals to that in in Figure 5 suggests that the top layer for these films is TiN while the bottom layer is a solid solution of Ti with nitrogen.

The titanium films nitrided by RTA at temperatures of 900 -1100 ^{0}C in this study were found to be unaffected by buffered HF after 5 minutes of etching, as indicated by no change in the sheet resistance. This confirms previous reports that TiN is not attacked by buffered HF.

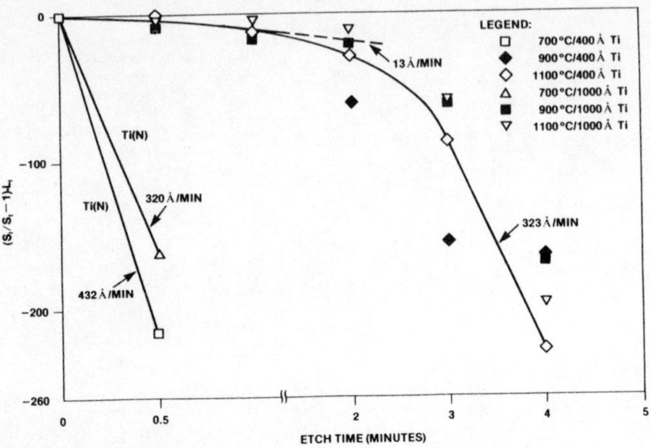

Figure 6: Normalized etch rate data (per equation (1) in text) for TiN formed on Si_3N_4 substrates by RTA for 20 seconds.

<u>Auger and TEM Analyses</u>

Figures 7 and 8 show the Auger depth profiles for various samples. The Auger analyses of TiN is complicated by the fact that the main Auger electron emission from nitrogen occurs at energy that completely overlaps the signal from titanium. Hence, the Auger analyses of TiN is semiquantitative, but much information can be gained by comparing the high and low energy transition peaks for Ti[4]. The depth profiles obtained by monitoring these two transitions of Ti, labelled 'Ti' and 'L' in Figures 7 and 8 reflect the Ti and Ti+N concentrations in the film as discussed in an earlier publication[4]. The depth profile data for 1000 Å Ti on SiO_2 shows evidence of thermal oxide breakdown (Figure 7). At the Ti-SiO_2 interface, the occurrence of a redox reaction leading to some silicide formation and release of oxygen into the Ti has been previously suggested[4]. The oxygen released by this redox reaction is snowploughed ahead, causing a large oxygen concentration peak in the middle of the TiN film which is formed by reaction of the Ti with the nitrogen ambient. The formation of TiN as evidenced by the Auger analyses has been discussed in a previous paper[4].

The large oxygen concentration in the titanium nitride formed over SiO_2 substrate, coupled with the etch rate results presented earlier, would indicate the formation of a titanium oxy-nitride throughout the initial Ti layer. The reduction of the SiO_2 layer by titanium increases with increasing anneal temperatures releasing larger amounts of oxygen in the film (Fig. 7). This behavior is not followed by the thin Ti over silicon oxide, as seen from Fig. 8, where some oxygen is still seen but is significantly less than other cases and indicates that the oxygen dissolution into the TiN for this case at 1100 ^{o}C anneal is highly suppressed. The etch rate data in Figure 5, indicating uniform etch rate for type 'B' films (Ti oxynitrides), do not reflect the varying oxygen/nitrogen concentrations in the films as suggested by the Auger depth profiles.

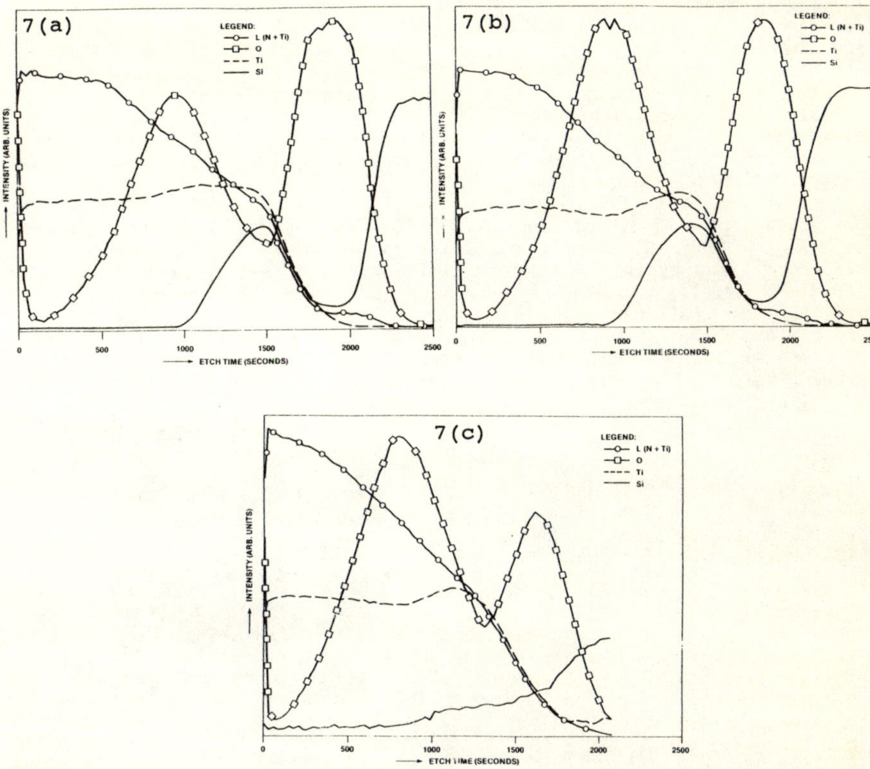

Figure 7: Auger depth profiles for TiN formed by 20 seconds RTA of 1000 Å Ti on SiO_2 at (a) 800 ^{o}C, (b) 900 ^{o}C, and (c) 1100 ^{o}C.

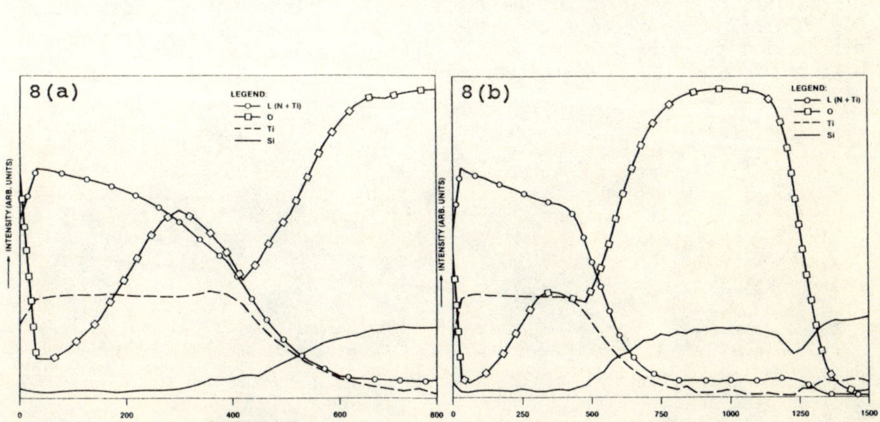

Figure 8: Auger depth profiles for TiN formed by 20 seconds RTA of 300 Å Ti on SiO_2 at (a) 900 ^{o}C, (b) 1100 ^{o}C.

The normalization of etch rates by equation (1) (used as the basis for the plot in Figure 5) was done with the assumption that the etch rate and the specific resistivity of the film are constant through the bulk. The lower etch rate and lower specific resistivity of oxygen-free TiN implied by our results will have compensating effects in the data normalization based upon equation (1) and hence it is suggested that the etch rate of Ti oxynitrides derived by using equation (1) is an average through the depth of the film and insensitive to the local oxygen concentration variations.

Cross-sectional TEM of samples of 1000 Å and 400 Å Ti on SiO_2 annealed at 1100 ^{o}C (Fig. 9) clearly shows the breakdown and consumption of the silicon dioxide by the nitridation reaction. The initial oxide thickness for these samples was ~800 Å and the figure shows that almost half the initial oxide has been reduced in the extreme case. The results further confirm the Auger analysis conclusions that the nitridation of Ti prevents the redox reaction from proceeding further and hence is more efficient in preventing the SiO_2 breakdown in the thinner Ti films annealed under identical conditions.

Figure 9: XTEM of 1000 and 400 Å Ti films on 900 Å SiO_2, annealed at 1100 ^{o}C for 20 seconds showing SiO_2 breakdown by Ti films.

Figure 10 shows the horizontal TEM images of Ti on SiO_2 samples annealed at three different temperatures. The figure shows an increase in the grain size from about 250 Å at 700 ^{o}C to about 2000 Å at 1100 ^{o}C, implying significant grain growth with increasing anneal temperatures. For reference, the grain size of the initial as deposited Ti films was ~250 Å. Microdiffraction patterns of the films formed at temperatures greater than 800 ^{o}C show the presence of cubic TiN with no residual Ti while the diffraction patterns for the 700 ^{o}C annealed films indicates a possible mixture of unreacted Ti with some cubic TiN.

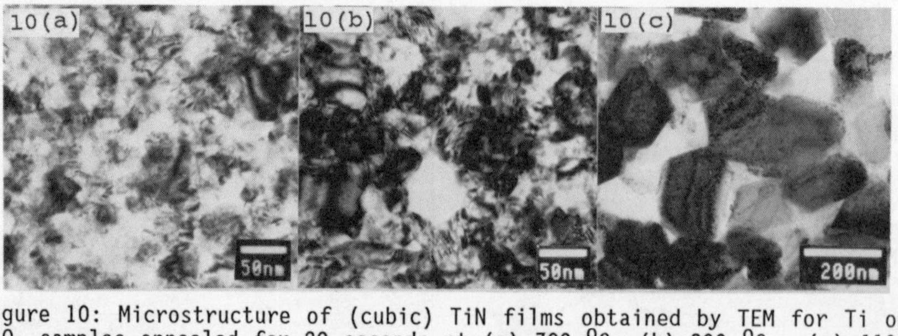

Figure 10: Microstructure of (cubic) TiN films obtained by TEM for Ti on SiO_2 samples annealed for 20 seconds at (a) 700 ^{o}C, (b) 900 ^{o}C., (c) 1100 ^{o}C.

The decrease in the resistivity of the films with higher annealing temperatures as discussed before could not have been due to increased oxygen incorporation (indeed this mechanism does not exist for TiN formed over Si_3N_4) and must be attributed mainly to the grain growth mechanism. Thus, the lowest sp. resistivity for TiN is achieved when the thinnest Ti layers are annealed at the highest temperatures (300 Å Ti, 1100 ^{o}C).

Model

From the discussion above, a model for the reaction of Ti with nitrogen over SiO_2 and Si_3N_4 substrates emerges. During the heat treatment, nitrogen diffuses into titanium while the silicon dioxide is reduced by the overlying titanium. The oxygen from the SiO_2 breakdown is released into the titanium layer leaving behind an oxygen depleted oxide region in contact with titanium. Ternary phase diagram of the Ti-O-Si system[6] suggests the formation of Ti_5Si_3 and TiO due to this redox reaction.

A comparison of TiO and TiN properties indicates same crystal structure (cubic, NaCl type) and very similar lattice parameters, 4.18 Å and 4.24 Å respectively. Additionally, both compounds are known to exhibit metallic conductivities[7]. Since Ti is a multivalent transition metal, a TiO_xN_y continuous solid solution of FCC structure by mixing TiO and TiN is very likely to form and is thermodynamically favored due to the entropy of mixing. The etch rate data presented earlier suggest that the nitridation reaction product of Ti on SiO_2 is indeed a single phase Ti oxynitride layer which is not bound by TiN and TiO on either side. This also implies that the nucleation of Ti-oxynitride occurs in a single step. The oxygen liberated by the interfacial redox reaction distorts the Ti lattice and must allow the nitrogen to diffuse deeper and nucleate TiO_xN_y throughout the bulk of the initial titanium film. In contrast, no such mechanism exists for the Ti on Si_3N_4 substrate which results in a TiN/Ti bilayer structure controlled by the diffusivity of the nitrogen through relatively pure (oxygen-free) TiN. In this regard, the nitridation of Ti in presence of dissolved oxygen impurity in the metal is another example of the well-known phenomenon of impurity enhanced diffusion and nucleation of new phases.

CONCLUSIONS

Insights into the formation mechanism of TiN have been made for films formed by rapid thermal nitridation of titanium films over SiO_2 and Si_3N_4 substrates. It has been shown that cubic, polycrystalline TiN can be nucleated at temperatures > 800 ^{o}C. At these high temperatures, the nitridation of < 1000 Å Ti over SiO_2 is essentially complete in about 20 seconds and results in a titanium oxynitride film rather than pure TiN. Nitridation of Ti on Si_3N_4 under similar conditions resulted in incomplete nitridation leading to a possible bilayer of TiN/Ti. The specific resistivity of the Ti oxynitride decreases with the formation temperature, reaching ~35 μ ohm-cm at 1100 ^{o}C anneals. The decrease in the resistivity is attributed mainly to the grain growth phenomena. Auger and TEM analyses show that the SiO_2 layer under Ti reacts with the metal to liberate oxygen into the titanium. The presence of the oxygen dissolved in Ti aids in virtually complete conversion of the initial Ti films to an oxynitride. The relative concentrations of the oxygen and nitrogen in the oxynitride product are dependent upon the formation temperature and the metal thickness. Increasing anneal temperature favors a larger amount of oxygen incorporation while thinner titanium films minimize the redox reaction.

ACKNOWLEDGEMENT

The authors greatly appreciate the assistance of Dr. P. Davies and Dr. B. Tracy of Intel Corporation for Auger and TEM analyses and many useful discussions.

REFERENCES

1. M. Wittmer and H. Melchior, Thin Solid Films, 93, pp. 397-405 (1982).

2. T. Tang, C.C. Wei, R. Haken, T. Holloway, C.F. Wan and M. Douglas, IEDM Technical Digest, IEEE, pp. 590-593 (1985).

3. J. E. Sundgren, Thin Solid Films, 128, pp. 21-44 (1985).

4. U. N. Mitra, P.W. Davies, R. K. Shukla and J. S. Multani, Proc. 5th Int. Symp. on Silicon Materials Sci. and Tech., Electrochem. Soc., 86-4, pp. 316-332 (1986).

5. P. J. Rosser and G. J. Tomkins, Proc. Layered Structures, Epitaxy and Interfaces Symp., edited by J. M. Gibson and L. R. Dawson, Materials Res. Society. (Publishers), p. 607 (1985).

6. R. Beyer, R. Sinclair & M.E. Thomas, Jour. Vac. Sci. & Technol., B2(4), p. 781, Oct.-Dec.(1984).

7. Encyclopedia of Chemical Technology, edited by Ed. Kirk-Othmer, 3rd edition, 5, p. 299, Wiley & Sons, Publishers, New York.

THERMAL STABILITY AND NITROGEN REDISTRIBUTION IN THE Ti/W-N BILAYER

F. C. T. SO*, C.-D. LIEN**, E. KOLAWA*, X.-A. ZHAO*** AND M-A. NICOLET*
*California Institute of Technology, Pasadena, California 91125
**Advanced Micro Devices, Sunnyvale, California 94088
***Permanent Address: Shanghai Institute of Metallurgy, Academy of Sciences of China, Shanghai, China

ABSTRACT

We investigate the thin-film reactions that take place in a $Ti/W_{60}N_{40}$ couple deposited onto Si and SiO_2. A partial redistribution of nitrogen from W-N into Ti occurs as a result of thermal annealing between 550°C and 850°C, leading to the formation of $Si/TiSi_2/TiN/W-N$ and $SiO_2/Ti-N-O/W-N$ structures on Si and SiO_2 respectively. A possible application of a Ti/W-N bilayer to realize the self-aligned $TiSi_2$ process with a W local interconnect is discussed.

INTRODUCTION

Tungsten nitrides (W-N) have recently attracted interest as diffusion barriers for contact metallization in VLSI technology [1,2] and as refractory Schottky gates to GaAs MESFETs [3,4,5]. Although W-N has many properties that make it highly favorable for device applications, some obvious limitations have been found. W-N was observed to adhere poorly to Au and Ag overlayers in contact structures [1,6]; also, the contact resistivities of W-N to n^+Si is unacceptably high for VLSI applications [7]. One simple way to alleviate such problems involves the inclusion of an interposed layer of another metal (e.g. Ti) between W-N and Si or Au to provide good contact resistivities and satisfactory adhesion. Under these conditions the thermal stability of W-N in the presence of Ti will be an important concern, since Ti has been used extensively as an intermediate adhesion layer and a primary contacting layer to Si [8,9].

In this work, we found that a $W_{60}N_{40}$ film decomposes in the presence of an adjacent Ti layer upon thermal annealing. Nitrogen from the W-N layer diffuses into Ti at 550°C or above. Such a redistribution of nitrogen is detrimental to the thermal stability of a <Si>/Ti/W-N/Al contact structure [10]. On the other hand, since the successful implementation of the self-aligned $TiSi_2$ process [9,11,12] is made possible with the diffusion of nitrogen from the reaction ambient into the Ti layer during silicidation, possible application of the Ti/W-N bilayer for the realization of the self-aligned $TiSi_2$ process is also described.

EXPERIMENTAL PROCEDURE

Commercially prepared and polished p-type <100>Si substrates of ~1-10 Ωcm resistivity ("<Si>") were used. Prior to loading into a sputter deposition system, the Si wafers were first oxidized slightly in an oxygen plasma for 10 min. and then rinsed in deionized water, followed by etching in a 10% HF solution for 3 min. Oxidized Si substrates were also used. Ti and W-N depositions were made by rf magnetron sputtering in a diffusion pump vacuum system with a base pressure better than 1×10^{-6} Torr. Titanium films of approximately 900Å were sputter-deposited in a 10 mTorr argon ambient at a rate of about 1100Å/min. A substrate bias of -50V was applied during Ti deposition. Following the Ti deposition, a W-N alloy film of about 1500Å thick was deposited by reactively sputtering W in a 50% N_2 and 50% Ar plasma. The total gas pressure for this sputter deposition is 15 mTorr.

Details of the W-N alloy deposition are reported in other work [1,13]. All as-prepared samples were subsequently annealed at temperatures from 600°C to 900°C in a vacuum of better than $5x10^{-7}$ Torr or in an argon ambient. Unless otherwise specified, the annealing duration for every temperature is 30 min. Backscattering spectrometry (BS) and Auger electron spectroscopy (AES) were used to monitor reactions and changes in the compositional profile of the <Si>/Ti/W-N and SiO_2/Ti/W-N samples. X-ray diffraction analysis was used to confirm the formation of various phases formed during annealing of the samples. In some cases, the top W-N layer was selectively etched off in an aqueous $KH_2PO_4 + K_3Fe(CN)_6$ + KOH solution [1,14] for more detailed analysis of the underlayers. Sheet resistance measurements were accomplished with a linear four-point probe setup.

The concentration of nitrogen in the as-deposited W-N alloy films was determined to be about 40 atomic percent by BS analysis of films deposited onto vitreous carbon substrates. The resistivity of the W-N layer is about 300 μΩcm. X-ray analysis indicates that W_2N is the dominant phase present in the as-deposited W-N films.

RESULTS AND DISCUSSION

1. <Si>/Ti/W-N

Figure 1 shows Auger depth profiles of samples of the configuration <Si>/900A Ti/1500A $W_{60}N_{40}$ before and after annealing in vacuum at 650°C and 800°C. It is evident that the nitrogen concentration in the W-N layer decreases as a result of heat treatment, accompanied by a progressive increase in the nitrogen content in the Ti layer. Nitrogen thus diffuses from the W-N into the Ti layer. The lowest temperature at which this nitrogen redistribution can be detected in a Ti/W-N bilayer is about 500°C [10]. The movement of nitrogen leads to the formation of a TiN_x layer below the W-N layer. However, according to Auger analysis, the Ti layer is only partially nitrided for annealing temperatures above 600°C. At these temperatures, Ti and Si also react, which leads to the formation of a $TiSi_x$ layer. The nitrogen front does not extend much into this $TiSi_x$ layer. Auger depth profiles show that the nitrogen level in the TiN_x layer varies little with annealing temperature from 650°C to 800°C. The $W_{60}N_{40}$ layer still contains a measurable amount of nitrogen after the 800°C annealing, although this amount is perceptibly lower than in the as-deposited sample and after annealing at 650°C [1,15].

The nitridation of Ti and partial consumption of Ti in forming a silicide resembles the outcome in the self-aligned $TiSi_2$ process where a Ti film on a silicon substrate is annealed in a nitrogen ambient above 600°C, as reported by numerous authors [11,12,16,17,18,19]. In the present case, it is simply the W-N film that acts as a source of nitrogen for the conversion of Ti to TiN_x.

The BS spectra for the as-deposited and the annealed samples of the configuration <Si>/Ti/$W_{60}N_{40}$ are shown in Figure 2. The W signal near its lower energy edge increases in height after the 650°C heat treatment. This is consistent with the AES analysis which reveals a partial loss of nitrogen from the W-N layer. The nitrogen is released preferentially from the W-N layer near the Ti interface for temperatures below 800°C. The BS data also supports the AES results which shows that the Ti evolves into a two-layer structure after heat treatment at 650°C. A silicide with a Si to Ti atomic ratio of about 2 forms between Ti and Si. The outer region of the Ti does not react with the Si substrate. The reason is the nitridation of this fraction of the Ti layer as revealed by Auger analysis. From the reduction in the signal height of the Ti, the nitrogen concentration in Ti can be estimated. The atomic fraction of N in Ti is 0.4 at 600°C, and rises slightly from 0.4 to about 0.55 at 700°C and above. X-ray analysis was

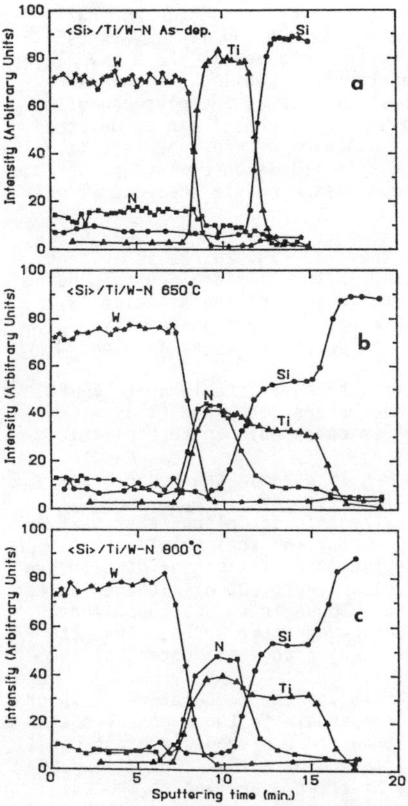

Fig. 1 AES depth profiles of samples of the configuration <Si>/900A Ti/1500A $W_{60}N_{40}$; (a) as-deposited, (b) 650°C 30 min. and (c) 800°C 30 min.

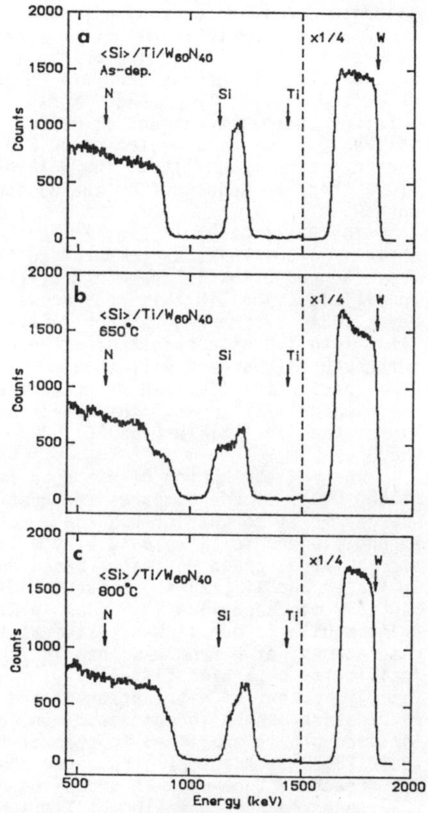

Fig. 2 BS spectra of samples of the configuration <Si>/900A Ti/1500A $W_{60}N_{40}$; (a) as-deposited, (b) 650°C 30 min. and (c) 800°C 30 min. The BS spectra are taken at a tilt angle of 20° relative to the incident beam with a backward detector angle of 10° relative to the incident beam.

carried out on annealed samples with the top W-N layer selectively etched off. The presence of $TiSi_2$ and TiN phases was confirmed for all samples annealed at 600°C or above. The formation of TiN is also evident from the appearance of the characteristic golden yellow color of TiN after the removal of the W-N layer. From Fig. 2, the TiN and $TiSi_2$ layers are estimated to be 350Å and 1600Å thick, respectively. The thicknesses of TiN and $TiSi_2$ remain more or less constant for annealing temperatures between 650°C and 750°C.

After an 800°C anneal, the W-N layer loses most of its nitrogen leading to a further increase of height throughout the whole W signal in the BS spectrum (Fig. 2c). A quantitative nitrogen depth profile obtained by the $^{15}N(p,\alpha\gamma)^{12}C$ nuclear reaction [20] shows that the W-N layer contains between

5 and 10 atomic percent of nitrogen after annealing at 800°C. At that temperature, the total amount of nitrogen in the W-N and TiN layers is reduced to about 70% of its original quantity contained in the $W_{60}N_{40}$ layer. Part of the nitrogen is probably released to the annealing ambient as described previously [1,15]. X-ray analysis indicates the presence of only α-W after thermal treatment at 800°C. No W nitride phases can be detected, unlike for samples annealed below 800°C. Whatever nitrogen is left is below our detection capability. The W layer thus formed is determined to be about 1100A thick as deduced from the BS spectrum, assuming the presence of only pure W.

The BS spectrum in Fig. 2c further indicates that the $TiSi_2$ layer shrinks in thickness as compared to that shown in Fig. 2b, as is evident from the reduction in the width of the step in the Si signal corresponding to silicide. The TiN thickness grows at the expense of the silicide layer (by > 200A) as the W-N further releases nitrogen. The transformation of $TiSi_2$ into TiN as a result of extended annealing in nitrogen at high temperatures has been well studied [17,19,27].

The $Si/TiSi_2$/TiN/W-N structure remains thermally stable up to 850°C, above which WSi_2 forms. This reaction temperature for W and Si is much higher than for <Si>/W ($\sim$650°C) [29], but is comparable to that of <Si>/W-N ($\sim$800°C) [30].

The redistribution of nitrogen from W-N into Ti is thermodynamically driven [10]. If one compares the heat of formation of TiN ($\Delta H_{TiN} \sim$ -80kcal/mole) to that of W_2N ($\Delta H_{W2N} \sim$-17kcal/mole), it follows that N is more strongly bound to Ti than to W by a large amount (63 kcal/mole, or 2.7 eV/N atom). Also, crude estimates based on published diffusivities of nitrogen in W [21] and Ti [22] show that the diffusion length Dt of nitrogen after 550°C 30 min annealing is $\geq$ 500A in Ti and $\sim$1000A in W. Nitrogen indeed is quite mobile in our Ti/W-N system at temperatures above 500°C. Thus the displacement of N from W-N into Ti is driven by a chemical potential and facilitated by a high diffusivity.

The extent of nitridation of the Ti film and the temperature at which nitridation occurs in our experiment are comparable to those observed when a Si/Ti couple is subjected to thermal treatment in a nitrogen ambient [16,17, 18,19,23,28]. Adams [17] reported the formation of a <Si>/880A $TiSi_2$/320A TiN structure from 600A Ti deposited onto Si after annealing at 800°C. Tang [23] observed the formation of TiN surface layers of $\sim$250A from a 1000A Ti on Si after a 675°C, 30 min heat treatment. It was also reported that same amount of nitrided Ti formed regardless of whether the Ti films are 1000A or 2000A thick after annealing at 600°C [16]. Delfino [28] and Brat [18] investigated the formation of $TiSi_2$ by RTA in nitrogen. Both obtained TiN surface layers of the order of 200A in thickness in the vicinity of 800°C. In our study, as described above, 360A of TiN is formed for annealing temperatures between 650°C and 750°C. The TiN thickness increases further to $\geq$ 560A after 800°C. That nitrogen moves in a Ti/W-N bilayer can be detected at temperatures as low as 500°C [10] by AES. Although the extent and amount of nitridation of Ti in a Ti/W-N couple may depend on the thickness and the nitrogen concentration of the as-deposited W-N layer, the fact that nitrogen redistributes in a Ti/W-N couple almost as readily as it diffuses into Ti from a gaseous ambient suggests that W-N releases nitrogen quite easily in the presence of Ti. In other words, W-N simply acts as a nitrogen container and as a source of nitrogen in the formation of TiN.

2. SiO_2/Ti/W-N

In the SiO_2/Ti/W-N system, nitrogen redistribution occurs in much the same way as in the former case. Fig. 3 shows how AES depth profiles evolve for a SiO_2/900A Ti/1500A $W_{60}N_{40}$ sample from the as-deposited state to one after annealing at 700°C in vacuum. As nitrogen diffuses into Ti, Ti also interacts with SiO_2. An intermediate oxygen-rich Ti-N-O layer is thereby

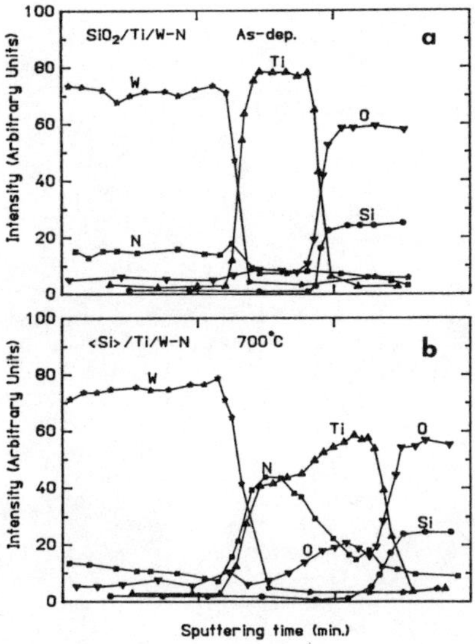

Fig. 3 AES depth profiles of a sample of SiO_2/900A Ti/1500A $W_{60}N_{40}$ before (a) and after (b) annealing at 700°C for 30 min.

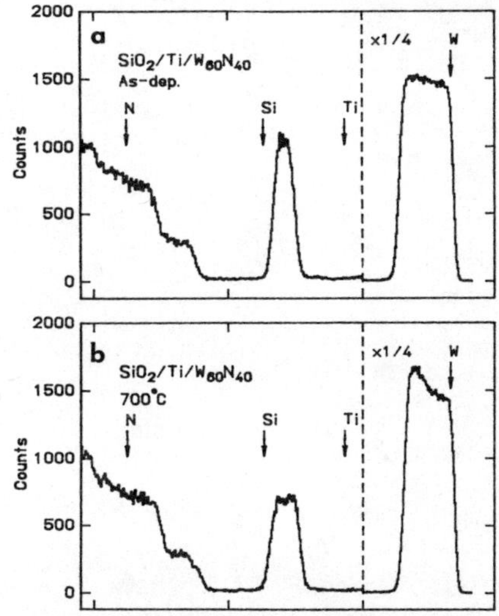

Fig. 4 BS spectra of samples of the configuration SiO_2/900A Ti/1500A $W_{60}N_{40}$; (a) as-deposited, (b) 700°C 30 min. The tilt and detector angles are the same as those described in the captions for Fig. 2.

formed. This observation is consistent with a previous report on the effect of RTA of Ti on SiO_2 in a nitrogen ambient [19]. Also, according to Ting et al. [24], Ti interacts insignificantly with SiO_2 up to 700°C, above which oxygen from a decomposed SiO_2 layer starts to dissolve in the Ti film. The oxygen-rich Ti layer in our annealed sample is probably caused by a dissolution of oxygen in Ti. BS analysis (Fig. 4) reveals a similar picture. The interaction between Ti and SiO_2 at 700°C is reflected in a slight broadening of the leading edge of the Si signal. (Notice that the signals generated at the Ti/W-N interface and the surface remain sharp steps.) However, one cannot ascertain from the spectra how much Ti or SiO_2 is consumed in such an interaction. X-ray diffraction analysis does not detect any oxide or silicide in this sample. The presence of Si in Ti is not detected by AES (Fig. 3). Although Ti appears to contain both nitrogen and oxygen in concentrations that vary with depth, as indicated by AES, the BS depth profile shows that the atomic fraction of Ti throughout the entire Ti-N-O layer stays very close to 0.45 (as calculated from the reduction in the height of the Ti signal of the annealed sample as compared to that of the as-deposited sample). From the spectrum of the sample annealed at 600°C (not shown), the average atomic fraction of Ti in the Ti-N-O reacted layer is about 0.6. Annealing at higher temperatures leads to an increase in the oxygen content of the Ti layer. However, the reaction between the SiO_2 substrate and Ti remains very limited up to at least 850°C. At this temperature, a pure Ti film on SiO_2 reacts rapidly, resulting in a titanium-rich silicide layer adjacent to the oxide, and a TiO_x layer above it [24]. That this does not occur here is probably due to the competing and rapid incorporation of nitrogen into Ti.

Possible application of Ti/W-N to the self-aligned $TiSi_2$ process: Formation of W local interconnect level

In a recent report on the self-aligned $TiSi_2$ process [23], a local interconnection (LI) level has been realized by the utilization of the TiN layer formed during silicidation. Fig. 5 shows the process steps that lead to the formation of local TiN interconnection lines between gates and source/ drain junctions. After the Ti-Si reaction step at 675°C in nitrogen, a LI pattern is defined photolithographically (Fig. 5b). A selective etch removes the unprotected TiN on SiO_2 and $TiSi_2$, leaving $TiSi_2$ intact at the gate, source and drain areas. After the resist strip, a TiN LI remains between the gate and junction (Fig. 5c). A final 800°C N_2 anneal reduces the $TiSi_2$ and TiN resistivities to their final values. By retaining part of the TiN formed during the initial silicidation step, a 0.11 μm thick LI of a sheet resistance of ∿10 Ω/sq results. The use of TiN to implement a LI has a number of advantages over the conventional buried interconnection schemes [23]. However, a major disadvantage is that W does not nucleate and grow on a TiN surface in the conventional self-aligned W LPCVD process advocated for via filling and the formation of second level interconnects [25].

As an alternative to the TiN LI technology just described, a Ti/W-N bilayer may be used to implement the self-aligned $TiSi_2$ process with the simultaneous formation of a W LI (see Fig. 6): After the sidewall formation process, source/drain anneal and the opening of contact holes (Fig. 6a), a blanket Ti/W-N bilayer is deposited in lieu of a single Ti layer. Upon annealing at 600-700°C, nitrogen will redistribute everywhere from the W-N layer into the Ti film, and TiN forms over the oxide; a $TiSi_2$ layer will also form, but over the oxide windows of the gate, drain and source areas, as in a typical self-aligned process. Following this heat treatment, the LI photoresist pattern is again laid down (Fig. 6b), and the W-N and TiN are then selectively etched off one after another, leaving a TiN/W-N LI running between source/drain contacts and gates (Fig. 6c). Subsequent annealing at 800°C will reduce the $TiSi_2$ and TiN resistivities further and also release nitrogen from W-N to form LI composed predominantly of cubic α-W.

To check that the electrical characteristics of the Ti/W-N bilayer are adequate for LI implementation, we have performed four point probe measurements for Ti/W-N bilayers on <Si> and SiO_2 as a function of annealing temperature. Fig. 7a shows the variation of the sheet resistance of a 900Å Ti/1500Å $W_{60}N_{40}$ couple on SiO_2 with annealing temperature. Sheet resistance decreases with annealing temperature above 600°C. An 800°C 30 min anneal reduces the sheet resistance to ∿2.5 Ω/sq. The sheet resistance of the TiN layer alone is 7.7 Ω/sq., as measured after the W overlayer is selectively etched off. The resistivity of the W layer (after the anneal) is therefore about 40 μΩcm, taking the thickness of W(WN) to be 1100Å as determined by BS. This resistivity value is only slightly higher than that reported previously for sputtered W-N and W layers annealed at 800°C [1], but is much higher than that reported for CVD W (8-15 μΩcm) [31], probably because of different microstructures and different degrees of incorporation of nitrogen and other impurities [13]. The resistivity of the TiN layer on SiO_2 is estimated to be about 77 μΩcm. Thus our Ti/W-N bilayer structure can potentially provide a LI three times more conductive than that composed of a single TiN layer.

The dependence of the four point probe resistance [32] on heat treatment for the Ti/$W_{60}N_{40}$ bilayer on <Si> is also shown (Fig. 7b). As is evident in the figure, the resistance diminishes as a result of annealing at 600°C and above as in the case of SiO_2/Ti/$W_{60}N_{40}$. At this temperature, both $TiSi_2$ and TiN form. Four point probe resistance obtained after annealing at 800°C is determined to be about 0.8 Ω/sq., with <Si>/$TiSi_2$/TiN contributing about 1.1 Ω/sq to the total value.

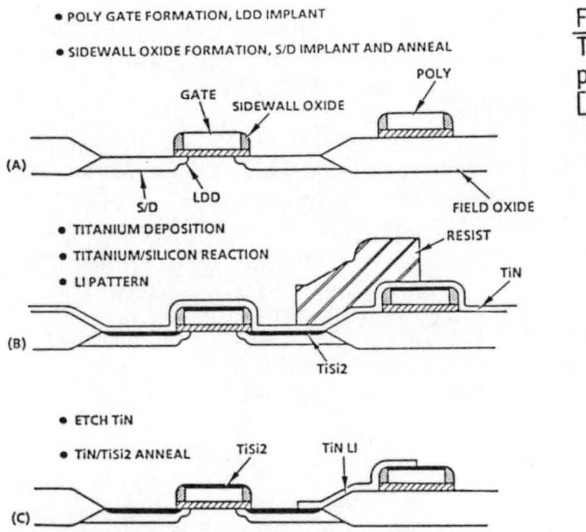

Self-aligned TiSi2 process showing the TiN LI patterning step.

Fig. 5 Conventional self-aligned $TiSi_2$ process showing the TiN LI patterning step, from Tang et al. [Ref. 23].

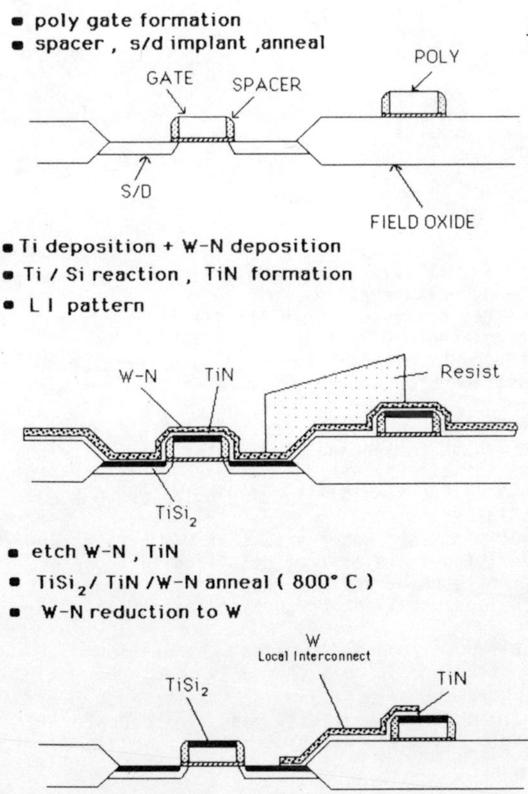

Fig. 6 Self-aligned $TiSi_2$ process with a W (WN) LI.

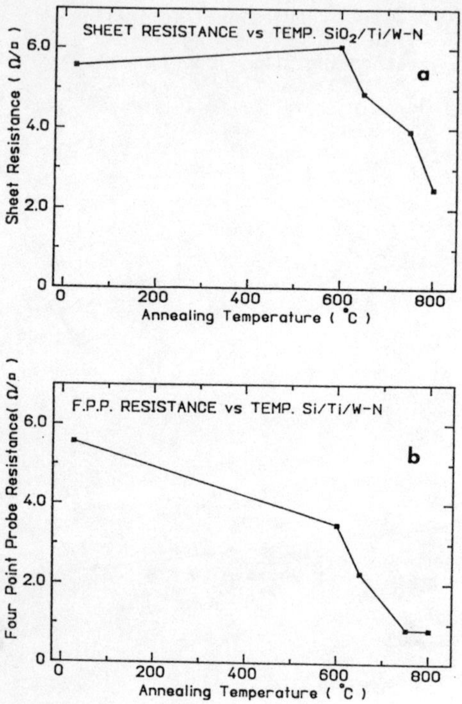

Fig. 7 (a) Sheet resistance of Ti/WN bilayers on SiO_2 as a function of annealing temperatures.
(b) Four point probe resistance of <Si>/Ti/WN as a function of annealing temperatures.

The implementation of the $TiSi_2$ self-aligned process with a W LI from a Ti/W-N bilayer offers the following advantages:

1) W LI provides a lower sheet resistance than its TiN counterpart because of the lower resistivity of the W(W-N) layer.

2) No additional photolithographic step is required to pattern the W LI. However, an additional deposition step and selective etch step for W-N is necessary.

3) A W LI that forms from the Ti/W-N bilayer can possibly act as a platform for nucleating LPCVD W growth for second level interconnects or the filling of vias. However, since not all the nitrogen is removed from the W layer even after annealing at 800°C, the feasibility of growing LPCVD W on such a W LI remains to be investigated.

4) Preliminary results (not discussed here) show that W-N acts as an excellent cap to protect an underlying Ti layer from oxidizing in argon or nitrogen annealing ambient. Recent work by Wei [26] has already shown that an overlayer of W on <Si>/Ti provides excellent protection against Ti oxidation.

One severe limitation of this novel process remains. As mentioned earlier, the Si/$TiSi_2$/TiN/W(WN) structure is only thermally stable up to 850°C. Above that temperature, W reacts with Si and forms silicide. This implies that the final heat treatment for resistivity reduction cannot exceed 850°C by furnace annealing. Rapid thermal processing may offer an alternative.

ACKNOWLEDGEMENTS

The authors take great pleasure in acknowledging the technical assistance from Rob Gorris. David N. Jamieson has helped tremendously in our data analysis. The assistance from Arlene Collinwood in preparing our manuscript is gratefully acknowledged. Financial support for this work was provided by the Army Research Office under contract number DAAG29-85-K-0192.

REFERENCES

1. H. P. Kattelus, E. Kolawa, K. Affolter and M-A. Nicolet, J. Vac. Sci. Technol. AB (6), p. 2246 (1985).
2. F. C. T. So, E. Kolawa, J. L. Tandon and M-A. Nicolet, J. Electrochem Soc., (in press).
3. N. Uchitomi, Y. Kitaura, T. Mizoguchi, Y. Ikawa, N. Toyoda and A. Hojo, in Extended Abstracts of the 16th Conference on Solid State Devices and Materials, Kobe, 1984, p. 383.
4. N. Uchitomi, M. Nagaoka, K. Nishibori, K. Shimada, T. Mizoguchi an N. Toyoda in the Fourth Workshop on Refractory Metals and Silicides for VLSI, San Juan Bautista, 1986.
5. N. Toyoda, N. Ucitomi, Y. Kitaura, M. Mochizuki, K. Kanazawa, T. Terada,, Y. Ikawa and A. Hojo, IEEE J. of Solid-State Circuits, 20, p. 1043 (1985).
6. E. Kolawa, F. C. T. So, J. Tandon and M-A. Nicolet, J. Electrochem. Soc. (in press).
7. F. C. T. So, (unpublished).
8. C. Y. Ting and B. L. Crowder, J. Electrochem Soc. 129, p. 2590 (1982).
9. C. Y. Ting, S. S. Iyer, C. M. Osburn, G. J. Hu and A. M. Scheweighart in Proceedings of the First International Symposium on VLSI Science and Technology (Electrochemical Society, New York, 1982) p. 224.
10. F. C. T. So, E. Kolawa, H. P. Kattelus, X.-A. Zhao and M-A. Nicolet, J. Vac. Sci. Technol. A4 (6), 1986.
11. C. K. Lau, in ECS Dig. Extended Abstracts, 83-1, p. 569 (1983).
12. M. E. Alperin, T. C. Hollaway, R. A. Haken, C. D. Gosmeyer, R. V. Karnaugh and W. D. Parmantie, IEEE Trans. on Electron Dev. ED32-2, p. 141 (1985).
13. E. Kolawa, F. C. T. So, X.-A. Zhao and M-A. Nicolet, presented in the Workshop on Tungsten and Other Refractory Metals for VLSI Applications, Palo Alto, 1986.
14. T. A. Shankoff and E. A. Chandross, J. Electrochem. Soc. 122, p. 294 (1975).
15. K. Y. Ahn, S. B. Brodsky, C. Y. Ting and J. Kim, J. Vac. Sci. Technol. (in press).
16. S. S. Iyer, C. Y. Ting and P. M. Fryer, J. Electrochem Soc. 132, p. 2240 (1985).
17. E. D. Adams, K. Y. Ahn and S. B. Brodsky, J. Vac. Sci. Technol. A3, 2264 (1985).
18. T. Brat, C. M. Osburn, T. Finstad, J. Liu and B. Ellington, J. Electrochem. Soc. 133, 1451 (1986).
19. A. E. Morgan, E. K. Broadbent and A. H. Reader, Proceedings of Material Research Society, Vol. 52, p. 298 (1986).
20. W. K. Chu, J. W. Mayer and M-A. Nicolet, Backscattering Spectrometry (Academic, New York, 1978).
21. J. Keinonen, J. Räisänen and A. Anttila, Appl. Phys. A 35, 227 (1984).
22. R. J. Wasilewski and G. L. Kehl, J. of the Inst. of Metals, 83, 94 (1954).
23. T. Tang, C. C. Wei, R. A. Haken, T. C. Holloway, L. Hite and T. Blake, in the Fourth Workshop on Refractory Metals and Silicides for VLSI, San Juan Bautista, 1986.
24. C. Y. Ting, M. Wittmer, S. S. Iyer and S. B. Brodsky, Proceedings of the Second International Symposium on VLSI Science and Technology (Electrochemical Society, Pennington, 1984), p. 397.

25. E. K. Broadbent, A. E. Morgan, J. M. DeBlasi, P. van der Putte, B. Coulman, B. J. Burrow, D. K. Sadana and A. Reader, J. Electrochem Soc. 133, 1715 (1986).
26. C. S. Wei, J. Van der Spiegel, M. Setton, J. Santiago, M. Tanielian and S. Blackstone, in Proceedings of the Materials Research Society, Vol. 52, p. 297 (1986).
27. H. Kaneko, M. Koyanagi, S. Shimizu, Y. Kubota and S. Kishino, IEEE Trans. Electron. Devices, ED-33, 1702 (1986).
28. M. Delfino, E. K. Broadbent, A. E. Morgan, B. J. Burrow and M. H. Norcott, IEEE Electron. Device Letts., EDL-6, 591 (1985).
29. M-A. Nicolet and S. S. Lau, in VLSI Electronics: Microstructure Sciences, edited by N. G. Einspruch (Academic, New York, 1983), Vol. 6; Materials and Process Characterization, edited by N. G. Einspruch and G. B. Larrabee (Academic, New York, 1983), Chap. 6.
30. H. P. Kattelus, F. C. T. So (unpublished).
31. See for example, S. Sachdev and S. D. Mehta, in the Proceedings of the 1985 Workshop on Tungsten and Other Refractory Metals for VLSI Applications, New Mexico, 1985, p. 161.
32. The term "four point probe resistance" is used here to distinguish our resistance values from the true sheet resistance of the Ti/W-N conducting layer on Si. No correction has been applied to exclude the contribution to the resistance measurements from the Si substrates. However, the 1-10 cm Si substrates used for our experiment is expected to be much more resistive than the conducting layers.

PROPERTIES OF REACTIVELY SPUTTERED WN_x FILMS

E. KOLAWA*, F. C. T. SO*, X.-A. ZHAO** AND M-A. NICOLET*
*California Institute of Technology, Pasadena, California 91125
**Permanent Address: Shanghai Institute of Metallurgy, Academy of Sciences of China, Shanghai, China

ABSTRACT

Tungsten nitride (WN_x) films were prepared by reactive rf sputtering of a tungsten target in N_2/Ar plasma. The resistivity, intrinsic stress and atomic composition of the films were studied as functions of various sputtering parameters: sputtering power, gas pressure, gas composition and substrate bias. The atomic percentage of nitrogen (x) of the WN_x films was found to decrease with sputtering power and to increase with the partial pressure of nitrogen. The intrinsic stress of WN_x is compressive. The dependence of film properties upon sputtering parameters for WN_x differs significantly from that of reactively sputtered TiN films, and may be attributed to the difference in the formation mechanisms of TiN and WN_x.

INTRODUCTION

Tungsten nitride (WN_x) thin films have been investigated recently as diffusion barriers in Si and GaAs contact metallizations [1-3], or as high temperature Schottky gates to GaAs MESFET [4-6]. WN_x can be formed easily and reproducibly by reactive sputtering from a W target in a nitrogen and argon gas mixture. Reactively sputtered TiN is the most widely investigated system [7-9]. Important properties of TiN from the point of view of diffusion barrier application (resistivity, intrinsic stress, impurity content, etc.) depend critically on the sputtering conditions.

Despite the fact that WN_x has already been considered for use in device applications, little is known about the film properties as a function of deposition parameters. In this study, we report on the influence of various sputtering process parameters, such as the power, total gas pressure, partial pressure of nitrogen, and substrate bias on the composition, deposition rate, resistivity and intrinsic stress of WN_x films.

EXPERIMENTAL PROCEDURE

Silicon wafers, carbon substrates and thin cover glass slides (18 x 18 x 0.160 mm^3) were used as substrates. Prior to loading into the chamber, silicon wafers and glass slides were processed through cleaning steps in ultrasonic baths of TCE, acetone and methanol. All WN_x films of this study were deposited by reactive rf sputtering using a planar magnetron W cathode of 7.5 cm diameter. The substrate holder was placed about 7 cm below the target and was neither cooled nor heated externally. The sputtering system is equipped with an oil diffusion pump and a chiller trap that yields a background pressure about 1×10^{-6} Torr prior to sputter deposition. Prior to striking the discharge, the total gas pressure of premixed N_2/Ar gas (subsequently referred to as an "initial" gas pressure) was adjusted with a variable leak valve and monitored with a capacitive manometer. Films were sputtered with a total gas pressure of 5, 10 and 15 mT. The relative partial pressure of nitrogen, defined as the ratio of the partial pressure of nitrogen to the total gas pressure, $p(N_2)/p(N_2+Ar)$, was varied from 0% to 70%.

A Sloan Dektac stylus instrument was used to measure the thickness of the films. Their resistivities were then determined from sheet resistance measurements with a four-point probe. The atomic composition of the films was determined by backscattering spectrometry. Carbon substrates were used for these measurements to increase the accuracy of the nitrogen/tungsten ratio derived from the backscattering spectrum.

The film stress was determined using the Dektac stylus instrument by measuring the resulting deflection profiles of the thin cover glass samples. Original deflections of the glass substrates were taken into account. Details on the procedure are given in reference [9].

RESULTS AND DISCUSSION

The crystallographic structure of sputter-deposited WN_x depends on the concentration of nitrogen [10] in films. At low nitrogen concentration (below $\sim$15at.%), the films contain a mixture of αW (BCC) and βW (cubic; A15). An amorphous phase is observed in the range of $\sim$15 ÷ 30% nitrogen. Above that, a single fcc W_2N phase is shown by x-ray but with an increased lattice constant up to 4.28A (bulk 4.126A).

Fig. 1 shows the dependence of the nitrogen concentration of the WN_x films on the dissipated sputtering power for 3 different initial total gas pressures: 5mT, 10mT and 15mT. The percentage of nitrogen in the admitted sputtering gas is kept at a constant level of 40% for each deposition. The nitrogen concentration decreases continuously as the sputtering power is varied from 100 to 600W. There is no sharp transition in the nitrogen concentration of the deposited WN_x films as power changes, although x increases more rapidly below 300W for the case of 10 and 15mT. Sputtering power clearly has a strong influence on the stoichiometry of the WN_x. For 10mT initial total gas pressure, the nitrogen level in the films can be changed 10 fold (from 5% to 50%) by varying the power from 100W to 600W.

Another important feature that is immediately noticable is that increased nitrogen levels can be incorporated in the films when the initial total gas pressure rises. This effect of initial total gas pressure on the composition of the film is most pronounced at 600W sputtering power. For 5 and 10mT gas pressure, the nitrogen level in the films are almost below the detection limit of backscattering spectrometry (BS). However, with 15mT, more than 20 percent of nitrogen is present in the films. This observation contrasts that of the TiN case in which the N/Ti ratio changes negligibly over a wide range of initial total gas pressure [7]. On the other hand, changing the initial total gas pressure influences the deposition rate of the WN_x films insignificantly (Fig. 1). For all three gas pressures, the deposition rate increases almost linearly with sputtering power with the same slope.

Figure 2 shows the dependence of the nitrogen concentration in the WN_x film on the relative nitrogen partial pressure for an initial constant total gas pressure of 10mT and two sputtering powers as parameter. Increasing the initial nitrogen partial pressure (p_{N2}) increases the nitrogen level in the films for both 200W and 400W sputtering power. This agrees well with previous work [1,10]. Sharp increases in nitrogen concentration are observed when p_{N2} is raised beyond 30%. Such increase appears real, and is reflected in the change of deposition rate in the case of 400W: rapid increase in nitrogen concentration in films as p_{N2} changes is accompanied by relatively rapid decrease in deposition rate. This correlation between nitrogen incorporation and deposition rate is not so obvious in the 200W case. Although the deposition rate of WN_x decreases with the amount of relative nitrogen partial pressure during sputtering, the decrease is not as abrupt as in the case of TiN. In the TiN case, there exists a threshold in the nitrogen initial partial pressure above which TiN can be deposited. Passing through this threshold always results in a very abrupt change of sputtering rate.

Figure 3 shows the resistivities of the WN_x films as a function of film compositions. The resistivity values are taken from all our films deposited under every sputtering conditions discussed in this paper for which the

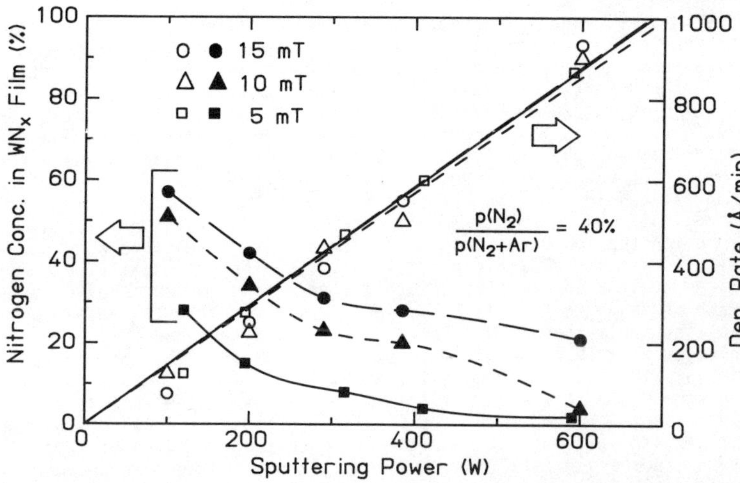

Fig. 1 Nitrogen concentration in the WN_x films and deposition rate for samples sputter-deposited with a fixed 40% N_2/60% Ar gas mixture at 5, 10 and 15mT of initial total gas pressure.

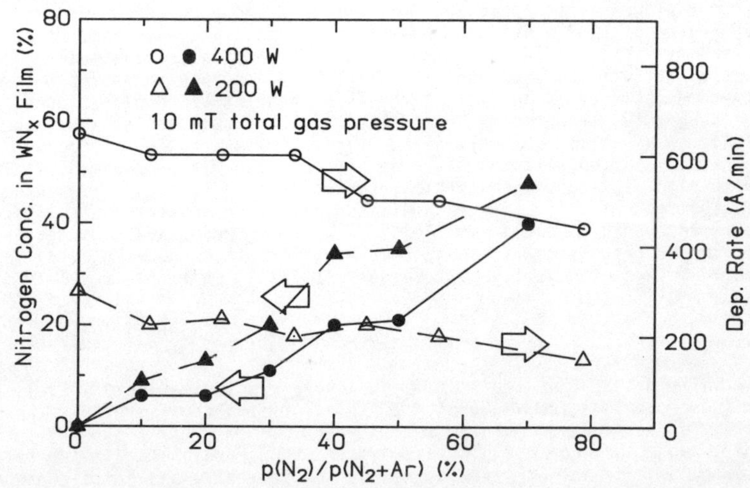

Fig. 2 Nitrogen concentration in the WN_x films and deposition rate as a function of initial relative partial pressure of nitrogen in the gas mixture with a fixed initial total gas pressure of 10mT for samples deposited with 200W and 400W sputtering power.

substrate bias was zero. Some data points were obtained from previous work by Kattelus et al. [1]. It is clear that the resistivity of WN_x increases monotonically with the nitrogen content in the films, regardless of the sputtering parameters. The resistivity rises slowly with nitrogen concentration up to about 30at.% ($x < 0.43$). Above that, the increase in resistivity steepens. This transition in resistivity also corresponds to the transition from amorphous WN_x to polycrystalline WN_x as x increases. For films containing about 50% nitrogen, resistivities are in the range of 2000 μΩcm (not plotted in Fig. 3). All the amorphous WN_x films have resistivities in the range of ~200 μΩ cm. The fact that the resistivity correlates with the nitrogen concentration regardless of the sputtering conditions used suggests that the microstructure of the WN_x films is dictated principally by the amount of nitrogen present in the film. Such a correlation is in marked contrast with what was observed for TiN [7-9]. Near-stoichiometric TiN obtained under different sputtering conditions have resistivities that can vary by an order of magnitude. The above observation seems to indicate that the formation mechanism of TiN and WN_x may be quite different.

The stress of all W and WN films is compressive. The variation in stress level with composition of the films is shown in Figure 4. As for the resistivity, stress is more a function of the nitrogen content of the films than of the sputtering conditions. As the nitrogen concentration increases, the compression of films decreases and reaches its lowest value for amorphous samples near 15 ÷ 20at.% nitrogen concentration. Past this point, stress rises again. Beyond about 30at.% nitrogen, the films are polycrystalline and consist of the W_2N phase. According to previous work [10], increasing the nitrogen concentration in W_2N increases the lattice constant from the bulk value of 4.126A to 4.28A at about 45at.% of N. This fact suggests that the excess nitrogen atoms are incorporated into W_2N lattice and that the increase of compressive stress originates from lattice expansion. In this respect, W_2N is similar to TiN [7,8].

Figure 5 depicts the change in composition, resistivity and compressive stress for two WN_x deposition conditions at a fixed initial total pressure of 15mT (25% nitrogen partial pressure, 400W dissipated power, and 40% nitrogen partial pressure, 200W dissipated power) versus negative substrate bias. With an increase in the negative bias the concentration of nitrogen in the films and their resistivities decrease. This result is in good agreement with the fact that resistivity is determined mainly by concentration of nitrogen in the films (Figure 3). In terms of internal stress, the two samples show opposite dependence on negative substrate bias. For the film with low nitrogen concentration, the compression rises as the negative bias increases; for the film with high nitrogen concentration, the trend is opposite. This outcome is as expected from Fig. 4. This response to a negative substrate bias differs radically from that observed for TiN films [8,9]. There, an abrupt transition is observed in the internal stress, the resistivity and the film color in a very narrow range of negative bias values (10V).

CONCLUSIONS

We believe that the results presented here are indicative of reactive sputtering mechanisms for WN_x that are different from those of TiN. The following facts argue for such an interpretation:

(i) When the rf power is increased, all other parameters being fixed, the nitrogen content in the WN_x film decreases monotonically and can approach zero (i.e. pure W is deposited) without short discontinuity in the deposition rate. That is not so for TiN.

(ii) The deposition rate of WN_x depends primarily on the dissipated power, and little on total gas pressure and composition. For TiN, large changes in deposition rate are noted with changes in the total pressure and gas composition at fixed power.

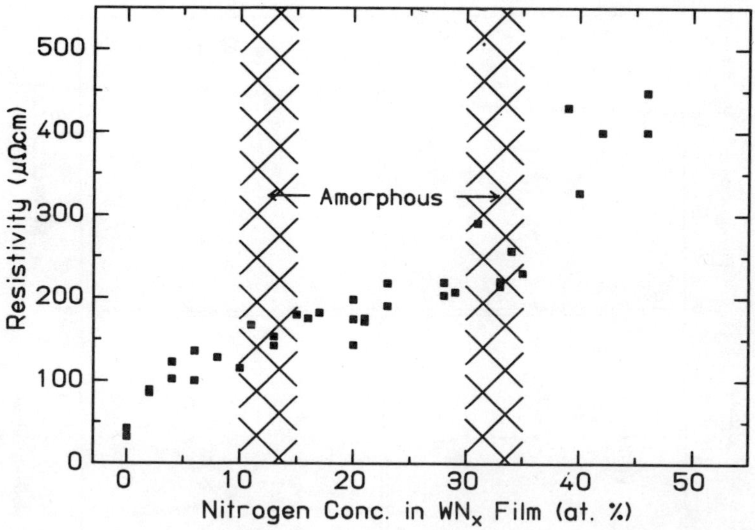

Fig. 3 Resistivity of WN_x films as a function of their nitrogen concentration. The data include all our films deposited under all different sputtering conditions.

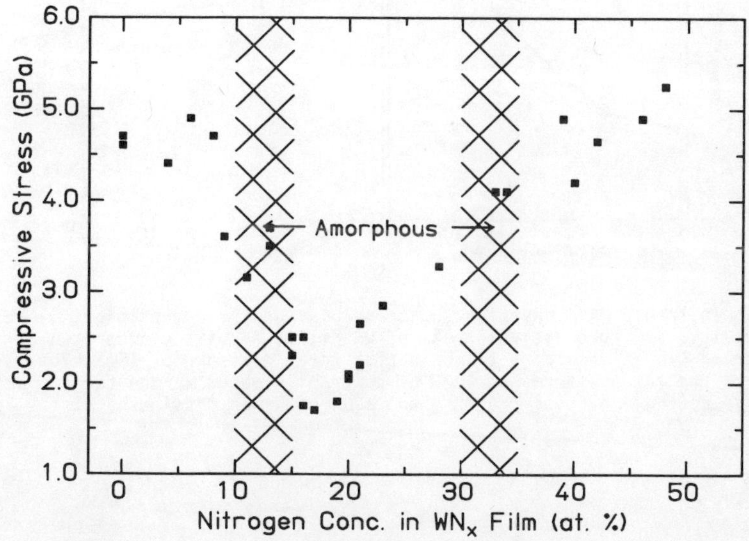

Fig. 4 Compressive stress of WN_x films as a function of their nitrogen concentration. The data include all our films deposited under all different sputtering conditions.

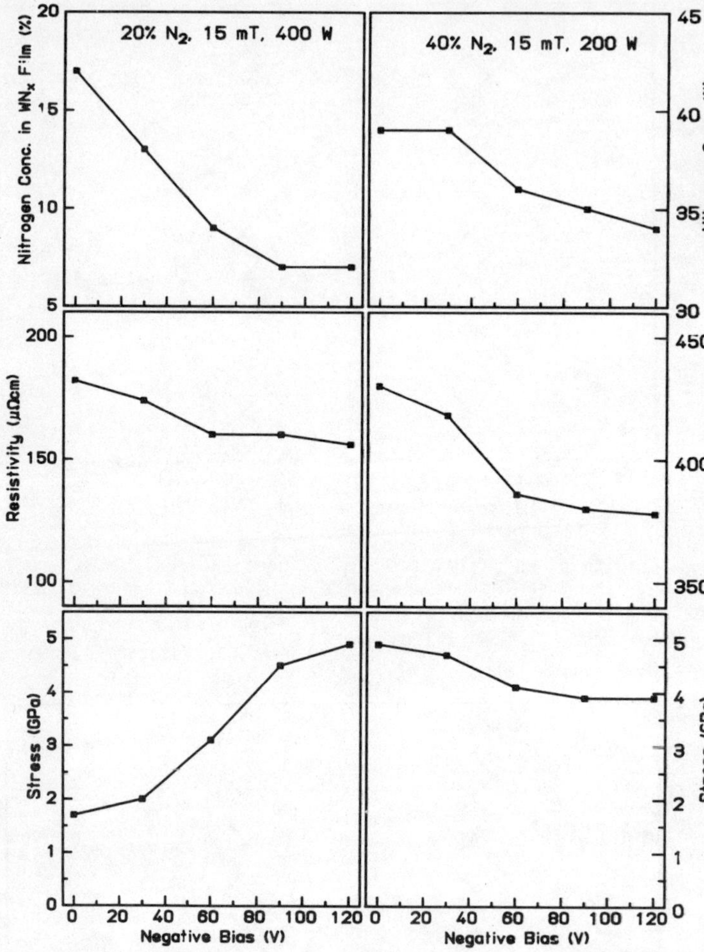

Fig. 5 Influence of a negative substrate bias on the composition, resistivity, and compressive stress of WN_x films deposited under two sputtering conditions for a fixed initial total pressure of 15mT: 25% relative nitrogen partial pressure and 400W of power dissipated at the target, and 40% relative nitrogen partial pressure and 200W of power dissipated at the target.

(iii) A negative dc bias of the substrate changes the resistivity of WN_x little; for TiN, this bias is critical. The same thing is true for the compressive stress.

The reactive sputtering characteristics of WN_x are attractive for industrial application because of the monotonic dependence of the film parameters on the sputtering conditions.

ACKNOWLEDGEMENTS

The authors gratefully acknowledge financial support from the Army Research Office under contract number DAAG29-85-K-0192. Technical assistance from R. Gorris, help in data analysis from D. Jamieson, and manuscript preparation from A. Collinwood are also gratefully acknowledged.

REFERENCES

1. H.P. Kattelus, E. Kolawa, K. Affolter and M-A. Nicolet, J. Vac. Sci. Technol. A 3, 2246 (1985).
2. F.C.T. So, E. Kolawa, J. Tandon and M-A. Nicolet, to be published in J. Electroc. Soc.
3. E. Kolawa, F.C.T. So, J. Tandon and M-A. Nicolet, to be published in J. Electroc. Soc.
4. N. Uchitomi, Y. Kitoura, T. Mizoguchi, Y. Ikova, N. Toyoda, A. Hojo in Extended Abstracts of the 16th Intl. Conf. on Solid State Devices and Materials, Kobe, pp. 383-386 (1984).
5. N. Uchitomi, M. Nagaoka, K. Nishibori, K. Shimada, T. Misoguchi and N. Toyoda in the Fourth Workshop on Refractory Metals and Silicides for VLSI, San Juan Bautista (1986).
6. K.Y. Ahn, S.B. Brodsky, C.Y. Ting and J. Kim, J. Vac. Sci. Technol. (in press).
7. J.E. Lundgren, B-O. Johansson and S.-E. Karlsson, Thin Solid Films 105, 353 (1983).
8. S. Kanamori, Thin Solid Films (in press).
9. H.P. Kattelus, J. Tandon, C. Sala and M-A. Nicolet, J. Vac. Sci. and Technol. A 4, 1850 (1986).
10. K. Affolter, H.P. Kattelus and M-A. Nicolet, MRS Symp. Proc. Vol. 47 (MRS, Pittsburgh, 1985), p. 167.

TiW AS A BARRIER METAL FOR SMALL CONTACTS

S. SAITO, K. MATSUDA, K. NISHIZAWA, AND K. SAKIYAMA
SHARP Corp. Integrated Circuits Group,
2613-1 Ichinomoto, Tenri-City, Nara, 632, Japan

ABSTRACT

Titaninm Tungsten (TiW) films have been used for bipolar devices because they exhibit both good adhesion to silicon dioxide [1] and serve as an excellent diffusion barrier Aluminium Silicon(AlSi). Films of TiW which are deposited by dc magnetron sputtering are easy to control and exhibit good stepcoverage. In this paper, the interaction of the AlSi/TiW/Si trilayer structure is studied by means of X-ray diffraction, sheet resistance, junction leakage current. The interaction of AlSi/TiW is found to occur at annealing conditions of 500°C for 60min, however, TiW and Si do not react even at 500°C. Low contact resistances for TiW films on n^+Si and p^+Si are demonstrated for annealing of up to 500°C for 60min.

INTRODUCTION

In recent years, a need has arisen for metallization systems that are suitable for use in very large scale integration (VLSI) processes. This is particularly true for contacts because as the size of the contact becomes smaller(~1μm) the junction depth also scales and becomes considerably shallower(~0.3μm) . In general, the following problems need be solved for the contacts : (1) junction spiking [2][3] (2) Si precipitation [3][4] (3) discontinuity of metal at contact hole. In order to eliminate these problems, many barrier metals have been extensively studied and utilized. The barrier metal requires: (1) low resistance and good ohmic contact to both p^+Si and n^+Si substrates, (2) good barrier effect between Al and Si, (3) conformal stepcoverage.

Recently, some promising barrier metals such as Ti,TiN[5][6],and CVD-W[7] have reported. However, the use of these materials often makes processing complicated and unstable. The purpose of this study is to develop a simple barrier process for VLSI contacts.

EXPERIMENTS

The multilayer structure of AlSi(Si-1wt%)/TiW(Ti-10wt%) was continuously deposited by dc magnetron sputtering onto Si (100) wafers. Before the AlSi deposition, the wafers were preheated. The base pressure was better than 5×10-7torr, the Ar gas pressure was 8mtorr and sputtering power was kept 1kw for TiW and 7kw for AlSi. The AlSi/TiW multilayer was patterned by both wet etching (AlSi ; H_3PO_4 , TiW ; 31% H_2O_2) and 2 step dryetching(AlSi;BCL_3+CCL_3F+He, TiW;CCL_3F+O_2). After patterning the AlSi/TiW, wafers were annealed in furnace for 60min.

RESULTS AND DISCUSSION

Film Quality Of TiW

The TiW film density, compositions, intrinsic stress, wet etching rate, film structure, resistivity are evaluated (Table I). All film quality except intrinsic stress is found to be independent of the sputtring conditions. On the other hand, the intrinsic stress apparently a direct

Table I Film quality of TiW

Resistivity	64.5 $\mu\Omega \cdot cm$
Density	16.2 g/cm^3
Compressive Stress	2 ~ 7 ×10E-9dyn/cm^2
Ti Concentration	5.6 wt%
Etch Rate (31%, H_2O_2)	150 A/min
Film Structure	Colummar
Reflectivity (436nm)	50%

function of the sputtering power (0.5kw~2kw) and the Argon gas presssure (3mtorr~14mtorr). When a TiW film was deposited using a lower pressure and higher power, the compressive stress in the film become higher.

Next of all, the stepcoverage of the AlSi/TiW film is also evaluated. Wafers were prepared with PSG(1.2μm) which was subsequently dry-etched to form vias. Next the AlSi(0.9μm) and the TiW(0.6μm) films were deposited independently. The film thickness at three points (T1,T2,T3), shown in Fig. 1, was measured. The TiW film was fonud to be thicker than the AlSi film at both the minimum thickness of the bottom of via hole(T2) and on the side walls of the vias(T3). At the point of H/D=0.6 (aspect ratio), the AlSi film was observed to be thicker than the TiW film (ref. Fig. 2). Because the stepcoverage of AlSi film was poor (little deposition on the side walls of the vias), the effective via hole size for AlSi becomes larger than that for TiW. These results indicates that TiW films are easy to control and have conformal stepcoverage. Therefor, we consider that TiW is one of several materials which are applicable for small contact.

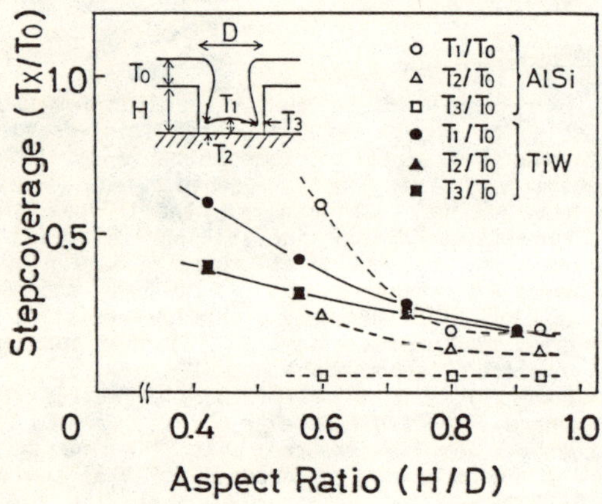

Fig.1 Stepcoverage of TiW(0.6μm) and AlSi(0.9μm) on via hole pattern. The via hole depth was 1.2μm.

Barrier Effect Of TiW

For macroscopic evaluation, X-ray diffraction analysis and sheet resistance measurements were perfomed. The X-ray diffraction spectra taken from the AlSi/TiW/Si thin film system are shown in Fig 2. The deposited films AlSi(0.5μm)/TiW(0.3μm) on Si substrates were annealed at 470°C and 500°C for 60min. From the X-ray diffraction spectra, the WAl_{12} peaks are observed after annealing at 500°C for 60 min. However, any siliside formations is not observed in these diffraction spectra. That is, the TiW/Si interface is staible up to 500°C.

The same samples were used to study the dependence of the sheet

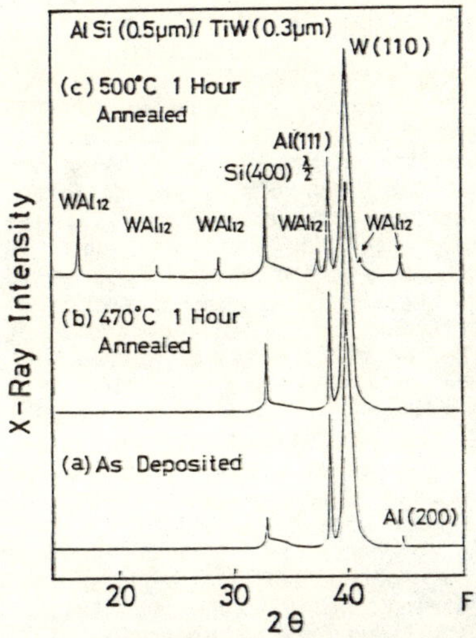

Fig.2 X-Ray(Cukα) diffraction spectra of AlSi/TiW/Si.

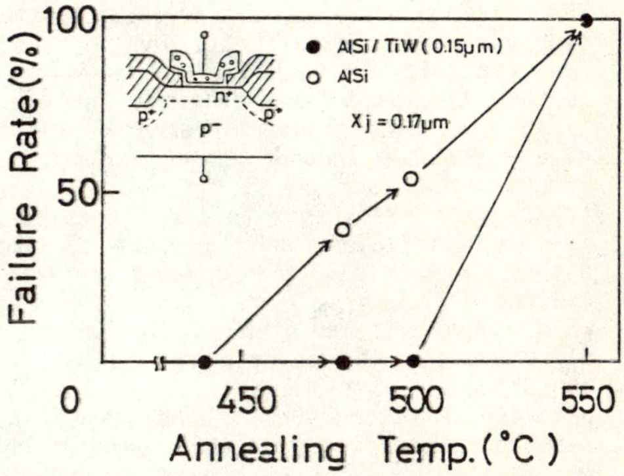

Fig.3 Failure rate dependence on annealing temperature. N^+P diode of whitch current over 1×10^{-11}A at 5 volts is defined as a failure sample.

resistance on annealing temperature. The sheet resistance was fonund to remarkably increase with annealing temperatures over 500°C. This data supports the diffraction date and indicates the formation of WAl_{12}.

For microscopic evaluation, the junction leakage current was measured for samples annealed at 440°C, 480°C, 500°C and 550°C. After the Arsen implantation (80kev with 5×10^{15}ions/cm^2) into P-type Si (junction area;400×400μm^2), the samples were activated at 950°C for 30min in N2. The junction depth was 0.17μm for these condition. After etching 100×100μm^2 via holes, AlSi(0.6μm)/TiW(0.15μm) and AlSi(0.6μm) were deposited to form electrodes of diodes. These diodes leakage current was less than 1×10^{-11}A. However, if the interaction of AlSi/TiW/Si has occured ,resulting in Al spiking of shallow junction, the junction leakage current would have been much greater. In this experiment, a leakage current in excess of 1×10^{-11}A at 5 volts reverse bias was defined as a failed barrier. The same sample structure was annealed at 440°C, 480°C, 500°C and 550°C for 60min. The relationship between the failure rate and the annealing temperatures is shown in Fig. 3. Only the AlSi(without TiW) film begins to fail at temperature of 440°C and 480°C. On the other hand, the AlSi/TiW flim do not fail up to 500°C.

Based on these results, we found that TiW films are effective for the prevention of Al spiking and can serve as an excellent diffusion barrier between AlSi and Si layers for temperatures up to 500°C.

Contact Resistance

In general, on heavily doping Si, the barrier width decreases, tunnering current becomes important. In this case, specific contact resistance may be written as

$$Rc \approx \exp\left[\frac{4\pi\sqrt{\varepsilon m^*}}{h}\left(\frac{\Phi B}{\sqrt{N_D}}\right)\right]$$

where ε is dielectric permittivety of Si, m* is the effective mass of charge carrier, h is Plank′ s constant, ΦB is barrier height and N_D is the doping concentration. Therefore, contact resistance is strongly related to (1) ΦB (2)N_D. It is also influenced by (3)metal-Si interface state.

With respect to interface states, M.J.Kim et al.[9] have reported that the dry etching can be a major cause of increased contact resistance. Therefore the effect of surface treatment and the annealing were also studied in this experiment. The contact resistance was measured using the four probe Kelvin method after annealing at 440°C for 60min. Two different surface treatment were selected for use after via hole etching(RIE). The first treatment a 1% Buffered Hydrofloric Acid (BHF) wet etching (STD clean) and the second was an isotropic plasma etching which was used to remove about 30nm of Si before the 1% BHF wet etching. The dependece of the contact resistance to these surface treatment and annealing conditions is shown in TableII. With respect to samples which were lightly doped with either BF_2 or As, the contact resistance of those samples which were treated using only the STD clean was much higher than that of the samples which were plasma etched 30nm and then wet etched. Annealing is an effective way to decrease the contact resistance. Howerer, the effect of annealing is considerably less than that of the surface treatments. Detailed investigations of the TiW/Si interface were conducted using Auger Electron Spectroscopy (AES). The etchig products of RIE(for example polimerization film of carbon and fluorine) often may not be observed. These products are tought to be removed during the photoresist striping by a plasma and a chemical cleaning with H_2SO_4 and BHF. We determined that a damage region of 20nm (into the Si substrate) existed, this finding is based on life time measurements in the Si after RIE. Using both the AES analysis and the life time measurements, it can be determined that Si surface damage caused by RIE is a major mechanism which can cause an increase in the contact resistance.

The effect of carrier concentration on contact resistance is considered next. An N^+ diffusion layer was formed by As^+ implantation (80kev; 8×10^{14},1×10^{15},2×10^{15},5×10^{15} and 8×10^{15}ions/cm^2). The carrier concentration was evaluated using SUPREMII. The P^+ diffusion layer was also formed by BF_2 implantation (65KeV; 8×10^{14},1×10^{15},2×10^{15} and 5×10^{15} ions/cm^2). The differencial sheet resistance method was used to measure the carrier concentrations. All of the samples were activated at 950°C for 30min. The contact resistances for both Pch and Nch correlates with the value of $1/(N_D)^{1/2}$.(Fig.4) The tunnering current was found to be larger than the current flow between the TiW film and the Si substrate. The slops of two lines are independent of contact area. And the carrier concentration and the junction depth are different for all samples. This is because ion implantation at vatious combinations oh dose, species, and ion energy were

TableII Effect of surface treatrent

Surface Treatment	Impurity	N_D [10^{20}cm^{-3}]	N2 Annealing	Rc ($1\mu m^2$) [Ω]
STD Clean (1% BHF)	As	10.2	——	28
	As	10.2	440°C,1H	22
	As	7.8	——	800
	As	7.8	440°C,1H	140
	BF2	7.4	——	370
	BF2	7.4	440°C,1H	76
300Å Plasma Si Etch + STD Clean	As	10.2	——	21
	BF2	7.4	——	42
	BF2	7.4	440°C,1H	42

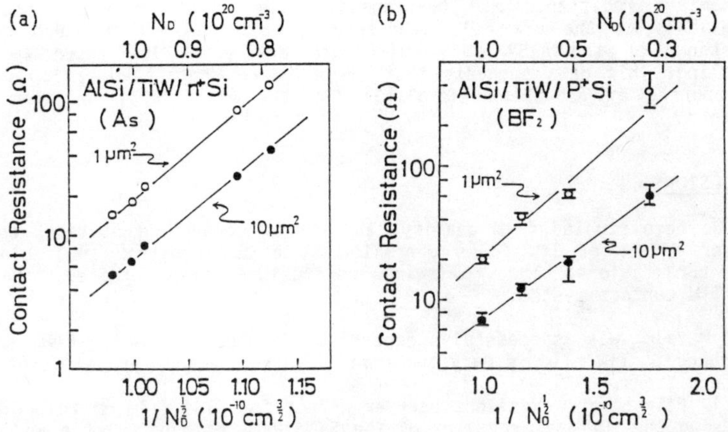

Fig.4(a)(b) Contact resistance as a function of carrier concentration at the surface. Carrier concentration is (a) calcurated by SUPREM II (b) measured by differenctial sheet resistance method.

used for the preparation of these samples. However, Fig.4 shows that the contact resistance is not related to the junction depth, rather it is dependent only on the surface carrier concenration.

The dependence of the contact resistance on annearing temperature is shown in Fig 5. The contact resistance was found to be stable up to 500°C for 60min even for very small contact size(1.7μm^2). Although the Al/W reaction take place at 500°C, the TiW film behaves as an excellent barrier even at the temperature up to 500°C. The dependence of contact resistance on the annealing time at 440°C is shown in Fig.6. Without the underlayer TiW film, the contact resistance increaces with annealing time.An increace

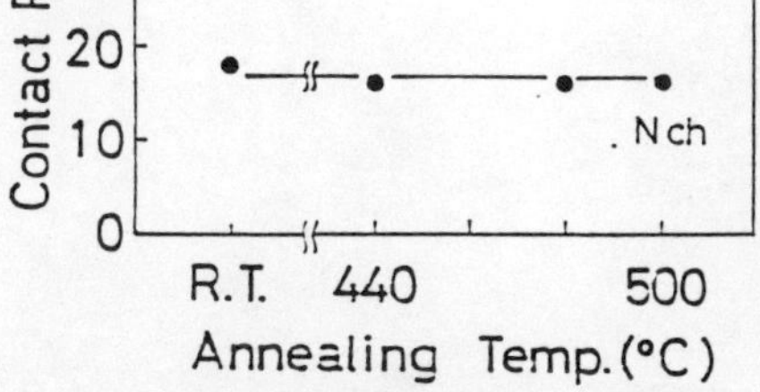

Fig.5 Contact resistance dependence on annealing temperature of AlSi/TiW structure.

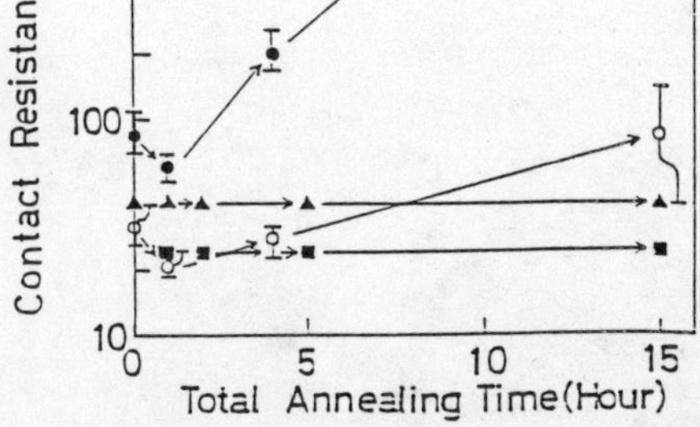

Fig.6 Contact resistance dependence on annealing time at 440°C.

of contact resistance may be caused by Si atoms in AlSi that grow epitaxially at the via hole. Conversely, both the Pch and Nch contact resistance of the AlSi/TiW/Si structure are still low after 15 hours annealing. This data suggests that there is no contact migration problem as reported by J.G.Chern et al.[3] for the AlSi/TiW/Si trilayer metal contacts.

CONCLUSIONS

We have studied film quality, ability to control deposition, and the barrier effect of TiW films deposited by a dc magnetron sputtering for VLSI applications. The following conclusions are obtained, for the AlSi/TiW contact system.

(1) TiW film was successfuly deposited without any major film quality change. TiW film is easy to control by choosing proper sputtering conditions.
(2) TiW film has an excellent barrier effect for AlSi film up to 500°C.
(3) Using the surface treatment of the Si plasma etching, stable and low resistance cantact can be obtained. One such example is the 1μm^2 contact resistance of 18Ω (N_D=1.02×10^{20}ions/cm^3) for Nch and 42Ω (N_D=0.74×10^{20}ions/cm3) for Pch.

Thus, a simple process with reliable barrier characteristics can be attained using the TiW interlayer for annearing temperatures up to 500°C.

REFERENCE

1. P.G.Ghate, J.C.Blair, C.R.Fuller, and G.E.Mcguire, Thin Solid Filmes, 53, 117 (1978).
2. D.Pramanic and A.N.Saxena, Solid State Tech. 127, 131 (1983).
3. J.G.J.Chern, W.G.Oldham and N.Cheung, IEEE Trans. Electron Devices, 32, 1341 (1985).
4. H.M.Naguib and L.H.Hobbs, J. Electrochem. Soc.,125,169 (1978).
5. T.Maeda,T.Nakahara,et al., VLSI Symposium, 8,(1985).
6. I.Suni, M.Biomberg, J.Saarilahti, J. Vac. Sci. Tech., A(3), Nov/Dec, 2233 (1985).
7. R.A.Levy, M.L.Green, P.K.Gallagher and Y.S.Ali, J. Electrochem. Soc., 133, 1905 (1986).
8. S.E.Babcock, J. Appl. Phys., 53, 6898 (1982).
9. M.J.Kim, D.M.Brown, S.S.Cohen, et al., IEEE Trans. Electron Devices, 7, 1328 (1985).

THE STABILITY OF THE ELECTRICAL AND STRUCTURAL PROPERTIES OF ELECTRON BEAM DEPOSITED AND SPUTTER DEPOSITED Ta AND Mo SCHOTTKY CONTACTS TO GaAs

A. K. KULKARNI, M. G. THOMPSON AND G. H. WHIPPLE
Department of Electrical Engineering, Michigan Technological University, Houghton, MI 49931

ABSTRACT

The stability of the electrical properties of Mo and Ta Schottky contacts to GaAs has been investigated by measurement of current-voltage and capacitance-voltage characteristics on as-deposited samples and samples annealed at 350°C, 450°C and 550°C for 30 minutes. E-beam deposited Mo contacts improved upon annealing, whereas, sputter deposited Mo contacts degraded. The e-beam and sputter deposited Ta contacts were very stable up to 450°C and became unstable after annealing at 550°C. The changes in the electrical properties are compared with the corresponding changes in the structural properties as determined by Auger depth profile plots.

I. INTRODUCTION

GaAs metal-semiconductor transistors (MESFETs) are emerging as the most important electronic devices for microwave and high speed digital applications. GaAs integrated circuits are often realized by a self-aligned gate process. In this process, the Schottky gate contact must maintain its stability during post ion-implantation annealing around 800°C. Refractory metals appear to be potentially suitable materials for this self-alignment process.

Refractory metals, i.e. W, Ti, Ta and Mo and their alloys Ti:Pt, TiW, TiW:Cr, and Ti:Cr:Pt, were initially considered as diffusion barriers to prevent the outdiffusion of gallium in the top gold layers to improve the reliability of Schottky and ohmic contacts to GaAs [1,2]. The refractory metals Ti, W and Ta turned out to be unsuitable because of problems related to interdiffusion, oxide formation and degradation of the interface [3-5]. W. J. Devlin has shown barrier heights determined from capacitance-voltage measurements of 0.7eV on as-prepared Au:Mo sputtered contacts and 0.82eV on an annealed (400°C, 23 hrs.) contacts [6]. On the other hand, the barrier height determined from current-voltage characteristics increased from 0.54eV to 0.67eV upon annealing. The ideality factor improved from 1.15 to 1.06 upon annealing. The Mo layer has been shown to peel off around 450°C due to mismatch of thermal expansion coefficients of Mo and GaAs [7]. Further heating of Mo contacts in arsenic over pressure at temperatures of 800°C to 920°C has resulted in compound formation, i.e. Mo_5As_4 [8].

The barrier height of Ta-GaAs contacts has been reported to be 0.85eV [9]. S. D. Mukherjee et al. have reported partial oxidation of Ta at 425°C (1 hr.) anneal and complete oxidation of Ta after 550°C (1 hr.) anneal [5]. Recently J. R. Waldrop et al. have reported barrier heights of 0.7eV on e-beam evaporated and sputtered Ta contacts and 0.8eV on sputtered Mo contacts to GaAs [10]. In this paper, we report the changes in the electrical and structural properties of Mo and Ta contacts to GaAs due to annealing.

II. FABRICATION AND PROCESSING

The GaAs substrates used in this study were n-type (N_D = 3 x 10^{17} cm^{-3}, ρ = 0.0064 ohm-cm, Si doped), < 100 > oriented. The substrates were cut into samples of size 4mm x 8mm and chemically etched in a caros etch [H_2SO_4:H_2O_2:H_2O (5:1:1)] for 2 minutes and a bromine-methanol etch for 5 minutes. After etching, the samples were dried in nitrogen and placed into an ultra high vacuum system (base pressure: 10^{-9} torr) for ohmic contact deposition.

The back ohmic contacts were AuGe (12% Ge by weight) deposited by e-beam evaporation to obtain 500nm thick AuGe films. The AuGe contacts were annealed at 400°C in forming gas (90% N_2 + 10% H_2) for 5 minutes to obtain contact resistances on the order of 10^{-5} ohm - cm^2 [11]. After ohmic contact formation, the front Schottky contacts (dia. = 0.5 mm) were deposited through shadow masks by either electron beam evaporation or rf sputtering. The geometry of a typical sample is shown in Fig. 1.

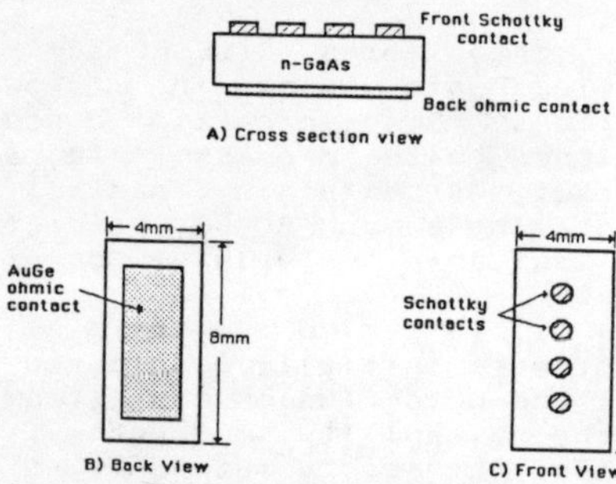

Fig. 1. Geometry of Mo and Ta Schottky diode samples.

Prior to e-beam evaporation, the vacuum system was pumped to a base pressure of 10^{-9} torr. Typical pressures during deposition were 10^{-7} - 10^{-6} torr. The Mo and Ta evaporation charges were made up of cut pieces of 99.97% and 99.99% pure rod, respectively. The deposition rate of Mo was 9.0 nm/min and that of Ta was 22.5 nm/min as determined with a Kronos QM-311 Thickness Monitor. It was desirable to have a relatively thick metal layer to withstand probing during electrical measurements. Depositions of 150 nm resulted in films of high quality that did not crack or peel off during subsequent anneals.

Mo and Ta contacts were also fabricated by rf sputtering. The target materials were 99.95% pure and the sputtering gas was high purity argon (99.999%) at a pressure of 15 mtorr. The sputtering system was diffusion pumped and equipped with a liquid nitrogen cold trap to minimize oil contamination. Prior to introducing argon into the chamber, a base pressure of 8 x

10^{-7} torr was achieved. The target surfaces were sputter cleaned immediately before deposition. A deposition sputtering power density of 0.62 w/cm^2 was used throughout this study. This power density represents a trade-off between deposition rate and sputter damage to the GaAs surface. The deposition rate of Mo and Ta were 7.6 nm/min and 3.5 nm/min, respectively. The sputtered contacts were deposited to a thickness of 150 nm ± 20 nm and adhered well to the GaAs surface.

III. EXPERIMENTAL PROCEDURE AND RESULTS

Electrical Characteristics

The electrical properties of the Schottky diodes were obtained by measuring current-voltage (I-V) characteristics and capacitance-voltage (C-V) characteristics at room temperature. The I-V characteristics were initially observed on a Tektronix 577 Curve Tracer to qualitatively assess properties such as forward cut-in voltage, series resistance and reverse breakdown voltage. Detailed I-V data was taken over several decades of current using a Keithly 225 Current Source and a Keithly 181 Nanovoltmeter. This data was used to determine the diode quality factor (n) and the saturation current (I_S) of the Schottky diode samples. Capacitance measurements were made at 1MHz as a function of applied reverse bias voltage with an MSI Electronics 864 Capacitance Meter. C-V data was used to determine the diffusion potential (ϕ_o) and the doping concentration (N_D) beneath the contact.

Typical I-V data for Schottky diodes in this study showed sharp forward cut-in characteristics and relatively soft reverse breakdown characteristics in the range of 2-4 volts. This soft breakdown region may be attributed to excess leakage current arising from the high field region near the contact edge since no guard ring was employed. Also, tunneling current could be a dominant component due to surface damage during contact deposition and lack of surface passivation. The diodes were subjected to 30 minute isochronal anneals in N_2 atmosphere at 350°C, 450°C and 550°C.

Semi-logarithmic plots of forward I-V data were used to determine the diode quality factor (n) and the reverse saturation current (I_S) from the thermionic emission theory. Typical data for sputtered Mo and Ta contacts is plotted in Fig. 2. and the results of several samples are listed in Table I.

The diffusion potential of Mo and Ta contacts were determined from C-V data assuming an inverse square law model for the depletion capacitance of a Schottky diode. The difference between the Schottky barrier height (ϕ_B) and the diffusion potential (ϕ_o) corresponds to the energy difference between the Fermi level and the conduction band edge. For our relatively heavily doped substrates, this difference is negligible and $\phi_B \doteq \phi_o$. The experimental values can be compared to an expected value of 0.94V given by the 2/3 energy gap approximation. Barrier heights determined from C-V data are summarized in Table I.

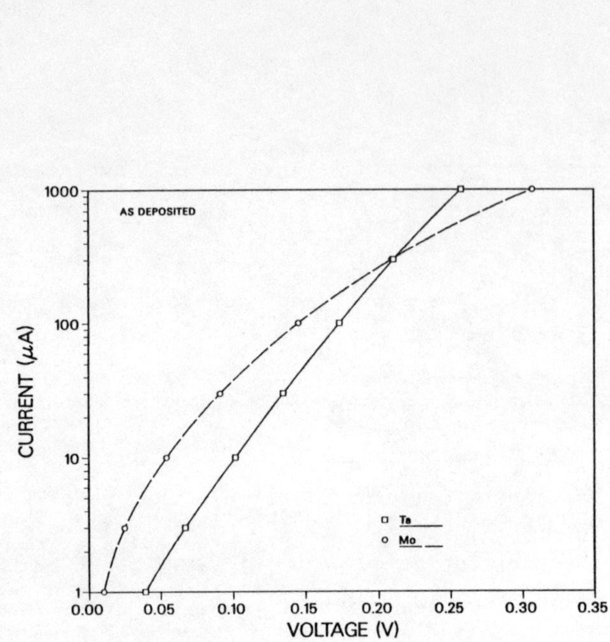

Fig. 2. Typical forward I-V characteristics of as-deposited sputtered Mo and Ta Schottky diodes.

Table I

Electrical Properties of E-beam Deposited and Sputter Deposited Mo and Ta Contacts to GaAs

SAMPLE	Anneal Temp. (^{o}C)	Barrier Height Φ_B(ev)	Doping Concen. $N_D(x10^{17}cm^{-3})$	Cut-In Voltage V_γ(V)	Breakdown Voltage V_{BR}(V)	Ideality Factor (n)	Saturation Current I_s(A)
E-beam Mo	(as-dep)	1.3	33.2	0.6	2.5	-	-
"	350^oC	1.325	14.7	0.4	1.8	-	-
"	450^oC	1.35	2.78	0.4	1.2	-	-
"	550^oC	1.20	1.64	0.2	0.8	-	-
Sputtered Mo	(as-dep)	0.98	3.3	0.11	2.4	1.76	3.9×10^{-8}
"	350^oC	0.92	2.5	0.15	2.4	1.48	1.0×10^{-6}
"	450^oC	1.13	13.3	0.13	2.4	1.40	1.0×10^{-6}
"	550^oC	3.84	23.3	0.20	1.5	1.44	1.9×10^{-7}
E-beam Ta	(as-dep)	1.10	6.97	0.6	3.0	1.70	8.3×10^{-11}
"	350^oC	1.05	7.82	0.4	2.5	-	-
"	450^oC	0.975	7.82	0.33	3.0	1.49	3.9×10^{-11}
"	550^oC	x	x	0.3	3.0	-	-
Sputtered Ta	(as-dep)	0.86	3.7	0.22	2.4	1.30	9.0×10^{-8}
"	350^oC	1.08	2.2	0.37	2.1	x	x
"	450^oC	1.29	1.4	x	x	x	x
"	550^oC	x	x	0.29	x	2.41	3.9×10^{-7}

- No data
x Could not be measured

Structural Properties

Compositional depth profiles of refractory metal-GaAs interfaces were obtained by Auger electron spectroscopy (AES) with a PHI 545 Auger electron spectrometer. Data reduction and conversion of Auger peak heights into depth profiles was accomplished by computer methods. Depth profiles of several samples, both as-deposited and thermally aged were obtained to study the structural degradation at the interface of metal and GaAs.

Figure 3 shows the AES depth profile of an as-deposited (e-beam) Mo-GaAs contact. The excessive amounts of carbon and oxygen seen here are undesirable and may have caused the poor electrical properties of as-deposited samples. Figure 4 compares the depth profiles of sputtered Ta-GaAs contacts annealed at 350°C and 550°C. The increased amount of oxygen (≃20%) for 550°C annealed contacts indicates oxide formation which has led to the degradation of the electrical properties. Annealing at 550°C for 30 minutes might have cured the sputter damage of the GaAs surface resulting in a slightly more abrupt interface and less diffusion of Ta into GaAs.

IV. DISCUSSION

As seen in Table I the barrier heights and doping concentrations of as-deposited (e-beam) Mo-GaAs contacts are higher than expected values. Upon annealing the observed values approach expected values. On the other hand, the barrier heights and doping concentrations of as-deposited (sputter) Mo-GaAs contacts show close to expected values. Upon annealing high values are observed for these contacts. This opposite trend in e-beam and sputter deposited Mo samples could be due to changes in the interfacial reactions. The Ta-GaAs contacts gave close to expected values of barrier heights and doping concentrations in both e-beam and sputter deposited samples and only small changes were observed in these values upon annealing up to 450°C. However, no measurements could be made on these contacts once they were annealed at 550°C for 30 minutes. This is essentially due to complete oxidation of these films at 550°C as reported by other investigators [5] and also observed in AES depth profiles here. The ideality factors reported here for sputtered Mo contacts are typically 1.4 to 1.8 and the saturation currents are very high (in μA). The thermionic emission model used to determine these values may not be appropriate because of the heavy doping (N_D = 3 x 10^{17} cm^{-3}) of the GaAs substrates.

The results reported here on refractory metal contacts to GaAs are somewhat different from our previous work on gold based contacts to GaAs [12]. In gold based contacts to GaAs, both the barrier heights and doping concentrations increased substantially upon annealing. In this respect, refractory metal contacts appear to be more stable than gold based contacts.

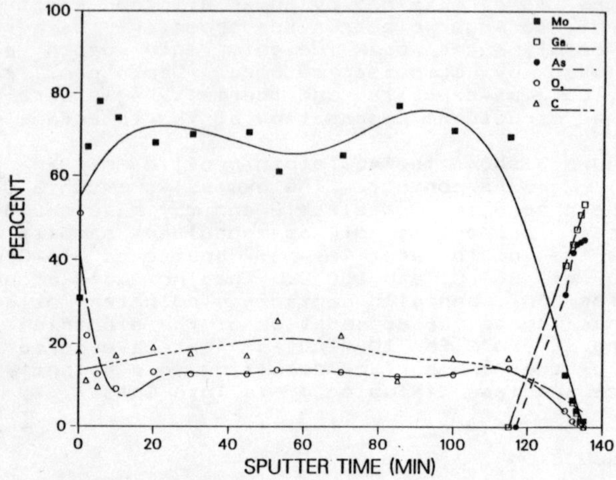

Fig. 3. Auger depth profile of e-beam as-deposited Mo-GaAs contact.

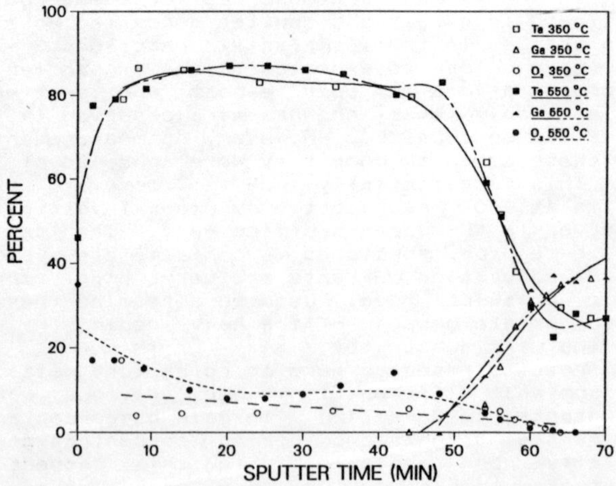

Fig. 4. Auger depth profiles of sputtered Ta-GaAs contacts after annealing at 350°C and 550°C for 30 minutes.

REFERENCES

1. H. M. Day, A. Christou and A. C. Macpherson, J. Vac. Sci. Technol. 14, 939 (1977).

2. W. T. Anderson, A. Christou, J. F. Giuliani and H. B. Kietrich, IEEE Trans. Ind. Elect. 29, 149 (1982).

3. A. Lahav and M. Eizenberg, Appl. Phys. Lett. 45, 256 (1984).

4. C. J. Palmstrom and D. V. Morgan, in Reliability and Degradation, edited by M. J. Howes and D. B. Morgan (John Wiley and Sons Ltd., New York, 1981), pp. 207-215.

5. S. D. Mukherjee, C. J. Palmstrom and J. G. Smith, J. Vac. Sci. Technol. 17, 904 (1980).

6. W. J. Devlin, Electron. Lett. 16, 93 (1980).

7. S. D. Mukherjee, D. V. Morgan, M. J. Howes, J. G. Smith and P. Brook, J. Vac. Sci. Technol. 16, 138 (1979).

8. K. Suh, H. K. Park and K. L. Moazed, J. Vac. Sci. Technol. B1, 365 (1983).

9. S. M. Sze, Physics of Semiconductor Devices, 2nd ed. (Wiley Interscience Inc., 1981), p. 291.

10. J. R. Waldrop, S. P. Kowalczyk and R. W. Grant, J. Vac. Sci. Technol. 21(2), 607 (1982).

11. A. K. Kulkarni and J. T. Lukowski, J. Appl. Phys. 59(8), 2901 (1986).

12. A. K. Kulkarni and B. M. Post, Thin Solid Films 123, 1 (1985).

ACKNOWLEDGEMENT

This research was supported in part, by the National Science Foundation under Grant No. ECS-8409946.

PRACTICAL PREPARATION OF SPUTTERED REFRACTORY METAL DIFFUSION BARRIERS

RONALD S. NOWICKI
Integrated Processing, 722 Kenley Way, Sunnyvale, CA 94087

ABSTRACT

Sputtered thin films of refractory metals have not always performed well as diffusion barriers, which has led to some controversy whether these barriers can be used with any degree of predictability. This concern has been of great interest to users who require higher device reliability under rigorous operating conditions, e.g. aerospace and defense. For example, the value for the energy of activation which governs the interdiffusion rate of aluminum or gold through RF diode sputtered $Ti_{33}W_{67}$ was determined to be ca. 0.6 eV. This could have profound consequences on electronic devices that are used in aircraft and missile systems and may explain the need for continuous field testing of such systems, i.e. unpredictable barrier failures could preclude their use for "Hi-Rel" applications. In this review, it is shown that thorough characterization of a given refractory metal barrier system must include data concerning pre- and post-deposition treatment. In particular, deposition technique will be shown to be a strong determinant of barrier properties. This view will be supported by historical data which supports this author's model for the deposition parameters which appear to optimize diffusion barrier properties in refractory metals.

INTRODUCTION AND REVIEW

The use of refractory metals as thin film "diffusion barriers" in current semiconductor technology has been well-documented [1]. In some structures, the barrier is used as a Schottky barrier contact directly to n or p type silicon and usually has an overlying film of solderable or bondable metal, e.g. aluminum or gold, which also provides minimal resistivity interconnections.

In other structures, the barrier is employed to minimize or reduce the interdiffusion between an underlayer such as aluminum, and an overlayer, such as gold [2]. Also, barriers have been employed as adhesive layers to silicon or ceramic dice in conjunction with a solderable or bondable overlayer such as gold, silver or nickel. This structure allows the attachment of the die to a package or carrier by "eutectic" or solder bonding.

Over the years, these materials have not always performed well as diffusion barriers [3]. This has led to some controversy as to what deposition parameters actually govern barrier properties. For example, the value for the energy of activation which determines the diffusion rate of aluminum or gold through relatively pure rf diode sputtered $Ti_{33}W_{67}$ thin films was determined to be 0.6 eV [4]. Thus, following the Arrhenius relationship, at room temperature the interdiffusion would be complete in ca. three years. If the Ti-W were used as a contact to Schottky diodes, electrical instabilities of the diodes could occur even with room temperature storage! When one considers that much of the early work with refractory metal diffusion barriers was done in vacuum systems having typical "base" pressures in the high 10^{-3} Pa (10^{-5} Torr) range [5], and that arrival rates of oxygen or nitrogen at this pressure are several atomic layers per second, it is not surprising that difficulties in the reproducibility of resultant film properties have arisen. Additionally, early workers did not have the analytical tools at their disposal to investigate resultant film purity, etc. Nevertheless, it was surprising to find near-stoichiometric molybdenum oxynitride produced by the rf sputter deposition of Mo in pure argon at base pressures of ca. $1.3X10^{-3}$ Pa ($1X10^{-5}$ Torr) [6]. This discovery would probably not have been made without the aid of nuclear microanalysis. More recently, analytical tools such as Auger, ESCA and SIMS have reached the sensitivities and resolution needed to perform quantitative analysis of trace film impurities with good confidence.

The realization that the barrier properties of refractory metal thin films are strongly dependent upon the impurities introduced into these films during deposition has slowly evolved [7]. As stated above, the major difficulty in achieving this realization has been the problem with analysis of the low concentration (0.1 - 1 at%) of impurities such as oxygen or nitrogen needed to improve the barrier properties. It appears that these impurities preferentially reside within the grain boundaries of the refractory metal and thereby impede the diffusion process. Since the grain boundary volume is determined to a large extent by crystallite size, one can approximate the concentration of impurity atoms needed to saturate the grain boundaries. For example, assuming one monolayer coverage of all grains, for a crystallite size of approximately 30 nm, it would require less than 10 at% of an impurity to fill the grain boundary voids. One can then understand the difficulties with the implementation or control of this variable in practical deposition technique. Thus, the aura of "black art" concerning the barrier properties persists up to recent time. However, Holloway and Nelson [8], and more recently Dalal et al. [9] discovered that the barrier properties of several refractory metal thin films improved with the controlled introduction of oxygen or water vapor into the films during the evaporation procedure. Holloway postulated that the improvement in barrier properties was due to "capping" of the grain boundaries.

With sputter deposition, there generally exists a "shadow region" near the periphery of the deposition zone, where film darkening can be observed. This darkening is associated with a "gettering" phenomenon, wherein gaseous trace impurities in the vacuum system are chemically combining with the deposition material through "reactive sputtering." Thus, residence time within this zone can determine the degree of impurity incorporation in a resultant film, i.e., greater residence time results in greater impurity content for substrates moving through the "shadow region", as in an in-line or multipass deposition mode.

This "gettering" effect may actually improve the barrier properties of refractory metals deposited with significant time in the shadow region, similar to the improvement reported by Dalal [9] for thermally-evaporated films. Although the titanium/platinum/gold metallization has been reported [10,11] to be metallurgically stable on beam lead devices, Tisone and Cruzan [12], and later Kanamori [13] reported failure of the titanium as a diffusion barrier to platinum at ca. 360° C. It is surmised as highly probable that this non-consistency of barrier properties may be the result of variable incorporation of nitrogen and oxygen impurities in the titanium between depositions, or within the titanium deposition itself. It is a well-known phenomenon in electron-beam evaporation to observe an initial rise in pressure during the onset of evaporation of refractory metals, followed by a decrease in pressure as the evaporation proceeds. Thus, "gettering" of gaseous impurities from the internal components of the vacuum chamber, due to out-gassing, may result in the actual formation of titanium oxynitrides during deposition. The realization of this phenomenon may have encouraged early workers to begin additional studies geared toward the improvement of the barrier properties of titanium by the controlled introduction of impurities [14].

More-recently, it is becoming increasingly apparent that any study involving diffusion barriers in practice is incomplete without data concerning pre- and post-deposition treatment, i.e. such treatment can also affect barrier effectiveness. For example, Chang and Quintana [15] studied the out-diffusion of silicon through RF sputtered tungsten in the presence and absence of a "native oxide" layer on the silicon (they employed in situ sputter cleaning to remove the oxide). It was found that the silicon out-diffusion was definitely suppressed by the removal of the "native oxide." The suppression of silicon out-diffusion into an overlying aluminum metallization by removal of the "native oxide" was also observed by Learn and Nowicki [16].

The profound effect of post-deposition treatment on barrier properties was aptly demonstrated by Chang and Chu [17] in the platinum/nickel/platinum metallization on oxidized silicon. They reported that the out-diffusion rate of nickel through the platinum at 400° C could be greatly reduced by carrying out the heat treatment in an oxygen ambient. Similar studies by Holloway and Nelson [8] were reported earlier for the chrome/gold metallization on silicon. Chang and Chu also reported [18] a reduction in out-diffusion rate for chromium in the chromium/platinum metallization and for silicon in the silicon/platinum system. Morimoto, et al. [19] reported the fabrication of stable (to 950° C) molybdenum silicide contacts to doped silicon by arsenic implantation through the molybdenum followed by heat treatment in nitrogen at 1000° C. When the heat treatment was carried out in hydrogen, the resultant electrical characteristics of the device were severly degraded. Van Gurp, et al. [20], however, reported the failure of molybdenum silicide as a diffusion barrier between silicon and aluminum at 500° C under a hydrogen-nitrogen ambient. The resultant unacceptable shifts in the contact barrier height were produced by silicon out-diffusion and aluminum in-diffusion through the barrier. Thus, the literature provides much evidence that the barrier properties of refractory metals can be quite variable depending on deposition technique and pre- and post-deposition treatment. Hopefully, this variability will be addressed in the fabrication of the new VHSIC devices now being used in "Hi-Rel" applications [21].

EXPERIMENTAL PROCEDURES FOR EFFECTIVE DIFFUSION BARRIER PREPARATION

A. The substrates are cleaned in a manner so as to remove "native" oxides, etc. After loading the substrates into the deposition system, an in situ bakeout to ca. 250° C for ca. 10 min. is recommended. This should be done ideally in the load chamber rather than the deposition chamber. If the wafers cannot be heated under vacuum, they should be kept under laminar flow using a heated air source. After this preheat cycle, the wafers are transferred to the deposition chamber where they are given a mild sputter, or glow discharge clean if the devices allow such treatment. This clean should remove any residual oxide or monolayer-scale residues in contact areas, etc. This step usually results in a small burst of pressure due to removal of such residues. Therefore, the deposition chamber pressure should be monitored after this step. Ideally, the deposition should begin at base pressures less than 8X10^{-7} Torr [22]. Sputter deposition of the refractory metal diffusion barrier should be initiated in pure argon to enhance adhesion to the substrate. After ca. 50 nm of metal has been DC or Rf sputtered, the gas is switched to a mixture of ca. 25% nitrogen in argon. The plasma color will usually change from the characteristic violet to crimson under some conditions. Deposition of the nitrided barrier is continued until at least 100 nm have been deposited. Ideally, deposition is done under RF or DC bias to enhance step coverage of the barrier over steep topography [23]. After the nitrided metal has been sputtered, the gas mix is switched back to pure argon, and ca. 50 nm of the barrier metal is sputtered. Gas flow is then stopped, and the residual nitrogen is pumped from the deposition chamber (nitrogen can reduce film reflectance in aluminum alloys [22]). When the base pressure has reached the high-10^{-7} Torr range (HVAC pump "throttled"), deposition of the aluminum alloy can begin in pure argon to the desired thickness, preferably under RF or DC substrate bias. The nitrided barrier can usually be etched in the same etchant as the pure refractory metal or alloy [24]. As with any metallization, the intrinsic film stress should be optimized to minimal values. The deposition pressure vs. stress relationship [25] is a useful tool for such optimization.

B. Another frequently-used technique for improving barrier properties is to expose the deposited film to atmosphere [8,26]. This procedure probably oxidizes the grain boundaries (and the surface) of the refractory metal. One problem with this technique is that the adhesion of an overlying metal such as gold is usually marginal. However, this difficulty can be overcome by an in situ sputter etch to remove the surface oxide immediately prior to the sputter deposition of this overlayer. This procedure reduces some of the effectiveness of the barrier, but can be useful in single-source deposition systems.

SUMMARY

In light of the recent requirements for extremely highly-reliable thin film devices in aerospace and defense systems, it has become important to fully characterize and optimize these films for physical properties such as stress and barrier effectiveness. The deposition procedures outlined in this paper and those found in the references cited should allow the preparation of such films.

REFERENCES

1. M.-A. Nicolet, Thin Solid Films, 52 (1978) 415.
2. T.S. Liu, W.R. Rodrigues de Miranda and P.R. Zipperlin, Solid State Technol., 23 (1980) 71.
3. J.M. Harris, E. Lugujjo, S.U. Campisano, M.-A. Nicolet and R. Shima, J.Vac. Sci. Technol., 12 (1975) 524.
4. R.S. Nowicki, J.M. Harris, M.-A. Nicolet and I.V. Mitchell, Thin Solid Films, 53 (1978) 195.
5. C.R. Fuller, private communication.
6. R.S. Nowicki, W.D. Buckley, W.D. Mackintosh and I.V. Mitchell, J. Vac. Sci. Technol., 11 (1974) 675.
7. J.E. Baker, R.J. Blattner, S. Nadel, C.A. Evans, Jr. and R.S. Nowicki, Thin Solid Films, 69 (1980) 53.
8. P.H. Holloway and C.C. Nelson, Thin Solid Films, 35 (1976) L13.
9. U.S. Patent 4,215,156 (1980) to H.M. Dalal, M. Ghafghaichi, L.A. Kasprzak and H. Wimpfheimer.
10. M.P. Lepselter, Bell Syst. Tech. J., 45 (1966) 233.
11. J.H. Swafford, I.E.E.E. Trans. on Parts, Hybrids, Packag., PHP13 (1977) 208.
12. T.C. Tisone and P.D. Cruzan, J. Vac. Sci. Technol., 12 (1975) 677.
13. S. Kanamori, Thin Solid Films, 75 (1981) 19.
14. U.S. Patent 3,879,746 (1975) to P.R. Fournier.
15. C.C. Chang and G. Quintana, J. Electrochem. Spectr., 2 (1973) 363.
16. A.J. Learn and R.S. Nowicki, Appl. Phys. Lett. 35 (1979) 611.
17. C-A. Chang and W.K. Chu, J. Appl. Phys., 52 (1981) 512.
18. C-A. Chang and W.K. Chu, Appl. Phys. Lett. 37 (1980) 161.
19. M. Morimoto, E. Nagasawa, H. Okabayashi and M. Kondo, "An Mo Gate 4K Static Ram Fabricated Using a Novel Direct Contact Technology," presented at the International Electron Devices Meeting, Washington, D.C., December, 1981.
20. G.J. Van Gurp, J.L.C. Daams, A. van Oostrom, L.J.M. Augustus and Y. Tamminga, J. Appl. Phys., 50 (1979) 6915.
21. A. Oscilowski, Electron. Eng. Times, July 14, 1986, p. T6.
22. R.S. Nowicki, J. Vac. Sci. Technol., 17 (1980) 384.
23. R.S. Nowicki and B. Schiefelbein, Proc. 1985 Workshop: "Tungsten and Other Refractory Metals for VLSI Applications," Ed. R.S. Blewer (Materials Research Society, Pittsburgh, 1986) p. 341.
24. R.A.M. Wolters and A.J.M. Nellissen, Solid State Technol., 29(1) (January, 1986) 131.
25. D.W. Hoffman and J.A. Thornton, Thin Solid Films, 45 (1977) 387.
26. R.S. Nowicki and R.P. Henderson, original authors of article presented by B.E.P. Beeston and A.J. Aronson, Microel. Manuf. Test., 9, No.6 (1986) p.33.

APPENDIX

LIFETIME/INTERDIFFUSION CALCULATIONS

In practice, some degree of confidence with regard to diffusion barrier lifetime can be calculated with the Arrhenius equations:

$$Ea = k \ln \frac{\tau_2}{\tau_1} \left[\frac{1}{T_2} - \frac{1}{T_1} \right]^{-1}$$

Where,

Ea = energy of activation (in eV)

$k = 8.62 \times 10^{-5}$ (Boltzmann's)

τ_1 and τ_2 are times to visible degradation

T_1 and T_2 are temps. (°K) corresponding to times of visible degradation

Thus, if degradation is seen at 7 min. (τ_1) @ 325°C (T_1) and 1020 min (τ_2) @ 150°C (T_2),

$$Ea = 8.62 \times 10^{-5} \ln \frac{1020}{7} \left[\frac{1}{423} - \frac{1}{598} \right]^{-1}$$

$$Ea = 0.62 \text{ eV}$$

$$\tau = \text{Const. } e^{Ea/kt}$$

→If degradation is seen at 10^4 h @ 125°C,

$$(\text{for } Ea \sim 0.62),\ \tau(300°C) = 10^4\ e^{\frac{0.62}{8.62 \times 10^{-5}} \left[\frac{1}{573} - \frac{1}{398} \right]}$$

$$\tau\ (300°C) = 41 \text{ h.}$$

Re: F. H. Reynolds, Proc. IEEE 62, 212 (1974).

D. S. Peck and C. H. Zierdt, and others, in same issue.

DEVELOPMENT OF REFRACTORY METALS AND SILICIDES TARGETS, AND THEIR CHARACTERISTICS

Mutsumi Suzuki
Head Office: Nippon Mining Co., Ltd.
8-10, Toranomon 2-chome, Minato-ku, Tokyo, Japan

High purity refractory metals and silicide targets have been developed to meet a wide range of requirements for device fabrication processes. These targets have been evaluated by a number of users and satisfactory results were obtained. It is believed that sputtering will remain useful in the future, particularly as further improvements are made in sputtering equipment.

[1] Production Method of High Purity Refractory Metals and Silicides Targets

Nippon Mining (NMC) has three major methods which are used to produce sputtering targets for refractory metals and silicides. Chemical refining and an Electron Beam (E.B.) melting process are used for the production of Mo, W, Ti and Ta targets. A chemical refining and sintering processes is utilized to make the $MoSi_x$ and WSi_x targets. Finally, a special melting process has been developed to produce $TiSi_x$ targets [6].

(1) Mo Target

This target was developed for use in the manufacture of Mo gate electrodes which will be used to replace the current polysilicon or $MoSi_x$ gate electrodes in a polycide structure. For this application the target purity must be better than six-nines. Fe, Cr, Ni and other heavy metals must be removed from the target because these elements will cause adverse effects on voltage threshold, etching uniformity, etc. Examples of other "problem" elements, which must also be removed, are the alkali metals (ie: Na and K) which are sources of mobile charge and the radioactive elements (ie: U and Th) which can cause soft errors by emitting α-particles.

A new process for producing a high purity Mo target has been jointly developed by NMC and Nippon Telegraph and Telephone (NTT) [2], [3] (Fig. 1-1, 1-2). The principle method used to remove U is chemical refining while Na and K are removed by E.B. melting. Fe, Ni and Cr are removed from the target material by both chemical refining and E.B. melting.

Chemical refining of Mo is accomplished by crystallization of Molybdenum Hydrate (MoO_3 $N[H_2O]$) from a Molybdic acid solution (H_2MoO_4). Na, K and other light elements are completely eliminated by the E.B. melting process. This purification is accomplished under very high vacuum conditions (more than 10^{-4} torr) and high temperatures (more than 2,610 degrees C, the melting point of molybdenum). Residual impurities contained in the Mo target are as shown in Table I.

Raw Material (Powder)

↓ Chemical Refining

High Purity Metal

↓ Sintering

Sintered Compact

↓ E.B.Melting

E.B.Ingot

↓ Hot Working & Machining

High purity Target

Fig.1-1 Production Flow Chart Of Metal Targets

Molybdenum Powder (3N)

↓ Dissolution by HNO_3

H_2MoO_4

↓ Crystalization

$MoO_3 \cdot N[H_2O]$

↓ Drying

MoO_3

↓ Reduction by H_2

High purity Mo powder

Fig.1-2 Mo Chemical Refining

Table I

[Impurity Levels of NMC's Targets]

	Total Purity	U	Na	K	Fe	Ni	Cr	O
Mo	6N	<1	<10	<30	<0.01	<0.05	<0.01	30
W	6N	<1	<10	<30	<0.01	<0.05	<0.01	30
MoSix	5N	<1	<100	<100	<1	<1	<1	200
WSix	5N	<3	<100	<100	<1	<1	<1	200
TiSix	4N5	<1	<10	<30	30	10	<1	150
Ti	4N5	<1	<10	<30	10	10	2	600
W-Ti	5N	<3	<100	<100	2	1	<1	600
Ta	6N	<1	<10	<30	<0.01	<0.05	<0.01	60

Unit: U,Na,K ------- ppb

Fe,Ni,Cr,O --- ppm

(2) W Target

The application of the high purity W target to VLSI fabrication is similar to that of the Mo target. The requirements for the W target are also similar to those of the Mo target. The W target is produced by a process which is almost the same as that which is used to make the Mo target. The major difference is that the chemical refining is conducted by crystallization of WO_3 from an ammonium tungstate solution. It is very difficult to fabricate the W ingots into a sputtering target because of the hardness of pure W metal. Because of this NMC is also employing a sintering process to produce W targets with five-nine purity (Fig. 1-1, 1-3).

WO_3 Powder (3N)

↓ Dissoloution by NH_4OH

$(NH_4)_{10}W_{12}O_{41}$

↓ Add HNO_3 and Crystaization

$WO_3 \cdot N[H_2O]$

↓ Drying

WO_3

↓ Reduction by H_2

High purity W powder

Fig.1-3 Tungsten Chemical Refining

Raw Ta_2O_5 powder

↓ Dissolution by HF

H_2TaF_7

↓ Add KCl and Crystalization

K_2TaF_7 powder

↓ Reduction by Na

Ta powder

Fig.1-4 Tantalum Chemical Refining

(3) Ti Target

The Ti target is used for deposition of Ti metal on the gate of a salicide structure prior to silicidation by lamp annealing. Pure Ti is also used as the starting material of both the $TiSi_x$ and the Ti-W targets. The Ti target is produced by chemical refining and application of the E.B. melting process. Ti can be purified by various methods. Distillation of $TiCl_4$, thermal decomposition of Ti Iodide, and the fused-salt electrolysis methods are all satisfactory for this purification process. NMC has attained Ti with Fe less than 5 ppm and Oxygen less than 300 ppm by employing fused-salt electrolysis method. For manufacturing the Ti target, NMC employs a cold press process to produce the starting ingots for the E.B. melting without the use of a sintering process.

(4) Ta Target

Atsugi Electrical Communication Laboratories (A.E.C.L.) of NTT has stated that Ta_2O_5 thin films can be utilized as a high dielectric material for use in thin film capacitors if the Ta target is sufficiently pure. By analyzing the characteristics of leakage currents observed in thin film capacitors made using Ta_2O_5, NTT concluded that these leakage currents were caused by impurities in the thin film. After reducing alkali metal and heavy metal contaminants, the final objective is to eliminate other residual refractory metals, such as Mo and W, because the oxides of these metals are electrically conductive. The process used to produce these ultra high purity Ta targets is a combination of chemical refining and E.B. melting which was jointly developed by NTT and NMC Central Research Laboratories (NMC C.R.L.). In the chemical refining of Ta, crystallization of Potassium Fluorotantalate (K_2TaF_7) from Fluoro-Tantalum-Acid (H_2TaF_7) solution is made by adding Potassium Chloride (KCl) to the solution. Ta ingots are easily fabricated by rolling at room temperature, in contrast with W [5] (Fig. 1-1, 1-4).

(5) WSix and MoSix Targets

These targets were developed for use on the gate electrode of VLSI polycide structures such as the 256KB DRAM. Because of the application, films deposited from these targets must be of very high purity and must have very low U, Th, Na and K concentrations. In addition, a uniform molar ratio of Si to Mo or W in target is essential. For producing MoSix and WSix targets, a sintering method is general applied. Previously it has been difficult to obtain sufficiently high purity, high density and well sintered targets in the general market. High performance silicide targets were developed by NMC by introducing a special reaction and sintering method of silicide formation (Table 1). The deviation from the designed molar ratio of target is within 0.2 (about 1%) (Fig. 1-5).

(6) Ti-W Target

NMC has also developed a special sintering process to produce Ti-W targets for application as a diffusion barrier in contact holes (Fig. 1-6). The use of high purity Ti and W powders as starting materials has allowed NMC to successfully manufacture high purity targets as shown in Table 1.

High purity Si Block
↓
Crushing
↓
Milling
↓
High purity Si powder   High Purity Mo or W powder
↓→→→→→→↓←←←←←↓
↓
High Purity Silicide
↓
Crushing
↓
Sintering
↓
High purity Silicide Block
↓
Machining
↓
High purity Silicide Target

Fig.1-5 Production Flow of MoSix and WSix

High Purity W powder   High purity Ti powder
↓→→→→→↓←←←←↓
↓
Mixing
↓
Sintering
↓
Ti-W Ingot
↓
Machining
↓
Ti-W Target

Fig. 1-6 Production Flow of Ti-W

(7) TiSix Target

The typical application of $TiSi_x$ is as a gate electrode or interconnect material for VLSI with 0.8μm or 0.5μm design rules because its resistivity can be as low as 13$\mu\Omega$cm to 15$\mu\Omega$cm. NMC has developed a new special melting process to produce $TiSi_x$ targets in contrast to the older sintering method.

[II] Evaluation of NMC's Targets and Some Comments on Their Applications

NMC's targets have been widely evaluated and used by various laboratories and plants in Japan. The following are some excerpts from the papers written by leading researchers and comments on these targets.

(1) Mo and W Targets

Although the melting points of Mo and W are as high as 2,610 degrees C and 3,387 degrees C, respectively, these metals are easily oxidized and sublimate by oxidation to MoO_3 or WO_3. In order to avoid these unfavorable chemical changes, some countermeasures must be taken for VLSI production processes. One of the more notable countermeasures for the Mo gate process is a two-step deposition developed by Drs. H. Oikawa and T. Amazawa of A.E.C.L., NTT [1-A]. During the sputtering process, Mo will penetrate into SiO_2 if the substrate is heated to a temperature higher than 200 degrees C. To prevent this undesired penetration into the oxide, the temperature of the substrate must be held at room temperature at the start of sputtering. After 30 Angstroms of Mo have been deposited, sputtering can be continued while the temperature of the substrate is raised. NTT also claims that they can coat the Mo gate with a thin MoO_2 film using the interfacial oxidation method [1-B]. With regard to the W gate process, the Wet-Hydrogen Process was developed by Drs. S. Iwata and N. Yamamoto of the Central Research Laboratories, Hitachi, Limited [1-C]. This process is used to the regrow SiO_2 films which are damaged during an ion implantation.

High quality thin films can be obtained by sputtering in a clean chamber, obtaining a low base pressure in the chamber, control of residual gases (particularly atmospheric species), and appropriate heating of the substrate (higher than 450 degrees C in case of Mo). Dr. H. Oikawa was able to produce thin films with resistivities as low as 6.3$\mu\Omega$cm, a value which is close to the bulk value obtained using high purity Mo target and load lock sputtering system. The characteristics of the Mo gate structures produced by sputtering NMC's high purity target are suitable for practical applications. However, while the resistivity of the Mo films are more than twice that of Al films, they may still be applicable as a VLSI interconnect because the films are free from hillock and electromigration problems [3].

The characteristics of W gate structures made by sputtering NMC's high purity target combined with a proprietary annealing technique proved to be comparable to those of traditional polySi gates as was demonstrated by Dr. N. Yamamoto, et al. of Hitachi, Limited and Mr. H. Miyazaki, et al. of NMC. [7]. One of the practical applications which should be noted is that the best results were obtained when shields, made of materials having good adhesion to Mo or W, were placed around targets and wafers to prevent the generation of Mo or W flakes which might drop onto the wafers.

(2) MoSix and WSix Targets

The range of permissible sputtering process parameters for the deposition of WSix are narrower than those for pure metals, i.e., control of a suitable Ar pressure, sputtering power, temperature of the substrates, and the distance between wafer and the target. For WSix films, those films with to high a molar ratio of Si to W of films (Si rich) will blister where as those with to low a molar ratio will peel off from the wafer. Therefore, the molar ratio of target must be carefully controlled to obtain the most suitable ratio in the deposited films. The resistivity of the MoSi 2.2 and WSi 2.4 films produced from NMC's targets are approx. 75$\mu\Omega$cm and 60$\mu\Omega$cm respectively. These MoSix and WSix targets have already been successfully applied to the production of devices manufactured using 2μm to 1μm design rules. High purity and low O_2 are necessary to produce films with low sheet resistivity. The effect of these improvements is widely accepted, however, the mechanisms of how film resistivities vary with sputtering conditions is not yet clear.

Most of the parameters associated with sputtering conditions and the target production processes needed to decrease the generation of particles have been established. However, the details of these mechanisms have yet to be studied. Dr. S. Sawada, et al. of NMC contend that the main cause for the generation of particles is as follows. Voids in the target or an uneven density in the target combined with the difference in sputtering yield for W and Si generate uneven sputtered surface on the target. In time, this uneven surface turns into small conical protrusions in the target. These protrusions burst, possibly by their own thermal energy or by an arc discharge in the plasma. When the protrusions burst, the result is undesirable particles which have more than 0.5μm diameters. One countermeasure to this problem is sufficient sintering of the target. This process has been implemented by NMC in order to obtain a higher density target and to tighten the coupling among particles in the target. This additional sintering results in a more planar sputtering surface on the target.

(3) Ta Target

Using a high purity Ta target jointly developed by NTT and NMC, the leakage current of Ta_2O_5 films grown by reactive sputtering was drastically lowered as expected [5]. This result proves that the NMC Ta target is applicable to devices which are much more highly integrated than the 4M DRAM, having planar capacitors and without trenches. For these applications, the technology used to grow high quality oxide films developed by A.E.C.L. of NTT is also essential.

[III] Closing

(1) As minimum line widths become narrower, from 0.8μm to 0.5μm, the deposition of uniform films onto an uneven surfaces becomes even more difficult. So, to extend the application of sputtering, an improvement in step coverage is essential. To this end, numerous studies are being conducted, including; controlling the direction of bundles of sputtered particles, etching overhangs at the corner of steps by bias sputtering, changing the configuration of targets, etc. The results of these studies seem to be favorable at the present.

Sputtering, as compared to CVD, has advantage of high productivity, a clean deposition system, easy of operation, and a lower cost. With the development of new targets and improvements in tne capabilities of sputtering systems, sputtering will continue to be main production method in the age of more integrated VLSI using 0.8μm and 0.5μm design rules.

(2) The successful development of NMC's targets is credited to the devoted efforts of following scientists. Mr. N. Mori of the NMC Head Office whose main contribution was to develop the E.B. melting process. Messrs. H. Miyazaki and I. Kyono et al. of NMC C.R.L. who developed the chemical refining methods of Mo, W, Ti, and Ta metals. Dr. S. Sawada of NMC Isohara Plant who developed the sintering method for the WSix, MoSix, Ti-W, W targets etc. and also developed the special melting process for TiSix. Mr. Y. Nakamura et al. of NMC C.R.L. whose main contribution was to develop analysis method for low level impurities in targets. Future new targets will be also developed by these scientists.

ACKNOWLEDGEMENT

The author wishes to thank Drs. H. Oikawa of NTT, N. Yamamoto of Hitachi, Ltd. and N. Hosokawa of Anelva Corp. for their advice and encouragement.

REFERENCES

[1] Extended Abstracts of the 15th Conference on Solid State Devices and Materials, Tokyo, 1983

(A) T. Amazawa and H. Oikawa, pp. 229 - 232
(B) H. Kyuragi and H. Oikawa, pp. 237 - 240
(C) N. Yamamoto, S. Iwata, N. Kobayashi and T. Terada, pp. 217 - 220

[2] T. Amazawa, H. Oikawa, N. Shiono and N. Honma, Extended Abstracts of the 16th (1984 International) Conference on Solid State Devices and Materials, Kobe, 1984, pp. 269 - 272

[3] H. Oikawa and T. Amazawa, Proceedings of Third International Symposium on VLSI science and technology, Vol. 85 - 5 (Electro-Chemical Society, Toronto, Canada, May 12 - 17, 1985)

[4] N. Mori and S. Kambara, Proceedings a Symposium Held at AMAX Materials Research Center "Physical Metallurgy and Technology of Molybdenum and its alloy", pp. 53 - 57 (1985)

[5] C. Hashimoto, H. Oikawa and N. Honma, Extended Abstracts of the 18th (1986 International) Conference on Solid State Devices and Materials, Tokyo, 1986, pp. 253 - 256

[6] H. Miyazaki, I. Kyono and S. Sawada, JST Reports, Vol. 2, No. 3, Autumn 1986, pp. 30 - 38

[7] N. Yamamoto, H. Kume, S. Iwata, K. Yagi, N. Kobayashi, N. Mori and H. Miyazaki, J. Electrochem. Soc., Vol. 133, No. 2, Feb. 1986, pp. 401 - 407

Reactive Ion Etching of Refractory Metal Films

Reactive Ion Etching of Refractory Metals for Gate and Interconnect Applications

R. J. Saia and B. Gorowitz
General Electric Corporate Research and Development Center, P.O. Box 8, Schenectady New York, 12301.

ABSTRACT

Refractory metals have been extensively investigated for use as gates or as interconnects, primarily due to their low resistivity and high temperature stability [1-7]. Recent advances in CVD tungsten technology have also provided the means to produce metal plugs in contacts and deep vias. This paper describes reactive ion etch (RIE) processes that can be used for the fabrication of molybdenum (Mo) gates, molybdenum/titanium-tungsten (Mo/TiW) interconnects, and CVD tungsten (W) plugs. Etching was done in a batch loaded parallel plate RIE system. The CVD tungsten plug etchback process uses photoresist planarization for both selective and non-selective tungsten, and a plasma chemistry that etches photoresist and tungsten at the same rate. An analysis of plasma etch chemistry , using the results of emission spectroscopy, is presented.

EXPERIMENTAL

An Anelva Model 503 reactive ion etch system was used for the Mo, Mo/TiW, and W plug etching described here. This parallel plate system is manually loaded and has a capability of etching six 100 mm diameter wafers per batch. RF power, at a frequency of 13.56 MHz, is provided to the lower 42 cm diameter water cooled electrode. The wafers being etched are placed in holes in a plate (carbon or quartz) which covers this electrode. The system is equipped with a cryo-pumped high vacuum system capable of reducing background pressures to the 10E-6 torr range between runs.

Mo and Mo/TiW films were deposited primarily by dc planar magnetron sputtering in a MRC Model 603 system. For comparison purposes, some molybdenum was also deposited in a Varian 3140 S-Gun system. The selective [8,9] and non-selective [10] CVD W deposition systems are described elsewhere. A thin (500 A) layer of sputtered Mo was used as an adhesion layer for the non-selective CVD W [10].

An optical emission spectrometer was used to monitor the plasma through a quartz window in the RIE chamber, and a laser interferometer was used to determine etching endpoints of metals and photoresist over silicon dioxide.

MOLYBDENUM GATES

Using chlorine based gases such as CCl_4 [11-13] or Cl_2 [14], mixed with high percentages of oxygen, it is possible to etch molybdenum anisotropically with selectivities to silicon dioxide as high as 80:1. The main products of etching with these mixtures are molybdenum oxychlorides ($MoOCl_3$ and $MoOCl_4$). Figure 1 shows that when oxygen is added to CCl_4, the relative intensity of Cl atom signals (256 nm) increases. Percentages of oxygen as high as 80% are typically used. This plasma chemistry etches Mo with high selectivity to a number of films including silicon dioxide, silicon nitride, silicon, tungsten, titanium silicide, and aluminum. These materials ordinarily etch to some extent in chlorinated plasmas, but they form a passivating oxide surface layer under the conditions described here. The etch rate of Mo increases with pressure while those of photoresist and silicon dioxide decrease (see Figure 2). As a result, the pressure used for this process is typically in the 200 - 250 mtorr range. The etch rate of Mo was found to be affected by the type of metal deposition system that was used. DC magnetron sputtered Mo etched at a slower rate than S-Gun sputtered Mo (see Figure 3). Secondary- ion-mass spectroscopy (SIMS) depth profiles showed that the amount of incorporated oxygen present in the Mo film sputtered using an S-Gun was higher than that deposited in the magnetron system (see Figure 4). It is believed that the added oxygen enhances the formation of the volatile molybdenum oxychloride etch products. An SEM photo of an etched submicron Mo gate is shown in Figure 5.

MOLYBDENUM / TITANIUM TUNGSTEN INTERCONNECTS

Molybdenum interconnects (5000 A thick) with a thin (800 A) underlying layer of TiW (30 atomic % Ti) have been shown to produce good ohmic contact with n+ silicon, p+ silicon, and polysilicon. They also have excellent

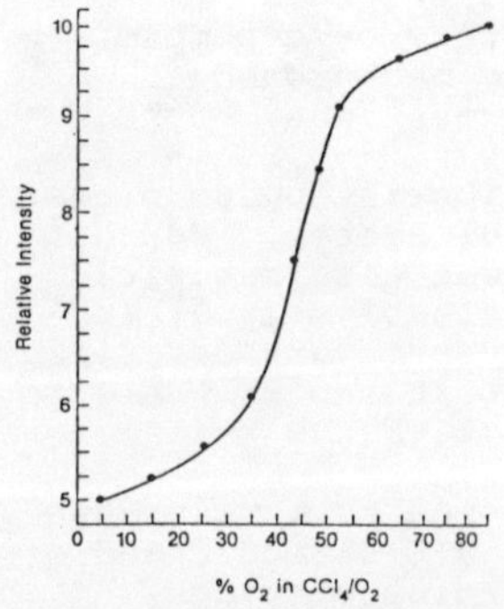

Figure 1. Relative intensity of Cl atom signals (256 nm) vs. percentage of O_2 in CCl_4/O_2, monitored by emission spectroscopy.

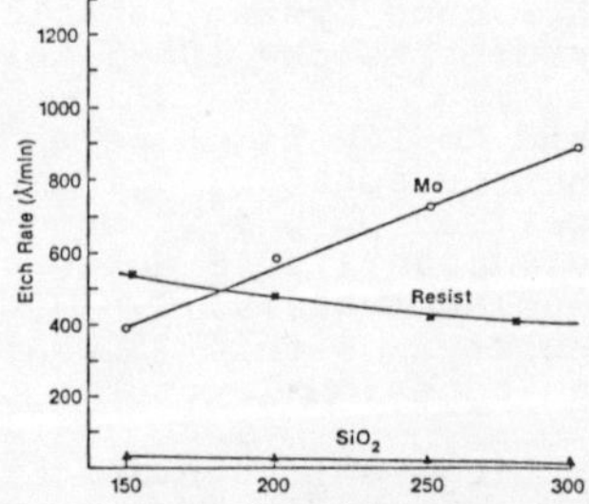

Figure 2. The etch rate of Mo photoresist and silicon dioxide vs. pressure. Etching parameters: flow - 100 sccm, 75% O_2 in CCl_4/O_2; power - 300 W (0.25 W/cm2); quartz cathode cover.

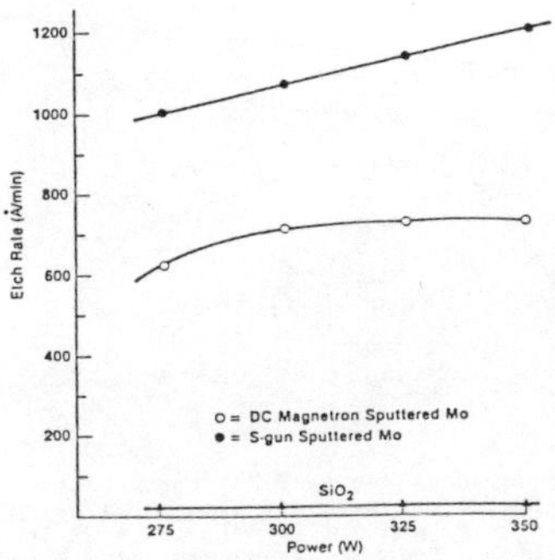

Figure 3. Etch rate of Mo films from two different sources vs. rf power. Conditions are the same as in Fig. 2, at 250 mtorr.

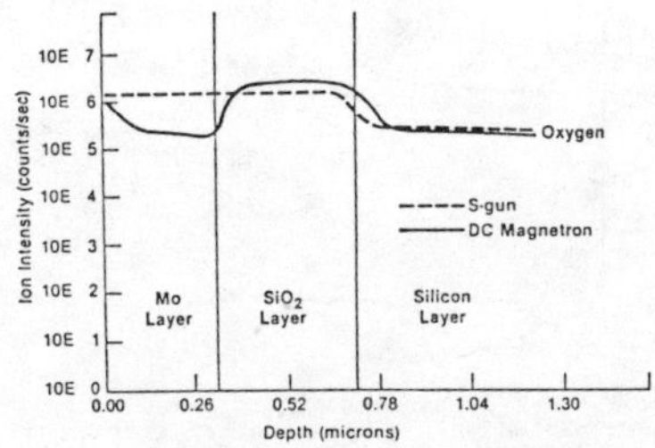

Figure 4. Secondary-ion-mass spectrometry depth profile of oxygen in Mo-film samples deposited from two different sources.

thermal stability at temperatures up to 650 C, and they do not suffer from the spiking and electromigration problems associated with aluminum [1,15]. One step reactive ion etching of this bi-layer metalization can be performed using a chloro/fluoro- gas mixture such as SF6/Cl2. As Cl2 is added to a SF6/Cl2 mixture, the etch rates of Mo and TiW first increase, then decrease. Optimum conditions were obtained at 10% Cl2 / 90% SF6 (see Figure 6). Linewidth losses were also reduced by the addition of Cl2 to SF6, and by using lower power (see Figure 7). The addition of Cl2 to SF6 was found to reduce the emission signals of atomic F in the discharge, presumably by the formation of FCl [16]. The reduction in the generation of free F, or a sidewall passivation film of FCl compounds, may explain the lower linewidth loss. The etch rate of Mo and the selectivity of Mo to photoresist were found to increase by almost a factor of two by adding oxygen to SF6/Cl2. Figure 8 shows that the relative intensity of F atom signals also increases with O2 addition, and then decreases. Above 30% oxygen, the

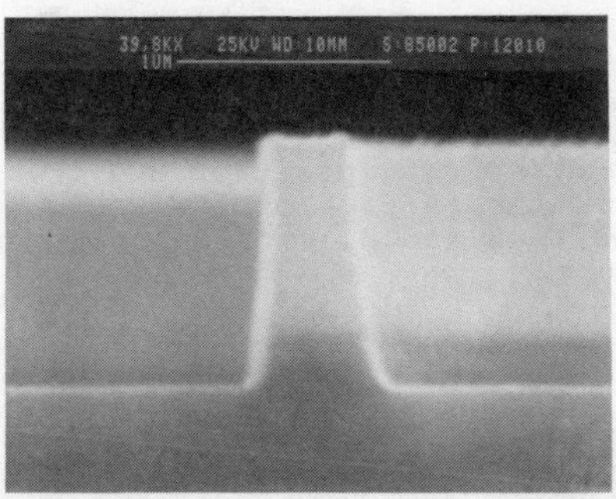

Figure 5. SEM photo of a Mo gate etched in CCl4/O2. Photoresist has not been removed.

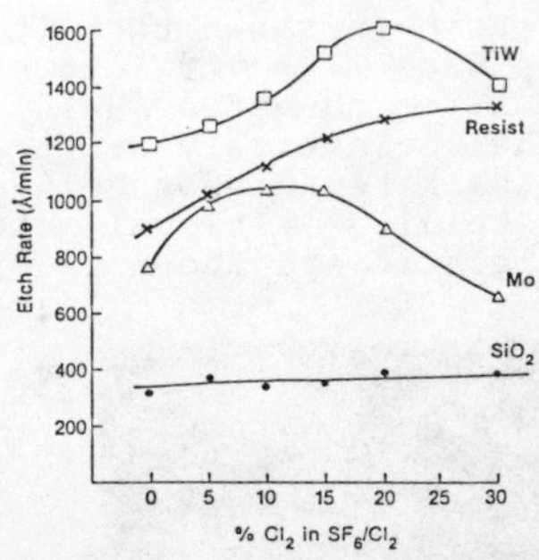

Figure 6. The etch rates of Mo, TiW, photoresist, and silicon dioxide vs. % Cl2 in SF6/Cl2. Etching parameters: flow - 30 sccm; pressure - 65 mtorr; power - 250 W (0.21 W/cm2); quartz cathode cover.

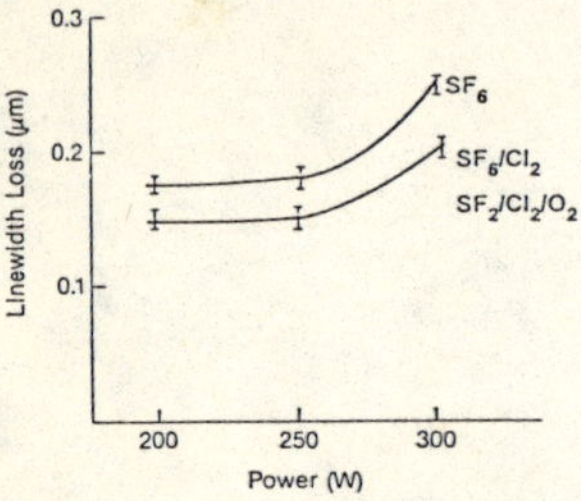

Figure 7. Mo/TiW Linewidth loss measured from SEM photos vs. rf power after etching in SF6, SF6/Cl2, and SF6/Cl2/O2.

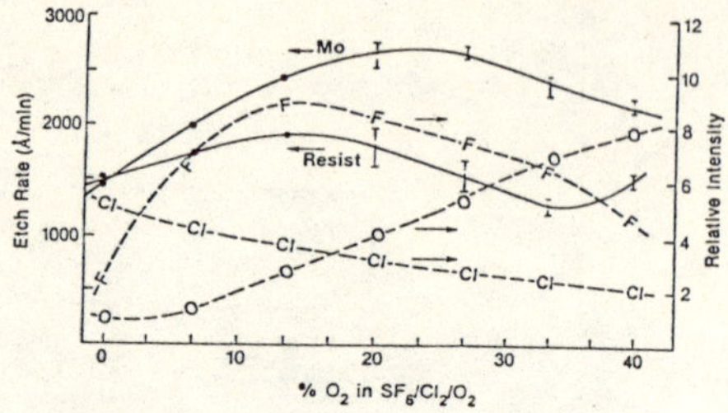

Figure 8. Etch rates of Mo, photoresist, and the relative emission signals of F (704 nm), Cl2 (256 nm), and O (777 nm) vs. % O2 in SF6/Cl2/O2. Etching parameters: flow - 60 sccm, ratio of Cl2:SF6 = 0.17; pressure - 65 mtorr; power - 200 W (0.16 W/cm2); quartz cathode cover.

photoresist etch rate increases excessively. Optimum selectivity of Mo to photoresist is obtained with 25-30% oxygen added. An SEM photo of an etched Mo/TiW line is shown in Figure 9.

CVD TUNGSTEN PLUG ETCHBACK

In multilevel metal structures, step coverage of the films overlying the first level of metal can be improved by planarizing the topography. This is sometimes accomplished by coating the dielectric films with photoresist, baking the resist to level it, and then etching the dielectric and the resist simultaneously. After this planarization, the depth of vias on the same chip may vary by as much as one micron. Metal step coverage in such deep holes can be poor with standard sputtering techniques and can lead to discontinuities or poor reliability. A typical method of improving step coverage is to taper the via sidewalls during etch. The diameters of vias of different depths on the same chip however are difficult to control using a tapered sidewall process.

It has been shown that it is possible to completely fill deep vias with straight sidewalls, using chemical vapor deposited tungsten [8-10]. The tungsten can be deposited either selectively or in blanket form. Planarization etchback techniques using both types of deposition and RIE to clear residual metal from the surface of the dielectric are shown in Figures 10 A and B. In the batch

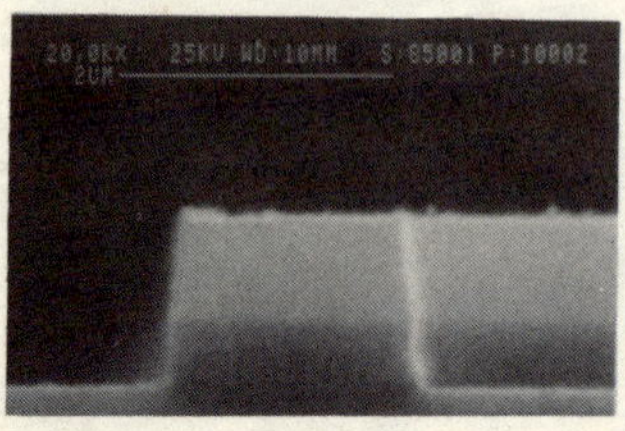

Figure 9. SEM photo of a Mo/TiW line etched in SF6/Cl2/O2. The photoresist has not been removed.

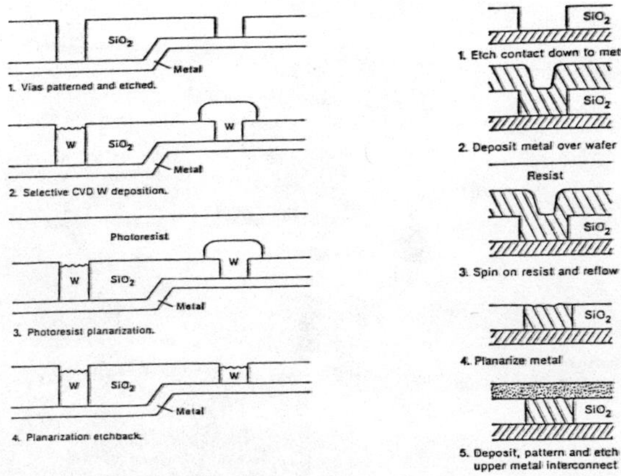

Figure 10. (A-left) Sequence of forming tungsten plugs in vias of varing depths using selective CVD W deposition and photoresist etchback. (B-right) Similar sequence using blanket CVD W.

loaded reactive ion etcher, a 1:1 etch rate of W and of the planarizing photoresist layer can be obtained by varying the pressure using CCl2F2/O2 (see Figure 11). Because of the large consumable area provided by a carbon cathode cover in the RIE system, the etch rate uniformity in this process is +- 2.2% across a 100 mm wafer. Trenching at the periphery of vias filled with non-selectively deposited W was first seen using the CCl2F2/O2 process (see Figure 12A). It was first thought that this was due to a slightly faster etch rate of the thin (500 A) Mo adhesion layer that was used. The cause however, was later traced to the non-selective W deposition process, since the same trenching was observed when the Mo adhesion layer was not used. It is believed that the first 1000 A of W contained residual oxygen, since the CVD reactor was not loadlocked. After the addition of a pre-processing step using a WF6 purge at low temperatures to remove moisture from the deposition chamber, trenching after W etchback was no longer observed (see Figure 12B).

Laser interferometry was used to monitor the etching of photoresist during selective W "nailhead" etchback, or of resist and W using blanket metal deposition. The photoresist planarization and plasma etchback technique is much better suited for use with selective CVD W deposition. In this case, since the area of W deposited is very small (2-5% of the surface area of the wafer), photoresist planarization makes up for non-uniformities in the deposited film thickness. Also, if the deposition is not completely selective, the etchback can clean up excess W from the dielectric surface.

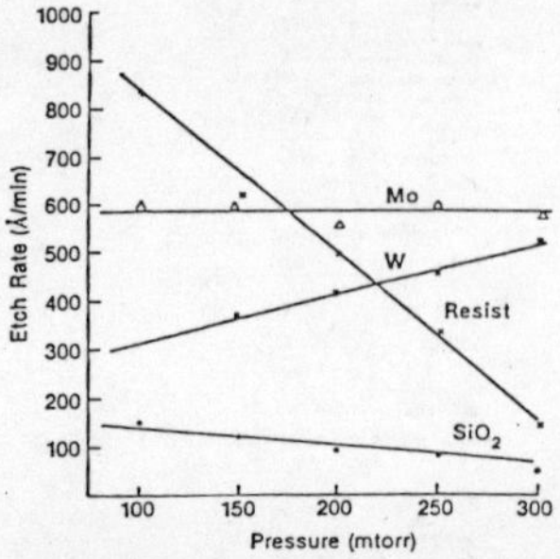

Figure 11. Etch rate of W, Mo, SiO2, and photoresist vs. pressure. Etching parameters: flow - 60 sccm, 75% O2 in CCl2F2/O2; power - 300 W (0.25 W/cm2); carbon cathode cover in a reactive ion etcher.

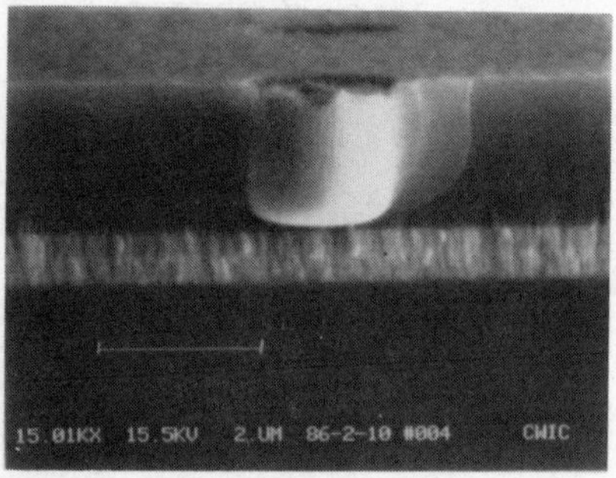

Figure 12. (A-left) SEM photo showing voids present along the sidewalls of a via filled with non-selective CVD W, after planarization / etchback.
(B-right) SEM photo showing a via filled with W, without voids along the sidewalls. Planarization / etchback conditions were identical to those used in (A). An oxygen gettering step was added before film deposition in the non-selective CVD W process.

SUMMARY / CONCLUSIONS

Etching of Mo gates and Mo/TiW interconnects can be performed using CCl4/O2 and SF6/Cl2/O2. The selectivity of Mo to silicon dioxide is much higher using chlorine and oxygen containing gases than when using fluorine based gas mixtures. The addition of Cl2 to SF6 was found to reduce Mo linewidth loss while the addition of O2 to this mixture increased the selectivity of Mo to photoresist by a factor of two.

Vias between the first and second levels of metal in multilevel metal systems can be filled with plugs using non-selective and selective CVD W followed by photoresist planarization and plasma etchback. With photoresist planarization, nucleation sites or nailheads associated with non-uniform selective CVD W can be eliminated. The implementation of a moisture absorbing step before non-selective W deposition eliminated sidewall trenching.

ACKNOWLEDGEMENTS

The authors wish to thank Jim Schermerhorn, Kevin Durocher, and Len Douglas for their assistance in dry etching, Jim Norton for providing the SEM photos, and Ron Wilson, Bob Stoll, and Dave Woodruff for tungsten deposition.

REFERENCES

1) M. J. Kim, D. M. Brown, S. S. Cohen, P. A. Piacente, and B. Gorowitz, IEEE Tans. Elect. Devices, ED-32, No. 7, 1328 (1985).

2) H. J. Geipel, Jr., N. Hsieh, M. H. Ishag, C. W. Koburger, and F. R. White, IEEE Trans. Electron Devices ED-27, 1417 (1980).

3) P. L. Shaw, IEEE Trans. Electron Devices ED-26, 631 (1979).

4) B. L. Crowder and S/ Zirinsky, IEEE Trans. Electron Devices ED-26, 369 (1980).

5) T. P. Chow and A. J. Steckl, Appl. Phys. Lett. 36, 297 (1980).

6) M. Zhany, J. Z. Li, I. Adesida, and E. D. Wolf, J. Vac. Sci., Technol b, Vol. 1, Iss. 4, 1037 (1983).

7) T. P. Chow and A. J. Steckl, IEEE Trans. Electron Devices ED-30, 1480 (1983).

8) R.H. Wilson, R.W. Stoll and M.A. Calacone, "Highly Selective, High Rate Tungsten Deposition", Proceedings of the Workshop on Tungsten and Other Refractory Metals for VLSI Applications, October 7-9, 1985 / Alburquerque, New Mexico, Edited by Robert S. Blewer, Materials Research Society, Pittsburgh, Pa., page 35, 1986.

9) R.W. Stoll and R.H. Wilson, The Electrochemical Society Meeting, Extended Abstracts, Vol. 86-2, paper 354, p. 531, San Diego, CA., Oct. 1986.

10) D.W. Woodruff, R.H. Wilson and R.A. Sanchez-Martinez, "Adhesion of Non-selective CVD Tungsten to Silicon Dioxide", Proceedings of the Workshop on Tungsten and Other Refractory Metals for VLSI Applications, October 7-9, 1985 / Alburquerque, New Mexico, Edited by Robert S. Blewer, Materials Research Society, Pittsburgh, Pa., page 173, 1986.

11) Y. Kurogi and K. Kaminura, Jpn. J. Of Appl. Phys., Vol. 21, No. 1, 168 (1982).

12) K. Hirata, Y. Ozaki, M. Oda, and M. Kimizuka, IEEE Trans. Electron Dev., ED-28, No. 1, 1323 (1981).

13) B. Gorowitz and R. J. Saia, General Electric TIS Report 82CED249 (1982).

14) T. Watanabe, U.S. Patent No. 4 478 678 (1984).

15) M.J. Kim, D.M.Brown, S.S Cohen, P.A. Piacente, and B. Gorowitz, IEEE Trans. Electron Devices. ED-32, No. 7, 1328 (1985).

16) D. Bhogeswara Rao and R. E. Gleason , The Electrochemical Society Meeting, Extended Abstracts, Vol. 86-2, paper 311, p. 466, San Diego, CA., Oct. 1986.

PLASMA ETCHING OF CVD TUNGSTEN FILMS FOR VLSI APPLICATIONS

Ching-Hwa Chen*, Laura C. Watson**, Donald W. Schlosser***
*Lam Research Corp., 47655 Warm Springs Blvd., Fremont, CA 94539
**Genus Corp., 515 Ellis St., Mt. View, CA 94043
***Integrated Device Technology, 3236 Scott Blvd., Santa Clara, CA 95052

ABSTRACT

Etching of tungsten composite films was investigated in single wafer planar plasma etchers using SF_6 plasmas. Processes were developed to etch tungsten for both interconnect and plug applications. Tungsten etching was found to be enhanced by ion bombardment and flourine atom concentration. Good anisotropy, one micron per minute etch rates, and good uniformity were obtained in the Lam AutoEtch 790 system. Addition of CHF_3 was found to adequately passivate sidewalls during interconnect etching so that undercutting of the composite film was eliminated.

INTRODUCTION

CVD tungsten has become an increasingly popular material for use in solving some of the challenges presented by today's high density, small geometry device structures. In particular, CVD tungsten solves the problem of filling high aspect ratio contact and via structures because of its superior step coverage. The goal of this work was to develop a tungsten etch process exhibiting good anisotropy, fast etch rates, and good uniformity, to be used in both plug etch back and interconnect technologies.

One problem with CVD tungsten films historically has been the uniform etching of the films. CVD blanket tungsten is a composite film which requires an adhesion layer for good adhesion to oxides. Identifying a process that etches the adhesion and tungsten layers at the same rates has been difficult. In this study, the Genus model 8402 reactor was used to deposit the CVD blanket tungsten films. The films are a composite of a tungsten silicide adhesion layer, a bulk tungsten layer, and a tungsten silicide cap which improves chemical resistance and promotes photoresist adhesion.

It is known that tungsten can be etched using flourine containing gases such as CF_4 and SF_6 plasmas.[1-6] It is believed that tungsten etching is primarily controlled by flourine atoms generated in the discharge of these plasmas. The addition of O_2 to CF_4 increases the tungsten etch rate considerably and exhibits a maximum at about 30% O_2 addition. On the other hand, the tungsten etch rate gradually decreases with increasing O_2 percentages in SF_6 plasmas. In this work, the etching characteristics of tungsten in SF_6 plasmas were investigated in a range of conditions relevant to plasma processing.

EXPERIMENTAL

Wafers were <100> silicon deposited with 5000 angstroms of thermally grown oxide followed by 10,000 angstroms of Genus blanket tungsten. The wafers were patterned with a metal test mask using positive photoresist.

The Genus blanket tungsten used in the study is a composite film. Because tungsten does not adhere readily to oxide surfaces, the surface is prepared with an argon plasma etch followed by a hydrogen plasma etch. A 1200 angstrom thick layer of tungsten silicide is deposited as a 'glue' layer for the tungsten. The bulk tungsten layer is then deposited by reduction of WF_6 by SiH_4 and H_2. A 200 angstrom tungsten silicide "cap" is deposited as a final step to prevent oxidation of the wafers as they are removed from the process chamber which is maintained at 450°C.

The significance of the composite tungsten film structure lies in the different and difficult-to-control etch rates of the silicide and tungsten layers. In the past, researchers have found that overetching of the silicide layer is difficult to avoid because of the lack of a well-defined etch endpoint for the bulk tungsten and the high etch rate of tungsten silicide. Developing a process that etches both materials in a 1:1 ratio is especially important to the success of the plug etch-back technology; overetching the silicide will leave a void between the tungsten plug and the walls of the contact structure.

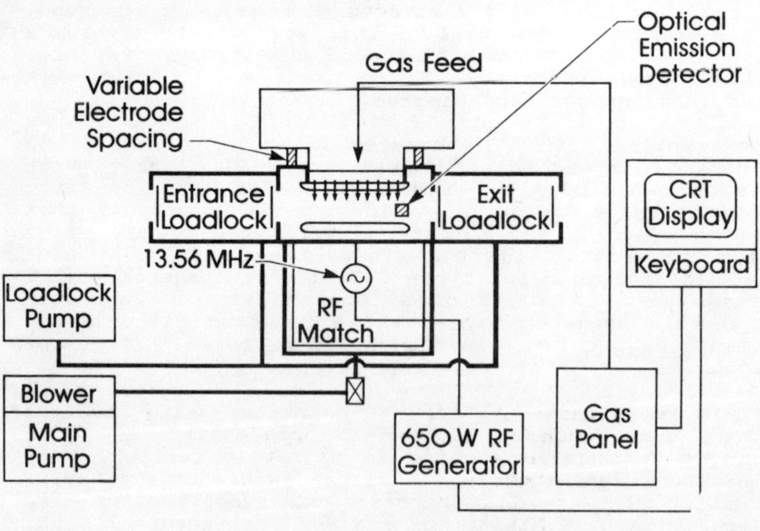

Figure 1. LAM Research Autoetch 790

The etch study was conducted on two systems, the Lam Research AutoEtch 490 and 790, both of which are parallel-plate, single wafer diode reactors. (See figure 1.) The 490 applies power to the top electrode and the 790 applies power to the bottom electrode. We use the terminology of plasma etcher (PE) for the 490 and reactive ion etcher (RIE) for the 790. Both units are equipped with entry and exit vacuum loadlocks which permit the process chamber to remain at low pressure, free from contamination from the environment. Both systems apply power at 13.56 MHz. An adjustable electrode gap allows flexibility in process development; the chemistry and electrode spacing can be matched to provide optimum uniformity across the wafer. Process recipes, gas flows, wafer transport, gap, and endpoint monitoring are microprocessor-controlled and can be programmed for multistep in-situ etching processes.

RESULTS AND DISCUSSION

SF_6 chemistries were used to etch the blanket tungsten films. Figure 2 shows that the tungsten etch rate increases as partial pressure of SF6 is increased. Figure 3 compares the tungsten etch rate at different power levels on both systems. On both systems the tungsten etch rate increases with power and percentage of O_2 in the SF_6 plasma. The RIE configuration was found to be more efficient because the tungsten surface is bombarded by more energetic ions, resulting in a much faster etch rate. These results indicate that the tungsten etch is enhanced by ion bombardment and increased fluorine concentration. Addition of O_2, up to 80% partial pressure, to SF_6 dramatically increases the etch rate. O_2 is known to increase the F and SOF_2 concentrations, etc. The dominate compound formed in the presence of O_2 is WOF_4; in the absence of O_2, it is WF_6.[7] It was found that one micron per minute etch rates can easily be attained in the single wafer diode reactive ion etcher.

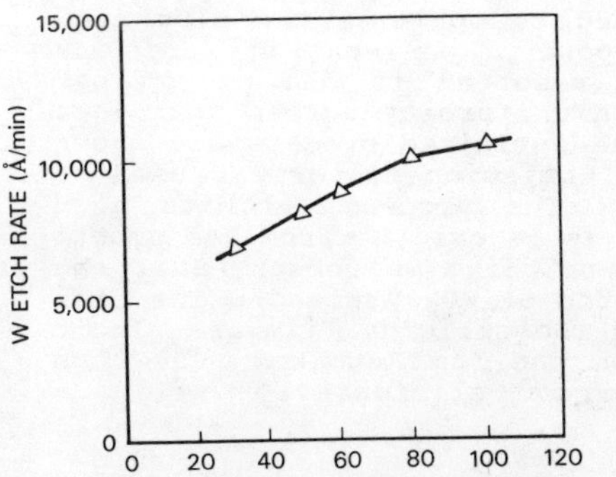

Figure 2. Tungsten etch rate vs. SF_6 pressure in SF_6/He plasma

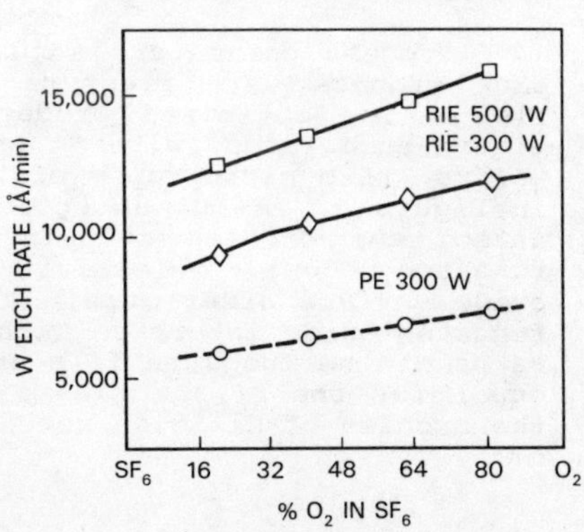

Figure 3. Tungsten etch rate vs. O_2 concentration in SF_6 plasma

Figures 4 and 5 show that the tungsten etch rate is not strongly dependent on pressure, gap spacing or changes in DC bias. The etch rate of the oxide substrate was found to decrease dramatically with increasing pressure, due to the drop in DC bias. This suggests that the role of ion energy for the tungsten etch is not the same as that involved in oxide etching. Etch rate uniformity was found to be a strong function of gap spacing. Adjustment of the gap spacing significantly affects gas flow pattern and plasma confinement. The Lam AutoEtch 790 offers flexible gap adjustment which allows for process uniformity optimization.

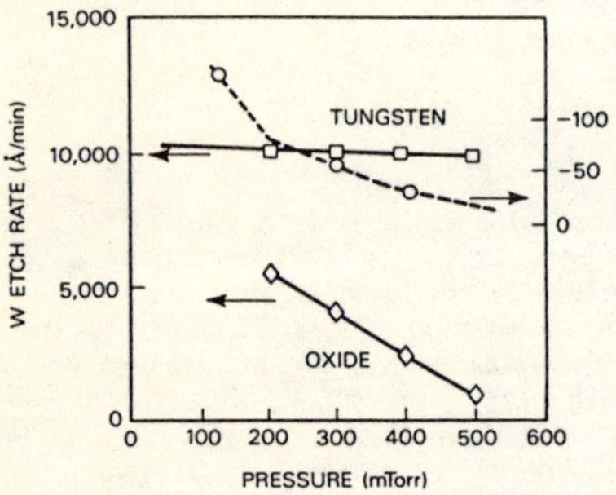

Figure 4. Tungsten and oxide substrate etch rates vs. pressure in $SF_6/He/O_2$ plasma

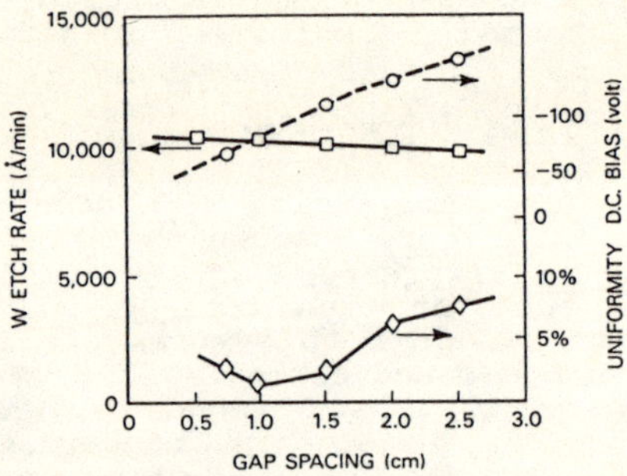

Figure 5. Tungsten etch rate uniformity vs. gap spacing in $SF_6/He/O_2$ plasma

PLUG ETCH BACK PROCESS

Process parameters adjusted to optimize the plug etch back process were pressure, power, gap spacing, and gas flow. The optimized process resulted in one microm per minute etch rates, with 5% etch uniformity across four inch wafers. SEM micrographs of the optimized process are shown in Figure 6. A 703 nm optical emission filter was used to detect the process endpoint. The process exhibits good tungsten to oxide selectivity, as is evident from the smooth oxide surface after etch. No notching was observed at the tungsten-oxide interface with the SF_6/O_2 process. The etch rates of the tungsten film and the silicide film are almost equal for the SF_6/O_2 plasma in the Lam AutoEtch 790, thus the problem of silicide notching was eliminated.

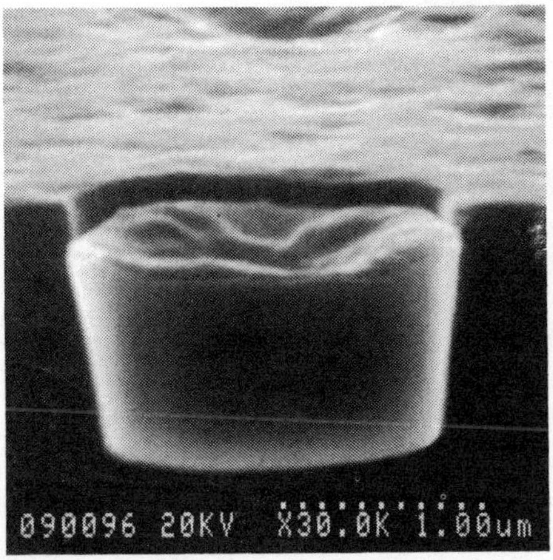

Figure 6. Tungsten plug etch back with SF_6/O_2 plasma

Figure 7. Anisotropic W/WSix etch can be obtained with sidewall passivation

INTERCONNECT ETCH PROCESS

To achieve anisotropic etching, the vertical etch rate of the film must far exceed the lateral etch rate. The lateral etch rate can be suppressed by a number of mechanisms, each of which is specific to the film being etched and the chemistry of the plasma. It is known that the tungsten etch rate is enhanced by ion bombardment and that good anisotropy may be obtained without a resist mask using flourine chemistry. However, intense ion bombardment severely damages photoresist. Sidewall passivation by either resist erosion or addition of another passivant is necessary to obtain vertical sidewalls.

SF_6/He chemistry was studied on both the plasma and reactive ion configured etchers for anisotropic tungsten applications. In the plasma etcher, severe undercutting was observed at both the tungsten and silicide layers. Using the reactive ion etch configuration, anisotropic etching of the tungsten was achieved, but the silicide layer was severely undercut. Sidewall passivation by resist erosion was not sufficient to protect the silicide sidewall from undercutting in the flourine environment. To retain anisotropy of the silicide layer, an additional passivation source, CHF_3, was used. CHF_3 generates an additional carbon source for sidewall passivation. The tungsten etch rate decreased with the addition of CHF_3. Figure 7 shows the anisotropic profile of the tungsten composite films etched by SF_6/CHF_3/He plasmas. Figure 8 shows that etch rates of resist and tungsten increase with power density. Figure 9 shows the etch rate of tungsten in SF_6/CHF_3 mixtures in both the plasma and reactive ion etch systems. The addition of CHF_3 slows etching in the plasma system because the system does not have much polymer sputtering capability. The tungsten etch rate in the reactive ion etch system is less affected by polymer build up due to its ion bombardment capability.

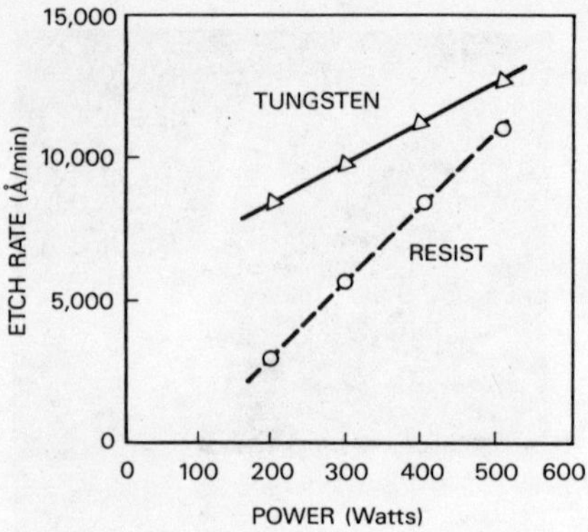

Figure 8. Tungsten and resist etch rates at different power levels in SF_6 plasma

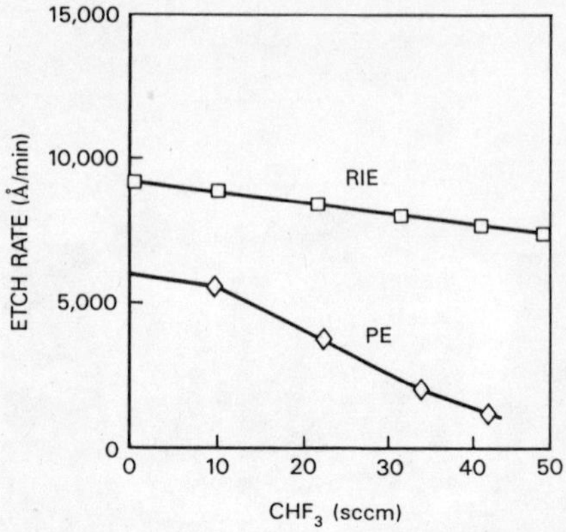

Figure 9. Tungsten etch rate vs. CHF_3 in SF_6 plasma

CONCLUSION

In this study plasma etching of the composite blanket tungsten films was demonstrated for both interconnect and plug applications. The etch system used to successfully etch the film was the Lam AutoEtch 790, a reactive ion etch system with variable gap spacing. The tungsten etch rate was found to be enhanced by ion bombardment and higher concentrations of fluorine. Tungsten plug etch back can be performed in SF_6/O_2 environment without notch formation. Anisotropic interconnect etching of composite tungsten over adhesive silicide can be achieved through sidewall blocking passivation mechanisms.

ACKNOWLEDGEMENTS

We wish to thank S. DeOrnellas for helpful discussion and revisions of this paper. We would also like to thank J. Tucker, T. Yep, W. Rust, and N. Zetterquist for supporting this work. L. Reed's, J. Collins' and T. Summan's help on the experimental work and SEM operation was greatly appreciated.

REFERENCES

1. K. Maeda and K. Fujino, Denki Kagaku, 43, 22, 1975.
2. R. Kumar, C. Ladas, and G. Hudson, Solid State Technology, 19(10) 54, 1976.
3. C. Tang and D.W. Hess, J. Electrochem. Soc. 131, 115, 1984.
4. H.F. Winters, J. Vac Sci. Technol., A3, 700, 1985.
5. S. Adachi and N. Susa, ibid 132, 2981, 1985.
6. J.N. Randall and J.C. Wolfe, J. APPL Phys. Lett. 39, 742, 1981.
7. G. Turban and A. Picard, ISPC-7, 1073, 1985.

REACTIVE ION ETCHING OF SUBMICROMETER SIZE FEATURES IN TUNGSTEN THIN FILMS

TIMOTHY J. WHETTEN, NELLIE Y. WHETTEN AND EDWARD D. WOLF
National Research and Resource Facility for Submicron Structures and School of Electrical Engineering
Knight Laboratory, Cornell University, Ithaca, NY 14853

ABSTRACT

The ability to etch tungsten thin films anisotropically is critical for high density tungsten interconnect fabrication. We have used contrast enhanced photolithography to define 0.6-1.0 micrometer wide 1.3 micrometer high lines with vertical walls in photoresist. An anisotropic, high aspect ratio, reactive ion etching process is described in which the lines are transferred into the tungsten. The dependences of source gas, masking material, tungsten grain size, pressure and power on etching anisotropy, etched surface roughness, and etching rate are discussed. The final etch profile will be described in terms of masking material, sidewall erosion, grain size and etching anisotropy.

INTRODUCTION

Tungsten is an attractive material for use in metallization applications for VLSI because of its excellent high temperature properties, electromigration resistance, and low resistivity. As an interconnect material, tungsten will have applications where electromigration is a concern, where there will be subsequent high temperature processing (>600°C), or in other situations where aluminum alloys are not suitable. Tungsten may play an important role in self aligned silicide (SALICIDE) structures. It is an excellent x-ray absorber, and as such is being used in x-ray mask applications.

High resolution patterning of tungsten thin films is required for all of these applications. High aspect ratio structures (where the film thickness is at least the same order of magnitude as the lateral dimensions) are often required in order to obtain acceptable interconnect resistivity and small lateral dimensions simultaneously. Reactive ion etching has been demonstrated to etch 200 nm wide structures of high aspect ratio in tungsten. [1,2] Because of the combination of directional ion bombardment and chemical reaction between the plasma species and the tungsten surface, good selectivity can be obtained between the tungsten film and both the masking material and the underlying substrate by an appropriate choice of etching parameters and gasses.

Optical lithography has long been the workhorse of the microelectronics industry. Although electron, ion beam and x-ray lithography have yielded excellent results for submicrometer geometries, low throughput, expensive capital equipment investment and various other problems continue to preclude their extensive use in production. With the use of a contrast enhancement layer on conventional photoresist, optical lithography can probably satisfy the submicrometer requirements of VLSI production for the short term future (without any additional capital investment!).

PROCESS

Contrast enhancement material (CEM) is a photo-bleachable dye which is initially opaque. When an aerial image of a mask is incident on the CEM, the regions of highest light intensity are bleached out faster than the lower intensity, dark areas of the mask. It is possible to completely expose the underlying photoresist in the light areas before the CEM is bleached through in the dark areas (see Figure 1). Therefore, the CEM effectively forms an in situ contact mask when exposed. The contrast level of the aerial image is enhanced, thereby improving the resolution and sidewall verticality of conventional photoresist.

We have used General Electric's Altilith Contrast Enhancement Material CEM-420 and its associated products with Shipley's Microposit S1400 series resist. Positive photoresist is spin-coated on HMDS-primed wafers, and then soft-baked according to standard process (90 ^{o}C, 30 min. in a convection oven). A barrier layer of polyvinyl alcohol (GE's CEM-BC5) is then spun on to prevent the formation of an interfacial layer between the CEM and photoresist. CEM-420 is spun on and this bilevel system is exposed using a Canon FPA 141 (4:1 reduction) projection aligner at an energy level of 144 mJ/cm^2. The CEM is stripped with CEM-SA15 (85% toluene, 15% anisole). The photoresist is then developed according to standard procedure; we used Shipley's Microposit 351 Developer, diluted 1:5 with deionized water.

The photoresist was then used directly as a mask for the reactive ion etching process. Alternatively, 60 nm of chrome was evaporated onto the photoresist/tungsten surface as a liftoff mask. The photoresist was then stripped, leaving chrome where tungsten was exposed. The chrome was then used as the etching mask.

Reactive ion etching of the tungsten films was performed in a modified Applied Materials parallel plate diode reactor with 13.56 MHz excitation and a plate spacing of 7 cm. CF_4, O_2 and SF_6 were used as the source gases is this set of experiments.

Our requirements for this etching process were (in order of importance) to obtain (1) very little, or no, undercut of the masking material, (2) high selectivity between the etching of tungsten and both the masking material (photoresist or Cr) and the underlying substrate (SiO_2 or Si), and (3) as high an etching rate as possible without sacrificing the first two conditions.

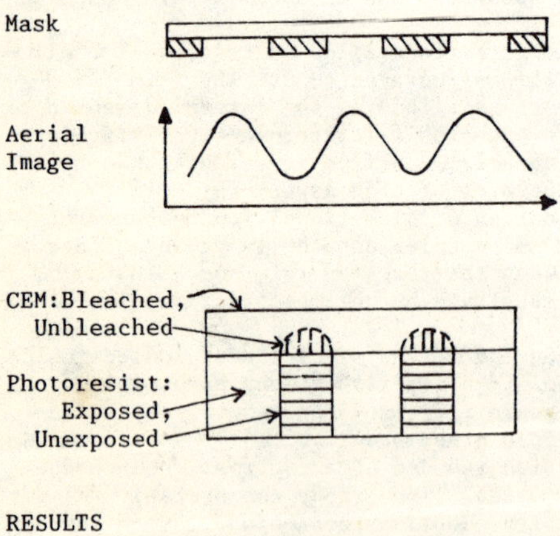

Figure 1. Formation of an in situ contact mask in the CEM layer due to the bleaching of the CEM by the incident aerial image.

RESULTS

An SEM micrograph of photoresist patterned with CEM is shown in Figure 2. Besides the obvious improvement in resolution and verticality of the sidewalls over conventional photoresist processing, the use of CEM also increases process latitude by increasing the depth of focus. As an example, for equal 1 um lines and spaces the CEM process has been shown to produce

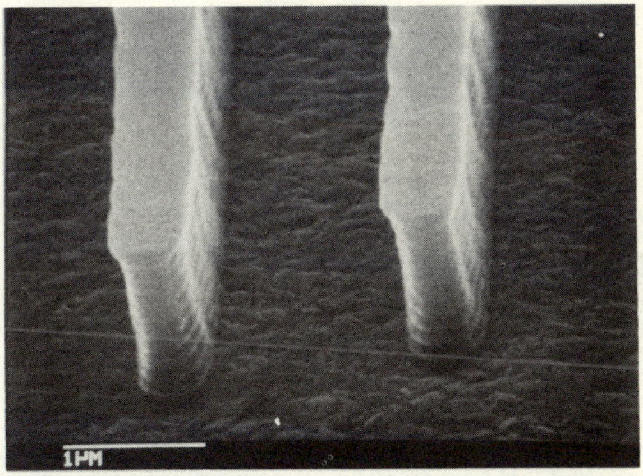

Figure 2. An SEM image of photoresist on tungsten. The photoresist was exposed with CEM.

essentially identical profiles in the resist at ± 3.0 μm of best focus [3]. This is almost a doubling in depth of focus. As mentioned earlier, the CEM process requires no additional capital investment and adds only 3 spin steps to the standard photolithography process.

The CEM process gives the best results with a Na-based developer. It still does enhance contrast with a metal ion free (MIF) developer but not quite to the dramatic degree that it does with a Na-based developer.

The tungsten films described in this investigation were sputter deposited in a CVC RF sputter deposition system. The grain size was determined to be 30 nm by x-ray diffraction. The grains are highly oriented about the [100] direction and have a columnar structure. Both the alpha and the metastable beta phases were observed. The resistivity of the films varied from 20-200 X 10^{-6} ohm-cm depending on the amount of beta phase present. Although the films were etched without any heat treatment, annealing the films to transform the beta phase to alpha phase was observed to reduce the resistivity.

Etching with either pure SF_6 or SF_6-O_2 mixtures gave unacceptable etching profiles. Large mask undercuts were often obtained and the sidewall smoothness was poorer than that for the CF_4-O_2 mixtures. This was probably due to the large amount of free fluorine in SF_6 plasmas. For CF_4 and CF_4-O_2 mixtures, it was generally found that the best etching profiles were obtained at low pressures (15 mtorr) and high power (0.25-0.5 W/cm^2). Adding oxygen to pure CF_4 improved the etching profile slightly.

The dependence of etching rate of W, Si and SiO_2 on oxygen content in the etching source gas is shown in Figure 3 for 15 mtorr total pressure. A maximum in the selectivity of W etching over both Si and SiO_2 is obtained at 29% O_2 flow. The flow rate of CF_4 was kept constant at 30 sccm. These results are consistent with previous work [4,5]. The maximum in the etching rate of tungsten and silicon has been observed to be at concentrations close to that of the maximum in the free F concentration.

The etching profile of 600 nm thick tungsten, etched with CF_4-29% O_2 at 0.25 W/cm^2 is shown in Figure 4. Chrome was used as the etching mask and SiO_2 is the underlying substrate (now exposed in the etched regions of the

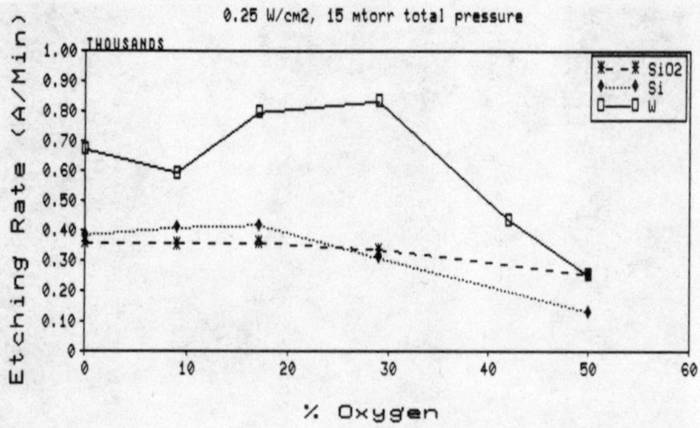

Figure 3a. Etching rate as a function of oxygen content in the source gas. The flow of CF_4 was held constant at 30 sccm and the pressure was constant at 15 mtorr.

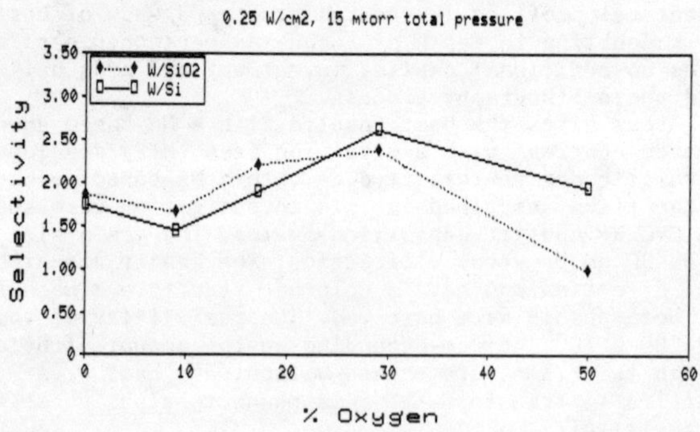

Figure 3b. Ratio of the etching rate of tungsten to Si and SiO_2. The etching parameters were the same as those for Figure 3a.

film). There is no observable undercut of the chrome mask and the wall profile is 15 degrees from vertical. We tentatively attribute the roughness of the sidewall to preferential chemical etching along the grain boundaries. Very large grain CVD tungsten exhibited roughness on the order of the grain size as well. We have etched (110) single crystal tungsten and have observed no such sidewall roughness.

Photoresist can be used as an etching mask although it etches much faster than the tungsten, limiting the thickness of the tungsten film that can be etched. Figures 5 and 6 show etching profiles of tungsten films etched with a photoresist mask. In obtaining the structure shown in Figure 5, no oxygen was used in the etching gas. The vertical etching rate of the photoresist was 100 nm/minute and there was a loss of 16 nm/minute from each

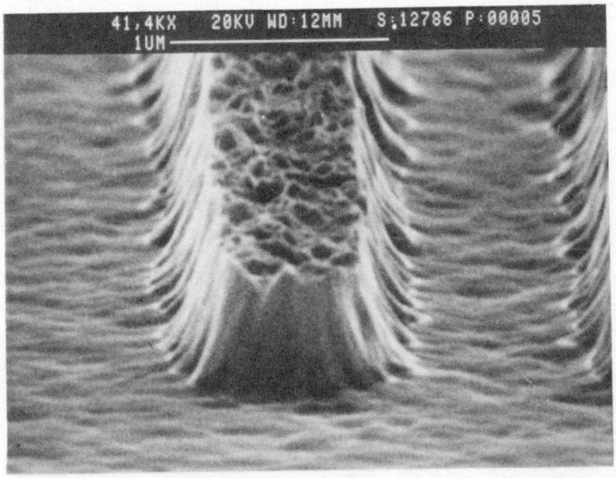

Figure 4. An SEM image of tungsten etched with a chrome mask. The film was etched at 0.25 W/cm^2, 15 mtorr and 29% O_2.

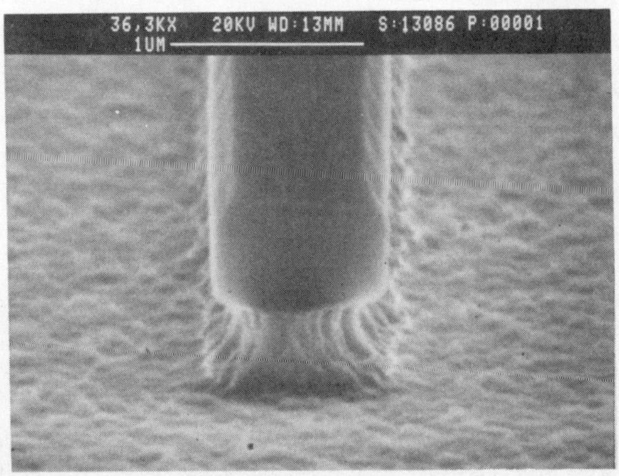

Figure 5. An SEM image of tungsten etched with a photoresist mask. The film was etched at 0.25 W/cm^2, 15 mtorr and no O_2.

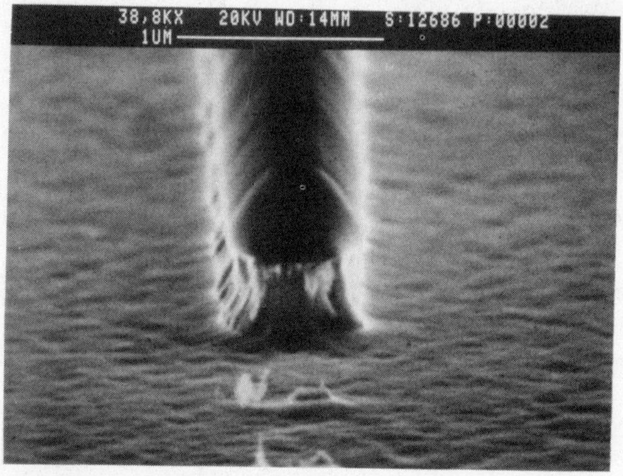

Figure 6. An SEM image of tungsten etched with a photoresist mask. The film was etched at 0.25 W/cm^2, 15 mtorr and 29% O_2.

sidewall. Slightly undercut sidewall profiles were obtained with roughness of the same order of magnitude as with a chrome mask. The etching rate of the tungsten was 30% less than that with a chrome mask (and no O_2), due to either the consumption of etching gas species such as F radicals or sputtering of carbon from the photoresist onto the tungsten surface during etching.

Figure 6 shows the structure obtained when a photoresist mask is used with the same etching parameters as in Figure 4 with a chrome mask (29% O_2). The walls are nearly vertical although there is some undercut and the etching rate of the tungsten was reduced by the presence of the photoresist, this time by 25%. In this case the photoresist etched more isotropically. The original photoresist line was 1.5 micrometers wide and 1.5 micrometers tall with vertical walls. The final resist profile is a trapezoid 0.5 micrometers tall and 0.5 micrometers wide at the base. Thus the resist was etched 0.5 micrometers from each side while it was etched 1.0 micrometer vertically.

The (110) single crystal tungsten was observed to etch at a rate 1/2 that of sputter deposited films and CVD films. Work by Winters [6] and more recently by Bensaoula et. al. [7] suggests that damage induced by ion bombardment increases the reaction probability of tungsten from a very small value to almost unity. The grain boundaries in the sputter deposited and CVD films may provide similar sites for adsorbtion and reaction. Preferential adsorbtion and/or reaction of plasma species at the grain boundaries would also explain the sidewall roughness observed and the development of surface roughness of the film during the etching process. In addition, there may be a crystallographic dependence on the etching rate.

CONCLUSIONS

We have demonstrated the ability to fabricate submicrometer dimension high aspect ratio structures using optical lithography for pattern definition and reactive ion etching for pattern transfer. Contrast enhanced lithography using CEM allows us to generate vertical wall structures in photoresist which were then used either as a liftoff mask or directly as a mask for reactive ion etching. We are investigating potential methods for stabilizing the photoresist so that the excellent resist profile that is obtained using CEM is maintained during the etching process. One possibility is to deep UV cure the resist after development. Another possible method is to use an aqueous surface treatment to thermally stabilize the resist [8].

If a metal mask is required for RIE of the underlying W, the same CEM process that produced the RIE etching mask can also be used for the lift-off process. Alternatively, the barrier layer can be omitted to encourage intermixing of the CEM and the photoresist to form an undercut profile, ideal for lift-off of a metal mask.

There is room for further work to understand the mechanisms for sidewall roughening and to improve both the sidewall profile and amount of undercut. Improved sidewall characteristics might be achieved by changing the tungsten deposition technique to obtain a grain structure of small, more nearly equiaxed, grains instead of the long columnar grains observed in these films, as well as further improvements in the etching techniques.

ACKNOWLEDGMENTS

This work was performed in part at the National Research and Resource Facility for Submicron Structures which is supported in part by the National Science Foundation under Grant ECS-8200312. Support was provided in part by SRC through Cornell Microscience and Technology. The authors would like to thank Dr. Robert Blewer at Sandia National Laboratories, Dr. Lynn Rathbun and Richard Bjoyko for their assistance with deposition and analysis of the tungsten thin films and General Electric Corporation for CEM material.

REFERENCES

1. John N. Randall and J.C. Wolfe, Appl. Phys. Lett. 39, 742 (1981).

2. M.L. Schattenburg, I. Plotnik and Henry I. Smith, J. Vac. Sci. Technol. B. 3, 272 (1985).

3. B.F. Griffing and P.R. West, General Electric CRD Report No. 82CRD252, 1982.

4. C.C. Tang and D.W. Hess, J. Electrochem. Soc., 131, 115 (1984).

5. C.J. Mogab, A.C. Adams and D.L. Flamm, J. Appl. Phys., 49, 3796 (1978).

6. Harold F. Winters, J. Vac. Sci. Technol. A, 3, 700 (1985).

7. A. Bensaoula, J.A. Strozier, Jr., J. Yu, A. Ignatiev and J. Wolfe, presented at the 33rd National Symposium of the American Vacuum Society, Baltimore, MD, 1986 (unpublished).

8. John J. Grunwald and Allen C. Spencer, "A novel aqueous surface treatment to thermally stabilize high resolution positive photoresist images," Advances in Resist Technology and Processing III, C. Grant Willson, editor, Proc. SPIE 631, 62 (1986).

REACTIVE ION ETCHING OF AN Al-Si/TiW MULTILAYER

N.TAKENAKA, D.KIMURA, K.NISHIZAWA, AND K.SAKIYAMA
SHARP Corp.Integrated Circuits Group, 2613-1 Ichinomoto,Tenri-City, Nara, 632, Japan

ABSTRACT

Titanium tungsten films are known to be useful as a contact barrier metal as well as being effective in suppressing metal hillocks in VLSI interconnection applications. The films used in this study consist of a sputtered aluminum-silicon layer on top of a sputtered titanium tungsten layer. The titanium tungsten film is reactive ion etched using a CCl_3F/O_2 gas mixture. The typical titanium tungsten film etch rate is 85nm/min with an etch selectivity to SiO_2 of 6:1. An anisotropic profile is achieved in etching an Al-Si/TiW multilayered film.

INTRODUCTION

One important aspect of progress in VLSI device fabrication is the need to shrink contact sizes. In order to decrease electrical resistance in smaller contacts, a multilayer metallization using a barrier metal has been effective. The several types of materials reported for this use[1][2], titanium tungsten is one of the most promising due to its good characteristics and easy in processing. Titanium tungsten can be etched in a CF_4 plasma or in a CCl_4/O_2 mixture[3]. However, the etch selectivity to SiO_2 is poor in a CF_4 plasma, and CCl_4 gas can be a health hazard. Furthermore, in the case of a multilayered structure, undercut is significant[4][5]. In this report a CCl_3F/O_2 gas mixture is used to obtain an anisotropic etch profile and good selectivity to SiO_2.

EXPERIMENTAL

The etching apparatus used was a conventional parallel plate, cathode coupled reactive ion etching system with a 13.56MHz RF power supply. The etching samples were prepared by continuous deposition of a 300nm thick TiW underlayer then a 600nm thick Al-Si (1%) top layer on thermal SiO_2. These samples were etched with a Shipley MP1400-27 resist mask. The etching conditions of the top Al-Si layer were held constant, using a BCl_3:CCl_3F:He (=13:3:2) gas mixture. Pressure was 15Pa and power density was 0.34W/cm^2.

RESULTS AND DISCUSSION

< 1 > Etching Characteristics Of A TiW Monolayer

Figure 1 shows the etch rate dependence on the oxygen partial pressure in a CCl_3F/O_2 gas mixture. Total gas flow rates for CCl_3F/O_2 are held at 50sccm. The SiO_2 etch rate weakly depends on the oxygen partial pressure in CCl_3F/O_2. On the other hand, the etch rate of TiW increases dramatically with an increasing O_2 percentage in CCl_3F/O_2. A maximum etch rate peak is seen at about 50% O_2. This etch rate is a four-fold increased over that of a pure CCl_3F gas.

Thus, the selectivity of TiW to SiO_2 is a maximum at 50% O_2 in CCl_3F/O_2.

Figure 2 shows the etch rate dependence on RF power density. Both TiW and SiO_2 etch rates increase monotonically with increasing power density. However, the selectivity of TiW to SiO_2 decreases with increasing power.

Figure 3 shows the etch rate dependence on gas pressure. The TiW etch rate remains nearly constant and the SiO_2 etch rate decreases slightly with increasing gas pressure.

The optimum TiW monolayer etching condition is then determined : 50% O_2 in CCl_3F/O_2, a pressure of 10Pa, and a power density of 0.27W/cm2.

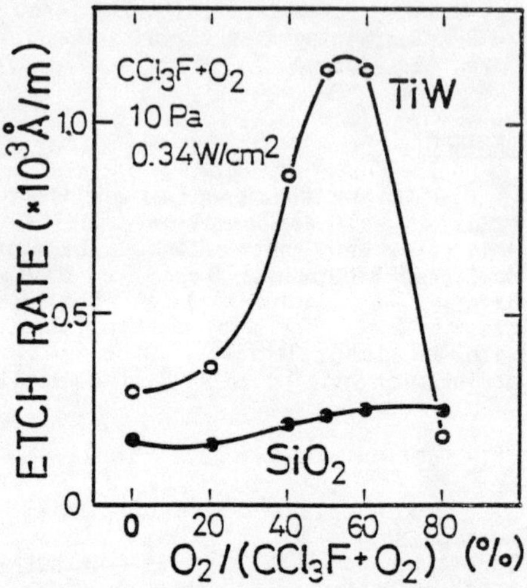

Figure 1. The etch rate dependence on the oxygen partial pressure in a CCl_3F/O_2 gas mixture.

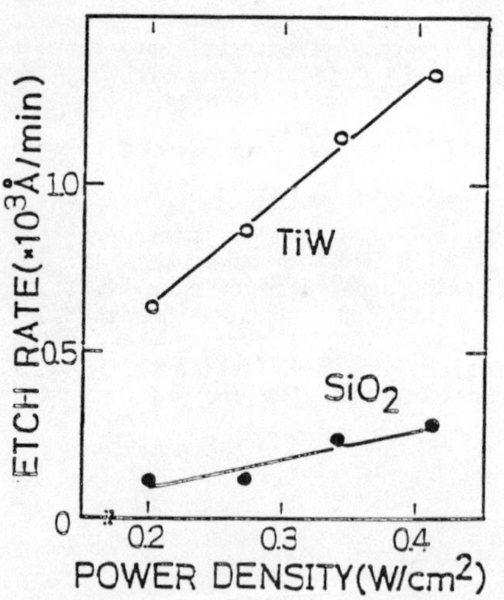

Figure 2. The etch rate dependence on RF power density. (CCl_3F/O_2= 25/25sccm; 10Pa)

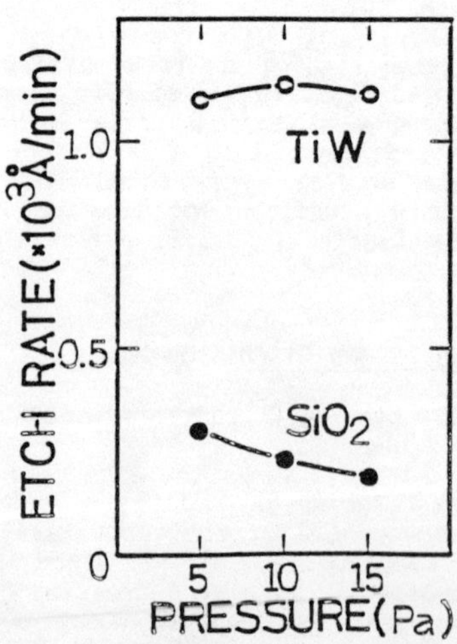

Figure 3. The etch rate dependence on gas pressure. (CCl_3F/O_2= 25/25sccm ; 0.34W/cm2)

< 2 > Etching Characteristics Of An Al-Si/TiW Multilayer

2-1) Undercut: An undercut of the underlayer in a multilayer structure such as Al-Si/TiW is a problem and can occur even if each film can be etched anisotropically when etched as a monolayer. When the Al-Si/TiW film is etched using $BCl_3/CCl_3F/He$ to etch Al-Si film then CCl_3F/O_2 to etch TiW film, a deposited film can be observed on the sidewall. Such deposited film may protect the sidewall and suppress the undercut. In order to investigate the formation mechanism of the deposited film, the following experiments were carried out. The etched profiles examined by SEM were the TiW monolayer, Al-Si monolayer, and Al-Si/TiW multilayer. The three etching conditions used for the three samples are listed with the results in table I. In the case of 2 step etching, the deposited film was observed for all etched materials.

etching condition / etched layer	Al-Si etching $BCl_3/CCl_3F/He$	TiW etching CCl_3F/O_2	2 step etch step 1:Al-Si etching step 2:TiW etching
TiW layer	——	X	O
Al-Si layer	X	——	O
Al-Si/TiW multi layer	X	——	O

Table I. Deposited film dependence on etched materials and gases
deposited film O: YES
× : NO

Because CCl_3F gas is common for all etching gas conditions and He gas is inert gas, BCl_3 and O_2 gases are considered to be related to the formation of sidewall deposition. The deposited film can be removed easily with O_2 plasma and ultrasonic cleaning.

2-2) Etching residue: The Al-Si/TiW multilayer 2 step etching causes etching residue, where each layer is etched using its standered etch conditions.
Residue observations are;

a) The residue is observed even after the 1st step (Al-Si etching).

b) The etching sample deposited with substrate heating has less density and larger size of residue islands than the one without substrate heating.

c) The etching sample with heat treatment (for example N_2 annealing) after deposition has less density and larger size of residue islands than the one without heat treatment after deposition.

d) The long overetching time for Al-Si films may reduce the density of the eching residue islands.

From above observations, Al-Si/TiW interface reaction it is hypothesized that a occured during heat cycle and causes products. We believe these products remain as the etching residue after 2nd step (TiW) etching.

The etched pattern profile is shown in Figure 4. The two step etching is carried out for the Al-Si/TiW multilayer etching. The 1st step is Al-Si etching using $BCl_3/CCl_3F/He$ gases, and the 2nd step is TiW etching using CCl_3F/O_2 gases. The anisotropic etching profile is attained for an Al-Si/TiW multilayer film.

Figure 4. The pattern profile after Al-Si /TiW etching. No resist mask.

CONCLUSIONS

By using a CCl_3F/O_2 gas mixture for titanium tungsten etching, an anisotropic profile is achived in etching an Al-Si/TiW multilayer film. A maximum etch rate is seen at about 50% O_2, and the etching selectivity of TiW to SiO_2 is maximum at the same O_2 gas concentration. When bottom layer TiW is etched using CCl_3F/O_2 after top layer Al-Si etching using $BCl_3/CCl_3F/He$, deposition occurs on the sidewall. We consider that BCl_3 and O_2 gases cause the formation of the sidewall deposition. The reaction between Al-Si and TiW interface causes residue. The residue decreases with prolonged Al-Si overetching.

REFERENCES

1. I.Suni, M.Biomberg, et al., J. Vac. Sci. Tech., A(3), Nov/Dec,2233 (1985)
2. R.A.Levy, M.L.Green, P.K.Gallagher et al., J. Electrochem. Soc., 133, 1905 (1986)
3. P.M.Schaible, G.C.Schmartz, J.Electrochem.Soc.,March 730 (1985)
4. M.Y.Tasai, H.H.Chao, et al., J.Electrochem.Soc., 128, 2207 (1981)
5. E.C.Whitcomb, A.B.Jones, Solid State Tech., April, 121 (1982)

Recent News Papers

THE HF PARTIAL PRESSURE AS A SELECTIVITY DETERMINANT DURING THE SELECTIVE DEPOSITION OF TUNGSTEN

L.F.Tz. Kwakman[1], W.J.C. Vermeulen[2], E.H.A. Granneman[1], M.L. Hitchman[3]

1) Advanced Semiconductor Materials Europe bv., Rembrandtlaan 2, Bilthoven, The Netherlands
2) Twente University of Technology, P.O Box 217, Enschede, The Netherlands
3) Univerity of Strathclyde, 295 Cathedral street, Glasgow, Great Britain

ABSTRACT

A simple gasflow model is used to calculate the HF partial pressure during W-deposition inside a conventional horizontal, hot-wall LPCVD reactor, both in between the wafers and in the annular zone between wafers and wall.When comparing experimental results with this model, it can be shown that selectivity loss on an oxide surface is directly related to the local HF concentration: at the onset of selectivity loss, tungsten nuclei are formed on the oxide surface of which the number per unit area is related to the local HF concentration.
Furthermore, a kinetic model for the tungsten nucleation on SiO_2 surfaces is presented.

INTRODUCTION

Because of its excellent electrical and thermal properties, Tungsten is considered to be a promising candidate material for contact and diffusion barriers, low resistance gates and interconnect lines in VLSI circuits.The present W-LPCVD process uses WF_6 and H_2 as source gases.In the absence of hydrogen, when WF_6 is reduced by the silicon substrate, the process is completely selective, but only thin layers (100-1000 Å) can be formed, due to the self limiting character of this reaction [1,2]. Also, under appropriate experimental conditions, the hydrogen reduction reaction gives a selective tungsten deposit, due to the fact that the dissociative adsorption of hydrogen preferentially takes place on metal surfaces. However, selectivity in this case is normally lost when W layers thicker than a few thousand Angstroms are deposited. The ultimate layer thickness which can be grown selectively, depends on many factors such as process conditions, reaction by-products, or system cleanliness.

In this study the effects of the reaction by-products SiF_4 and HF, on selectivity were examined. A relation between the number of W nuclei on SiO_2 (selectivity loss) and the local HF concentration inside the reactor was established. Furthermore, a kinetic model was developed which describes a possible growth mechanism of tungsten on silicon dioxide in terms of the HF and WF_6 concentrations and the deposition time and temperature.

EXPERIMENTAL PROCEDURE

Tungsten films have been deposited in a conventional ASM horizontal hot wall LPCVD reactor. The deposition gases WF_6 (99.99 %), Ar (99.9995 %) and H_2 (after purification by a palladium diffuser >99.9999 %) were injected into the reactor from the door side. The quartz reactor chamber (1450 mm long, inner diameter 200 mm) was evacuated with a rootsblower/ mechanical pump combination having an effective pumping speed of 280 m^3/hr.The temperature inside the reactor was calibrated and controlled within +/- 0.5 °C of the setpoint (300-400 °C).The total pressure in the reactor was controlled by a throttle valve located in the pumpline.The substrates were 4", >1 Ohm.cm, p-type <100> oriented silicon wafers, either uncoated or coated on both sides with 2000 Å plasma oxide. The bare silicon wafers received a 20 seconds dilute HF preclean followed by a rinse and spin dry, prior to the tungsten deposition. By doing so, the displacement reaction (WF_6 + Ar) is well controlled with respect to the Si-W interface quality and the resulting limiting W-thickness. The substrates were positioned in a quartz boat (500 mm long) which in turn was inserted in the quartz tube in the conventional way. Loading and unloading were done at 250 C under a flow of nitrogen.After each run, the quartz tube and boat were in-situ plasma-etched to remove any tungsten deposition.

Tungsten was deposited by a two-step process: after a pump down of the system and a 30 minutes heat-up of the wafers to the proces temerature of 400 °C, 20 sccm of WF_6 and 600 sccm Ar were monitored into the system at a total pressure of 0.4 Torr for 10 minutes. After this, the system was pumped down again, and 50 sccm WF_6 together with 1500 sccm H_2 were introduced in the system for 6-12 minutes.Finally, after a cooldown and a pump purge cycle, the run was terminated by a N_2 backfill to atmospheric pressure.

The resulting samples were analyzed on thickness and resistivity by means of a Dektak and a four-point probe, respectively. After subtraction of the W-thickness resulting from the WF_6/Ar reaction (≈ 200 Å), the deposition rate of the hydrogen reduction reaction was calculated. The oxide samples were examined on the presence of W nuclei with an optical microscope and a SEM.

RESULTS

When tungsten layers in excess of 2000 Å were deposited on a full silicon waferload (25 wafers), selectivity loss was observed on the boat and tube, starting from the first wafer position, getting worse downstream.When the waferload consists of sets of three wafers grouped together, two oxide wafers with one silicon wafer in between, another W-pattern was observed on the quartz boat (see fig. 1): W deposits were found on the boat only in between the wafers, with a density related to the waferspacing. Also on the oxide-coated wafers tungsten growth was observed, but only on the waferside facing a silicon wafer. This selectivity loss, which was observed under an optical microscope as tiny W-dots on the oxide, was minimal at the edge of the wafer, but increased dramatically towards the centre of the wafer. When we tried to deposit tungsten on a full oxide wafer load, so that <u>no</u> silicon or W was present in the system, exposures to the WF_6/H_2 flow up to 45 minutes did not lead to selectivity loss at all. Furthermore, the degree of selectivity loss was found to depend on the silicon waferload, waferspacing and position, and deposition rate and time.

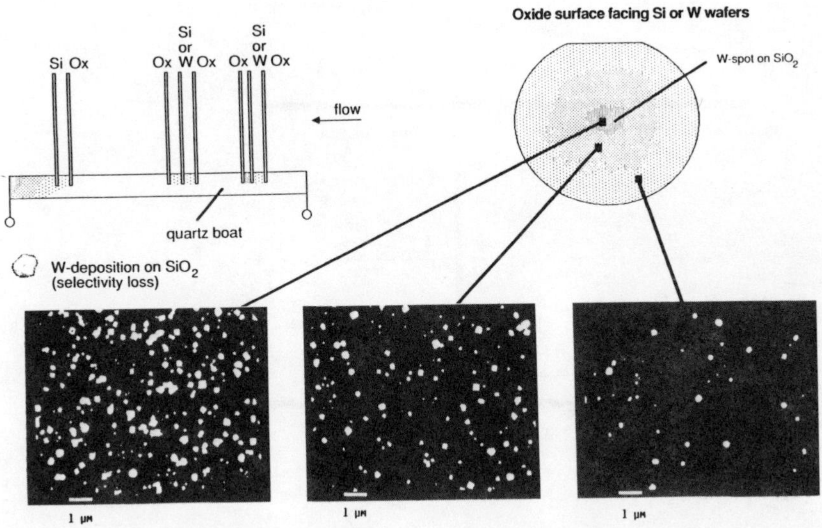

figure 1. Selectivity loss on quartzboat and wafers

These experimental observations all indicate that selectivity is strongly affected by the presence of a reaction byproduct. In experiments in which silicon wafers were replaced by tungsten covered wafers, the observed selectivity loss was equivalent, which means that HF is the main agent inducing selectivity loss (and not SiF_4).

We will now quantify the relation between the local HF concentration and the tungsten growth on SiO_2. We first calculate the HF concentration in between two adjacent wafers (one bare silicon and one oxide coated wafer) and in the annular zone between the wafers and the tube wall under the assumptions that: i) the deposition rate uniformity of W on silicon wafers is 100% both within wafer and from wafer to wafer, ii) the mass transport between two adjacent wafers is by diffusion only, iii) HF will mix quickly with the convective flow in the annular zone, iv) in the annular zone the mean convectionspeed is constant and v) there is no HF backdiffusion in the annular zone. After solving the one-dimensional diffusion equation, the HF concentration between two adjacent wafers is given by (see figure 2):

$$C_{HF}(r) = \frac{\text{constant} * R}{d * D_{HF,H_2}} * (r_w^2 - r^2) + C(r_w) \qquad (1.1)$$

With R = deposition rate of tungsten on silicon (cm/min)
d = waferspacing (cm)
D = diffusion coefficient of HF in H_2 (cm^2/min)
r, r_w = radius , wafer radius (cm)
$C(r_w)$ = boundary value ($r = r_w$); this is the (z-dependant) HF concentration in the annular zone between wafers ar. tube. It depends on the HF produced upstream.

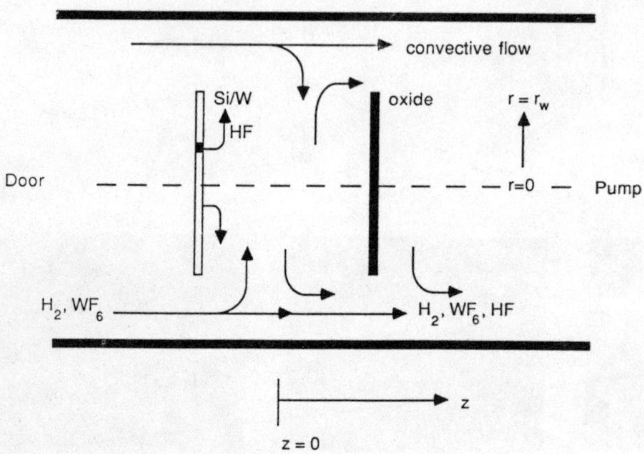

figure 2. A schematic of the reactor geometry

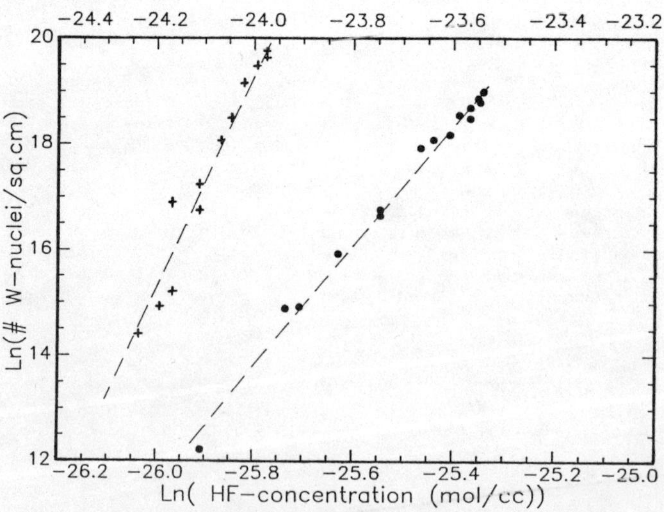

figure 3. Relation between the calculated HF concentration and the measured W nuclei density.

Equation (1.1) shows that the HF concentration decreases parabolically from the center($r=0$) towards the edge of the wafer ($r=r_w$).

Secondly, various sets of experiments were evaluated in more detail: most experiments yielded oxide substrates either free of any W nuclei (complete selectivity) or with a very high W nuclei density (coalescence or complete selectivity loss). Only very few oxide wafers were covered with a reasonable distribution of W nuclei. From SEM photographs taken at different radial positions (see fig.1) a W nuclei density distribution is deduced (total counts in the range 0.01 - 4.0 nuclei/um^2)

Figure 3 shows the results for two oxide wafers originating from different runs. In this graph ln $C_{HF}(r)$ as calculated from (1.1) is plotted versus ln(# W nuclei(r)) ($0 < r < r_w$). Straight lines result with slopes 12 and 22. This suggests a power dependance between the number of W nuclei and the HF concentration. Note that the actual value of the slope depends critically on the accuracy of the calculation of the HF density. A small variation in $C_{HF}(R_w)$ leads to a large variation in the slope. Since the slope 12 corresponds with a waferpair at the entrance (door) side of the system, it is probably more accurate than the other slope, which corresponds with a pair further downstream. The high values of the slopes indicate that only a slight variation in the HF concentration will result in a very large variation of the number of W nuclei.

NUCLEATION OF W ON SIO_2, A KINETIC MODEL.

The following, relatively simple mechanistic model is proposed for the W nucleation on SiO_2 surfaces:

W can be selectively deposited by the Hydrogen reduction reaction:

$$WF_6\,(g) + 3\,H_2\,(g) \longrightarrow W\,(a) + 6\,HF\,(g) \tag{2.1}$$

The selectivity of the process is explained in terms of the inability of hydrogen to dissociate on the oxide surface. In the case that HF is present, a way of generating atomic hydrogen on SiO_2 is by the dissociative adsorption of HF:

$$HF\,(g) + 2\,S \rightleftharpoons SF(a) + 2\,H(a) \tag{2.2}$$

S represents a free surface site, and SF(a) a fluorine atom chemisorbed onto the surface site. A likely candidate for S is a Si atom. This is not unreasonable since the Si-F bond is very strong. ($\Delta G \simeq -370$ kJ/mol).
On tungsten, H is known to be mobile. It is postulated that the activation energy for the hydrogen reduction of WF_6 ($E_\alpha \simeq 69$ kJ/mol) corresponds to the diffusion of H over the W surface [1].On SiO_2 however, H will be much less mobile, because of the likelihood of H bonding to Si or O atoms, yielding primarily hydroxyl groups. Thus in this case, W growth by the hydrogen diffusion mechanism is unlikely; instead the formation of a W nucleus of critical size, consisting of a cluster of n W- atoms (typically n=2 to 6), might be the rate limiting step.

After this nucleus of critical size has formed it can act as a nucleation site for further W growth by hydrogen dissociating on its surface. So, the nuclei formation rate J_w can be expressed as:

$$J_w = d[W_n]/dt = k_1(T) * [W_{cr}] \quad (2.3)$$

and the formation rate of a nucleus of critical size as:

$$d[W_{cr}]/dt = k_2(T) * [W]^n \quad (2.4)$$

where $[W_{cr}]$ is the concentration of W nuclei of critical size, [W] the concentration of atomic W formed on the oxide, and $k_i(T)$ are kinetic rate constants. The formation of a W nucleus of critical size is now depending on the reduction of WF_6 by adsorbed H. In order for this to happen though, a WF_6 molecule has to adsorb on a surface area where sufficient H atoms are present in close vicinity and suitable oriented to reduce it. Since the binding enthalpy of HF (~ 560 kJ/mol) is larger than that of OH in a Si-OH structure (< 460 kJ/mol), it is likely that the adsorbed H will be released in order to form HF. The rate of formation of a single atom will be given by:

$$d[W]/dt = k_3(T) * [WF_6(g)] * [H(a)]^6 \quad (2.5)$$

i.e. corresponding to the reaction:

$$WF_6(g) + 6\,H(a) \longrightarrow W(a) + 6\,HF(g) \quad (2.6)$$

The HF formed can, of course, be readsorbed and dissociate according to equation (2.2), but the HF liberated from (2.6) will be much less than that generated at the opposite silicon wafer, and so HF from reaction (2.6) is neglected.
Using the equilibrium (2.2) we can relate [H(a)] to [HF(g)] through an equilibium constant. Assuming all conversion rates to be small, we finally find from (2.2)- (2.5) for the number of tungsten nuclei:

$$[W_n] = K(T) * [WF_6(g)]^n * [HF(g)]^{6n} * t^{n+2} \quad (2.7)$$

where K(T) is the overall rate constant including $k_1(T)$, $k_2(T)$ and $k_3(T)$ Equation (2.7) shows that the number of W nuclei formed on the SiO_2 depends on the WF_6 and HF concentrations, the deposition time and temperature.
A comparison between figure 3 and formula (2.7) indicates that n has a value between 2 and 4.

CONCLUSIONS.

Experiments have shown qualitatively as well as quantitively, that a volatile species (HF), which is generated uniformily and at a constant rate at the W substrate during the WF_6 /H_2 reaction, acts as a 'selectivity killer' during the selective deposition of tungsten. The experimental results are in good agreement with the presented model for W nucleation on SiO_2. Based on our experimental results, equation (2.7) will lead to a time dependence of the form

$$[W_n] = c * t^{4-6}$$

This time dependence is in good agreement with the experimental results obtained by McConica et al. [3].

Although we strongly believe that HF is the species causing selectivity loss, in principle one cannot rule out other candidates. One of the other species which is often stated to cause selectivity loss is WOF_x. However, it is more likely that this compound is formed by the reactions between HF, WF_6 and the SiO_2 substrate - thus being directly related to the HF concentration - rather than that it is desorbed from the W-substrate. Since the picture in this sense is not complete, further experiments need to be done, for example by adding HF from an external source. Preliminary results do indicate indeed that the addition of HF to the WF_6/H_2gas mixture affects the selectivity of the W-process adversily.

ACKNOWLEDGEMENTS

The authors would like to acknowledge the EEC for financial support of this work under contract no. MR-11-ASM.

REFERENCES

1. E.K. Broadbent and C.L. Ramiller, J. Electrochem. Soc., 131, 1427 (1984)
2. H.H. Busta and C.H. Tang, J. Electrochem. Soc., 133, 1195 (1986)
3. C.M.McConica and K.J. Cooper, in Proc.," Workshop on Tungsten and other refractory Metals for VLSI Applications",197 (1985)

THE CONTACT PROPERTIES TO $TiSi_2$ AND THE ADHESION WITHIN SUB-MICRON CONTACT HOLES OF ETCHED BACK CVD W/ADHESION LAYER FILMS

R.C. ELLWANGER, J.E.J. SCHMITZ, R.A.M. WOLTERS AND A.J.M. VAN DIJK
Philips Research Laboratories
5600 JA Eindhoven, The Netherlands

ABSTRACT

A CVD tungsten deposition and etch back process was developed which renders the contact hole coplanar with the oxide surface and forms little, if any, keyhole within the plug. The etch back is performed without the incorporation of a sacrificial planarizing layer. This deposition scheme, in combination with differing adhesion layers, was studied with respect to contact resistance and chemical stability to $TiSi_2$, as well as to adhesion within submicron contact holes. All layerings studied exhibited low values of contact resistance, and only adhesion layers of pure Ti showed a measurable level of interfacial fluorine (AES). In addition, W growth directly atop $TiSi_2$ was shown feasible under certain conditions. A large variation in adhesion performance was observed for the layerings studied. This ranged from layerings which "popped" out of contact holes during etch back, to layerings which allowed the plug, itself, to be severed during sample cleaving.

INTRODUCTION

The selective deposition of tungsten via low pressure chemical vapor deposition has been reported to be a potentially promising technique for the fabrication of very large and ultralarge scale integrated circuits (1-3). The reported applications range from Schottky diode barriers (4), current shunts (1,5,6), diffusion barriers (5,7), plasma etch barriers (8), and contact or via fills (9-11). Amongst these applications is perhaps the most urgent need, a planar filling of contact and via holes. This is demonstrated by the fact that present integrated circuit designs require contacts of submicron diameters and aspect ratios of the order of unity.

The direct application of selective growth for contact hole filling has, however, proven problematic. The growth of tungsten atop of single crystal Si has been demonstrated to produce tunnel defect structures, which can extend to beyond one micron depths into the substrate (12, 13). Attempts to alleviate the potential reliability problems posed by such defect structures include the growth of tungsten onto regions where silicide was first formed. The silicide would allow for self-aligned, low resistance contacts; whilst the W would act as a contact fill and diffusion barrier.Such systems were studied by Broadbent et al. (14). In their work it was found that a selective W growth directly atop $TiSi_2$ consumed the Si from the upper 40-50 nm of the silicide. This left behind a layer of Ti which contained a large concentration of fluorine and oxygen. Moreover the W layer which formed was completely non-adherent. When grown atop PtNiSi, the W encroachment along the silicide-silicon interface led to void formation along the perimeter (14).

Another approach which utilizes the conformal nature of CVD tungsten films is that of non-selective or "blanket" deposition followed by an etch back technique. CVD tungsten layers exhibit poor adhesion to oxide (15) and hence in such schemes an adhesion promoting layer is incorporated. This layer may either be sputtered (15-17) or chemically grown in situ prior to the W growth (18-21). A proper choice of adhesion layer should not only promote the adhesion of W, but also act as a barrier to any detrimental interaction at the substrate (Si or silicide) during tungsten growth. In the above cited works the two adhesion layers studied were sputtered Ti or CVD grown tungsten silicide. The literature to date has not reported results of the adhesion of the etched back structure within one micron, high aspect ratio contact holes. Either trenches (20), non-etched back contact holes (18), or contact holes of diameter well greater than one micron (19,21) are reported. In addition only Smith et al. (16) report contact resistance values. They used a sputtered adhesion layer of Ti and reported low contact values atop $TiSi_2$. They also reported low values of contact resistance when selective growth of W was performed above $TiSi_2$. This is in possible disagreement with the work of Broadbent et al. (14), but no analytic data was presented to clarify this point.

In this paper we describe a CVD blanket tungsten deposition and etch back process which renders the contact hole coplanar with the oxide surface, whilst producing little or no keyhole formation within the plug. The etching was done without the use of a sacrificial resist layer. We report the effect upon contact resistance to $TiSi_2$ of this W deposition process in combination with several adhesion layers. The contact resistance of a selective W growth directly atop $TiSi_2$ is also reported. Interfacial stabilities were studied using Auger Electron Spectroscopy. The adhesion of the etched-back plug within one micron to sub-micron contact holes was also studied for several different adhesion layers. A prefered CVD W/adhesion layer stack is given.

EXPERIMENTAL

100 mm diameter Si wafers of basically two groups were used for these experiments, the first being Kelvin contact resistance structures of various geometries. Such structures received implants through 55 nm of gate oxide of either $3x10^{15}$ B^+ cm^{-2} at 40 keV, or $5x10^{15}$ As^+ at 100 keV, and were annealed at either 900°C, 30 min or 975°C 30 min, respectively. Titanium was sputter deposited to a thickness of 40 nm and rapid thermal annealed (RTA) in N_2 at 650°C, 10 sec. The titanium nitride over oxide, as well as the thin TiN layer formed over the silicied contacts, was removed by etching in a solution of 1:1:5 H_2O_2: NH_4OH:H_2O. A second RTA of 850°C, 10 sec in N_2 was

employed resulting in a self-aligned bilayer of 6-8 nm TiN on top of 35 nm of $TiSi_2$. A similar bilayer structure is reported in the literature (22).

For the second group, TEOS grown oxide was deposited on non-patterned Si substrates to thicknesses varying between 0.7 and 0.9 µm. Contact holes were defined and etched using standard lithographic and dry etch techniques. These contact holes varied in base diameter from 0.8 to 1.1 µm.

At this point all substrates received the appropriate adhesion layer material as outlined in Table I. All substrates aside from the WSix adhesion layer, and the selective W depositions received first an in situ 1500 W, 150 V sputter etch which removed the equivalent of 8 nm of thermal SiO_2. The TiW and TiW-N adhesion layers were sputtered deposited to a thickness of 100 nm. The Ti and W sputtered adhesion layers were deposited to a thickness of 20 nm each. The CVD W reactor was cooled to 200°C, the sputtered adhesion layers were then loaded and the reactor was heated while pumping with an N_2 purge to 430°C. WSi_x was grown to a thickness of 50 nm using SiH_4 WF_6, and H_2 at 430°C. The selective W layers were grown using H_2 and WF_6, also at 430°C.

All tungsten layers (and also the WSi_x) were grown in a Genus 8402 reactor. With the exception of the selective W samples a two step process was used. This consisted of a W growth to 0.4 µm using WF_6, SiH_4, and H_2, and then continuing to a thickness of 1.0 µm with WF_6 and H_2 alone. Substrates on which contact holes had been defined were then etched-back in an Electrotech 600 etch system using a chemistry of SF_6 alone.

The Kelvin structures received a 0.5 µm thick W deposition, followed by a 0.2 µm aluminum coating and were then photolithographically defined and etched. The defined aluminum was then used to mask the blanket W/adhesion layer structure during an SF_6 etch.

Auger Electron Spectroscopy was performed using non-patterned Si substrates having been processed identical to Kelvin structure substrates from the point of titanium salicide formation, through to tungsten growth. AES was performed in a PHI 600 spectrometer using a 3 KeV, 1 µA, 60° incident beam rastered over 0.01 mm^2. Sputter depth profiling was accomplished using a 2 KeV, 4 µA, 60° incident Ar^+ beam rastered over 4 mm^2.

RESULTS AND DISCUSSION

W deposition and etch back

As described above, the deposition process incorporates two steps. We observed that SiH_4 has no effect upon W step coverage during the first 0.4 µm deposition into aspect ratio 1.0 µm diameter contacts or trenches. At this point however, a noticeable difference is observed, with H_2 reduction demonstrating a much greater degree of continued conformality as compared to a SiH_4-H_2 reduction (23). The initial SiH_4 containing step is used for growth rate enhancement. In addition it appears that the SiH_4 chemistry is very effective in the suppression of fluorine incorporation in underlying titanium containing layers (24).

The etch process was developed to allow for selectivity between the keyhole and the bulk of the W front. In experiments in which 0.7 µm deep, 1.0 diameter contact holes were filled with a W layer of only 0.8 µm thickness and then etched back, this selectivity was calculated to be on the order of 3:1. Such selectivity accompanied by a conformal deposition does not require a sacrificial layer for etch-back. Fig. 1 is an SEM of an etched-back W layer in a 0.7 µm deep, 0.9 µm diameter hole. Coplanarity of the W surface with the oxide surface is observed. Aspect ratios up to 0.8 have been studied.

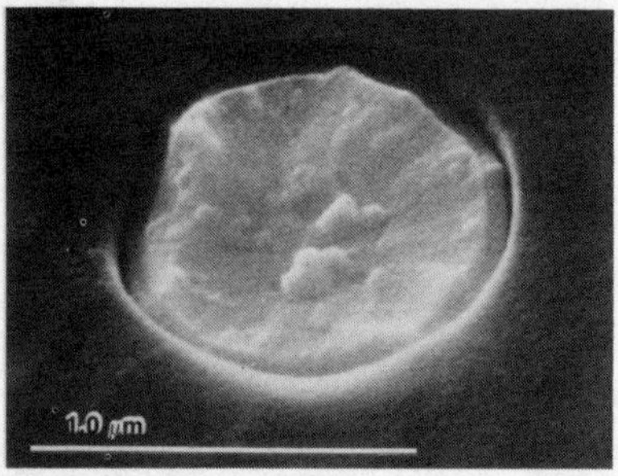

Fig.1- SEM micrograph of W contact filling using etch-back 0.9 µm diameter, 0.7 µm deep contact.

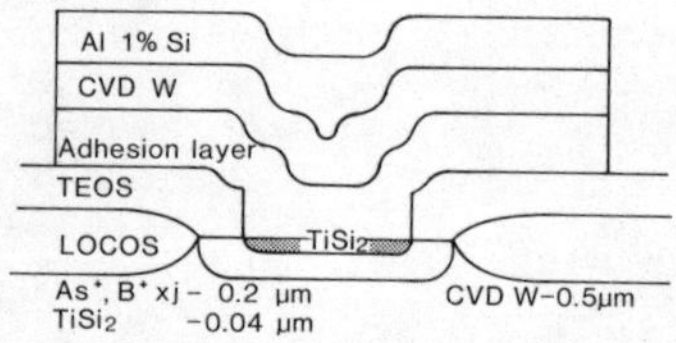

Fig.2- Cross-sectional schematic of the layer definition used in contact resistance measurements.

Contact resistance

Fig. 2 shows how the various layers were defined upon the Kelvin structures used for contact resistance measurements. The $TiSi_2$ was formed in the contact opening and the CVD W/adhesion layer was defined by the aluminum as a blanket strip. With such a structure one isolates the adhesion layer / $TiSi_2$ / Si interfaces. Table I lists the various layerings studied as well as the corresponding contact resistance to both n+ and p+ Si. It is shown that for the four sputtered adhesion layers, as well as for the CVD WSi_2 adhesion layer, neither the adhesion layer nor the CVD W processing effects an increase in the measured contact resistances.

Table I - Various layerings for which contact resistance measurements were performed. Al1%Si/Si and Al1%Si/TiW/$TiSi_2$/Si are included for comparison.

Layering	N + ($5x10^{15}$, 100 keV	P + ($3x10^{15}$, 40KeV)
Al1%Si/CVD W/W/Ti /$TiSi_2$/Si	$1.7x10^{-7}$ Ω-cm^2	$3.6x10^{-7}$ Ω-cm^2
Al1%Si/CVD W/Ti /$TiSi_2$/Si	2.4 "	3.5 "
Al1%Si/CVD W/TiW /$TiSi_2$/Si	2.1 "	4.1 "
Al1%Si/CVD W/TiW-N/$TiSi_2$/Si	1.9 "	3.5 "
Al1%Si/CVD W/WSi_x /$TiSi_2$/Si	2.0 "	4.1 "
Al1%Si/Sel.W/ /$TiSi_2$/Si	2.6 "	4.1 "
Al1%Si /TiW /$TiSi_2$/Si	1.9 "	3.3 "
Al1%Si /Si	$4.4x10^{-6}$	5.1 "

The CVD W/adhesion layer layerings were examined by AES. The only instance in which fluorine was detected at the adhesion layer/$TiSi_2$ interface was the layering which used 20 nm Ti layer only (Auger detection limit for fluorine is about 0.5 at.%). In this case approx. 4 at.% fluorine was measured. This interaction is much less than what was observed by Broadbent et al (15). We attribute this to the addition of SiH_4 (25) in the present case. At the 4% fluorine level no effect was observed in the contact resistance.

Selective W (H_2 + WF_6) directly atop $TiSi_2$

In Table I it is also shown that the contact resistance of a selective W deposition (H_2 reduction alone) to the $TiSi_2$/Si contact is, for n+ and p+, of the same value as the other layerings studied. These films were 80 nm thick and demonstrated good adhesion to the $TiSi_2$ layer. Fig. 3a is an Auger depth profile of this layering. The fluorine signal is at the noise level, again meaning at or below 0.5 at.%. TEM cross section has verified that the TiN layer formed during $TiSi_2$ anneal in N_2 was still present as a continuous layer after W deposition.

In the study of Broadbent et al (15) this TiN layer was removed prior to selective W deposition. Fig. 3b shows the Auger profile of a selective deposition upon a $TiSi_2$ layer upon which the upper TiN layer was removed in a solution of $NH_4OH:H_2O_2:H_2O$. This profile is nearly identical to that shown by Broadbent, high O and F (appr 20 at %) densities can be seen. For this deposition a total lack of adhesion was observed.

It thus appears that W can be grown selectively atop $TiSi_2$ without degrading the contact properties. This is dependant upon the presence of an upper TiN layer on to the silicide surface prior to W growth. However, the question of growth atop of TiN is not fully answered. In the work of Broadbent no nucleation of W was found to occur atop a TiN surface. Hence it was removed. In our study we observed no difficulties in film growth on this TiN layer. The critical issue here is not yet known.

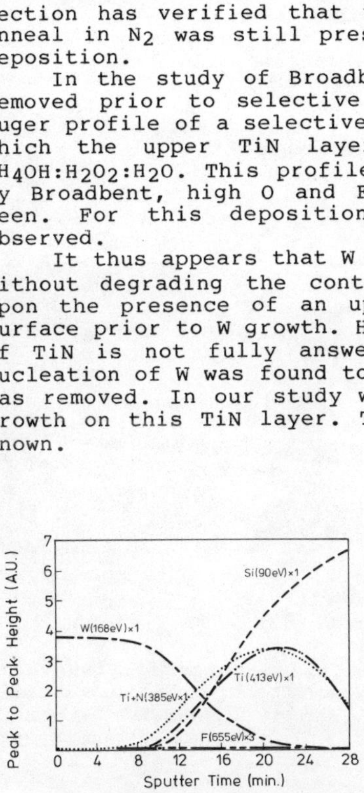

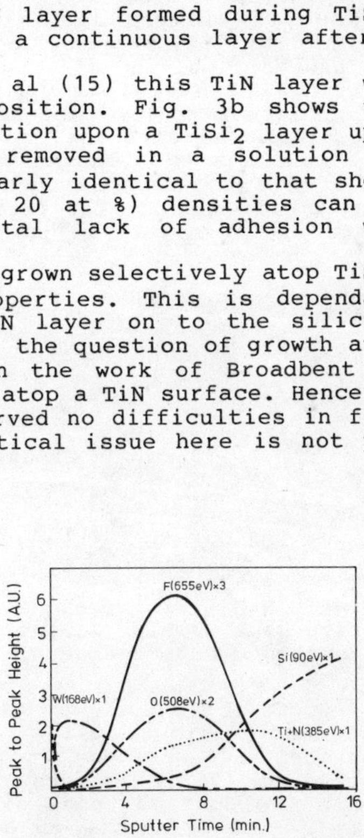

Fig.3- Auger depth profile of
a) W/TiN/$TiSi_2$/Si structure. b) W/$TiSi_2$/Si structure.

Adhesion Properties

CVD WSi_x is technologically attractive as an adhesion layer because of the in situ nature of the deposition. However, it did not provide adhesion at the one micron diameter contact level. Fig. 4 is an SEM micrograph of an etched-back W/WSi_x layering with contact holes of 1.0 µm diameter, 0.8 µm depth. Some plugs remain intact inside the hole, others intact outside of the contact, and still others were not to be found. Of primary importance was to determine if the loss of adhesion was caused by the etch-back process, or if it was related solely to the adhesion properties of the WSi_x.
To evaluate this, we terminated the etch-back process both prior to endpoint and after endpoint for several samples groups. SEM inspection of the contact topography followed. Fig. 5 is an SEM micrograph of a 1.0 µm diameter, 0.8 µm deep contact prior to endpoint. A continuous film is seen above the oxide. Because this film is thin the contact periphery can be seen. No attack of the WSi_x or of the $SiO_2/WSi_x/W$ interface can be observed. In fig. 6 we show a micrograph of a contact which has been overetched. The WSi_x layer was etched by the SF_6 at a lower rate (within the contact hole) than the W plug. No enhanced etching can be seen at the W/WSi_x interface. The WSi_x is, however, pulling off from the oxide sidewall. Just prior to clearing the WSi_x film from the oxide surface all plugs were intact inside the contact holes and no attack of the plugs was observed. Upon clearing the oxide surface of WSi_x a variation of some plugs to all plugs are relieved from the contact hole (25). This occurs, although, the WSi_x etch rate is less than that of the W plug. The "popping" of the plugs, hence, appears to be related solely to the adhesion properties of the WSi_x.

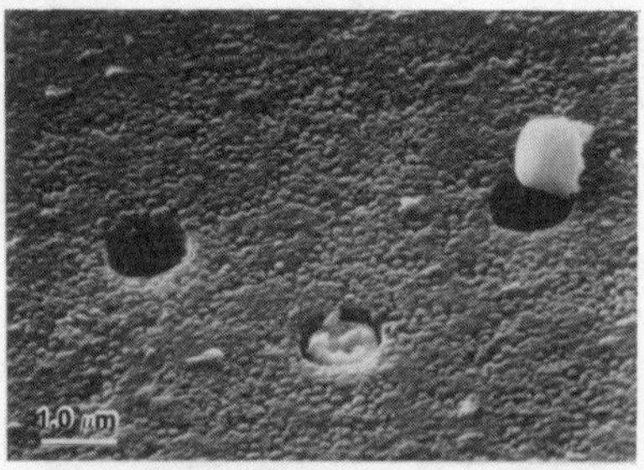

Fig.4- SEM micrograph of etched-back W/WSi_x. 1.0 µm diameter, 0.8 µm contacts.

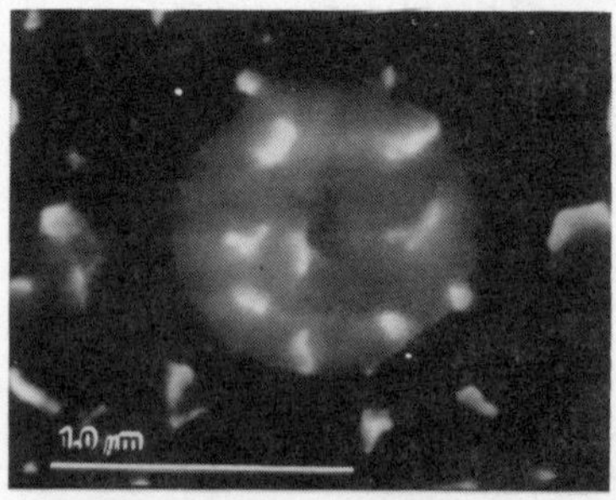

Fig.5- SEM micrograph of W/WSi_x. Thin, continuous surface layer is present.

Sputtered layers of TiW and reactive sputtered TiW-N were also studied for adhesion performance. These films are easily deposited and have been previously shown compatable with MOS processing (26). In both cases all plugs remained intimate within the contact holes after completion of the etch step. A further quantification of adhesion performance appeared to be the resultant plug topography along a cleave. Substrate wafers were broken by stylus pressure to the wafer periphery. Cleaves which occurred within high density contact patterns were found

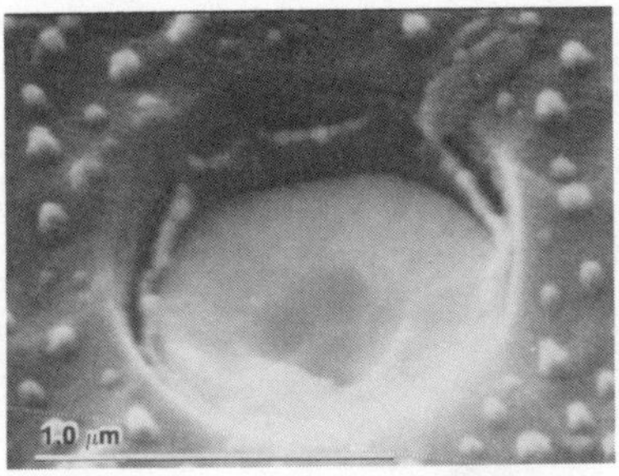

Fig.6- SEM micrograph of an overetched W/WSi_x plug.

Fig.7- SEM micrograph of a W/WSi_x plug along cleave.

and then observed in the SEM. WSi_x/W plugs were, as expected, fully gone after such a process. Even in the instance where a 100 nm layer was left upon the surface and with a non-critical aspect ratio (1.1 µm diameter, 0.6 deep) the W and WSi_x are both removed from the contact hole. This is shown in fig. 7. The plug can be seen here having been rotated 90°. No WSi_x can be discerned. Fig. 8 is an SEM micrograph of the TiW-N/W structure. For such layers the plug was always missing, but the TiW-N layer remained in tact. TiW/W layers, however, demonstrated adhesion adequate to allow the plug to be cleaved. Fig. 9 shows this for a 1.0 µm diameter, 0.7 µm deep contact. Of further interest is fig. 10. Here the structure has been overetched to a degree that less than 200 nm of TiW is left vertically along the plug sidewall. Still the SiO_2/TiW/W interfacial adhesion is adequate for the plug structure to have remained in tact.

Fig.8- Micrograph of cleaved W/TiW-N plug. Note step coverage of sputtered layer.

Fig.9- SEM micrograph of W/TiW plugs. 1.0 µm diameter, 0.7 µm deep contacts.

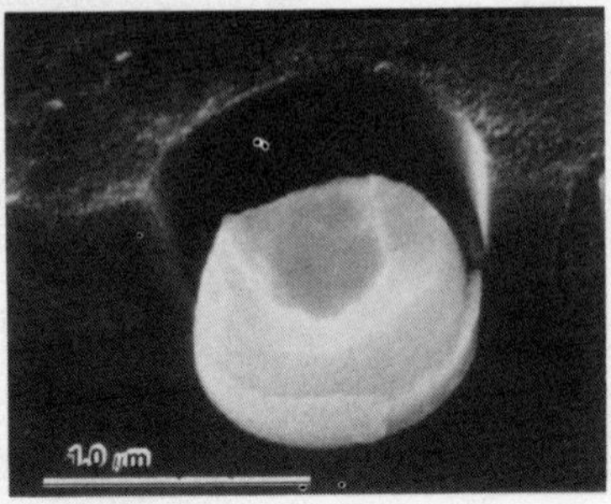

Fig.10- SEM micrograph of overetched W/TiW plug along cleave.

SUMMARY AND CONCLUSIONS

We have described a blanket W deposition and etch-back process which renders the contact hole co planar with the oxide surface and for which little or keyhole formation may be observed. We have further presented contact resistance data for a variety of adhesion layers to $TiSi_2$ upon which W was grown. It was shown that in all cases, the $TiSi_2$/Si interface is the determining factor for contact resistance and that the subsequent processing had neglicable effect upon this value. As determined by AES, pure titanium was the only layer for which fluorine was detected, when combined with blanket W processing.

Tungsten silicide (CVD) demonstrated unsuitable adhesion properties for a one micron and submicron contact dimension. Adhesions layers of sputtered TiW performed well in all aspects studied. Additionally sputtered TiW is a technology step for which much IC experience already exists. It is our conclusion that sputtered TiW is a best choice adhesion layer film for a blanket CVD W technology.

Further, it was shown that a H_2 reduction of WF_6 atop titanium silicide structures can be accomplished without the degradation of the silicide, providing a TiN top layer is present. Upon removing this layer the $TiSi_2$ was attacked rendering a non-adherent W layer as well as a high F and O concentration in the $TiSi_2$ film. Further examination is required in order to understand reported differences in selective tungsten growth upon the $TiN/TiSi_2$ salicide structure.

ACKNOWLEDGEMENTS

The authors express appreciation to: P. Oosting and W. v.d. Wijgert for AES; J. Bos for etching; H. v. Houtum, L. Bos, T. Nellissen and H. Godon for assistance in substrate preparation; and T. Nellissen for some contact resistance measurements. We also gratefully acknowledge E. Broadbent, H. v. Houtum and H. Kalter for helpful discussions.

REFERENCES

1. P.A. Gargini and I. Beinglass, IEDM Tech. Dig., 54 (1981).
2. E.K. Broadbent and C.L. Ramiller, J. Electrochem. Soc., 131, 1427 (1984).
3. K.L. Saraswat, S. Swirhun and J.P. McVittie, in "VLSI Science and Technology/1984", K.E. Bean and G.A. Rozgonyi, Editors, p. 409, The Electrochemical Society Softbound Proceedings Series, PV 84-7, Pennington, NJ (1984).
4. C. Crowell, J. Savace and S. Sze, Trans. Metall. Soc. AIME, 233, 478 (1965).
5. N.E. Miller and I. Beinglass, Solid State Technol., 25, 85 (1982).
6. W.A. Metz, J.J. Szluk, G.W. Miller and H.O. Hayworth, IEEE Trans. Electron Device Lett., edl-6, 372 (1985).
7. S. Swirhun, K.C. Saraswat and R.M. Swanson, ibid., edl-5, 209 (1984).
8. J. DeBlasi, C. Werkhoven, T. Bril and P. Mayo, Abstract 517, p. 756, The Electrochemical Society Extended Abstracts, Vol. 84-2, New Orleans, LA, Oct. 7-12, 1984.
9. T. Moriya, S. Shima, Y. Hazuki, M. Chiba and M. Kashiwagi, IEDM Tech. DIg., CH 1973-7/83, 550 (1983).
10. R.S. Blewer and V.A. Wells, Proc. First Int'l IEEE VLSI Multilevel Interconnect Conference, p. 153, IEEE Cat. no. 84CH1999-2, New Orelans, LA, June 21-22, 1984.
11. R.H. Wilson, R.W. Stoll and M.A. Calcone, Proc. Second Int'l IEEE VLSI Multilevel Interconnect Conference, p. 343, IEEE Cat. no. 84CH2197-2, Santa Clara, CA, June 25-26, 1985.
12. W.T. Stacy, E.K. Broadbent and M.H. Norcott, J. Electrochem. Soc., 132, 444 (1985).
13. M.L. Green and R.A. Levy, J. Electrochem. Soc., 132, 1243 (1985).
14. E.K. Broadbent, A.E. Morgan, J.M. DeBlasi, P. van der Putte, B. Coulman, B.J. Burrow, D.K. Sadana and A. Reader, J. Electrochem. Soc., 133, 1715 (1986).
15. D.W. Woodruff, R.H. Wilson and R.A. Sanchez-Martinez, "Tungsten and Other Refractory Metals for VLSI Applications", Ed. R.S. Blewer, p. 173, Materials Research Society, Pittsburg, Penn., 1986.
16. G.C. Smith and R.B. Jucha, Proc. Third Int'l IEEE VLSI Multilevel Interconnect Conference, p. 403, IEEE Cat. no. 86CH2337-4, Santa Clara, CA, June 9-10, 1986.
17. S. Mehta, S. Mittal, A. Haranahalli and D. Ranadive, ibid., p. 418.
18. C. Fuhs, E.S. McInerney, L. Watson, N. Zetterquist, "Tungsten and Other Refractory Metals for VLSI Applications", Ed. R.S. Blewer, p. 257, Materials Research Society, Pittsburg, Penn., 1986.
19. W.A. Metz and E.A. Beam, ibid., p. 249.
20. G.C. Smith, Proc. Second Int'l IEEE VLSI Multilevel Interconnect Conference, p. 350, IEEE Cat. no. 84CH2197-2, Santa Clara, CA, June 25-26, 1985.
21. G. Higelin, C. Wieczorek and V. Grewal, Proc. Third Int'l IEEE VLSI Multilevel Interconnect Conference, p. 443, IEEE (at. no. 86CH2337-4, Santa Clara, CA, June 9-10, 1986.

22. M. Delfino, E.K. Broadbent, R.E. Morgan, B.J. Burrow and M.H. Norcott, IEEE Trans. Electron Device Lett., edl-6, 591 (1985).
23. J.E.J. Schmitz et al., to be published.
24. R.C. Ellwanger et al., to be published.
25. Such range depends upon depth and size of contact hole, as well as the underlying topography.
26. R.A.M. Wolters and A.J.M. Nellissen, Solic State Technol., 29 (2), 131 (1986).

MECHANISMS FOR WHISKER GROWTH IN THIN FILMS

John E. Sanchez, Jr.
Xerox Palo Alto Research Center, 3333 Coyote Hill Road
Palo Alto, California 94304

ABSTRACT

Electromigration (e-m) induced damage and failures continue to be a major reliability concern in VLSI technology. The trend for decreased feature sizes as well as for wafer scale interconnections will continue to present greater challenges for metallization materials, systems, and processes. As an example, multi-level metallization schemes have led to a new class of e-m failures, namely interlevel shorts due to whiskers[1-3]. These whiskers, along with voids, hillocks, and mounds are manifestations of the basic electromigration process in aluminum, which is mass flux due to the effective force on metal lattice atoms from electron momentum transfer during current flow[4]. This brief paper will review the observations and mechanisms of whisker growth in thin films, and apply the whisker growth phenomena as an effective mechanism for mechanical stress relief and vacancy creation during electromigration stressing.

HISTORICAL

It is interesting that the first observations of "whiskers" on the surfaces of thin films occurred as a result of short circuit failures of capacitors in military electrical equipment during WWII[5]. However a systematic study of the nature and growth of these whiskers was not published until after the war. The whiskers were found to grow spontaneously in lead, tin, cadmium, aluminum, silver, gold, indium, and bismuth films[6-10] under compressive stress. Later, whiskers were found[11,12] to have grown after electromigration stressing of aluminum films deposited on SiO_2-silicon substrates. Beyond circuit reliability concerns, whisker formation generated much interest in the metallurgy literature of the 1950's. This interest was due to the remarkable strength of the whiskers[13], as well as due to the controversy surrounding the proposed mechanisms for whisker formation[14]. The conditions under which whisker growth was observed can be generalized to the following;

1) films are under significant compressive stress, due to the deposition conditions (typically electroplating), "thermal", or applied;
2) temperatures are at least half the melting point, $T/Tm \simeq 0.5$;
3) growth occurs at the base, not at the tip;
4) growth rate $\simeq$ 1 cm/year;
5) whiskers are straight, with angular kinks;
6) other stress relief/deformation mechanisms are impeded in regions of whisker growth[15,16].

However, understanding of whisker growth conditions and mechanisms has yet to be applied to films undergoing e-m stressing.

MECHANISMS FOR WHISKER GROWTH

Whisker growth mechanisms in thin films under compressive stresses have been been accepted for some time, and these mechanisms[14-17] are consistent with the observations listed above. These mechanisms rely on non-conservative dislocation glide or climb which produces step height increases where the dislocation meets the surface, although other purely diffusional mechanisms are plausible. However, all mechanisms produce excess vacancies equal in magnitude to the volume of the whisker produced on the surface, resulting in vacancy concentrations greater than in thermal equilibrium, as these dislocations reproduce and glide/climb to the surface. These excess vacancies must diffuse away from their dislocation source for continued whisker growth. A likely mechanism among several similar possible is described below, however direct confirmation of the precise (dislocation) mechanism operative for a given whisker must be made by direct observation, as by transmission electron microscopic imaging.

Amelinckx, et al.[17] describe a mechanism involving a screw dislocation pinned at its ends, Fig. 1, climbing and gliding under applied stress into a helix, with the helix axis intersecting the surface. Each loop of the helix can be thought of as an equivalent "disk" of extra material that will create a disk of burger's vector (i. e., ~ lattice spacing) step height as each loop reaches the surface. The forces attracting the loops to the surface include "image forces"[17], the effective negative surface energy due to surface reactions[14,16] (oxidation), and the compressive stresses in the bulk[7,8,15]. Continued accumulation of these loop steps at the surface produce the observed whiskers, and the whisker growth rate is given by the combined rate of loop generation and loop migration to the surface. The rate of whisker growth is estimated to be ~ 0.5 to 1 μ / hour, however increasing the applied stress gave growth rates up to ~ 1 μ / second[15]. As the loops are created and migrate, they emit large numbers of vacancies, on the order of ~ 10^7 to 10^9 / second. As pointed out[7,14-17], this excess vacancy concentration will limit the rate of loop generation and non-conservative glide/climb. In turn, this excess concentration of vacancies will be determined by the rate of diffusion of vacancies away from the loop-helix complex to sinks at the surface or to internal sinks. In fact, from mass balance, the rate of whisker growth is determined by the rate of vacancy generation and migration away from the helix complex.

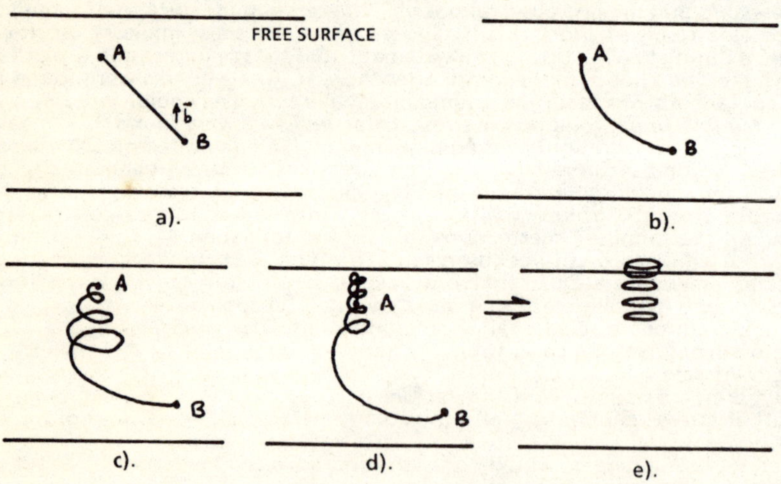

Figure 1. Helix dislocation mechanism for whisker growth, after Amelinckx, Bontinck, Dekeyser, Seitz, 1957. a) dislocation with screw character ($\vec{b}$ = burgers vector) pinned at points A and B; b) glide/climb of dislocation, emitting vacancies; c) loop formation and multiplication, with further vacancies created; d) loop migration to surface, with surface step height increased by loop burgers vector; e) loops are geometrically equivalent to disks of "extra material".

This mechanism is typical of most of the mechanisms for whisker formation, in that the dislocation mechanism is aided or driven by compressive stresses in the film, and large numbers of vacancies are emitted in the bulk as the whisker grows. It will be shown that the former process may be critical for enhancing electromigration induced damage in aluminum test stripes.

ELECTROMIGRATION DAMAGE

E-m damage is essentially a diffusion phenomenon, mass flux due to the force of conduction electron monmentum transfer (in aluminum) to lattice ions. However, substantial vacancy creation is necessary to produce the magnitude and number of damage artifacts, such as hillocks, voids, and whiskers associated with electromigration that must necessarily form after significant sustained mass flux. To illustrate, the number of vacancies N_v "comprising" an e-m induced void of size ~ 2 μ^3 is

$$N_v = (2\,\mu^3)(6.0 \cdot 10^{10} \text{ lattice sites}/\mu^3) = 1.2 \cdot 10^{11} \text{ vacant sites.}$$

In comparison, the equilibrium number of vacancies N_e at 125°C in an aluminum test stripe 100 μ by 1 μ by 2 μ, where the equilibrium vacancy fraction[18,19] at 125°C is $5.0 \cdot 10^{-9}$, is given by

$$N_e\,(125°C) = (200\,\mu^3)(5.0 \cdot 10^{-9})(6.0 \cdot 10^{10} \text{ sites}/\mu^3) \simeq 6.0 \cdot 10^4 \text{ vacancies.}$$

Thus, the number of new vacancies necessary for the formation of a 2 μ^3 void exceeds the equilibrium value on the order of $N_v/N_e = 2.0 \cdot 10^6$.

The significance of vacancy creation during sustained mass flux is illustrated in another fundamental way. The activation energy associated with self-diffusion is comprised of two terms, the energy of migration of a vacancy (atom) and the energy of creation of the vacancy[20]. Typically, and especially for aluminum, the vacancy creation energy term dominates the migration energy term, so that variations in effective activation energy during aluminum electromigration may be more indicative of the mechanisms for vacancy creation rather than vacancy (atom) migration. Thus, the "grain boundary diffusion activation energies" reported[21,22] for many electromigration kinetic studies may indicate that internal defects such as dislocations and grain boundaries are the origins of the necessary new vacancies, consistent with the decreased energy required at these defects for vacancy creation, and not necessarily indicative solely of the diffusion-migration path.

These two points, the significant lack of "equilibrium" vacancies necessary for observed electromigration damage and the significance of the vacancy creation energy term in the overall diffusion activation energy, illustrate the important role of vacancy creation mechanisms during electromigration. It should be emphasized that sustained vacancy creation must lead to surface growths or features[20]. These features then decorate the localized sites of creation of vacancies that become involved in e-m induced mass flux, but are not necessarily indicative of overall e-m mechanisms or paths as proposed elsewhere[23,24]. Hillocks may form where grain boundaries intersect the surface, where vacancies were formed on the surface and then migrated along the boundary into the film, for example if the film is under compressive stress. This flux is normal to the e-m force and mass flux. Then, these vacancies may diffuse through the grain bulk parallel to the e-m force, producing mass flux and electromigration. Thus, in this example the overall sequence for electromigration was surface vacancy creation, boundary migration, and bulk diffusion.

WHISKER GROWTH DURING ELECTROMIGRATION

Typical conditions during electromigration testing include elevated temperatures (> 200°C, T/Tm ≃0.5 for Al), where the thermal expansion mismatch between (Si) substrate and aluminum film can lead to compressive stresses on the order of 100 MPa[25]. The stressed film may and will plastically deform in many ways, such as hillock formation[26], Nabarro-Herring[26] or Coble[27] creep, dislocation glide and/or climb[28], whisker formation, etc.. However, during electromigration stressing those deformation mechanisms that also create new vacancies to replace those swept away by electromigration induced mass flux will be favored. Thus, hillock formation and whisker growth become qualitatively more likely during electromigration testing. This bias to whisker formation may not be pronounced if other sites for vacancy creation are operative, such as grain boundaries, or if plastic deformation mechanisms substantially reduce the compressive stress level.

Increased control of e-m damage and improved interconnect reliability has been achieved through alloy additions (i. e., Cu from 0.5 to 4.0 a/o) to aluminum. Further, the use of refractory metal or compound underlayers for aluminum spike control is common, usually in combination with alloyed aluminum. Both approaches, however, may reduce the effectiveness of grain boundaries as internal sources of vacancies. In the former case, Cu may migrate to and decorate the grain boundaries, reducing the boundary migration rate (due to vacancy flux across the boundary) or reducing the rate of vacancy-emitting boundary dislocation climb. In the latter case, impurity metal atoms from TiN, W, TiW, and Ti film underlayers may rapidly diffuse along grain boundaries to "harden" the boundaries in a similar manner. However, the result in both cases is that boundaries may no longer be efficient sources of vacancies, requiring other localized vacancy creation mechanisms to become significant. These "hardened" boundaries will also reduce the rates of plastic deformation due to the above mechanisms, leaving a higher compressive stress in the film. This situation is analogous to the last general observation for whisker growth, that is that other deformation mechanisms are impeded. (A more detailed study of the effects of grain boundary diffusion, grain growth, and boundary dislocation structures on e-m mechanisms will be presented elsewhere[29].)

A qualitative model for the conditions and mechanisms for whisker growth in thin films can now be assembled. During electromigration testing at elevated temperatures, vacancies are swept away by the mass flux and sink at surfaces and internal sites, such as boundaries and boundary triple points, creating a net vacancy undersaturation. Further, due to the typical compressive stresses in the film, mechanisms for stress relief via removal of mass from the bulk to the surface are operative. For relatively pure (weak) films, internal grain boundaries serve as convenient and efficient sites for vacancy creation and deformation, leading to hillock formation, diffusion "creep", boundary slip via boundary dislocation climb, and low stress levels. However, these efficient sources of vacancies lead to rapid electromigration failures via void formation. Alloying additions (Cu) successfully reduce the failure rates and increase the effective "activation energy" for electromigration by "hardening" the grain boundaries. The boundaries may also be "hardened" by diffusing species from barrier metal underlayers, reducing the participation of boundaries in mass flux and stress relief, leading to higher stresses and/or allowing other mechanisms to become significant. Whisker formation mechanisms involving bulk dislocation generation and motion are relatively unaffected by boundary "hardening", and thus serve as both as effective mechanisms for stress relief deformation and vacancy creation for e-m mass flux. Whisker formation is thus expected be be a significant failure mode in double level metal interconnect structures[3] in alloyed aluminum films with refractory metal film or compound underlayers.

CONCLUSIONS

Accepted mechanisms exist for rapid whisker growth in compressed thin films, similar to the conditions of films undergoing e-m stressing.

The conditions of films undergoing e-m stressing are consistent with previous whisker observations and mechanisms of formation, namely high compressive stresses and a diffusion mechanism (e-m induced mass flux) to accomodate the creation of excess vacancies.

Alloying additions and refractory metal underlayers can eliminate or reduce the effectiveness of other stress relief and vacancy creation mechanisms, making whisker growth a more likely process for accomodating stress and e-m induced mass flux. Whisker formation is an expected mode for e-m induced short circuit failures in these metallization systems.

REFERENCES

1) J. R. Lloyd, *Int'l. Rel. Phys. Symp.*, 208, 1983.
2) J. R. Lloyd, *Int'l. Rel. Phys. Symp.*, 48, 1984.
3) J. M. Towner, *Int'l. Rel. Phys. Symp.*, 81, 1985.
4) H. B. Huntington, *Thin Solid Films*, **25**, 265, 1975.
5) K. G. Compton, A. Medizza, S. M. Arnold, *Corrosion*, **7**, 327, 1951.
6) S. E. Koonce, S. M. Arnold, *J. Appl. Phys.*, **24**, 365, 1953.
7) J. Franks, *Acta Met.*, **6**, 103, 1958.
8) G. Sines, *J. Phys. Soc. Japan*, **15 #7**, 1199, 1960.
9) Th. Werner, R. Tidecks, B. D. Johnson, *J. Crys. Growth*, **73**, 467, 1985.
10) I. A. Blech, P. M. Petroff, K. L. Tai, V. Kumar, *J. Crys. Growth*, **32**, 161, 1975.
11) I. A. Blech, E. S. Meieran, *J. Appl. Phys.*, **40#2**, 485, 1969.
12) R. W. Thomas, D. W. Calabrese, *Int'l. Rel. Phys. Symp.*, 1, 1983.
13) C. Herring, J. K. Galt, *Phys. Rev.*, **85**, 1060, 1952.
14) J. D. Eshelby, *Phys. Rev.*, **91**, 775, 1953.
15) R. M. Fisher, L. S. Darken, K. G. Carroll, *Acta Met.*, **2**, 368, 1954.
16) F. C. Frank, *Phil. Mag.*, **44**, 854, 1953.
17) S. Amelinckx, W. Bontinck, W. Dekeyser, F. Seitz, *Phil. Mag.*, **2**, 355, 1957.
18) R. O. Simmons, R. W. Baluffi, *Phys. Rev.*, **117#1**, 52, 1960.
19) B. von Gerard, H. Peisl, R. Zitzman, *Appl. Phys.*, **3**, 37, 1974.
20) J. Friedel, *Dislocations*, Pergamon Press/Addison-Wesley Publishing, 1964.
21) H.-U. Schreiber, *Sld. St. Elect.*, **28#11**, 1153, 1985.
22) H.-U. Schreiber, *MRS Symp. Proc.*, Vol. 71, Spring 1986, Palo Alto.
23) J. R. Lloyd, *Thin Solid Films*, **91**, 175, 1982.
24) J. M. Towner, *Solid State Tech.*, 197, October 1984.
25) P. H. Townsend, H, VanderPlas, *MRS Spring Symp.*, 1985, San Francisco.
26) P. Chaudhari, *J. Appl. Phys.*, **45#10**, 4339, 1974.
27) R. L. Coble, *J. Appl. Phys.*, **34**, 1679, 1963.
28) M. Murakami, T.-S. Kuan, I. A. Blech, *Prep. Prop. Thin Films*, K. N. Tu, R. Rosenberg, editors, Academic Press, 1982.
29) J. Sanchez, Jr., to be published.

Tungsten Deposition in Porous Silicon for the Formation of Buried Layer Conductors

ROBERT S. BLEWER, SYLVIA S. TSAO AND MAREN TRACY
Sandia National Laboratories, P.O. Box 5800, Albuquerque, NM 87185

ABSTRACT

It has been well established in recent years that silicon can be converted into highly porous material under the appropriate anodic etching conditions. Metallization of such a porous skeleton could be technologically interesting for device applications, especially if the resistance of the metal path could be made sufficiently low. We report results on the feasibility of forming buried conductors under single crystal silicon using low pressure chemical vapor deposition of tungsten on the internal surfaces of porous silicon. The infiltration of tungsten into the silicon depends on its porosity, and on tungsten deposition process parameters such as deposition duration, total deposition pressure, and carrier gas species. The resulting tungsten/porous silicon layers were characterized by transmission and scanning electron microscopy, by Rutherford backscattering and X-ray diffraction and by four-point probe resistivity measurements.

INTRODUCTION

Recently, there has been interest in buried conductors in silicon as a means for achieving multilevel metallization to increase device speed and radiation hardness. A conductive grid underneath radiation-sensitive devices could be an effective shield against the collection of radiation-induced carriers in sensitive circuit nodes [1,2]. As was proposed in a recent paper [3], one method for forming such a buried layer underneath a single crystal silicon overlayer is low pressure chemical vapor deposited (LPCVD) tungsten infiltration into porous silicon. This approach is analogous to the FIPOS (Full Isolation by Oxidized Porous Silicon) technique for Silicon-on-Insulator (SOI) except that tungsten, instead of oxygen, is incorporated into the porous material. Porous silicon is first formed underneath silicon islands which will eventually contain microcircuits. Tungsten is then deposited to form a conductive layer wherever porous silicon has been formed. This approach offers potential advantages such as little or no damage to the overlying crystalline silicon, low processing temperature (<275°C), self-alignment of conductors, and use of only conventional CVD equipment.

A practical consideration in this approach is the distance that tungsten is able to penetrate into the pores. In this paper, we summarize results on the dependence of the penetration of tungsten into porous silicon on the microstructure of the porous silicon and on the tungsten deposition conditions [4].

POROUS SILICON FORMATION

Two types of porous silicon specimen geometries were used, distinguished by whether the porous silicon (PS) layer was formed at the surface, or was buried underneath single crystal Si islands [5]. Most of the measurements were made in the first geometry. In this case, a PS layer, 15 to 25 um thick, was formed in a p-type, boron-doped, (100) oriented, virgin silicon wafer with 0.01 Ω-cm resistivity as shown in Figure 1. Sets of samples were prepared, each set under a different anodization condition.

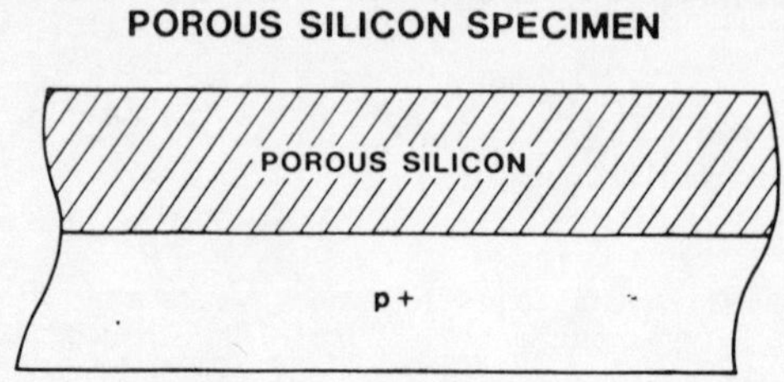

Figure 1. The PS surface layers, 15-25 μm thick, were themselves formed in p-type, boron-doped, (100) silicon wafers with 0.01 Ω cm resistivity. Hydrofluoric acid concentrations and current densities during anodization were varied to obtain different PS densities and pore sizes. All wafers were stored in dilute HF solution to minimize pore surface reoxidation, and then rinsed in de-ionized wafer just prior to the tungsten deposition.

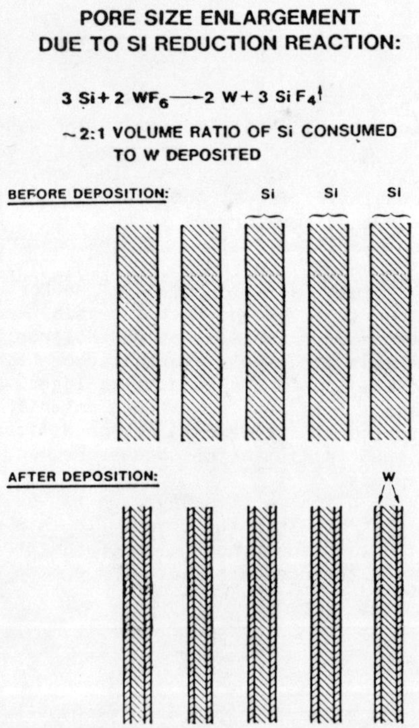

Figure 2. The surface layers of the pores in porous silicon are converted to tungsten using WF_6 gas. The reaction self-limits when the tungsten thickness reaches 10 nm - 30 nm (Ref. 7).

The hydrofluoric acid (HF) concentration was varied from 10 to 30 wt. %, and the current density was varied from 10 to 100 mA/cm^2.

The anodization conditions and resulting densities and average pore sizes of the different sets of PS samples are summarized in Table I. The porous silicon density obtained from weight loss measurements increased with increasing HF concentration and decreasing current density. The average pore diameter, deduced from cross-section transmission electron microscopy (TEM), decreased with increasing PS density, in agreement with data reported by Beale, et. al. [6].

Table I. Porous silicon formation conditions and resulting densities and pore diameters of sets of specimens.

Specimen	HF conc. wt. %	Current Density, j mA/cm2	$\rho_{Porous\ Si} / \rho\ Si$	Average Pore Diam. nm
A	10	15	0.31	8.4 ± 2.4
B1	20	148	0.16	15.0 ± 2.5
B2	20	59	0.36	11.0 ± 1.0
C1	30	148	0.42	9.3 ± 2.4
C2	30	59	0.58	7.5 ± 1.5

TUNGSTEN DEPOSITION

The reduction of WF_6 by Si ($2WF_6 + 3Si \rightarrow 2W + 3SiF_4$), one of two reactions typically used in LPCVD tungsten technology [7], is an attractive first step in porous silicon metallization. First, the reaction is chemically selective with respect to silicon. Second, since the ratio between the volume of silicon consumed to that of tungsten deposited is roughly 2:1, there should be no blockage of pores due to the tungsten deposit. In fact, this reaction should enlarge the pores as shown in Figure 2, and enhance penetration of the source gas (WF_6) far into the pores. Third, if the distance between pores is short enough, it should be possible to consume all of the silicon between adjacent pores and thus form pure, though porous, tungsten. It should be noted that once the pores are coated with tungsten, additional tungsten can be deposited in a second step, using hydrogen reduction of WF_6 [7], to densify the deposit, increase mechanical integrity, and reduce electrical resistance.

The extent of tungsten penetration into the porous silicon layer was determined from cross-section transmission electron microscopy (TEM) and scanning electron microscopy (SEM) as shown in Figure 3. For the specimen shown in Figure 3(a), tungsten penetrated 4.7 µm into the 23 µm thick porous silicon layer after 20 minutes of the silicon reduction reaction at 243°C. The bright and dark bands correspond to the tungsten-porous silicon (WPS) layer and the unreacted porous silicon (PS) layer, respectively. Under typical flow conditions for the silicon reduction reaction (11 sccm WF_6 flow rate and 0.210 torr total pressure), the penetration of tungsten into the porous silicon increased with time but appeared to saturate at long times, with the apparent saturation level increasing with increasing porosity. At 8 torr (stagnant) partial pressure of WF_6, tungsten penetration into the

POROUS TUNGSTEN LAYER FORMATION

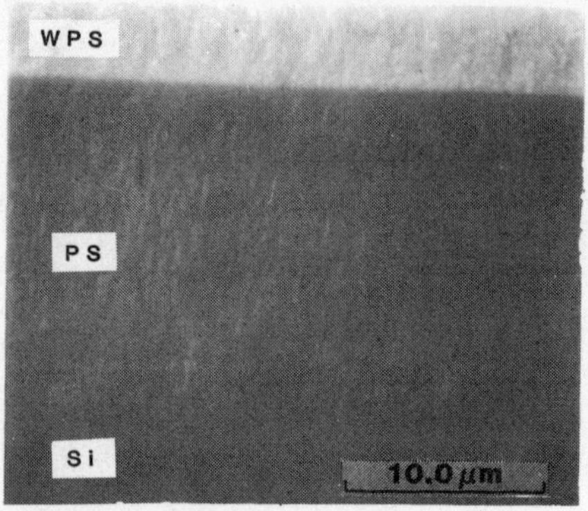

Figure 3(a) Cross-section SEM of specimen C1 after 20 min of the Si reduction reaction at 243°C. The bright and dark bands correspond to the tungsten-porous silicon and unreacted porous silicon layers, respectively.

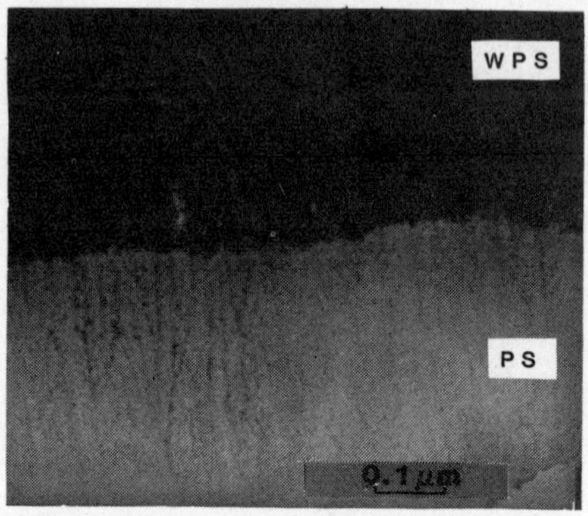

Figure 3(b) Cross-section TEM of specimen B2 after 20 min of Si reduction reaction at 238°C. Note the columnar structure of the porous silicon (lower layer) and the tungsten/porous-silicon (upper layer).

RUTHERFORD BACKSCATTERING SPECTRUM OF WPS LAYER

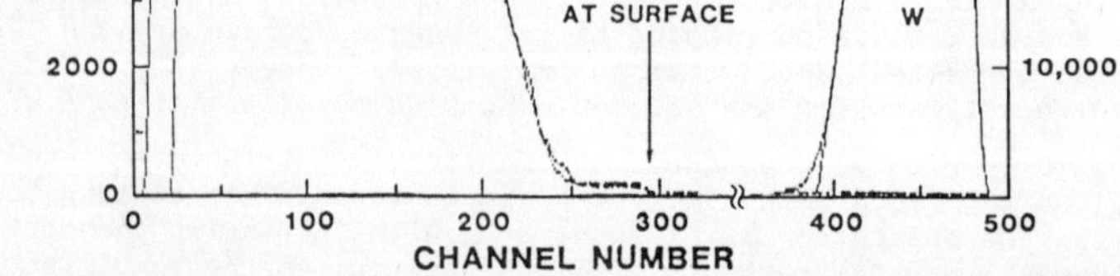

Figure 4. RBS spectrum of a porous silicon sample after one minute exposure to WF_6 gas.

porous silicon increased from 5μm to > 15μm. The degree of penetration was also increased by using helium rather than argon as the carrier gas. In contrast, the tungsten deposit on a non-porous silicon control wafer self-limited in thickness at only about 10 nm.

Figure 4 shows a depth profile by Rutherford backscattering spectroscopy (RBS) on a reacted sample [8]. There is a uniform concentration of tungsten throughout the reacted (WPS) layer, terminating in a sharply delineated rear interface between the reacted and unreacted PS, as confirmed by cross-section transmission electron microscopy (See Figure 3(b).) The sharpness of the WPS/PS interface is also evident from cross-section SEM micrographs, such as Figure 3(b) in which the backscattered electron image is much brighter in the reacted PS. From such micrographs, we can extract well-defined tungsten vertical penetration depths from the geometric surface of the wafer.

Depletion of silicon (down to ~6%) and the presence of oxygen (as much as 56%) in the WPS layer are evident in Figure 5; PS layers contain typically ~5% oxygen [9]. The high oxygen content in the WPS layer is probably due to water in the PS layer remaining after the predeposition rinse and spin dry and can result in formation of beta phase tungsten, as determined by x-ray diffraction. Comparison of the RBS data with cross-section scanning electron microscope (SEM) measurements of the reacted layer thickness indicates a very high degree of porosity remaining in the film.

TUNGSTEN INFILTRATION UNDER EPI ISLANDS

Preliminary studies have been made of buried tungsten layer formation. In these studies, n-/p+ layer structures were first formed in a p- wafer as indicated in Fig. 5. Openings were etched in the n-layer and PS was formed in the p+ layer [3,5]. Figure 6(a) is a cross-section SEM micrograph of a 49μm wide silicon island under which PS has been formed and subsequently exposed to 8 Torr of stagnant WF_6 for 20 min. In this example, tungsten penetrated 20μm from both edges of the island. Islands less than 40 μm wide were fully isolated by the WPS layer during the same deposition run, as shown in Figure 6(b). Mechanical integrity of the islands was maintained. We expect the lateral tungsten penetration to be limited only by the lateral extent of the PS itself.

SCHEMATIC DIAGRAM OF TUNGSTEN PENETRATION INTO PSL UNDERNEATH Si ISLAND

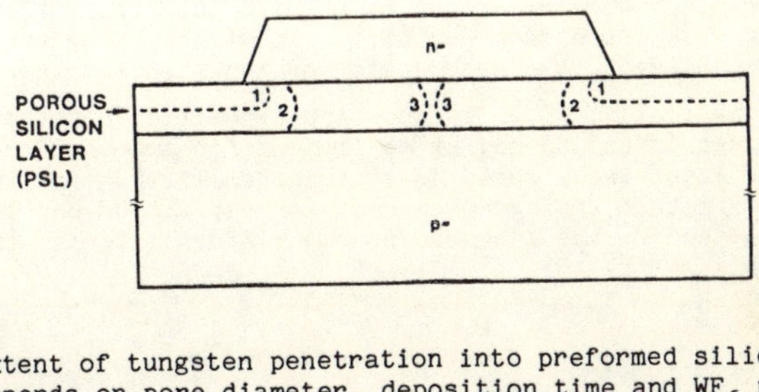

Figure 5. Extent of tungsten penetration into preformed silicon pores depends on pore diameter, deposition time and WF_6 partial pressure, and can thus be tailored to the desired application.

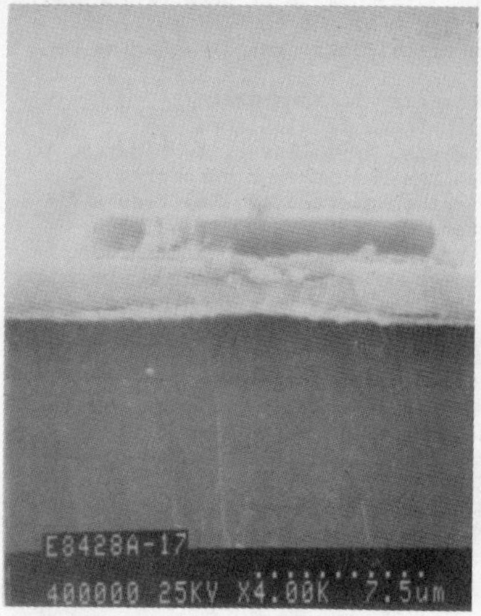

Fig. 6(a). Cross-section SEM micrograph of a single crystal silicon island 49 μm wide which was not fully isolated by metallized porous silicon. Tungsten penetrated 20 μm from the island edges during the 20 min. high-pressure Si reduction deposition.

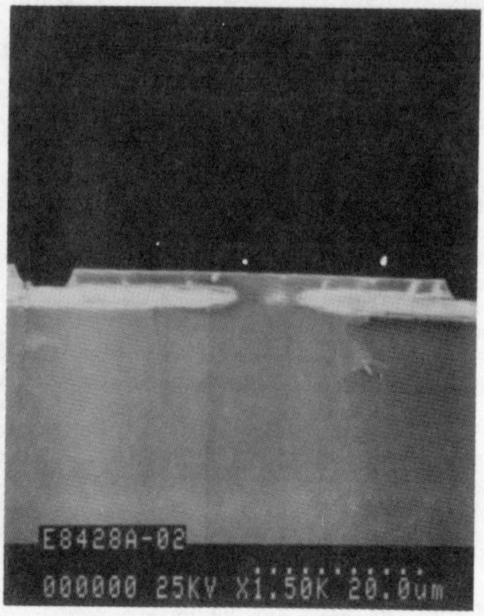

Fig. 6(b). Cross-section SEM micrograph of a silicon island 18 μm wide which has been fully isolated by metallized porous silicon.

SUMMARY

We have demonstrated that buried conductors can be formed under single crystal silicon islands by LPCVD tungsten infiltration of buried layer porous silicon. The Si reduction reaction is very attractive for porous silicon metallization since it does not result in pore blockage. At the typical flow conditions used for tungsten LPCVD by the Si reduction reaction, the penetration of tungsten into porous silicon increased with time but appeared to saturate at long times with the apparent saturation level increasing with increasing porosity. At higher (stagnant) partial pressures of WF_6, however, the penetration of tungsten into porous silicon is increased dramatically. The Si reduction deposition can therefore metallize very deep pores and may be sufficient for some buried conductor applications not requiring a fully dense tungsten layer. Preliminary experiments confirm that the hydrogen reaction deposition can be used to fill up pores to densify the tungsten-porous silicon material and to decrease its resistivity.

ACKNOWLEDGEMENTS

We thank C. E. Case for the invaluable assistance in the preparation of the porous silicon specimens, N. A. Creager and F. P. Ganyard for the scanning electron microscopy, A. Datye and J. Aubin for the cross-section transmission electron microscopy, and P. S. Hund and T. R. Guilinger for their help in the initial phases of this work. This work supported by the U. S. Department of Energy under contract number DE-AC04-76-DP00789.

REFERENCES

1. M. R. Wordeman, R. H. Dennard, and G. A. Sai-Halasz, Technical Digest, IEEE, International Electron Devices Meeting, 40 (1981).
2. G. A. Sai-Halasz, M. T. Wordeman and R. H. Dennard, IEEE Trans. Elect. Devices ED-20. 725 (1982).
3. S. S. Tsao and R. S. Blewer, Semiconductor-on-Insulator and Thin Film Transistor Technology 1985, edited by A. Chiang, M. W. Geis and L. Pfeiffer (Materials Research Society, Pittsburgh), Vol. 53, 199 (1986).
4. S. S. Tsao, R. S. Blewer and J. Y. Tsao, Appl. Phys. Lett. 49, (7) 403 (1986).
5. L. Nesbit, Technical Digest, IEEE IEDM, 800 (1984).
6. M. I. J. Beale, N. G. Chew, M. J. Uren, A. G. Cullis and J. D. Benjamin, Appl. Phys. Lett. 46, 86 (1985).
7. R. S. Blewer, Solid State Technology 29 (11), 117 (Nov. 1986).
8. RBS Measurements were performed by Charles Evans and Associates, Redwood City, CA.
9. L. G. Earwater, J. P. G. Farr, P. E. Grzeszczyk, and I. Sturland, Nucl. Instrum. Methods Phys. Res. B9, 317 (1985).

Panel Discussion Summary

RELIABILITY OF DEVICES WITH TUNGSTEN ELEMENTS

Panel Discussion

Reported by: **Sunil Mehta** *(Panel Chairman)*
Cypress Semiconductor
3901 North First Street
San Jose, CA 95134

Panel Members: **Robert Blewer**, Sandia National Laboratories
Nathan Cheung, University of California-Berkeley
Ronald Nowicki, Integrated Processing
James Roland, Hewlett-Packard
Rama Shukla, Intel Corporation

INTRODUCTION

The purpose of the panel was to review the available data on the effect of tungsten on device reliability and to discuss the potential reliability concerns related to the use of tungsten in VLSI applications.

SUMMARY OF RELIABILITY DATA

A comprehensive review of published data on the reliability of devices with tungsten elements was presented *[R. Blewer]*. It was mentioned that in the case of tungsten in contacts, junction leakage has a strong bearing on device yield and reliability. It has been reported by Moriya and Itoh *[1]* that junction leakage can be completely suppressed by the use of a tungsten barrier, even for very shallow junction depths, as shown in Figure 1. Moriya and Itoh also reported contact electromigration data as a function of contact size, and showed that the mean time to failure (MTF) can be considerably increased by planarizing the contact with selective tungsten (Figure 2). The overall effect of tungsten on device reliability has been studied by Metz et al. *[2]* for LSI circuits with tungsten in the contacts. They found that, at a stress temperature of 125°C, there were no failures for 32,000 device-hours, which implies a projected failure rate of less than 267 FITs at a normal operating temperature of 70°C. Wells and Walko *[3]* have studied the reliability of tungsten when used as an interconnect material in a thermopile (80/20 Si-Ge). The reported MTF data, which may have relevance to VLSI applications, is given in Figure 3, and represents contact-resistance failure caused by formation of tungsten silicide which leads to voids at the interface.

The projected improvement in electromigration performance by the use of tungsten interconnect as a replacement for aluminum-based interconnects was discussed in depth *[N. Cheung]*. A comparison of the reported failure data on tungsten with the projected performance of Al-Si-Cu was presented. The reported MTF for tungsten at stress conditions of 4×10^6 A/cm^2 and 415°C is 1200 hours*[4]*, which is three orders of magnitude higher than the MTF of Al-Si(2%)-Cu(4%) extrapolated to the same

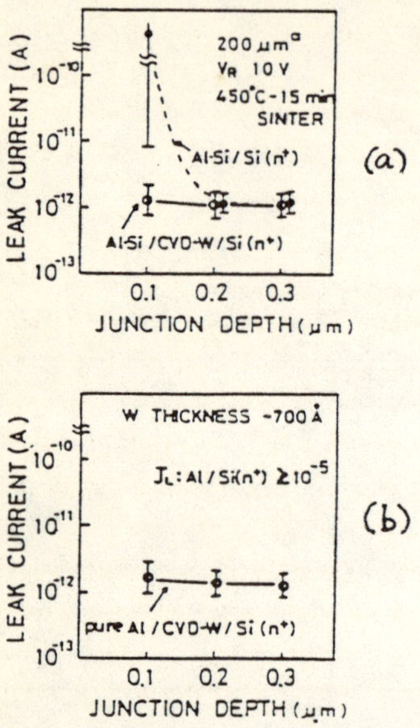

Figure 1. Influence of CVD tungsten on junction leakage. (a) With Al-Si, (b) With pure Al.

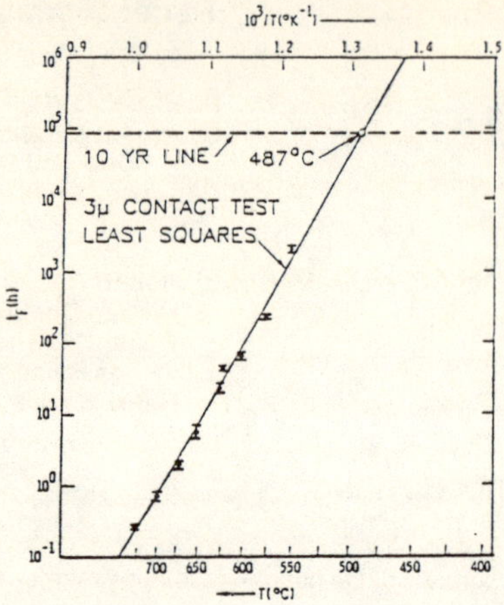

Figure 3. Electromigration failures of tungsten contacts to Si-Ge in a thermopile.

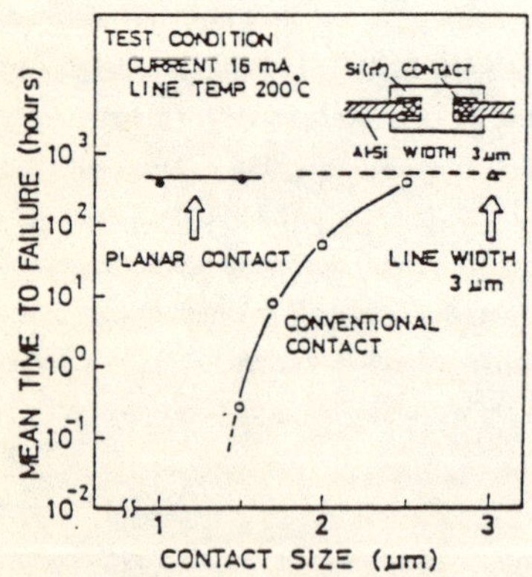

Figure 2. Electromigration performance of contacts with and without thick selective tungsten.

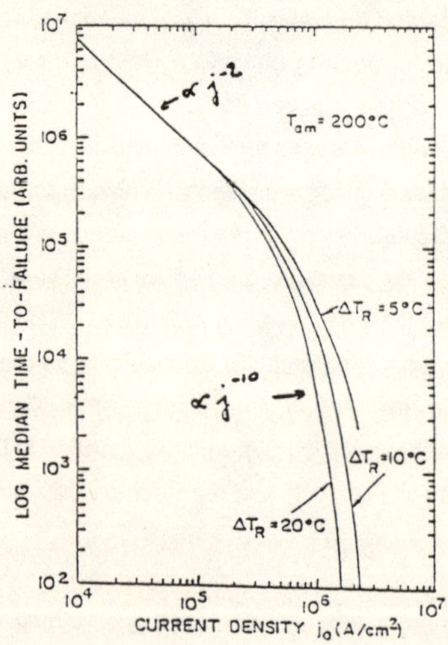

Figure 4. Effect of stress temperature on mean time to failure.

stress conditions (1.7 hours). The use of tungsten as a contact material was also discussed, and a number of points were emphasized in this regard. The extrapolated failure mode must be carefully determined (open vs. leakage failures). The effect of stress temperature on MTF must be accounted for, as shown in Figure 4. Lastly, the proper value of the current exponent must be used in the expression for MTF while performing any contact lifetime extrapolation.

Contact electromigration data on the use of tungsten as a barrier metal were reviewed *[R. Shukla]*. A comparison of the MTF with and without a tungsten barrier was done as early as 1982 by Gargini et al. *[5]* for 3 u x 3 u contacts. They reported that, for the junction depths studied (0.5 to 0.6 u), the use of tungsten completely eliminated leakage failures at the contacts. Open failures with or without a tungsten barrier can be interpreted as Al-Si line failures at the contact sidewalls. Since the leakage failure mode is often dominant for contacts without a barrier, open failures with tungsten should be compared to leakage failures without tungsten. Such a comparison is given in Figure 5. It can be observed that tungsten increases the mean time to failure by a large factor, especially for larger currents. A more recent study for smaller geometries (1.5 u x 1.5 u contacts, 0.3 u junction depth) has been done by Sachdev and Mehta *[6]*, using stress conditions of 8 mA and 200°C. This study essentially replicates the findings of Gargini et al., as shown in Figure 6. For both open-failure and leakage-failure modes, the tungsten contact barrier was found to increase the electromigration resistance by a substantial factor. In fact, up to 880 hours of stressing, there were no leakage failures and only one open failure for the structures with tungsten in the contacts.

The use of sputtered refractory metals as diffusion barriers, and the influence of the method of their preparation on reliability was discussed *[R. Nowicki]*. Thus, for example, in the case of sputtered tungsten or titanium-tungsten, the sputtering ambient plays an important role in determining barrier properties. This is illustrated in Figures 7 and 8, which show a comparison of TiW and nitrided TiW films, the latter having been sputtered in a nitrogen-argon ambient *[7]*. It can be observed that nitrided TiW is an adequate barrier under the stress conditions studied (300°C), while TiW is not. Step coverage is another important factor influenced by the deposition method that has an impact on device reliability. Hence, the deposition technique and the pre- and post-deposition treatment must be carefully characterized while attempting to optimize the reliability of a barrier-metal scheme.

A description of Hewlett-Packard's NMOS III technology (which employs tungsten as both Metal 1 and Metal 2) was presented *[J. Roland]*. For the first level of tungsten interconnect, this technology uses a glue layer that also functions as an etch stop and silicon-diffusion barrier. Details of the process can be found elsewhere in this volume *[8]*. Under the processing conditions used, the fabricated devices exhibited no lifetime failure due to the use of tungsten as an interconnect metal. No electromigration failures were detected even with extended stressing; the electromigration resistance of tungsten was estimated to be at least two orders of magnitude higher than that of aluminum.

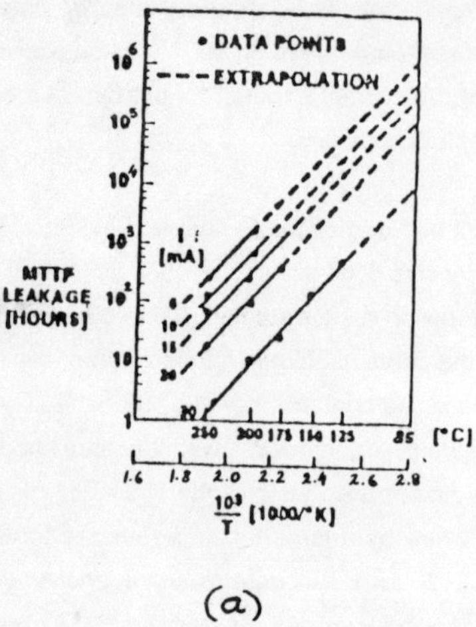

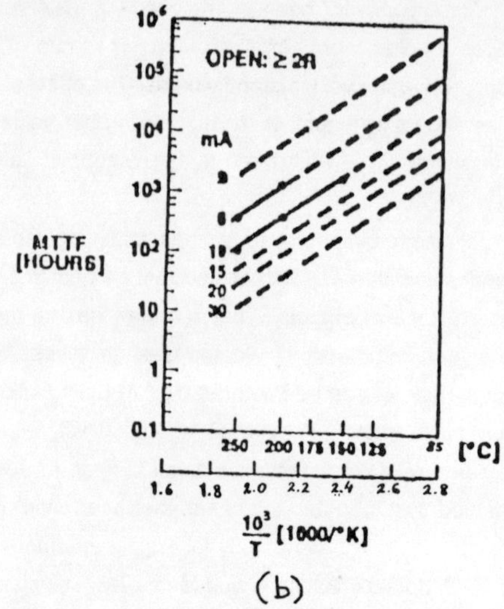

Figure 5. Electromigration failures versus temperature.
(a) No tungsten (leakage failures), (b) With tungsten (open failures).

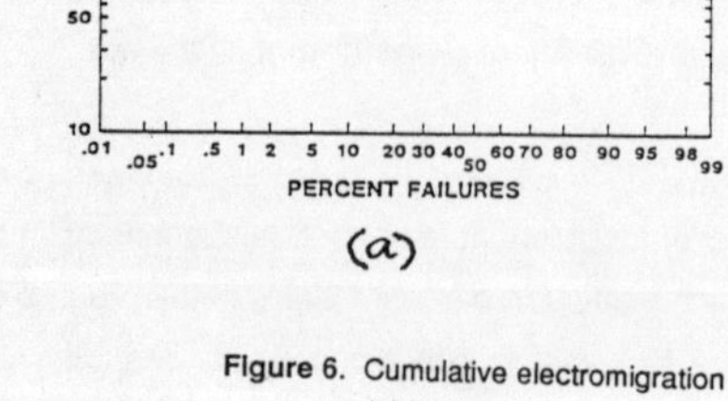

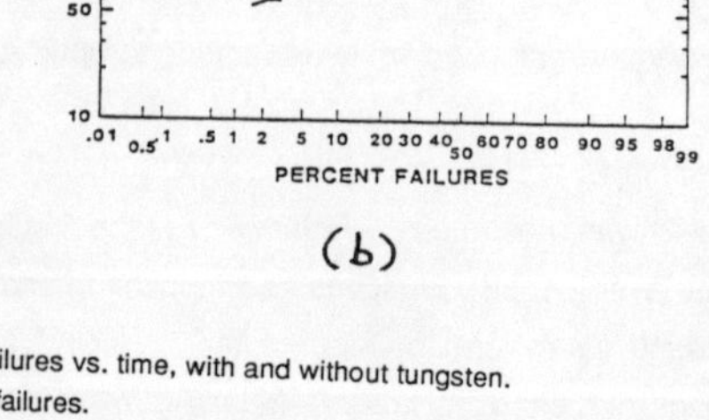

Figure 6. Cumulative electromigration failures vs. time, with and without tungsten.
(a) Leakage failures, (b) Open failures.

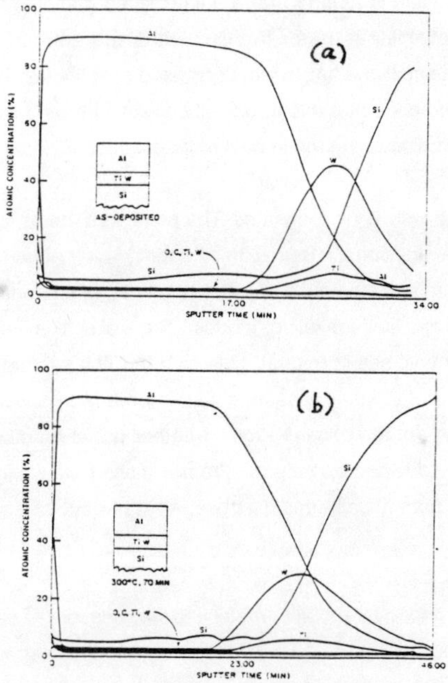

Figure 7. Auger depth profiles of Al/TiW/Si.
(a) No anneal, (b) 300°C anneal.

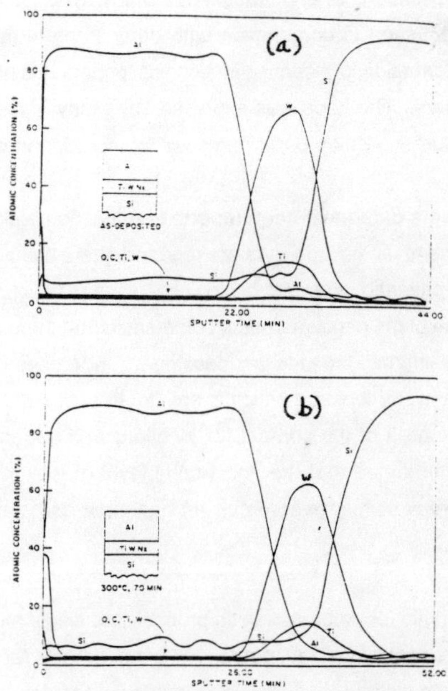

Figure 8. Auger depth profiles of Al/TiWN$_x$/Si.
(a) No anneal, (b) 300°C anneal.

DISCUSSION

Several concerns about the potential problems with CVD tungsten in VLSI devices were raised by the panel members and workshop participants. The following is a compilation of the significant points that were discussed.

A number of workshop participants felt that there is a dearth of device *yield* data for VLSI applications of CVD tungsten. One way to look at yield is as short-term reliability, so it should be discussed in conjunction with other (long-term) reliability issues. In this regard, the amount of information forthcoming in technical papers has been small and has in fact decreased over the last five years. The hope was expressed by many in the audience that this trend will be reversed, and that future workshop participants will be more forthright in discussing the impact of tungsten on yield.

Some data have been reported on junction leakage with CVD tungsten. The point was made that whenever leakage data are reported, there should be additional information on doping levels, junction depths and other processing conditions to truly understand the cause of junction failure. In response, one of the panel members commented that there are two leakage failure modes. The first is aluminum spiking into the junction, causing a sudden and dramatic rise in reverse-bias current. The second is transistor leakage which is smaller in comparison. In the latter case, it is very important to specify the area of the contact (or junction) and dopant type and concentration. Another panel member commented that the acceptable level of leakage is device-dependent. From a testing viewpoint, however, the reverse current generally rises very dramatically under stress, so it is easy to spot problems.

Junction leakage has been probably the single most important problem restricting the widespread use of selective CVD tungsten. Tungsten tunnels (or wormholes) are a cause for concern in this respect. Even without tunnels, high electric field strength at the tip of the tungsten encroachment (however small it may be) can cause leakage. Despite major efforts, there has not been much success in eliminating leakage at the field oxide edge. Short channel effects will become more important for smaller geometries and will also have to be considered.

When blanket tungsten is used as the first-level interconnect, it has been observed *[8]* that covering the tungsten films with dielectric before alloying almost always results in device failures unless a silicon-diffusion barrier is used underneath the tungsten. This emphasizes the point that one should test completed devices, rather than depositing the tungsten and testing the devices immediately. Back-end integration plays an important role in influencing junction leakage and other reliability issues.

Although tungsten is expected to have extremely high electromigration resistance owing to its refractory nature, it has proven difficult to quantify the extent of improvement relative to aluminum.

This is because it is not easy to make tungsten lines fail under stress. Even when the structures fail, the failures may be more related to localized melting than to electromigration failures. From a fundamental viewpoint, it would be useful to determine the electromigration resistance of tungsten quantitatively, and perhaps some ingenious structure could be devised to perform this measurement.

In regard to device reliability, a common concern has been the fluorine content of CVD tungsten. A recent study *[9]* indicates that the F content is higher at the tungsten/substrate interface than in the bulk tungsten film. A representative SIMS profile is shown in Figure 9. The F content is about 0.2 atomic percent in the bulk film, and decreases with temperature, as shown in Figure 10. These levels of fluorine are comparable to those measured in CVD tungsten silicide films (which are used extensively in polycide applications), and are not expected to be a problem for device reliability.

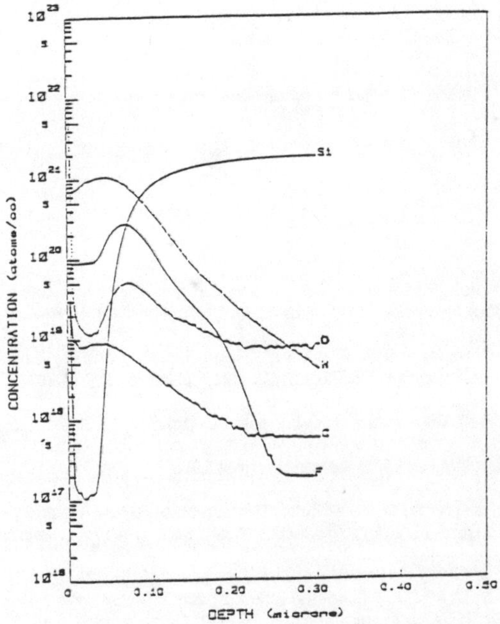

Figure 9. SIMS depth profile of CVD tungsten film deposited at 300°C.

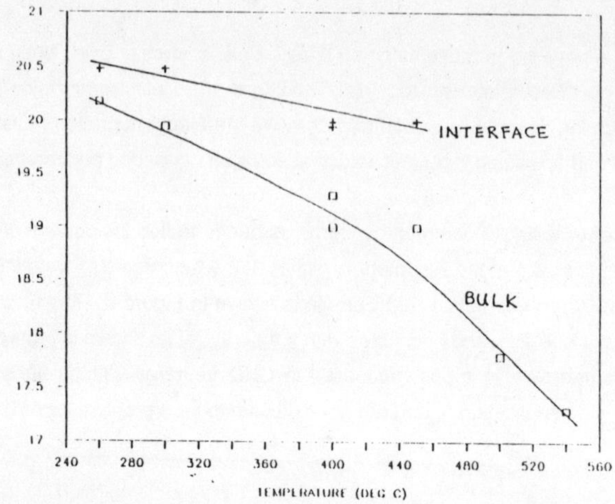

Figure 10. Effect of deposition temperature on fluorine concentration.

REFERENCES

[1] T. Moriya and H. Itoh, "Selective CVD of Tungsten and its Applications to MOS VLSI," *Workshop on Tungsten and Other Refractory Metals for VLSI Applications* (R. S. Blewer, Ed.), p. 21, 1985.

[2] W. A. Metz, N. J. Szluk and G. W. Miller, "Reproducibility and Reliability of Selective Tungsten Shunt and Contact Applications in CMOS LSI Circuits," *Ibid.*, p. 527, 1984.

[3] V. A. Wells and R. J. Walko, "LPCVD Tungsten Deposition on Si-Ge Alloy," *Ibid.*, p. 497, 1984.

[4] P. Merchant, *Hewlett-Packard Journal*, August 1982.

[5] P. A. Gargini, C. Tseng and M. H. Woods, "Elimination of Silicon Electromigration in Contacts by the Use of an Interposed Barrier Metal," *Proceedings of the 20th Annual Reliability Physics Symposium*, p. 66, 1982.

[6] S. Sachdev and S. D. Mehta, "Tungsten Interconnects in VLSI," *Workshop on Tungsten and Other Refractory Metals for VLSI Applications* (R. S. Blewer, Ed.), p. 161, 1985.

[7] R. S. Nowicki and B. Schiefelbein, "Studies of the Diffusion Barrier Properties of DC Magnetron Sputtered Titanium-Tungsten on Gallium Arsenide and Silicon," *Ibid.*, p. 341, 1985.

[8] J. P. Roland, "Device Reliability When Using Tungsten Interconnects," *This Volume.*

[9] C. Yang, S. D. Mehta and P. Davies, "Composition and Structure of CVD Tungsten Films," Presented at the *Materials Research Society Annual Meeting,* December 1986 (to be published in the proceedings).

DEVICE RELIABILITY WHEN USING TUNGSTEN INTERCONNECTS

JAMES P. ROLAND
Hewlett-Packard, Fort Collins, IC Division
3404 East Harmony Road, Fort Collins, CO 80525

Among the most advanced chips currently in production are those made with Hewlett-Packard's NMOSIII technology in Fort Collins, Colorado. A chip set made with this process is the heart of the HP9000 Computer. NMOSIII is a modified n-channel silicon gate process featuring three and one-half levels of interconnect -- diffusion, polysilicon with buried contacts and two levels of refractory metal. Features of the process that relate to the metal layer are shown in Figure 1. A more complete description of the process and designs in the process can be found elsewhere (1,2,3).

The use of zero overlap vias at Metal I to poly contacts allows significantly higher device density to be achieved. However, as is illustrated in Figure 2, the fabricated structure can look different from the designed structure because of the shrinkage of lines, the growth of contact holes, and the tolerances of alignment. A direct consequence of this design rule as it is used at Metal I is that the tungsten is allowed to come into contact with polysilicon, silicon and silicon dioxide. Therefore, when Metal I is patterned, the etch must be highly selective to these materials since they can be part of the source, drain or gate of the device. Furthermore, a dry etch is called for at this level because of the fine dimensional control required. Since a practical dry tungsten etch which is highly selective to silicon does not exist, a thin etch stop and glue layer was instituted below the Metal I layer as illustrated in Figure 3.

A class of materials evaluated for this application were silicides such as platinum silicide, palladium silicide, nickel silicide and chrome silicide. Both self-aligned silicides formed with existing silicon and co-deposited materials were evaluated. Because of subsequent processing required for the full NMOSIII process, a test of the practicality of these materials was undertaken. The test structure used was 250,000 diodes in parallel, and the treatment, intended to simulate worst case processing exposure, involved annealing at 500°C for 10 hours.

When heat treatment took place directly after the first metal layer was patterned, device characteristics were largely unchanged. However, when heat treatment followed the application and patterning of an oxide layer, diodes were leaky and in many cases totally conducting. Analysis of the devices revealed massive silicon outdiffusion. The silicon was found to reside within the tungsten line and could be easily detected tens of microns away from the diode from which it came. As a result of this study, the functionality of the tungsten underlayer was expanded so that it serves the purposes of glue layer, etch stop, and silicon diffusion barrier. Materials were developed which meet these requirements. Devices fabricated using these materials exhibit no known lifetime failure due to the use of tungsten as interconnect material.

References

1. J.W. Beyers, L.J. Dohse, J.P. Fucetola, R.L. Kochis, C.G. Lob, G.L. Taylor, and E.R. Zeller, "A 32b VLSI CPU Chip", IEEE International Solid State Circuits Conference, 1981. ISSCC Digest of Technical Papers, 1981, pp. 104-105.

2. J.M. Mikkelson, L.A. Hall, A.D. Malhotra, S.D. Seccombe, and M.S. Wilson, "An NMOS VLSI Process for Fabrication of a 32b CPU Chip", IEEE International Solid State Circuits Conference, 1981. ISSCC Digest of Technical Papers, 1981, pp. 106-107.

3. Hewlett-Packard Journal, August 1983 (Entire Issue devoted to NMOSIII).

Figure 1

Hewlett Packard NMOSIII Process

Layers of Interconnect

* Diffusion
* Polysilicon
* Metal I
* Metal II

Zero overlap vias at Metal I

Metal I

* tungsten
* 0.4 ohms/sq
* 1.5 micron line/1.0 micron space

Metal II

* tungsten
* 0.04 ohms/sq
* 5.0 micron line/3.0 micron space

Figure 2

Zero Overlap Vias

Tungsten contacts allowed to Source, Drain and Gate through polysilicon

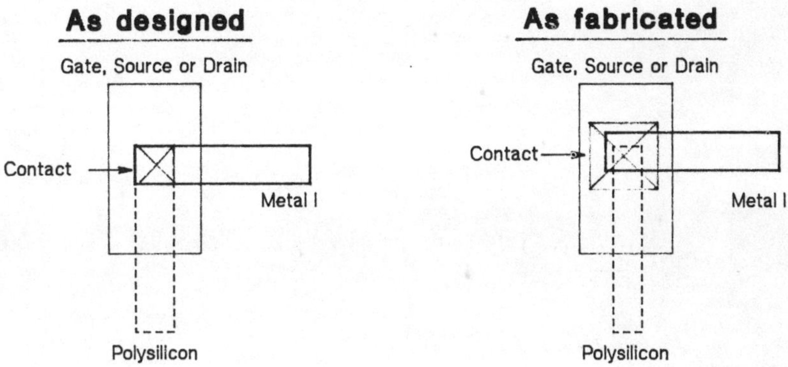

Figure 3

Metal I Structure

Tungsten: 0.4 ohms/sq

Glue layer/etch stop

Polysilicon or silicon or oxide

Author Index

Ali, Y.S., 85
Amano, J., 93

Balland, B., 169
Bartholomew, L., 131
Blewer, R.S., 235, 401
Boone, T., 85
Bradbury, D.R., 93
Broadbent, E.K., 101, 247
Brors, D.L., 147
Brown, J.M., 215, 227

Chen, C.-H., 357
Chen, J.Y., 9
Chow, R., 137
Creighton, J.R., 43

Davidson, B.A., 85
Deal, B.E., 27
DeBlasi, J.M., 265
Delfino, M., 265

Ellwanger, R.C., 283, 385

Feldman, L.C., 85
Flanigan, E.J., 199
Foster, R.F., 147
Fraser, D.B., 3

Georgiou, G.E., 227
Gralenski, N., 131
Granneman, E.H.A., 377
Green, M.L., 85, 227
Gorowitz, B., 349
Gutai, L., 265

Hara, N., 159
Harshbarger, W.R., 137
Heiny, M.A., 69
Hieber, K., 111
Hitchman, M.L., 377
Holschwandner, L.H., 187

Ichikawa, M., 257
Ikuta, T., 123
Inayoshi, K., 257
Inoue, S., 59
Iwata, S., 159

Kamins, T.I., 93
Kang, S., 137
Kao, D.-B., 27
Kimura, D., 371
Kobayashi, N., 159
Kolawa, E., 301, 311
Kulkarni, A.K., 325
Kurita, K., 257
Kusumoto, Y., 123
Kwakman, L.F. Tz., 377

Laderman, S.S., 93
Ledys, J.-L., 169
Lien, C.-D., 301
Liftshitz, N., 215
Liu, R., 227
Lubowiecki, V., 169

Maeda, M., 59
Matsuda, K., 319
McConica, C.M., 51, 69
Mehta, S., 411
Mihara, S., 275
Morgan, A.E., 101

Nakahara, S., 85
Nakayama, I., 123
Ng, S.L., 93
Nicolet, M.-A., 19, 301, 311
Nishizawa, K., 319, 371
Nowicki, R.S., 333

Ohba, T., 59
Ohyama, Y., 59

Park, K.-H., 275
Plossu, C., 169

Rana, V.V.S., 187
Ray Chiu, K.C., 177
Roland, J.P., 419
Rosner, S.J., 93

Sadana, D.K., 101
Sahin, T., 199
Saia, R.J., 349
Saito, S., 319
Sakiyama, K., 319, 371
Sanchez, J.E., 395
Sanchez-Martinez, R.A., 207
Sato, Y., 275
Schlosser, D.W., 357
Schmitz, J.E.J., 385
Sears, J.T., 199
Shukla, R.K., 291
So, F.C.T., 301, 311
Susoeff, M., 137
Suzuki, M., 339

Takakuwa, K., 123
Takenaka, N., 371
Taylor, J.A., 187
Thompson, M.G., 325
Towner, J.M., 247, 283
Tracy, M.E., 235, 401
Tsai, N.S., 187
Tsao, S.S., 401
Tseng, S., 147
Tsuchikawa, H., 275
Tsuzuki, N., 257

van der Putte, P., 77, 101
van Dijk, A.J.M., 385
Vermeulen, W.J.C., 377

Watanabe, K., 257
Watson, L.C., 357
Wendling, T.P.H.F., 111
Whetten, N.Y., 363
Whetten, T.J., 363
Whipple, G.H., 325
Wieczorek, C., 111
Williams, D.S., 215, 227
Wolf, E.D., 363
Wolters, R.A.M., 385
Woodruff, D.W., 207

Yamamoto, N., 159
Yoshida, M., 275
Yoshikawa, S.A., 291

Zetterquist, N.E., 177
Zhao, X.-A., 301, 311

Subject Index

adhesion, 187
  layers, 3, 169, 177, 187, 247, 385
  strength, 177
aluminum, 147, 187, 247, 265, 395
  containing layers, 93
annealing, 265
anodization, 401
APCVD, 131
Auger spectroscopy, 43
activation energy, 51, 123, 247, 333

bipolar devices, 257, 319
blanket CVD tungsten, 3, 131, 159, 169, 177, 187
buried
  layer conductors, 401
  tungsten layer, 401

CCl_3F, 371
CF_4, 363
chemicals, 27
cleaning procedures, 27
CMOS, 9, 169
cobalt, 101
  silicide, 101, 227
cold wall CVD reactor, 137, 147, 177
contact filling, 385
contact resistance, 257, 265, 385
contacts, 19, 27, 319
contrast enhancement, 363

depletion, 131
desorption, 43
diffusion barrier, 19, 85, 101, 265, 275, 283, 301, 319, 333
diffusivity, 19
discontinous, 85
displacement reaction, 85

electrical characterization, 169
electromigration, 247, 283, 395
emissivity, 59
encroachment, 101, 111, 137, 199
equipment, 123

fluorine, 257, 385
  compounds, 93
flow configuration, 123

gallium arsenide, 325
gas phase, 131
gate electrode, 169

high purity, 339
high rate, 147
high vacuum, 3
hydrogen
  fluoride, 51, 69, 377
  fluoride desorption, 77
  reduction of WF_6, 59, 77
  reduction of MoF_6, 199
  wet, 159

impurities, 27
integrated circuit, 111, 169
interconnects, 27, 247, 257, 283, 349, 357, 395
interface, 19
  W/Si, 85
interfacial fluorides, 187
infrared
  sensor, 147
  thermometer, 59

kinetic model, 377
kinetics, 77, 199

laminar jet, 123
latch-up, 9
layered conductors, 247
leakage current, 227, 265, 319
lightly doped drain, 9
local interconnect, 291, 301
LPCVD, 377, 401

mass spectroscopy, 43
metal gate, 9, 159
metallization, 111, 247, 265, 275
MoF_6, 199, 207, 215
molybdenum, 159, 215, 325, 349
  LPCVD, 199, 215
  silicide, 199
MOS, 159
multilayer interconnect, 247, 371

native oxide, 27, 187
nitrogen distribution, 301, 311
nucleation, 69
  autocatalytic, 51
  on TiN, 187

oxygen, 357, 363, 371

partial pressure, 377
phosphine, 77
photolithography, 363
plasma, 349
  etching, 357
plug etchback, 349, 357, 385
porous, 85, 215
  silicon, 401
precipitation, 275
pretreatment, 235

radiation hardness, 401
rapid thermal annealing, 291
rate
  limiting step, 77
  of initiation, 51
reaction
  area, 51
  kinetics, 77
  time, 59
reactive
  ion etching, 349, 363, 371
  sputtering, 283, 311
refractory metal, 9, 159, 325, 339
residues, 235

Schottky barrier, 325
selective
  deposition, 137, 147, 207, 215, 227, 235, 257, 377
  tantalum silicide, 111
selectivity
  breakdown, 123
  loss mechanism, 43, 377
  thickness, 59
self-aligned, 101
self-limiting, 85
semiconductor device, 27
SF_6, 357
shallow junctions, 207, 235
short circuit, 247, 283
silicide, 339
silicon, 27
  consumption, 123, 137, 147
  oxide, 69, 177, 235
  etching, 207
  reduction, 59, 85, 235
  tetrafluoride, 69, 215
sputter target, 339
stress, 3, 177, 311
subfluoride disproportionation, 43
sublimator, 111
surface
  analysis, 27, 43
  preparation, 27, 235
  temperature, 59

tantalum, 325
temperature programmed desorption, 69
thermal
  annealing, 275, 319
  nitridation, 291
  oxidation, 27, 159
  oxide, 51
  stability, 159, 265, 301, 325
thin film reactions, 301
titanium, 3
  containing layers, 93
  nitride, 187, 275, 283
  oxynitride, 291
  silicide, 385
titanium-tungsten, 319, 333, 349, 371, 385
transmission electron microscopy, 101, 215, 227, 275
traps, 9
tungsten, 3, 159, 349, 363
  CVD, 69, 77, 85, 93, 101
  interaction, 93
  nitride, 311
  nucleation on oxide, 51, 377
  nuclei, 51, 377
  silicide, 131, 177, 357
  sputtered, 93, 257
  thick films, 85
turbulent jet, 123
two-step process, 111

uniformity, 147

vacancies, 395
via filling, 257, 349
void formation, 101

x-ray diffraction, 319

whiskers, 111, 247, 395
work function, 9
wormholes, 215

MATERIALS RESEARCH SOCIETY CONFERENCE PROCEEDINGS

VLSI-I—Tungsten and Other Refractory Metals for VLSI Applications, R. S. Blewer, 1986; ISSN: 0886-7860; ISBN: 0-931837-32-4

VLSI-II—Tungsten and Other Refractory Metals for VLSI Applications, E.K. Broadbent, 1987; ISSN: 0886-7860; ISBN: 0-931837-66-9

TMC—Ternary and Multinary Compounds, S. Deb, A. Zunger, 1987; ISBN:0-931837-57-x